A geologic time scale 1989

This book (GTS 89) is the planned successor to *A Geologic Time Scale* by W. B. Harland, A. V. Cox, P. G. Llewellyn, C. A. G. Pickton, A. G. Smith and R. Walters published in 1982 (GTS 82). It adopts the same style and employs and develops similar methods, but it has been entirely reworked. The state of the art in 1989 is thus presented; but the data assembled provide a source of reference which will serve for some years.

The work develops and assesses a new calibration of the geologic time scale employing a new database (amended to 1988). The new scale is summarized on the book covers and on a wall chart to be published separately. The work has been coordinated in detail throughout; nevertheless each chapter can stand on its own.

Chapter 1 sets out the principles and summarizes the results.

Chapter 2 treats the chronometric scale which is defined numerically and applies mainly for Precambrian time.

Chapter 3 treats the chronostratic scale which is defined in rock. Both Chapters 2 and 3 summarize progress made towards international agreement concerning definition, nomenclature and classification.

Chapter 4 explains the principles on which the database of isotopic determinations has been refined by tests both stratigraphic and geochemical. Many of the GTS 82 age determinations have been rejected; the database now includes Cenozoic data; about 700 determinations are listed.

Chapter 5 calibrates the chronostratic scale, applying and developing the chronogram method to the data. This enables a time scale to be generated automatically and so allows comparison of the effects of the inclusion and exclusion of different classes of data.

Chapter 6 updates the magnetostratigraphic scale which in turn enables some refinement of the resulting time scale.

Chapter 7. Natural events are distinguished from the artifactual time scales and some global events are plotted linearly against the new time scale. This material has been adapted to the wall chart.

Appendices: (1) reproduces a list of the origins of stage and other names; (2) gives a system of stratigraphic abbreviations; (3) treats isotopic dating methods; (4) plots 125 chronograms; (5) plots the definitive magnetostratigraphic time scale; and (6) introduces the wall chart.

The list of nearly 1000 cited and other references amounts to a selected bibliography. There are both general and stratigraphic indexes.

The whole work results in a time scale for 1989. By showing its mode of construction the reader can assess how the scale may be modified as more critical data become available. The estimated uncertainty in the time scale values adopted is significantly less than in previously published estimates.

The cover

The panels on the cover summarize the time scale arrived at in the book.

The panel on the front cover (Figure 1.3) is plotted to a linear scale of approximately 20 million years (20 Ma) to a millimetre. To the left is the scale of chronometric divisions defined numerically; to the right are shown some chronostratic divisions of higher rank abstracted from the tabulation on the back cover.

The two panels on the back cover repeat the definitive scale (Figure 1.7), which is the principal conclusion of the book. The numerical values are the 1989 estimates. The divisions are spaced for typographic convenience.

The colour scheme of both panels is an approximation of the agreed international scheme (see Appendix 6).

The wall chart

A newly designed coloured wall chart, about 60 cm × 102 cm in size ('A geologic time scale 1989') is available as a separate publication, either on paper folded to A4 size or on unfolded laminated plastic. The geologic time scale is illustrated in three panels: a panel scale spanning all of geologic time and two others showing Phanerozoic and Quaternary details. The Quaternary panel shows an archeological classification, the $\delta^{18}O$ record and magnetostratigraphy together with other data. The Phanerozoic panel shows magnetostratigraphy, stages, epochs and periods as well as significant biologic, orogenic, continental and oceanic events.

A geologic time scale 1989

W. Brian HARLAND

Richard L. ARMSTRONG

Allen V. COX

Lorraine E. CRAIG

Alan G. SMITH

David G. SMITH

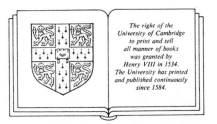

*The right of the
University of Cambridge
to print and sell
all manner of books
was granted by
Henry VIII in 1534.
The University has printed
and published continuously
since 1584.*

CAMBRIDGE UNIVERSITY PRESS

Cambridge

New York   Port Chester   Melbourne   Sydney

Published by the Press Syndicate of the University of Cambridge
The Pitt Building, Trumpington Street, Cambridge CB2 1RP
40 West 20th Street, New York, NY 10011, USA
10 Stamford Road, Oakleigh, Melbourne 3166, Australia

First published 1990

Printed in Great Britain at The Bath Press, Avon

British Library cataloguing in publication data

A Geological timescale 1989.
  1. Geological time
  I. Harland, W. B. (Walter Brian) II. Series
  551.701

Library of Congress cataloguing in publication data

A Geologic time scale 1989 / W. Brian Harland . . . [et al.].
    p. cm.
  Rev. ed. of: A Geologic time scale. 1982.
  Includes bibliographical references.
  ISBN 0 521 38361 7. – ISBN 0 521 38765 5 (paperback)
  1. Geological time. I. Harland, W. B. (Walter Brian), 1917–
II. Geologic time scale.
QE508.G3956 1990
551.7'01 – dc20   89-78175   CIP

ISBN 0 521 38361 7 hardback
ISBN 0 521 38765 5 paperback

v

Contents

Contents

Contents   vii

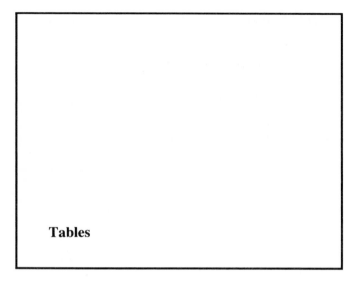

Tables

Figures

List of figures   ix

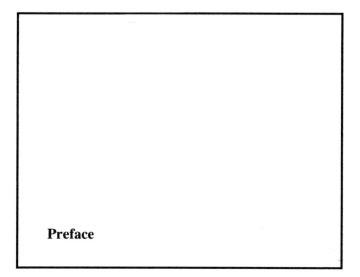

Preface

By the time that *A Geologic Time Scale* was published in 1982 (GTS 82), it had already been decided to undertake a revision of the book so as to incorporate newly available data as well as to remedy some shortcomings in it. Encouraged by sales of GTS 82, Cambridge University Press agreed to publish a new book.

The principal objective was to compile a new database for the project rather than to apply already published lists as had been done for GTS 82. Richard L. Armstrong's data compilation presented in 1976 (1978) had been used for the Paleozoic and Mesozoic scales in GTS 82 and the Hardenbol & Berggren 1978 Cenozoic scale had simply been incorporated unchanged into GTS 82. Consequently, GTS 82 was based mostly on data compiled before 1976. The new database in this work was updated to 1988.

Of the original six authors, Peter G. Llewellyn, Clive A. G. Pickton and Ronald Walters were no longer in a position to contribute to a new book. Peter Llewellyn, who did so much to promote GTS 82, had been given increased responsibility in British Petroleum (BP). Ron Walters, who retired from BP early through ill health, died in 1985. Clive Pickton was fully employed in the Deminex oil company.

BP continued to provide the main support for the new book. David Smith, Senior Stratigrapher in BP Research International at Sunbury, assumed responsibility for BP's part in the project. To ensure that a thorough revision of the book was carried out and the principal objective achieved, BP financed an 18-month Research Assistantship in the Department of Earth Sciences at the University of Cambridge. Under this grant, Lorraine E. Craig worked on the project from 1984 to 1986. She continued her work as an author while employed with Cambridge Arctic Shelf Programme as Senior Geologist.

Richard L. Armstrong, University of British Columbia, Vancouver, was an unwritten author of *A Geologic Time Scale 1982*, as his published list of isotopic dates was systematically employed throughout. He joined the authors principally to develop and refine the file of isotopic dates in Chapter 4 and the chronograms in Chapter 5 and to evaluate the modifications to the time scale in Chapter 6.

Allan V. Cox, formerly of the University of Stanford,

California, was to update his magnetostratigraphic chapter in GTS 82. He had already carried out considerable revision of the chapter before his death in January 1987. He remains an author of GTS 89 but cannot be held responsible for the work in its final form.

Alan G. Smith and W. Brian Harland continued with the work as before from the University of Cambridge.

The book is a joint work. All authors have criticized and contributed to successive drafts of each part, with the intention of producing a time scale that is more than a compilation. A unified study has been undertaken. Nevertheless, each has undertaken particular responsibilities as follows.

Chapters 1 to 3 were drafted by WBH. Chapter 3 is based on wide correspondence with international bodies by LEC who also revised most of these figures supplemented by critical input by others, especially DGS. Chapter 4 was drafted by RLA with the database compiled largely by RLA and LEC. Chapter 5 was drafted by AGS, while the chronogram work was done by LEC and AGS. Chapter 6 was begun by AVC but largely rewritten and expanded by AGS. After a more ambitious scheme was abandoned, DGS and AGS drafted Chapter 7, and the remaining figures were drafted by WBH with some input by DGS. Appendix 1 is unchanged from GTS 82 (appendix 2); Appendix 2 was developed from appendix 3 of GTS 82 by WBH; Appendix 3 continues the tables from appendix 1 of GTS 82 but has a new text drafted by RLA. The figures in Appendices 4 and 5 were computer designed by AGS and computer drawn by Steven Roberts, as were most of the figures in Chapters 5 and 6. The other figures were drawn for publication by Colin Yeomans with financial help from BP. A few were drawn by CUP, and two by RLA.

Coordination and assembly of the book was the responsibility of WBH with the help of Cambridge Arctic Shelf Programme staff – mainly Susan Morris who worked on the references at first, then all clerical work for two years by Isabella Antonio and the input and corrections for CUP of the final text and tables on disc by Ann Sparks.

The main work was done in the Department of Earth Sciences at the University of Cambridge where library facilities, office accommodation and much photocopying were provided by the University of Cambridge.

In the course of the work the procedures used in GTS 82 were followed and developed as outlined at the end of Chapter 1. This introductory chapter expresses the principles and reasoning behind the work and contains the resulting 'definitive' time scale that also appears on the back cover of the volume.

Chapters 2 and 3 began as one chapter as in GTS 82 but subsequent consideration led to the two time scales (chronometric and chronostratic respectively) being treated separately.

Chapter 2 begins with the 'unit of time' which appeared as Appendix 4 in GTS 82, and as it happens the chapter mainly concerns the Precambrian time scale.

Chapter 3 extends the Precambrian chronostratic scale using lunar data for the first part. For the revision of the Phanerozoic part it was policy to follow the conventions arrived at by the International Commission of Stratigraphy

(ICS), or that seemed likely to be agreed upon. When there was no such indication we made our own suggestions so that a relatively complete chronostratic scale is available for international use. Chairmen and secretaries of the many subcommissions of ICS responded generously to our own enquiries and their names are listed here. However, we must take responsibility for the scale as finally set out.

Thus Chapter 3 depended on help from K. Plumb (Precambrian Subcommission); J. Cowie (President of ICS, Chairman of Precambrian–Cambrian Boundary Working Group and Secretary Cambrian Subcommission); B. S. Norford (Cambrian–Ordovician Boundary Subcommission) and D. Price; C. H. Holland (Ordovician–Silurian Boundary Subcommission); R. B. Rickards (Silurian Subcommission); W. S. McKerrow (Ordovician–Devonian Subcommission); D. L. Dineley (Devonian Subcommission) and M. J. S. Rudwick; R. H. Wagner (Carboniferous Subcommission); W. H. C. Ramsbottom (Carboniferous Subcommission); J. M. Dickins (Gondwana Subcommission); K. S. W. Campbell (Permian Subcommission of Australia); Jin Yu-gan (Secretary Permian Subcommission); E. T. Tozer (Permian–Triassic Boundary and Triassic Subcommissions), D. Campbell and H. Campbell; A. Zeiss (Jurassic Subcommission) and A. Hallam; P. F. Rawson (Cretaceous Subcommission) and N. F. Hughes; T. Birkelund (Cretaceous Subcommission); K. Perch-Neilson (Cretaceous–Palaeogene Boundary Subcommission); D. G. Jenkins (Palaeogene Subcommission); J. Senes (Neogene Subcommission); R. P. Suggate (Quaternary Subcommission); J. E. Meulenkamp (Mediterranean Neogene Subcommission) and G. M. Richmond (Chairman, INQUA Working Group on Subdivision of Pleistocene).

J. W. Cowie and N. F. Hughes read late drafts of many of the chapters and Hughes also took upon himself much detailed work for the book as a whole as conscientiously as if he had been an author. The Quaternary Figure 3.17 was compiled by LEC and drawn by R. Khan of CASP from an ^{18}O climatic curve plotted especially for this work by N. J. Shackleton who also advised on the Quaternary time scale.

Chapter 3 is supported by Appendices 1 and 2. The list of (mostly) stage names with their origins is copied directly from Gregory & Barrett (1931) and has proved a useful supplement. Appendix 2 develops the three-letter stratigraphic abbreviations begun in GTS 82. Some respondents urged us not to change these from GTS 82 as they had been used in databases other than our own. The stratigraphic index attempts to list all stratigraphic names in the book.

Chapters 4 and 5 similarly began as one chapter but it was divided so that Chapter 4 developed the database and Chapter 5 derived the time scale from it.

In Chapter 4, after scrutiny of innumerable determinations, about 700 sample groups survived in the database (Table 4.2). This is therefore a quite original contribution. The chapter is supported by Appendix 3. We are indebted for help in the compilation of the material especially to A. G. Fisher, W. A. Berggren, P. F. Carr and N. J. Snelling (President of the Subcommission on Geochronology and Editor of the Geological Society publication). Proofs or preprints of their work were kindly provided in advance of publication in 1985 by W. A. Berggren and by N. J. Snelling (The Chronology of the Geological Record 1985) and were extensively used.

Chapter 5 records the application of the chronogram technique (as developed in GTS 82) to the data. It was necessary to agree the 127 stratigraphic stages early on so that the whole process could proceed in a logical fashion. A minor consequence was that further improvements to the chronostratic scale (Chapter 3) were not effected in the database. A major advantage was that a new scale could be produced automatically within a few hours, incorporating new data and varying the data input. From this it was found that glauconite determinations older than about 115 Ma gave systematically low values. In the final scale such data for ages in excess of 115 Ma were used as minimum estimates only.

Chapter 6 originally depended on a revision by AVC of his Chapter 4 in GTS 82. His death occasioned some delay in the whole project. It was some time before AGS could visit his laboratory in Stanford, California, to search through his archives and bring the data to Cambridge. St John's College, Cambridge, and CUP assisted with travel and research expenses for this visit and Stanford University provided computing and other facilities. The task then entailed familiarization and reworking the data. The help afforded in this by the former colleagues and students of AVC, particularly J. Tarduno, W. Harbert and D. Wilson, is acknowledged. Preprints, discussion and written comments were also provided by W. Berggren, T. Bralower, B. Haq, M. Lanphere, A. Mussett, W. Ogg and M. Steiner, with permission to quote their results. The time scale as produced from Chapter 5 was then subjected to further refinement in the light of magnetostratigraphic and ocean-floor spreading data. The penultimate draft of Chapter 6 benefited from peer review by E. Hailwood and D. Wilson. The 'definitive' magnetostratigraphic scale appears in twenty figures in Appendix 5. Figure 6.2 was redrawn from Berggren, Kent & Flynn 1985, figure 1, p.144.

Chapter 7 began ambitiously as a survey of geologic events and their bearing on the time scale, but was substantially reduced in scope. However, the pairs of figures 7.2 to 7.9 plot some selected events on linear time scales. Comments and suggestions were gratefully received from M. J. Hambrey, M. R. House, D. Rowley, D. W. Skelton and R. A. Spicer. The analogous figures in GTS 82, Chapter 5, were derived from a wall chart published simultaneously. However, for GTS 89 the figures were drawn for the book and a wall chart incorporating more data will appear as an independent publication.

At the end of the work we thank Cambridge University Press for having faith in our endeavours and for cooperating in the production of the book. We thank Caroline Roberts (Editor), and especially Karin Fancett (Sub-editor) without whose eagle eyes misprints, inconsistencies and infelicities on a much greater scale than appear here would have been perpetrated. Material for all the type was supplied on disc which Ann Sparks finalized for printing. Robert Scott joined CASP during the last stages of preparation for final submission to CUP. The work was typeset for CUP by the Paston Press of

Loddon, Norfolk, Robert Scott then completed the main stages of proof correction.

In conclusion, *A Geologic Time Scale*, though 'definitive' as of 1989, is in reality a transient time scale being no better than the data on which it is based. It will be superseded. However, we hope that the systematic work set out here will facilitate future revisions and focus attention on the opportunities for further work. For want of a better word we claim our method has been democratic in that the authors have worked together and also treated the data democratically without favouring some at the expense of others. We are certainly wrong in some of our conclusions but we claim to have made it clear how we arrived at them and that our conclusions cannot be far off the mark.

W. B. Harland (1)(3)
R. L. Armstrong (2)
L. E. Craig (3)(1)
A. G. Smith (1)
D. G. Smith (4)

(1) Department of Earth Sciences
    University of Cambridge
    Downing Street
    Cambridge CB2 3EQ
(2) Department of Geological Sciences
    University of British Columbia
    Vancouver
    British Columbia
    Canada V6T 2B4
(3) Cambridge Arctic Shelf Programme
    West Building
    Gravel Hill
    Huntingdon Road
    Cambridge CB3 ODJ
(4) BP Research International
    BP Research Centre
    Chertsey Road
    Sunbury-on-Thames
    Middlesex TW16 7LN

Postscript added to proofs in October 1989

The final copy and artwork were submitted to the press before the 28th International Geological Congress in Washington, July 1989, where the manuscript was available for inspection.

Our *definitive scale* (GTS 89) which appears here as Figure 1.7 were summarized and published as plastic-coated card in two formats by BP in July 1989 (for distribution at the Geological Congress). It was a prepublication of the time scale from this book with the same authorship.

For the Congress, a *1989 Global Stratigraphic Chart* published by the International Union of Geological Sciences was issued. This was compiled by J. W. Cowie and M. G. Bassett of the Bureau of the International Commission of Stratigraphy (ICS: IUGS). It appeared in the June issue of *Episodes* (as a tear-out unnumbered sheet Volume 12 No. 2). The *Phanerozoic* part was a chronostratic scale similar (but not identical) to our own. The Phanerozoic numerical values of this scale were taken from Snelling (1985). The *Precambrian* part was taken directly from the recommendations of the Subcommission on Precambrian stratigraphy (our Figure 2.2).

Cowie (former and current President of the International Commission of Stratigraphy) and Bassett (former secretary of ICS) had naturally been contacted by us and had received a draft of Chapter 3 for comment. It was therefore appropriate that the chronostratic scale used divisions mostly identical with those here; but not altogether. In addition to the named divisions as selected by us they offered alternative names without stating a preference as follows.

Stampian was given as alternative to Rupelian.
Latdorfian was given as alternative to Priabonian.
Selandian was given as alternative to Thanetian.
The alternative of merging Rhaetian into Norian is given.
Scythian was given as the 'Lower' Triassic stage.
Tatarian, Kazanian and Kungurian are continued but Changhsingian, Capitanian, Wordian and Roadian are given as alternatives.
'Upper Subsystem' was given for our Pennsylvanian. 'Lower Subsystem' was given for Mississippian. The boundary between them was taken at the initial Chokierian boundary (as in GTS 89 and GTS 82, but see additional note below on Carboniferous chronostratigraphy).
Alternatives for nearly all Carboniferous stages were given so that Soviet stages and non-Soviet stages appear in opposing columns throughout.
Ordovician stages were as in GTS 89. Alternative larger division names were offered but were not conterminous with the stage boundaries. Of these perhaps only Dariwilian might be noted which may be considered as an alternative to Dyfed.

We are not surprised by the above differences because it is evident that ICS must maintain an even-handed approach to recommendations until they are formally adopted and until then we prefer our unified scale.

For Cambrian time Cowie & Bassett presented a new sequence of stage names which we were not aware had been agreed by the Cambrian Subcommission. But, because J. W. Cowie has long been secretary of that Subcommission and because we have long awaited a lead from that body, we must take these names seriously even though no explanation has yet been published. They are as follows:

Cambrian	'Upper'	Trempealeauan
		Franconian
		Dresbachian (major part)
Cambrian	Middle	Dresbachian (minor part)
Cambrian	'Lower'	Mayaian
		Angaian
		Toyonian
		Botomian
		Atdabanian
		Tommotian
		Unnamed Division

For **Precambrian chronometry** the scheme given by Cowie, Ziegler & Remane (*Episodes* 12(2), 79–80) repeated here in Figure 2.2 was adopted by the Subcommission on Precambrian stratigraphy as a recommendation to the International Commission of Stratigraphy which accepted it at the International Geological Congress, Washington, July

1989. It thus only needs ratification by the International Union of Geological Sciences. This may not, however, be a pure formality because the scheme aroused some opposition as well as support. However, this classification (a chronometric one) was printed along with the chronostratic scale published by the IUGS as 1989 *Global Stratigraphic Chart* and was issued in *Episodes* 12(2) as explained above.

In this Precambrian scheme the terminal Precambrian division is not named and is referred to as 'Neoproterozoic III', it being left to the Working Group on the 'Terminal Precambrian system' to define and name it in due course. Definition will likely be in GSSP so further advancing the principle for a chronostratic Precambrian scale.

For **Cambrian chronostratigraphy** a GSSP for the initial Cambrian boundary has still not yet been decided; but, in papers presented to the Working Group in Washington, the point was made that both the Chinese and Siberian candidates may be difficult to correlate internationally because they are in carbonate facies, whereas Newfoundland, with a rich sequence of trace fossil zones in clastic rocks, may provide a greater correlation potential as well as a more complete succession. It was determined to come to a decision before the 1992 Congress.

For **Silurian chronometry**, M. A. Kleffner in 'A conodont-based Silurian chronostratigraphy' (*Geological Society of America Bulletin*, **101**, 904–912, July 1989) divided most of the Silurian Period (Telychian through Pridoli) into 48 'standard time units' as follows: Telychian 5; Sheinwoodian 4; Homerian $4\frac{1}{2}$; Gorstian 4; Kydfirduab $7\frac{1}{2}$; Pridoli 23. He used a graphic correlation method applied to 79 conodont species in 30 stratigraphic sections in Europe and North America.

Compared with our GTS 89 scale the duration ratios of the Wenlock and Ludlow epochs are similar but the Pridoli Epoch is much greater. If our tie-points be fixed then the Ludlow Epoch would be reduced to about half the Pridoli duration, i.e. Ludlow 5 Ma (GT 89 is 13.3 Ma) and Pridoli 10 Ma (GTS 89 is 2.2 Ma).

(Boucot (1975) estimated the Pridoli duration as half the Ludlow duration and Ludlow as slightly more than Gedinnian.)

For **Devonian chronostratigraphy** the initial Pragian boundary had been agreed (Chlupac & Oliver 1989 in *Episodes* 12(2), 109–113) at a GSSP at the base of bed 12 in the Velka Chuchle Quarry, in the northwest of Prague, Czechoslovakia. This was selected at a point coincident with the first occurrence of the conodont *Eognathodus sulcatus*.

The Subcommission on Devonian Stratigraphy, in a leaflet issued at the Congress entitled *Towards an International Language for the Devonian* (listing its 19 field and business meetings 1973 to 1989) gave the stage names and their definitions as in GTS 89.

They noted the initial Pragian, Eifelian, Frasnian and Famennian boundaries as already decided and plotted against columns of conodont zones, European ammonoid zones and North American ammonoid zones (from M. R. House, 1988, *Proceedings of Ussher Society*, **7**, 41–46). The initial Pragian boundary would be defined at or near the earliest record of *Eognathodus sulcatus* (as in our Figure 3.5). The initial Frasnian boundary would also be as in our text (p. 42) but to be defined in Belgium.

For **Carboniferous chronostratigraphy** Cowie & Bassett (1989) mentioned that a boundary between the two sub-systems, earlier than the initial Kinderscoutian boundary, i.e. initial Chokierian (which was being favoured by the Carboniferous Subcommission), coincided with neither the initial Pennsylvanian nor the initial Silesian boundaries.

We understand, however (fide W. H. C. Ramsbottom), that the initial Chokierian boundary will probably be selected for a GSSP and that it will not demonstrably conflict with the Mississippian–Pennsylvanian boundary so that these sub-period names are still in our opinion the leading contenders.

For **Triassic and Jurassic chronometry,** P. E. Olsen, S. Fowell, B. Cornet & W. K. Witte (IGC paper on Calibration of a Late Triassic–Early Jurassic time scale based on orbitally induced lake cycles) estimated at the Congress the duration of the Carnian Stage as 8 Ma (GTS 89 = 11.6 Ma); Norian and Rhaetian stages together as 21 Ma (15.4 Ma); and Hettangian as 2 Ma (4.5 Ma). Such changes would re-apportion this part of our scale with little further change. The revised Triassic–Jurassic boundary would be somewhat younger (205.5 Ma) than in GTS 89 (208 Ma).

For **Cretaceous chronometry** T. D. Herbert (IGC paper on Improvements of Cretaceous geochronology using the Earth's orbital pacemaker) from sediment cycles concluded durations as follows: Aptian 8 Ma (GTS 89 12.5 Ma); Albian 12 Ma (15 Ma) and Cenomanian 6.5 Ma (6.6 Ma). Such differences are well within the uncertainty of our chronogram calibration procedure.

For **Cenozoic chronometry** papers were presented at the Congress by C. C. Swisher (Single crystal ^{40}Ar–^{39}Ar dating and its application to the calibration of North American land mammal ages) and D. R. Prothero (Magnetostratigraphic calibration of Ar–Ar dating of Late Eocene–Oligocene terrestrial sequences of North America). He described precise re-determinations of ages and magnetic stratigraphy in western North America that were in progress leading to substantial revisions. GTS 89 has not used such mammalian stratigraphy to any great extent, but this work may significantly affect some magnetic reversal age assignments.

P. Andreieff presented a paper on Neogene chronology and chronostratigraphy: new data from Martinique and Guadeloupe (FWI) which gave Miocene ages in good agreement with GTS 89.

The stratigraphic position of the Priabonian–Rupelian boundary advocated by Premoli-Silva *et al.* in another IGC paper would shift its date to 35.0 Ma using the revised magnetostratigraphy of this study. This is close to the 34.5 Ma estimated in Premoli-Silva *et al.* (1988) and the 34.4 ± 0.6 Ma estimate of Glass *et al.* (1986), from independent data sets (see footnote to Table 6.6).

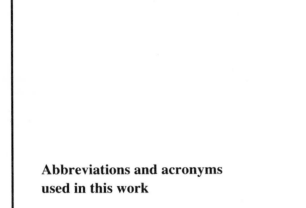

Abbreviations and acronyms used in this work

Organizations

AGI   American Geological Institute
BGS   British Geological Survey
BP   British Petroleum Company
CASP   Cambridge Arctic Shelf Programme
CGMW   Commission for the Geological Map of the World
CUP   Cambridge University Press
DNAG   Decade of North American Geology
DSDP   Deep Sea Drilling Project
GSC   Geological Survey of Canada
GSL   Geological Society of London
IAGA   International Association of Geomagnetism and Aeronomy
IAU   International Astronomical Union
ICS   International Commission of Stratigraphy
IGC   International Geological Congress
IGCP   International Geological Correlation Programme
INQUA   International Quaternary Association
IPOD   International Programme of Ocean Drilling
ISSC   International Subcommission on Stratigraphic Classification
IUGS   International Union of Geological Sciences
NZGS   New Zealand Geological Survey
UNESCO   United Nations Educational, Scientific and Cultural Organization
USGS   United States Geological Survey

Authors of this work are referred to in text as

RLA   R. L. Armstrong
AVC   A. V. Cox
LEC   L. E. Craig
WBH   W. B. Harland
AGS   A. G. Smith
DGS   D. G. Smith

Time scale publications (see **References** for further details)

PTS 64   The Phanerozoic Time-Scale (Harland *et al.* 1964)

PTS 71   The Phanerozoic Time-Scale – A Supplement (Harland & Francis 1971)
RLA 78   R. L. Armstrong 1978c
USGS 80   United States Geological Survey, Geologic Names Committee 1980 (Izett *et al.* 1980)
GSL 82   Geological Society of London (Snelling 1982)
NDS 82   Numerical Dating in Stratigraphy (Odin 1982b)
GTS 82   A Geologic Time Scale (Harland *et al.* 1982)
DNAG 83   Decade of North American Geology (Palmer 1983)
CGR 85
  (GSL 85)   The Chronology of the Geological Record (Snelling 1985b)
GTS 89   A Geologic Time Scale 1989 (this work)

Geoscientific concepts in this work

CMS   Chronometric scale
CSS   Chronostratic scale
FAD   First appearance datum
GCS   Geochronologic scale (*sensu lato*)
GCMS   Global chronometric (or geochronometric) scale
GCSS   Global chronostratic scale
GSSP   Global stratotype section and point
LAD   Last appearance datum
NCS   Natural chronologic scale
RCMS   Regional chronometric scale
RCSS   Regional chronostratic scale
RSSP   Regional (candidate) stratigraphic section and point
SSS   Standard stratigraphic scale replaced by GCSS
TCSS   Traditional chronostratic scale

Symbols

BP   Before Present (1950), also an oil company and sponsor of this work
SI   Système International d'Unités
Ga 10^9 years (Giga annum)
Ma 10^6 years (Mega annum)
ka 10^3 years (Kilo annum)
yr   year(s) (annum)

Stratigraphic abbreviations are listed in Appendix 2.

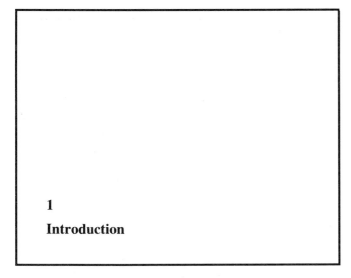

1

Introduction

1.1 Objective

A geologic time scale (geochronologic scale) is composed of standard stratigraphic divisions based on rock sequences and calibrated in years. It is thus (Figure 1.1) the joining of two different kinds of scale, a chronometric scale and a chronostratic scale. A chronometric scale (CMS) is based on units of duration – the standard second – hence a year. A chronostratic scale (CSS) is now conceived as a scale of rock sequences with standardized reference points selected in subsections, each particularly complete at and near the boundary and known as a boundary stratotype. The chronostratic scale is a convention to be agreed rather than

discovered, while its calibration in years is a matter for discovery or estimation rather than agreement. Whereas the chronostratic scale once agreed should generally stand unchanged, its evaluation will be subject to repeated revision. For this reason, no geologic time scale can be final and this particular attempt (GTS 89) must be qualified by '1989', its year of completion.

The concepts employed here have in the past been used in different combinations of words, for example **(standard) (global) (geo)chronostrat(igraph)ic (time) scale** is generally contracted to **chronostratic scale**. Other contractions may be clear enough in context. Such acronyms are shown in Figures 1.1, 1.2 and 1.4 and explained on p.xvi.

Regional chronostratic scales (RCSS) have gradually given rise (Figure 1.2) to the single global traditional stratigraphic scale (TCSS). This is being refined and standardized at global stratotype sections and points (GSSP) to give definition to the standard global chronostratic scale (GCSS); see Chapter 3. Regional points are competing for GSSP in this process. Chronometric scales were also first developed regionally and are being standardized as a single agreed standard chronometric scale (GCMS); see Chapter 2.

The calibration of any chronostratic scale in years yields what is commonly called a time scale (e.g. the title of this book, GTS). To distinguish such a calibration from other time scales they may be referred to generally as geochronologic (chronostratic & chronometric) scales (GCS).

Figure 1.2 shows the relationship of these entities.

Figure 1.2. Steps in the development of a time scale – GSSP = Global stratotype section and point; CMS = Chronometric scale; CSS = Chronostratic scale; GCS = Geochronologic scale; GCMS = Global chronometric scale; GCSS = Global chronostratic scale; TCSS = Traditional chronostratic scale; RCMS = Regional chronometric scale; RCSS = Regional chronostratic scale.

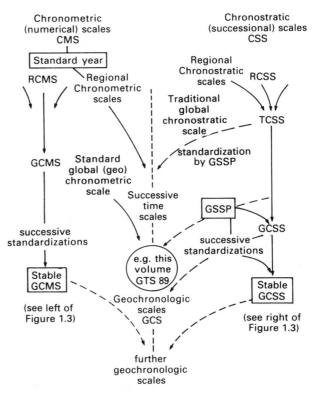

Figure 1.1. The making of a time scale.

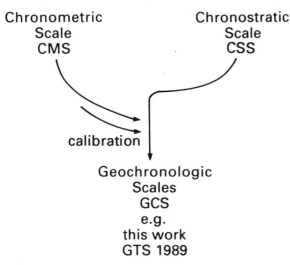

Figure 1.3. Outline chronometric and chronostratic time scales.

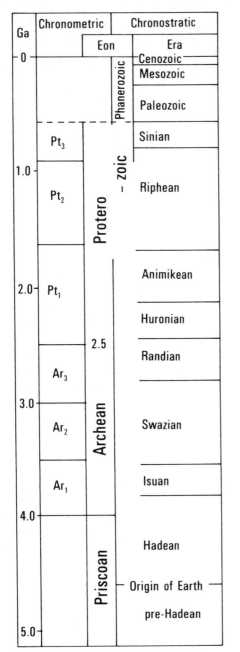

Figure 1.3 as an example illustrates side-by-side the evolving state of two such scales. Figure 1.4 identifies the logical steps in this process of calibrating sequences of natural events or natural chronologies (NCS).

1.2   The traditional chronostratic scale (TCSS)

The prodigious stratigraphic labours of the nineteenth century resulted in innumerable competing stratigraphic schemes. To impose some order the first International Geological Congress (IGC) in Paris in 1878 set as its objective the production of a standard stratigraphic scale. Suggestions were made for standard colours (Anon. 1882, pp.70–82), uniformity of geologic nomenclature (pp.82–4) and the adoption of uniform subdivisions (pp.85–7). There was also a review of several regional stratigraphic problems. In the

succeeding congress at Bologna in 1881, many of the above suggestions were taken substantially further, i.e. international maps were planned with standard colours for stratigraphic periods and rock types (e.g. Anon. 1882, pp.297–411) and annexes contained national contributions towards standardization of stratigraphic classification, etc. (pp.429–658).

In spite of this promising start, the IGCs did not have the continuing organization to carry these proposals through, except for the commissions set up to produce international maps. It was not until the establishment of the International Union of Geological Sciences (IUGS) around 1960 that the promise had a means of fulfilment, through the IUGS's Commission of Stratigraphy and its many subcommissions.

By 1878 the early belief that the stratigraphic systems and other divisions being described in any one place were natural chapters of Earth history was fading and the need to agree some conventions was widely recognized. Even so, the practice continued of describing stratal divisions largely as biostratigraphic units, and even today it is an article of faith for many that divisions of the developing international stratigraphic scale are defined by the fossil content of the rocks. To follow this through, however, leads to difficulties: boundaries may change with new fossil discoveries; boundaries defined by particular fossils will tend to be diachronous; there will be disagreement as to which taxa shall be definitive. So the traditional stratigraphic scale is of necessity evolving into a new kind of standard chronostratic scale.

1.3   Standardization of the global chronostratic scale (GCSS)

At the 1948 IGC one of the first attempts to standardize a stratigraphic boundary was made (the Pliocene–Pleistocene boundary by convention at the base of the Calabrian Stage in

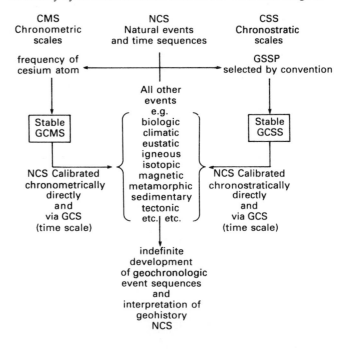

Figure 1.4. Steps in the calibration of sequences of natural events.

Italy). Such a decision had to be an agreed convention. It was agreed to standardize divisions at their boundaries only, and each boundary at only one locality. The international procedure to standardize such a boundary at a single point in a reference subsection was worked out by the Silurian–Devonian Boundary Working Group. Their procedure was first to agree upon the approximate position in the biostratigraphic sequence that would best fit existing usage and then to find a succession somewhere in the world where the Silurian–Devonian boundary was represented in fossiliferous rock with the best potential for correlation. If we take this procedure as a guide, the requirements for the standard **geo-** or **global chronostratic scale** (GCSS) follow.

A sequence of reference points in continuous subsections of uniform (marine) sedimentary facies selected with suitable potential for international correlation, state of preservation and access needs to be agreed. The precise reference point for each boundary is now known as the **global stratotype section and point** = GSSP (Cowie *et al.* 1986). It is then conceived as representing the point in time when that part of the rock was formed. Rock immediately below (formed before the point in time) or above (formed after it) should contain characters for correlation. Pairs of such points then define the intervening time span. The global chronostratic scale is ultimately defined by a sequence of GSSP.

The procedure has a significant consequence in the conception of chronostratic divisions. Before the standardization just described, the intervals were conceived as being the time equivalent of a rock unit that was already defined. Thus systems (series, stages or chronozones) were first described and the geologic periods (epochs, ages, chrons) were derived as the corresponding time intervals. The practice implied the assumption that the bases of such rock divisions are not diachronous. Even without that assumption, for a while 'body stratotypes' (type sections) were thought to be sufficient. The new procedure of defining boundary points effectively reverses the derivation. The time division (period, etc.) is now defined precisely by selecting initial and terminal points, while the corresponding rock formed in the interval (system, etc.) cannot be identified with certainty at its boundaries other than at the GSSP depending, as it does, on estimates of relative age by correlation. This generally yields a chronostratically well-dated main body of the rock division, but with uncertain initial and terminal boundaries away from the GSSP. To emphasize the primacy of time in such a time scale, Early, Mid and Late are used rather than Lower, Middle and Upper, for subdivisions of the primary named intervals.

Various names have been proposed for the newly standardized scale. The Geological Society of London (GSL) used **standard stratigraphic scale** (SSS), in contrast to the traditional stratigraphic scale (TSS) and regional stratigraphic scales (RSS) out of which it was evolving (George *et al.* 1967). The International Subcommission on Stratigraphic Classification (ISSC) referred to it as the **standard global chronostratigraphic scale** in the *International Stratigraphic Guide* (Hedberg 1976). Both the American Stratigraphic Code, and the ISSC *Guide* into which it grew, confused the matter somewhat. They divided the standard scale as described here into two categories: periods and systems. Their **geochronologic units** refer to periods, etc. and **chronostratigraphic units** to systems, etc. It is obvious that time and rock are different (e.g. as indicated by the words period and system), but when defined they both derive from the same standard reference points. The two apparently distinct disciplines, geochronology and chronostratigraphy in Hedberg's terminology, are thus different aspects of a single procedure.

It is both traditional and convenient to use a hierarchy of names for stratigraphic intervals (era, period, epoch, age, chron). The use of the hierarchy is largely a matter of habit but it has its uses in both economy of description and in describing events of different duration or uncertainty of correlation. The chronostratic divisions of any rank in the hierarchy are defined in the same way by GSSP. There is no difference in principle between a GSSP defining an initial chron or an initial era boundary. Indeed, the same GSSP may serve as the initial boundary for several ranks in the hierarchy. The ranks are then conterminous and the principle of conterminosity simplifies the use of a hierarchy.

The names for the spans are generally those favoured from classic sections. Once selected for the GCSS, however, they cease to have local reference and must be used internationally for the time span defined by the limiting points. It is convenient to retain familiar names but, when redefined at some distance from the eponymous locality, the local geologists must accept that the name has acquired a new meaning and possibly avoid its old use by renaming the original rock unit. For example, when Pridoli was accepted as an epoch name in the GCSS the original Pridoli Formation from which it came was renamed.

The above principles developed for the global chronostratic scale can be applied to regional chronostratic scales (RCSS) as a step in the process of correlation, but the multiplication of scales is not generally helpful. The work of standardization is considerable and need not be multiplied. Until such a global time scale is standardized, points in regional scales (RSSP) may be regarded as candidates for GSSP. The development of the global chronostratic scale is addressed further in Chapter 3.

1.4   The global chronometric or geochronometric scale (GCMS)

The proposal for a global chronometric scale is quite different. The scale is linear, i.e. it is compounded of units of equal duration. Therefore all that is essential is to define a standard unit – a second of time based on the cesium 'clock' – and so derive one year. In the same way that a linear scale of length is constructed from unit lengths and is so defined, the chronometric scale exists by virtue of the definition of a unit of duration (see Chapter 2).

A further convention is to compound the units into longer, named intervals. Such a scheme of millennia (10^3 yr), gigennia (10^9 yr), etc. is by no means essential but, as with the higher ranks of the chronostratic hierarchy, they may be convenient in general expressions of age. Unlike the chronostratic divisions they will be defined not by reference points in rock but by initial and terminal points, each defined by a finite number of units of duration BP (Before Present – by convention in ^{14}C determinations counted from 1950). These matters are taken further in Chapter 2.

There are those who think that there is some advantage in treating Precambrian history as sufficiently different from Phanerozoic history as to require the use only of named chronometric divisions for Precambrian time. This opinion derived from the general absence of fossils as it seemed in the Precambrian rocks, which is no longer so. Thus the Subcommission on Precambrian Stratigraphy of the IUGS agreed in 1976 that the boundary between Archean and Proterozoic should be defined at 2500 Ma exactly (but not yet ratified by the International Commission of Stratigraphy (ICS)); moreover, other subdivisions of Precambrian time have also been proposed along the same lines, as will be seen in Chapter 2. Our preferred alternative would be to extend the scheme of named chronostratic divisions backwards into Precambrian time (as is developed in Chapter 3). A parallel development of named chronometric divisions forward through Phanerozoic time is not proposed here but cannot be dismissed.

There is thus only one standard for a general chronometric scale (the second and hence the year). For the geochronometric scale **present** needs to be defined and the difference between BP and BC is generally irrelevant. The matter at issue is the naming of spans of time and their numerical definitions. Together these should provide a stable GCMS (see Chapter 2).

1.5   Statement of age

The two conventional scales outlined above (chrono-stratic and chronometric) do not in themselves enable us to date or to time-correlate rocks one with another. Their function is to provide agreed standards for expressing the ages of rocks. They reduce the number of ways in which geologic ages are stated (ideally to two: one by named intervals between defined events (GSSP) and one numerical). Both are conventions and neither is better than the other. The two do not and cannot define each other and so they are both needed. According to circumstances, some rocks can be dated chronometrically more precisely than chronostratically and for others more precise ages can be given chronostratically. Only if the conversion of one scale to the other were far more precise than it now is could any age usefully be given either in years or in named chrons.

Both scales are decided by convention and have therefore been referred to as artificial; but the word artificial has an unfortunate connotation of inferiority. The scales are artifacts and artifactual would better express their conventional nature.

Alternative terminology for the two ways of stating age (chronometric and chronostratic) have been **relative** and **absolute**, which is an unfortunate distinction because both are relative and neither is absolute. In German, **Phanomologische Alter** and **Chronometrische Alter** have been used (Englehardt & Zimmermann 1982, p.114). **Stratigraphic** is also unsuitable for CSS because in the wider sense of stratigraphy as geo-history both expressions for age are stratigraphic.

There is some similarity between the above pairs and McTaggart's (1908) **A series** and **B series**. These two concepts (argued by philosophers rather than geoscientists) have been applied to human experience of ever changing past, present and future (A series). This gives an objective sequence of events (whether conscious or historical) which relate to other events being earlier, coeval or later. Such events may be consolidated in a chronometrically calibrated sequence (B series, i.e. the **real time** of Mellor 1981).

1.6   Natural chronologies (NCS) and event sequences

Particular historical geological phenomena are commonly known as events. They are of many kinds, for example magmatic, sedimentary, biologic, tectonic, magnetic, eustatic, climatic, celestial and geochemical. They are the basis of stratigraphy in the full sense of geohistory and especially of event sequences with time-correlation. The current term **event stratigraphy** is redundant because stratigraphy has always depended on interpretation of strata in terms of events. Some events are more obvious, identifiable, widespread, predictable, or sudden than others. All are the subjects of investigation. Therefore all are in part matters of opinion – the substance of hypotheses liable to revision.

Interest in these natural phenomena motivates science. The time scales already discussed are tools only for the study of phenomena. The scales have no application without time-correlation, which is entirely dependent on the interpretation of natural phenomena. Events are such interpretations.

A GSSP may be conceived as relating to phenomena and therefore to interpreted events. For example, the tip of the 'golden spike' (the colloquial term for a GSSP) may separate two sand grains, one deposited before and the other after the designated point in time. In a uniform sequence the point has little significance as a special event – it is almost a non-event and so it can be the more readily agreed as a point in the conventional chronostratic scale.

Other events – a particular astronomical year – or (later) the perturbations of the cesium atom have been selected as the basis of the chronometric scale.

The distinction between time scale as a tool for the study of natural history and something expressing natural history itself has only slowly become recognized. It was previously assumed that most classes and subclasses of event would somehow fit into the divisions of the story that was gradually being revealed. Calibration of event sequences against independent time scales liberates the phenomena for investigation of their interplay. Some examples of classes of event sequence follow.

Sedimentation or magmatism (with subsequent alteration) yields bodies of rock that provide the most convenient descriptive units (formations) as introduced in Section 1.7.

Biologic evolutionary history, especially for Phanerozoic time, has given us not only the principal means of time-correlation but the basis of the unique progressive traditional stratigraphic scale. Definition of biozones, through biochrons, to chrons defined by GSSP, is only now slowly taking place. For this reason the figures in Chapter 3 list selected biostratigraphic events. In due course the distinction of GSSP from supposed extinction and other events will enable the more thorough investigation of biologic evolution. In this work, although there is no certainty as to the rate of biologic evolution, the approximate number of chrons has been used as a very rough and ready measure for the duration of each stage.

Tectonic events were once thought to punctuate and even separate the natural chapters of Earth history. The tendency to assume, rather than question, synchroneity of Earth movements is unfortunately perpetuated in the use of named episodes as time terms (e.g. Stille 1924). The meaning of Caledonian outside the North Atlantic Caledonide Orogen, or of Cimmerian outside the Tethyan region (see Sengor 1985) is not clear and stratigraphic precision is lost. The agreed time scale should be used in preference where a local tectonic term is not appropriate. Some such tectonic episodes are plotted in the figures in Chapter 7 as events and not in any way to substitute for a time scale. Ocean spreading is a more gradual sequence of events whose supposed uniformity through certain spans of time provides independent information to supplement biostratigraphic data in the interpolation and extrapolation of chrons and stages between tie-points as used in Chapter 6.

Magnetic reversals, in which mainly two alternate states are recorded, make a binary scale. Combined with other data for correlation and age assessment, magnetostratigraphy has become a powerful tool with global application. This is the subject of Chapter 6. Whereas the method enhances time-correlation, the sequence in time is not yet fully understood.

Eustatic changes are also global and almost geologically instantaneous, but they are inevitably confused by local and regional crustal controls of sea level. Even when eustatic changes are identified there may be different causes, from rapid transfer of water between the sea and continental ice sheets, through gradual changes in configuration of the ocean basins (by tectonic and gradational processes) to very slow changes in the mass and/or volume of the hydrosphere. When the contributing factors are better understood, eustatic change will enable some precise time-correlations to be made. However, the application of sea level to a time scale through predictable cycles is promising but still problematic.

Climatic change is also polygenetic and valuable in correlation (sedimentary facies, especially tillites). Relatively rapid climatic change is likely to be reflected regionally, if not globally, in the sedimentary record. Apart from climatic factors to do with atmospheric composition and the configuration and proportions of land and sea, there are what may be termed **celestial** factors.

Celestial phenomena are of four main kinds, some of which may prove to have predictive value and so ultimately serve to refine the time scale.

(i) The most obvious and certain of these is **Earth's daily rotation**, with the possibility of an annual scale based on summer and winter sedimentation rates even within one climatic regime. Biologic tissue accretion has also been so claimed. Such scales need to be corrected for the slowing down of Earth's spin.

(ii) Short-term fluctuations in **solar output** (e.g. sunspot 'cycles') are not always easy to distinguish from annual laminae. Even longer-term solar cycles have been suggested (e.g. Williams 1989) and solar output probably changed systematically throughout Earth history. It is possible also that solar output has varied irregularly from time to time. Solar output received on Earth as the 'solar constant' is the single most powerful factor in Earth's climate.

(iii) However, the best understood climatic fluctuation results from **planetary perturbations** of three kinds: **tilt** of Earth's spin axis; **eccentricity** of Earth's orbit; and **precession** of equinoxes. Together these yield a complex curve for the 'solar constant' at different latitudes, as first suggested by Croll (1875) and Milankovitch (1920, 1930). This was later pioneered for the Quaternary time scale by Zeuner (e.g. 1945), Emiliani (1965 *et seq.*) and by Evans (1971) until the process became well established not only for Quaternary time (e.g. Shackleton & Opdyke 1973, Imbrie & Imbrie 1979, Williams 1981). Such cyclic phenomena based on these planetary components with periods approximating to 40 ka, 100 ka and 20 ka respectively are already enabling effective refinement of the time scale.

(iv) Fourth amongst celestial phenomena are the major meteorite or bolide **impacts** leaving both chemical and possibly climatic (hence biologic) traces in the record. These may be random, yet of decreasing frequency in the evolution of the Solar System. It has, however, been agreed that some cyclicity may relate to Earth's passage through denser parts of the Solar System or nebula.

Radioactivity is the phenomenon, first applied to age determinations by chemical and later by isotopic analysis. The calibration of the chronostratic scale by decay rates is perhaps the main subject of this book, being addressed in Chapters 4 and 5 and in Appendix 3. The decay series is an atomic event sequence yielding apparent ages of the containing minerals and rocks. Isotopic dating is the most powerful method for calibrating the chronostratic scale. It is distinct also from the chronometric scale against which it is in turn calibrated.

The above classes of events, each with distinctive properties, all have some bearing on the time scale being calibrated by it. Events also support the scale through correlation. Such events give further promise, yet to be exploited in future time scales. A selection of events appears in the linear time plots in Chapter 7 (Figures 7.3, 7.5, 7.7 and 7.9).

1.7 Local rock units

Rock is the ultimate objective reference for both the study of the natural phenomena of geologic history and for the evidence of age. There is a well-established geologic convention for describing and classifying rock in named units, i.e. formations combined into groups, supergroups and complexes or divided into members and beds (e.g. Hedberg 1976).

All stratigraphic divisions as originally described were in effect local rock units even if they were intended to have regional or global significance. There is therefore most confusion in the eponymous areas (mainly in Europe) of the global chronostratic scale (GCSS). The original systems, series and stages were initially described as bodies of rock, and in many cases this usage persists explicitly. It is easier to proceed logically outside Europe where named rock units cannot be confused with named divisions of the GCSS.

1.8 Geochronologic scales (GCS)

To return to the point at which we began. A geologic time scale is really a dual scale: chronostratic and chronometric, fitted to each other more or less successfully.

Table 1.1. *Standard time scale of geochronology, on the basis of the Eocene Period for a time unit or geochrone (H. S. Williams 1893, p.295)*

Recent	1	
Quaternary		3
Pliocene	1	
Miocene		
Eocene	1	
Cretaceous	4	
Jurassic	3	9
Triassic	2	
Carboniferous	6	
Devonian	5	
Upper Silurian	4	45
Lower Silurian	15	
or Ordovician	15	
Cambrian	15	

Table 1.1 shows one of the earliest attempts at a geologic time scale, constructed in 1893 before methods of dating based on radioactivity were conceived. H. S. Williams (1893) was one of many who attempted a numerical scale. He used **geochrones** as his unit; his geochrone being the duration of a well-known period: the Eocene geochrone.

Wager (1964) summarized the history of numerical time scales in the *Phanerozoic Time-Scale* – a symposium dedicated to Arthur Holmes by the Geological Society of London. Indeed, Holmes' classic *Age of the Earth* (1913) developed one of the earliest time scales, which was prophetic in the measure of Phanerozoic time. Methods of dating based on radioactivity include both the earlier chemical analysis of radioactive and radiogenic element content developed and the later mass spectrometric isotopic analysis which has been widely applied since the 1950s. Such ages are now commonly referred to as isotopic ages or dates.

In this work the history of radiometric methods is not discussed but for convenience some of the resulting geochronologic scales are compared in Figure 1.5 and some of these are compared graphically in Figure 1.6. On the left-hand side the major chronostratic divisions (periods to sub-epochs) are tabulated and on the right-hand side the three-letter codes for stages are given (also listed in Appendix 2). This is not a complete list of stages, rather the initial stage of each division on the left is given so that the horizontal lines for the initial boundaries coincide. On those lines the values of selected time scales from 15 publications are given in separate columns in approximate order of publication. The references to each are at the back of this volume. A few explanatory comments on those scales may be helpful. The acronyms at the head of some columns are explained here and are also used elsewhere in this volume. They are also listed in the abbreviations list on p.xvi.

The first four columns with time scales by Holmes in 1937, 1947 and 1960 and by Kulp in 1961 show the evolution of scales based on chemical to those based on isotopic analysis.

PTS 64 was *The Phanerozoic Time-Scale*, a Geological Society of London (GSL) cooperative enterprise which developed a new database comprising 'items' (PTS) 1–337 (Harland, Smith & Wilcock 1964).

PTS 71, *The Phanerozoic Time-Scale – A Supplement*, developed the GSL project including a database with additional items (PTS 338–404, Harland & Francis 1971). It also included a substantial paper: *Towards a Pleistocene Time-Scale* by P. Evans. The pre-Pleistocene time scale in that volume was compiled by Lambert who considered the accumulated database (PTS 1–404) and rejected a majority of items as not reliable. He computed three scales for three different half-life values of rubidium; his extreme values are shown in the pre-Cenozoic column of PTS 71 (Lambert 1971, table 1, p.25). Berggren's 1969 scale for Cenozoic ages was adopted in PTS 71 (see Figure 1.5 this volume).

Van Eysinga's wall chart (1975) was widely distributed and the scale thereon is listed in the seventh column. His method of obtaining the values was not explained and is therefore assumed to have been by inspection of other scales.

In 1976, decay constants were standardized internationally by the IUGS Subcommission on Geochronology (Steiger & Jäger 1977 as well as in *Contributions to the Geologic Time Scale*, Cohee, Glaessner & Hedberg 1978) from IGC, Sydney, 1976. Pre- and post-1976 time scales are not strictly comparable because of the different decay constants. After 1978 the standardized and internally consistent set of decay constants have been used for all published scales. The first scale using the new constants was that by Armstrong (1978c) in the same volume.

The Geologic Names Committee of the US Geological Survey circulated a revised time scale in 1980 (Izett *et al.* 1980) that has received wide distribution on the back cover of the journal *Isochron/West*. Based on a somewhat larger collection of data, it eliminated the positive bias of its immediate predecessors, while maintaining a similar fit to the 1989 scale (see USGS 1980 column on Figure 1.5 and also Figure 1.6).

The first edition of our work (GTS 82) borrowed the database Armstrong had used for our newly systematized method of producing a time scale (Harland *et al.* 1982). Indeed, a developing database has been common to all attempts at making time scales. We adopted Berggren & Van Couvering's 1974 and Hardenbol & Berggren's 1978 ages for Cenozoic stratigraphic boundaries.

In parallel with the work towards GTS 82, *Numerical Dating in Stratigraphy* (NDS, edited by Odin, 1982b) was in progress. This substantial work in two volumes developed a largely new database listed as 'abstracts' (analogous to 'items' in PTS), NDS 1–251. The resulting scale is given in the eleventh column of Figure 1.5. This work was supplemented by separate papers as follows: Odin 1982a, Odin & Curry 1982, Odin & Kennedy 1982, and Odin *et al.* 1983. One notable feature of this work was the value for the initial Cambrian boundary, younger than had been used in earlier scales (Harland 1983b), and a general bias towards young Paleozoic dates. The database is heavily weighted in favour of K–Ar determinations on 'glauconies' for Mesozoic and Cenozoic time. The result, as illustrated in Figure 1.6, is a large and consistent negative bias when compared to our 1989 scale.

The Geological Society of London was at the same time working for a new time scale to update PTS 64. At a symposium meeting the preliminary scale (GSL 1982) was presented (Snelling 1982). After the meeting the work continued and proofs of NDS and GTS 1982 were made available. The results were eventually published by the GSL in the large volume *The Chronology of the Geological Record* (CGR edited by Snelling 1985b). This was a symposium volume in which at least six independent scales were argued in different chapters. These are distinguished in our Figure 1.5 by the letters (a) to (f). Snelling produced a composite scale (g) from the foregoing in a final chapter. The scales (a) to (f) were authored as follows: (a) Berggren, Kent & Van Couvering; (b) Berggren, Kent & Flynn; (c) Hallam, Hancock, La Brecque, Lowrie & Channell; (d) Forster & Warrington; (e) McKerrow, Lambert & Cocks; (f) Gale.

Figure 1.5. Comparison of earlier time scales with GTS 89.

Period & sub-period	Epoch & sub-epoch	Selected Initial Stages	This work	HAQ and VAN EYSINGA 1987	CGR (GSL) 1985	DNAG 1983	GTS 1982	NDS 1982	GSL 1982	USGS 1980	ARMSTRONG 1978c	VAN EYSINGA 1975	BERGGREN 1969 / PTS 1971	PTS (GSL) 1964	KULP 1961	HOLMES 1960	HOLMES 1947	HOLMES 1937
Quaternary	Holocene		0.01		(a)	0.01	0.01					0.01						
	Pleistocene		1.64	1.67	1.6	1.6	2.0	1.9	1.8–2.0	2.0			1.8	1.5–2.0	1	1	1	
Neogene	Pliocene 3/2/1	Zan	5.2	5.1	5.3	5.3	5.1	5.3	5.3	5.0		5	5.5	7	13	11	12	16
	Miocene 3	Tor	10.4		10.4	11.2	11.3						10.0					
	Miocene 2	Lan	16.3		16.2	16.6	14.4						14.0		25	25	26	32
	Miocene 1	Aqt	23.3	24	23.7 (b)	23.7	24.6	23		24		22.5	22.5	27	36	40	38	48
Paleogene	Oligocene 2	Cht	29.3		30.0	30.0	32.8	34		38		38	32.0	37–38	45	60	58	68
	Oligocene 1	Rup	35.4	36	36.6	36.6	38.0		36.5				36.0		52			
	Eocene 3	Prb	38.6	40	40.0	40.0	42	45		55	65	55	45.0	53–54	58	70		
	Eocene 2	Lut	50.0	50	52.0	52.0	50.5	53	58	68	90	65	49.0	65	63			
	Eocene 1	Ypr	56.5	55	57.8	57.8	54.9	65					53.5					
	Paleocene	Dan	65.0	66	66.4 (g 65)	66.4	65			96	96	80	65.0	88	110	135	127	108
Cretaceous	Gulf / Senonian	Con	88.5	97.5	95	88.5	88.5	95	95.5		123	100	95	100	135			
	Gallic	Cen	97.0	140	135	97.5	97.5	130	135	138	143	118	135	118		180	152	145
	K1 / Neocomian	Brm	132	160	152	124	125	150	154	205	162	141	195–205	136–146	181			
		Ber	146	184	180 (d)	144	144	(178)			182	160		157	200	225	182	193
Jurassic	Malm / Dogger	Oxf	157	210	205	163	163	204	230	240	212	195	235–245	172	230	270	203	227
	Dogger / Lias	Aal	178	230	230	187	188	229	245		234	230		192	280			
	Lias	Het	208	243	242	208	213	239	250	290	242	280	275–290	205				
Triassic	Tr 1	Crn	235	250	250	230	231	245	290	330	247	325	360–380	215	320	350	255	275
	Tr 2	Ans	241	270	270	240	243	290	325	360	269	345	405–430	225	345	400	313	313
	Tr 3 / Scythian	Gri	245	290	290	245	248	320	365	410	289	370	435–460	256.5	365	440	350	392
Permian	Zechstein / Lopingian	Lgt	250	305		263	263	360	409	435	307	395	500–530	280	390	500	430	431
	Rotliegendes / Guadalupian	Ufi	256	314	325	286	286	375	439	500	330	423	570–610	292.5	405	600	510	470
		Ass	290	325	354	296	296	385	464	570	341	435		312.5	425			
Carboniferous / Pennsylvanian	Gzelian	Kla	295	336	374	315	315	400	481		355	450		325	445			
	Kasimovian	Kre	303	352	391	320	320	418	499		367	500		337.5	500			
	Moscovian (Ste)	Mya	305	360	412	333	333	425	520		385	515		345	530			
	Bashkirian (Wes / Nam)	Vrk	311	376		352	352	438			396	540		359	~600			
Carboniferous / Mississippian	Serpukhovian	Che	318	390		360	360	455			416	570		370				
	Visean	Kin	323	410		374	374	470			432			395				
	Tournaisian	Pnd	333	440	425	387	387	475			440			435				
Devonian	D3 Late	Chd	350	449	442	408	408	495			446			445				
	D2 Mid	Has	363	460	454	414	414	530			454			500				
	D1 Early	Frs	377	480	461	421	421				465			515				
		Eif	386	500	470	428	428				477			540				
Silurian	Pridoli	Lok	409		492	438	438				492			570				
	Ludlow	Prd	411		513	448	448				500							
	Wenlock	Gor	424			458	458				510							
	Llandovery	She	430			468	468				524							
Bala	Ashgill	Rhu	439		570	478	478				545							
	Caradoc	Pus	443			488	488				575							
Dyfed	Llandeilo	Cos	464			505	505											
	Llanvirn	Llol	469			523	523											
Canadian	Arenig	Linl	476			540	540											
		Arg	493			570	570–590											
	Tremadoc	Tre	510				630											
Cambrian	Merioneth	Mnt	517				670											
	St David's	Sol	536															
	Caerfai	Tom	570															
Vendian	Ediacara	Won	590															
	Varanger	Sma	610															

Figure 1.6. Graphic comparison of time scales.

Deviations from 1989 Time Scale in Ma

The height of the dark areas gives the positive and negative deviations from the time scale presented in this book (GTS 89). On the right side of the figure are quantitative comparisons obtained by computing the percentage deviations at every point of comparison and then calculating the average deviation and net deviation as expressions of difference, and systematic bias between scales. This procedure normalizes the effect of differing numbers of points of comparison. All the comparisons are made on the basis of modern decay constants (Steiger & Jäger 1977). The fine dotted lines are 2% deviation bands. There is a general convergence with the new time scale in going from earlier to later proposals. The two earliest scales show large deviations but little systematic bias. Lambert (1971) picked somewhat old early Paleozoic boundaries, giving his scale (PTS 1971) a distinct positive bias. Armstrong (1978) likewise was positively biased, especially for Paleozoic time, but was less biased than Lambert's scale and only half as discrepant as previous scales. Except for Odin (NDS 1982), where a strong negative bias and large average discrepancy is evident, all 1980s scales show little net bias (less than 1%) and similar average discrepancies (of the order of 1.5 to 2.5%) and are thus of rather comparable quality, relative to the new scale. The DNAG (1983) and Haq and van Eysinga (1987) scales are the most similar to the new scale.

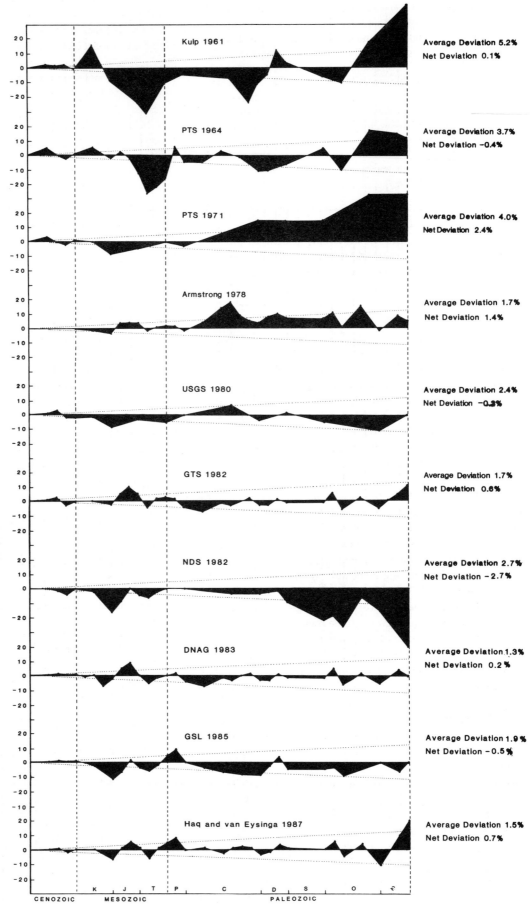

Kulp 1961 — Average Deviation 5.2% / Net Deviation 0.1%

PTS 1964 — Average Deviation 3.7% / Net Deviation −0.4%

PTS 1971 — Average Deviation 4.0% / Net Deviation 2.4%

Armstrong 1978 — Average Deviation 1.7% / Net Deviation 1.4%

USGS 1980 — Average Deviation 2.4% / Net Deviation −0.3%

GTS 1982 — Average Deviation 1.7% / Net Deviation 0.6%

NDS 1982 — Average Deviation 2.7% / Net Deviation −2.7%

DNAG 1983 — Average Deviation 1.3% / Net Deviation 0.2%

GSL 1985 — Average Deviation 1.9% / Net Deviation −0.5%

Haq and van Eysinga 1987 — Average Deviation 1.5% / Net Deviation 0.7%

CENOZOIC   MESOZOIC   PALEOZOIC

Points of Comparison

The *Decade of North American Geology* (DNAG 83) published its own scale (Palmer 1983) which was explicitly an eclectic one based largely on preprints of Berggren and coworkers for the Cenozoic scale. Much of the Cretaceous and remaining scale was modified on the basis of work to appear in NDS 82 or adopted intact from GTS 82.

A new 'time table' (Haq & Van Eysinga 1987) was a compilation using available sources but mainly based on Haq *et al.* (1987).

The last time scale column in Figure 1.5 is our own current scale produced, as explained below, from a newly assembled database and by methods improved from GTS 82.

Too late for detailed consideration here is the short review paper and coloured chart comparing 20 numerical time scales from 1917 to 1986 (Menning 1989).

Inspection of this material might lead one to think that there is not much to choose between most scales published since 1980. However, each new attempt strengthens and refines the available data by addition and removal. Some parts of the scale seem to be relatively stable, other parts will undoubtedly change significantly according to the accidents of the available geologic record. The applications throughout geoscience, however, make the effort worthwhile.

These time scales are made by interpolation between, and extrapolation from, tie-points that relate to particular rock samples in which a fortunate combination of circumstances allows, for example, isotopic determinations on rocks closely related to those with fossils that can be used to correlate with the stratotype. Methods, other than isotopic, of obtaining comparative durations are relative thicknesses, rhythmites, numbers of similar biozones and rates of ocean spreading. Other methods of correlation (besides biostratigraphic) are lithostratigraphic, paleomagnetic and paleoclimatic. Indeed, the best points are those in rocks with the most varied correlation potential. But it is a matter of chance where in the geologic columns on Earth such useful rocks are found. This leads to some parts of the combined scale being far more refined than others.

Some statement or qualification of uncertainty of each calibration is useful. There are several elements to be considered. Experimental error, usually expressed as standard deviation, informs only that the same rock sample gave or would give such a scatter of determinations. The environmental histories of the mineral in the rock and of the rock itself modify the closed system assumption on which the age calculations are based. Therefore it is well not to refer to isotopic dates as absolute ages but rather as **apparent**, being probably different from the **true** age. There are uncertainties that become evident when different samples of the same unit are analysed and uncertainties are introduced by interpolation between determined points. There are also uncertainties due to correlation precision of paleontologic methods, or whatever other methods are used.

There are structural uncertainties as to whether the relationships between rocks determined are as assumed. No way has yet been devised of expressing these uncertainties concisely. The expression ± is misleading as it generally refers only to the consistency of the analytical determinations expressed in the value given. One past cause of discrepancy in values reported for the same rock has been resolved by

international cooperation, namely, the agreement to use one internally consistent set of decay constants. Appendix 3 provides the means to convert older dates to those based on the conventional constants agreed in 1976.

It is the main purpose of this edition, as of the previous one, to show not so much our conclusions for a preferred time scale for 1982 and 1989 – which do not differ greatly from previous efforts – but rather to show clearly how such a scale has been constructed. Believing, as we do, that a scale is in need of frequent revision to take account of new results from many disciplines, we hope this work will facilitate future efforts by showing how it was made.

1.9   Procedures adopted in the construction of this time scale (GTS 89)

The **chronometric scale** is a purely numerical one. Its standardization and conventional divisions are outlined in Chapter 2.

In Chapter 3 the **chronostratic scale** is developed and additional information is given in Appendices 1 and 2.

The use of isotopic methods to calibrate rocks chronometrically and the assembly of a database are treated in Chapter 4 (and Appendix 3).

The principal task we set ourselves, however, is the **calibration of one scale against the other**, or to give numbers to chronostratic boundaries. This is done partly by chronograms, the subject of Chapter 5, and partly by using the ocean-floor anomalies to refine the chronogram estimates in Chapter 6. In this we follow a long line of attempts. Whether or not the resulting time scale is better than earlier ones (as it should be because we have more data), at least we claim to have systematically set out the procedures by which it has been constructed so that others could check our findings or vary the input to achieve a revised result. The steps taken were as follows.

(1) The **chronostratic scale** was constructed as outlined in Chapter 3. In the end this scale will be determined by a series of agreed international conventions concerning boundary points, names and classifications; we are now a long way on that road so that few major changes may be expected to what is set out in Chapter 3 except, of course, for Precambrian time.

(2) **A database of critical isotopic determinations** was assembled embodying our judgement on what results are acceptable experimentally and useful stratigraphically. There is no space here to argue the inclusion or exclusion of particular data, but they are all listed in Table 4.2 so that items can be deleted or added as appropriate.

(3) The **stratigraphic ranges** of these determinations were positioned in the chronostratic sequence and they therefore provided constraints on the calibration of chronostratic boundaries. In this stratigraphic matching of laboratory determinations we distinguished two kinds of uncertainty: **isotopic uncertainty** and **chronostratic uncertainty**. We treated these as follows (4) and (5).

(4) As in GTS 82, the **chronogram computation** took into account the experimental precision (isotopic uncertainty) of the determinations. Their construction is described in Chapter 5.

(5) In addition for this work we attempted to introduce a measure of **chronostratic uncertainty** into the calibration so as to make the fit of the boundary values more realistic. This takes into account a frequent

Figure 1.7. GTS 89 definitive time scale. The black dots mark tie-points, pseudo-tie-points or age modified by incorporating ocean-floor spreading data. Some of the intervals, especially in Rotliegendes, have been rounded.

Era	Sub-era / Period / Sub-period	Epoch	Stage / tie point ●	Age Ma	Stage abbrev.	Intervals Ma
Cenozoic	Quaternary or Pleistogene	Holocene		0.01	Hol	0.01
		Pleistocene		1.64	Ple	1.63
	Neogene / Ng	Pliocene 2	Piacenzian ●	3.4	Pia	3.6
		Pli 1	Zanclian ●	5.2	Zan	
		Miocene 3	Messinian ●	6.7	Mes	5.2
			Tortonian ●	10.4	Tor	
		2	Serravallian ●	14.2	Srv	5.9
			Langhian ●	16.3	Lan	
		1	Burdigalian ●	21.5	Bur	7.0
		Mio	Aquitanian ●	23.3	Aqt	
	Paleogene / Pg	Oligocene 2	Chattian ●	29.3	Cht	6.0
		Oli1	Rupelian ●	35.4	Rup	6.1
		Eocene 3	Priabonian ●	38.6	Prb	3.2
		2	Bartonian ●	42.1	Brt	11.4
			Lutetian ●	50.0	Lut	
		Eoc 1	Ypresian ●	56.5	Ypr	6.5
		Paleocene 2	Thanetian ●	60.5	Tha	4.0
		Pal 1	Danian ●	65.0	Dan	4.5
Cz TT Pg						
Mesozoic	Cretaceous	Seno-nian / Sen — Gulf	Maastrichtian ●	74.0	Maa	9.0
			Campanian ●	83.0	Cmp	9.0
			Santonian ●	86.6	San	3.6
			Coniacian ●	88.5	Con	1.9
		Gul	Turonian ●	90.4	Tur	1.9
		Gallic	Cenomanian ●	97.0	Cen	6.6
			Albian ●	112.0	Alb	15.0
		Gal	Aptian ●	124.5	Apt	12.5
			Barremian ●	131.8	Brm	7.3
		Neoc-omian / Neo	Hauterivian ●	135.0	Hau	3.2
			Valanginian ●	140.7	Vlg	5.7
	K1	Neo	Berriasian ●	145.6	Ber	4.9
	Jurassic	Malm J3 Mlm	Tithonian ●	152.1	Tth	6.5
			Kimmeridgian	154.7	Kim	2.6
			Oxfordian ●	157.1	Oxf	2.4
		Dogger J2 Dog	Callovian	161.3	Clv	4.2
			Bathonian	166.1	Bth	4.8
			Bajocian ●	173.5	Baj	7.4
			Aalenian	178.0	Aal	4.5
		Lias J1 Lia	Toarcian	187.0	Toa	9.0
			Pliensbachian	194.5	Plb	7.5
			Sinemurian	203.5	Sin	9.0
	J		Hettangian	208.0	Het	4.5
	Triassic	Tr3	Rhaetian ●	209.5	Rht	1.5
			Norian	223.4	Nor	13.9
			Carnian	235.0	Crn	11.6
		Tr2	Ladinian ●	239.5	Lad	4.5
			Anisian ●	241.1	Ans	1.6
		Scythian Tr1 Scy	Spathian	241.9	Spa	0.8
			Nammalian	243.4	Nml	1.5
Mz	Tr		Griesbachian ●	245.0	Gri	1.6
	Permian	Zechstein	Changxingian	247.5	Chx	2.5
			Longtanian	250.0	Lgt	2.5
			Capitanian	252.5	Cap	2.5
			Wordian	255.0	Wor	2.5
		Zec	Ufimian	256.1	Ufi	1.1
			Kungurian	259.7	Kun	3.6
		Rotliegendes	Artinskian ●	268.8	Art	9.1
			Sakmarian	281.5	Sak	12.7
	P	Rot	Asselian ●	290.0	Ass	8.5
Paleozoic	Carboniferous (Pennsylvanian)	Gzelian Gze	Noginskian ●	293.6	Nog	3.6
			Klazminskian	295.1	Kla	1.5
		Kasimovian	Dorogomilovskian	298.3	Dor	3.2
			Chamovnicheskian	299.9	Chv	1.6
		Kas	Krevyakinskian	303.0	Kre	3.1
		Moscovian	Myachkovskian ●	305.0	Mya	2.0
			Podolskian	307.1	Pod	2.1
			Kashirskian	309.2	Ksk	2.1
		Mos	Vereiskian	311.3	Vrk	2.1
		Bashkirian	Melekesskian	313.4	Mel	2.1
			Cheremshanskian ●	318.3	Che	4.9
			Yeadonian	320.6	Yea	2.3
	C2	Bsh	Marsdenian	321.5	Mrd	0.9
			Kinderscoutian	322.8	Kin	1.3
	C1	Serpukhovian	Alportian	325.6	Alp	2.8
			Chokierian		Cho	2.7

Era	Sub-era / Period / Sub-per.	Epoch	Stage / tie point ●	Age Ma/Ga	Stage abbrev.	Intervals Ma
Carboniferous	C2	Bashkirian	Marsdenian	321.5	Mrd	11.5
			Kinderscoutian	322.8	Kin	
	Mississippian	Serpukhovian	Alportian	325.6	Alp	10
			Chokierian	328.3	Cho	
			Arnsbergian	331.1	Arn	
		Spk	Pendleian	332.9	Pnd	
		Visean	Brigantian ●	336.0	Bri	40
			Asbian	339.4	Asb	17
			Holkerian	342.8	Hlk	
			Arundian	345.0	Aru	
		Vis	Chadian ●	349.5	Chd	
		Tournaisian	Ivorian	353.8	Ivo	13
C	C1	Tou	Hastarian ●	362.5	Has	
Devonian		D3	Famennian	367.0	Fam	15
			Frasnian	377.4	Frs	
		D2	Givetian	380.8	Giv	9
			Eifelian	386.0	Eif	
		D	Emsian	390.4	Ems	
		D1	Pragian ●	396.3	Pra	22
			Lochkovian ●	408.5	Lok	
Silurian		S4 Pridoli	Prd	410.7	Prd	2
		S3 Ludlow	Ludfordian	415.1	Ldf	13
		Lud	Gorstian	424.0	Gor	
		S2 Wenlock	Gleedonian	425.4	Gle	31
			Whitwellian	426.1	Whi	6.5
		Wen	Sheinwoodian	430.4	She	
		S1 Llandovery	Telychian	432.6	Tel	
			Aeronian	436.9	Aer	8.5
		Lly	Rhuddanian	439.0	Rhu	
Ordovician		Ashgill	Hirnantian ●	439.5	Hir	
			Rawtheyan	440.1	Raw	4
			Cautleyan	440.6	Cau	
		Ash	Pusgillian	443.1	Pus	
		Bala Caradoc	Onnian	444.0	Onn	
			Actonian ●	444.5	Act	
			Marshbrookian	447.1	Mrb	
			Longvillian	449.7	Lon	21 / 71
			Soudleyan	457.5	Sou	
		Bal Crd	Harnagian	462.3	Har	
			Costonian	463.9	Cos	
		Dyfed Llandeilo	Late	465.4	Llo3	
			Mid	467.0	Llo2	4.5
		Llo	Early	468.6	Llo1	
		Dfd Llanvirn	Late	472.7	Lln2	7.5
		Lln	Early	476.1	Lln1	
		Canadian Cnd	Arenig	493.0	Arg	17
O			Tremadoc	510.0	Tre	17
Cambrian		Merioneth	Dolgellian	514.1	Dol	7
		Mer	Maentwrogian	517.2	Mnt	
		St David's	Menevian ●	530.2	Men	19
		StD	Solvan ●	536.0	Sol	
		Caerfai	Lenian	554	Len	18
			Atdabanian	560	Atb	
Pz	€	Crf	Tommotian	570	Tom	16
Sinian	Vendian	Ediacara	Poundian	580	Pou	20
		Edi	Wonokan	590	Won	
		Varanger	Mortensnes	600	Mor	20
	V	Var	Smalfjord	610	Sma	
Z	Sturtian			/0.80	Stu	190
Riphean		Karatau		1.05	Kar	250
		Yurmatin		1.35	Yur	300
	Rif	Burzyan		1.65	Buz	300
Animikean				2.2	Ani	550
Huronian				2.45	Hur	250
Randian				2.8	Ran	350
Swazian				3.5	Swz	700
Isuan				3.8	Isu	300
Hadean		Early Imbrian		3.85	Imb	50
		Nectarian		3.95	Nec	100
		Basin Groups 1–9		4.15	BG1–9	200
	Hde	Cryptic		4.56	Cry	410

uncertainty of the reported chronostratic age of the samples isotopically determined and their time correlation. We show in Chapter 5, Table 5.1, how this additional uncertainty is applied systematically to all determinations. We applied a likely uncertainty (for example) of 2.5 Ma for Cretaceous ages. It should be possible in future to assess this uncertainty for each individual determination and so give each value a combined chronometric and chronostratic uncertainty.

(6) A personal computer was used to generate a chronogram from the database. The problem of how to use glauconites for time scale calibration was examined by **testing the consequences of inclusion and exclusion of data**. In summary, glauconite dates have been included in all the chronograms but, for boundaries older than 115 Ma, glauconite dates are treated only as minimum dates (Section 5.5). Perhaps more controversially illites were excluded from the chronograms altogether but are listed as supplementary data in Chapter 4 (Table 4.2).

(7) The next stage was to plot the chronogram ages and their ranges in Ma versus the chronostratic scale (in units of chrons) for each period (Figures 5.5 to 5.24), to achieve **graphic interpolation** of poorly defined boundary ages. For this a **chronostratic apportionment** was made by taking into account the relative duration of stages, etc. We assumed for certain moderate spans of time relatively uniform rates of evolution as measured by biozones. Others have recommended such apportionment procedures for time scale calibration (Van Hinte 1978a, 1978b, Westerman 1984). It was not done in GTS 82 and it could be done more thoroughly than we have done here.

(8) Visual inspection of the graphs showed that in many cases a smooth line could be drawn through the most precise chronogram values. These more precise values were regarded as 'tie-points' which fix the age of a particular stratigraphic boundary. Where the chronogram values had large uncertainties and did not form a smooth line, interpolation was adopted. This is often done intuitively ('eye-balling'), but was done here by linear interpolation between 'tie-points'. The only subjective decision lay in the initial selection of 'tie-points'.

(9) Further interpolations, based on the relative spacing of ocean-floor magnetic anomalies, were used to refine parts of the Cenozoic and later Mesozoic time scale.

(10) A **definitive time scale** was produced for publication here. The refinements possible with existing data, as well as modifications necessitated by new data, entail an indefinitely continuing series of revised numbers. As there is no finality or even temporary stability in this process the moment when the scale is fixed for publication is seemingly arbitrary. An internally consistent set of numbers was adopted for the charts, figures and tables in this book, and is based on the initial operations up to stage (9) above. That is to say the values are those from Chapters 5 and 6, as produced by chronograms in September 1988 on the data set from Chapter 4.

(11) In the remaining time before final submission for publication the authors were able to re-examine what had been done and to qualify some values. In particular, initial Cambrian chronometry continued to present distinct problems and was treated differently as outlined in Section 5.10. The overall conclusions are listed in the definitive time scale (Figure 1.7) and are summarized in the simplified scale (Table 1.2, p.12).

Time scales will be subject to continuing refinement. For example, if and when certain cyclic events (e.g. those resulting from planetary perturbations) become established (as is already the case in Quaternary stratigraphy) more precise scales may be possible. In the meantime here is a tool to calibrate geologic history and to measure rates of processes.

12

Table 1.2. *Simplified Phanerozoic time scale with rounded values*

EON	ERA (Durations in Ma)		PERIOD	SUB-PERIOD	EPOCH	(Some stages)	DURATION Ma	AGE Ma
Phanerozoic	Cenozoic 65		Quaternary 1.64		Holocene		0.01	0.01
					Pleistocene	Ple 3	0.12	0.13
						Ple 2	0.66	0.79
						Ple 1	0.85	1.64
			Tertiary	Neogene 22	Pliocene		3.5	5.2
					Miocene		18.3	23.5
				Paleogene 42	Oligocene		12.0	35.5
					Eocene		21.0	56.5
					Paleocene		8.5	65
	Mesozoic 180		Cretaceous 81	Gulf 32	Senonian		24	89
				Gallic 43		Tur/Cen	8	97
						(Alb 15)		
						(Apti 13)	35	
						(Bar 7)		
				K1 49	Neocomian		14	132
			Jurassic 61		Malm		12	146 / 157
					Dogger		21	178
					Lias		30	208
			Triassic 37		Tr3	(Rht/Nor 15) (Crn 12)	27	235
					Tr2		6	241
					Scythian		4	245
	Paleozoic 325	Late 164	Permian 45	Zecnstein		Lopingian	5	250
						Guadalupian	6	256
				Rotliegendes		(Kun/Art 13)		
						(Sak 13)	35	
						(Ass 9)		290
			Carbon-iferous 73	Pennsylvanian 33		Gzelian	5	295
						Kasimovian	8	303
						Moscovian	8	311
						Bashkirian	12	323
				Mississippian 40		Serpukhovian	10	333
						Visean	17	350
						Tournaisian	13	363
			Devonian 46		D3		14	377
					D2		9	386
					D1		23	409
		Early 161	Silurian 31		Pridoli		2	411
					Ludlow		13	424
					Wenlock		6	430
					Llandovery		9	439
			Ordovician 71	Bala		Ashgill	4	443
						Caradoc	21	464
				Dyfed		Llandeilo	5	469
						Llanvirn	7	476
				Canadian		Arenig	17	493
						Tremadoc	17	510
			Cambrian 60		Merioneth		7	517
					Saint Davids		19	536
					Caerfai	(Len 18)	34	
						(Atb/Tom 16)		570
	Sinian ~230		Vendian 40		Ediacara		20	590
					Varanger		20	610

2

The chronometric (numerical) scale

2.1 Essentials of the chronometric scale

The statement of age of an event on a chronometric scale is given as the number of units of time that lapsed between the instant of that event and a defined datum 'Present'. The scale is thus compounded by repetition of identical units and so is not a calendar date as might be derived from so many circuits of the Earth round the Sun in relation to the stars because each circuit varies in duration. Thus the standardization of the chronometric scale requires only that the unit of time be defined. However, two competing standards for this unit have been in use (as described below) and the IUGS had not decided which to follow (George *et al.* 1969). That the difference in numerical terms is insignificant for our purposes does not eliminate the need to state the observational principle by which geologic time is calibrated.

2.1.1 Mean solar second

Formerly the unit was the mean solar second, defined as 1/86 400 of the mean solar day.

2.1.2 Ephemeris second

The International Astronomical Union (IAU) recommended in 1957 that in astronomy and related sciences the ephemeris second be adopted as the fundamental invariable unit of time. It was defined as 1/31 556 925.9747 of the tropical year at 1900 January 0 days 12 hours ephemeris time. The IAU meeting in Prague in 1967 recommended its continuation in face of the competing standards for a second (references in George *et al.* 1969, especially Sadler 1968).

2.1.3 Atomic second

The following definition was adopted on 13 October 1967 at the meeting of the Thirteenth General Conference on Weights and Measures: 'The second is the duration of 9 192 631 770 periods of the radiation of the atom of cesium

133'. The frequency (9 192 631 770 Hz) which the definition assigns to the cesium radiation was carefully chosen to make it impossible, by any existing experimental evidence, to distinguish the atomic second from the ephemeris second based on the Earth's motion. No changes were anticipated in data stated in terms of the old standard to convert them to the new one (Weast 1969 *et seq.*).

The advantages claimed were that it is theoretically possible for anyone anywhere to build an atomic clock with a precision of ± 1 part per 10^{11} (or better), controlled by the cesium radiation, and calibration can be achieved in the laboratory in a few hours without the long time necessary to make astronomical observations. In practice, clocks are calibrated against broadcast time signals.

2.1.4 Conventions

Although the time standard was initially based on the year, the fundamental unit of time is the second (s). The Système International d'Unités (SI) of the General Conference on Weights and Measures also allows the use of the year (a). Geoscientists use years rather than seconds.

The conversion of years to seconds (useful for some physical calculations) is based on the above values. Thus 1 year = 31.56 megaseconds (1 a = 31.56 Ms), 1 thousand years = 31.56 gigaseconds (1 ka = 31.56 Gs), 1 million years = 31.56 teraseconds (1 Ma = 31.56 Ts) and 1 billion years = 31.56 petaseconds (1 Ga = 31.56 Ps). The presently available SI conventions at 10^3 intervals are:

10^{18}	exa	E	10^{-3}	milli	m
10^{15}	peta	P	10^{-6}	micro	μ
10^{12}	tera	T	10^{-9}	nano	n
10^{9}	giga	G	10^{-12}	pico	p
10^{6}	mega	M	10^{-15}	femto	f
10^{3}	kilo	k	10^{-18}	atto	a
10^{0}	unity	1			

Ages are given in years before present (BP) rather than BC. To avoid a constantly changing datum, present = AD 1950 (as in ^{14}C determinations), in round numbers the time of the beginning of modern isotopic dating research in laboratories around the world.

2.2 Chronometry for Earth history

Chronometric calibration of the age of geologic events is based fundamentally on two kinds of observation. (1) The Earth's motion is reflected in sedimentary phenomena such as varves, or biologic phenomena such as variations in growth; (2) isotopic decay rates are determined in the first place in real time in the laboratory using a clock.

The first category yields sedimentary patterns which may enable correlations to be made by distinctive signatures. Durations of successions of strata may be estimated in years or in cycles of years (solar or orbital). It is exceptional, however, to obtain such chronometric ages directly because of the need for the record to continue to the present as with tree rings.

The second category yields durations before present (BP), with varying degrees of precision and uncertainty, originally by ratios of elements and now by the analysis of isotopic ratios or radiation damage. This is the subject of Chapter 4. Events of any kind may be so calibrated as examples in the figures in Chapter 7 indicate, but the events we seek to calibrate are the initial boundaries of chonostratic divisions. These divisions are defined or discussed in Chapter 3 and their numerical calibration in Chapters 5 and 6.

2.3 Nomenclature for a geochronometric scale

Whereas the principal exercise in this volume is to define a chronostratic or successional scale and then to calibrate it numerically in years, a numerical scale can stand on its own so that ages of events are simply given approximately in years.

Years may be compounded to express a chronometric time scale with longer intervals. The time span of these intervals will probably be in round numbers to the base 10. Millennia or gigennia could be numbered as with centuries in human records.

It would be possible to set a zero datum at, say, 10 billion years BP so that all relevant history would be calibrated in forward time, but this has not been done. We use ka, Ma and Ga ages, i.e. durations back from the present, in each case BP being understood.

Two approaches to the problem have been made. One has been to fix the dividing points in the chronostratic scale at major historical events such as times of maximum orogenic activity. The other has been to approximate (manifestly rounded) numbers to perceived stages in the development of the Earth.

Some approaches to this problem are described below and are tabulated in Figure 2.1.

2.3.1 Geological Survey of Canada (GSC) publications

An example of the first approach was by Stockwell for the Geological Survey of Canada. In a succession of reports on new age determinations, he refined the classification of Precambrian rocks of the Canadian Shield. Stockwell's fourth report (1964) terminated his Archaean Eon at 2390 Ma, the Aphebian Era at 1640 Ma, the Paleohelikian Sub-era at 1280 and the Neohelikian Sub-era at 880 Ma. The Hadrynian Era would thus span 880 to 600 Ma, or to the initial Cambrian boundary. Thus the Proterozoic Eon was divided into Aphebian, Helikian and Hadrynian Eras. The names are from Greek roots: Aphebian is from 'aphebos' for old maturity; Helikian from 'helikia' for middle maturity; Hadrynian from 'hadrynes' for young maturity. Stockwell's numbers were successively modified for the GSC in 1973 and in 1982 (Harland 1983a), and even more recently by Okulitch (1988). The attempt was made to approximate the boundaries to a concept of terminal orogenic phases in the Canadian Shield (with an allowance for cooling ages) and then to define the divisions at exact numbers based on such estimates. The Archean–Proterozoic boundary, for example, was taken as 'the true age of the Kenoran Orogeny'. The Proterozoic divisions were defined at the Hudsonian and Grenville orogenies.

The evolution of Stockwells' original chronometric scheme to a possibly chronostratic definition of the same named divisions was evident in Douglas' paper for the GSC (1980) shortly before Douglas' death (see Harland 1983a).

However, Frarey (1981) rejected this chronostratic tendency and reaffirmed a chronometric scale as developed by Stockwell in the GSC and based on orogenic events to define boundaries.

2.3.2 United States Geological Survey (USGS) publications

An example of the second approach was by James (1972) for the United States Geological Survey. Precambrian W, X, Y and Z were defined with initial boundaries at 3.1 Ga, 2.5 Ga, 1.6 Ga and 0.8 Ga as tabulated in Figure 2.1. This scheme allowed for successively earlier divisions identified by the letters V, U, T, etc. as required. The divisions W, X, Y and Z were written fully as Weltian, Xenian, Yovian and Zedian. This was proposed in the USGS after Goldich (1968) had proposed another purely chronometric scheme dividing Precambrian time backwards from 600 Ma at 400 Ma intervals. He named these intervals backward Alpha–Precambrian (600 to 1000 Ma) to Theta–Precambrian (exceeding 3400 Ma). These scales are shown on Figure 2.1.

2.3.3 International proposals

With the prospect of national surveys from different continents each producing the most favourable numerical divisions for maps of their own territories, a move to produce an international standard was made in the Precambrian Subcommission of the International Commission on Stratigraphy (ICS).

Anticipating this, and hoping to preserve time-honoured stratigraphic names (e.g. Archean and Proterozoic) for the chronostratic scale in the sense in which they were first used (e.g. Alcock 1934), Harland (1975) proposed a totally artifactual scale, manifestly independent of Earth history (rather than approximately to reflect it) and for all geohistorical time. Using Latin roots (as most Greek roots had already been used in the chronostratic scale) the scheme summarized in Harland 1978 (p.20) and illustrated in Table 2.1 here was proposed.

This scheme was contrasted with the parallel need for a Precambrian chronostratic scale using the familiar

Table 2.1. *Global chronometric divisions (Harland 1975)*

	Late	
	-------- 0.5 --------	
Novo-time	Middle	
	-------- 1.0 --------	
	Early	
	----------------- 1.5 -----------------	
	Late	
Medio-time	-------- 2.0 --------	
	Early	
	----------------- 2.5 -----------------	
	Late	
	-------- 3.0 --------	
	Middle	
Antiquo-time	-------- 3.5 --------	
	Early	
	----------------- 4.0 -----------------	
Prisco-time	with scope for earlier divisions	

Figure 2.1. Comparison of chronometric schemes.

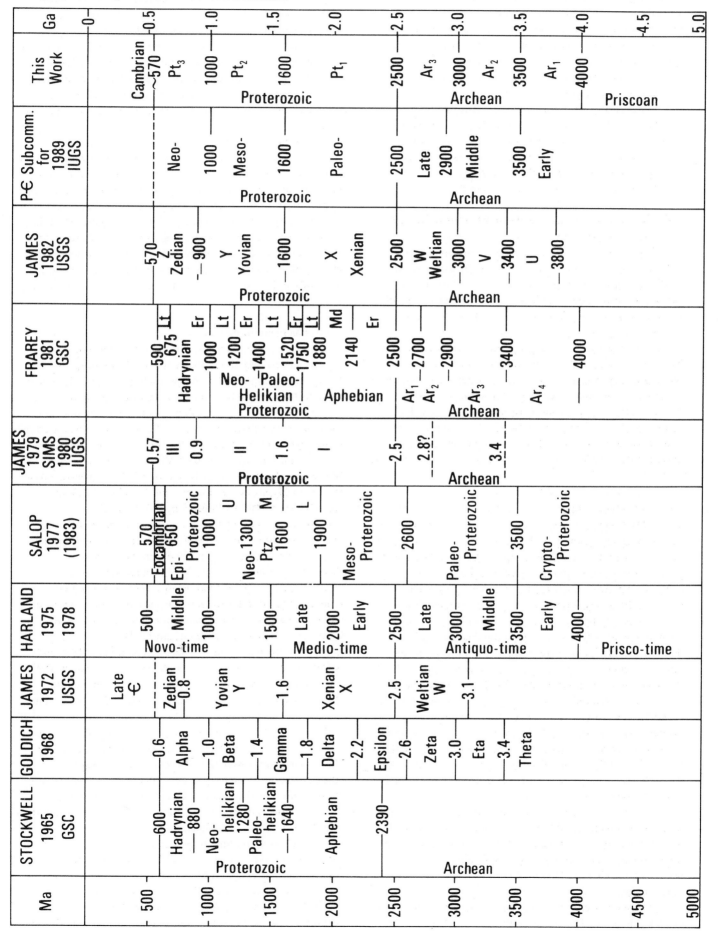

stratigraphic names. Moreover, the two time scales were argued each to be applicable to both Precambrian and Phanerozoic history and each having its different advantages and disadvantages. However, at that time only one scale was thought desirable for Precambrian history, the dual system was rejected by the Subcommission, and the familiar names Archean and Proterozoic were then agreed to be defined chronometrically rather than chronostratically. After much debate an Archean–Proterozoic boundary was defined at 2.5 Ga (Sims 1980). International proposals for the division of Archean and Proterozoic eons are described in Section 2.4. It should be noted that many other schemes have been proposed privately for international use (e.g. Salop 1983).

2.4 Chronometric divisions of Precambrian time

2.4.1 Priscoan time

The name Archean has hitherto been applied predominantly with rock in mind. In most attempts to divide Archean no division was proposed for time intervals earlier than the age of presently known or older rock, except Hadean which was used in a chronostratic sense. This usage has been followed here (see Chapter 3).

There remains a need to distinguish pre-Archean time. Not only did a great deal happen on Earth, though we depend on the Lunar record (see Chapter 3) for details of this, but the evolution of the Solar System and the whole cosmos preceded the formation of our planet. **Priscoan**, adopted in GTS 82, is continued. It is defined chronometrically as all pre-Archean time, encompassing the early history not only of the Solar System but of earlier cosmic events back to the beginning (e.g. to the 'Big Bang)'.

2.4.2 The Archean Eon

Archean originally referred to the oldest known rocks. **Archaeozoic** was defined as 'the earlier part of Precambrian time corresponding to Archean rocks' (Bates & Jackson 1980 – *AGI Glossary of Geology*). Archaeozoic had been introduced by Dana in 1872 as an alternative to Archean (Harland 1974). Nevertheless, in 1976, the Precambrian Subcommission treated Archean as a standard time division terminating at 2500 Ma, no initial boundary having been defined. Its relation to terrestrial rocks persists in the minds of most geoscientists because, in nearly every attempt to divide Archean time into the three divisions Early, Middle and Late (or 1, 2, and 3), numbers have been proposed such that the initial Middle Archean boundary approximates to or is proposed at 3500 Ma and the initial Late Archean boundary at around 3000 Ma. Moreover, the oldest rocks so far recorded from southern West Greenland or from southern Africa or Australia are commonly referred to as 'earliest Archean'. Even the use of Eon or Era implies a distinct span of time.

In these circumstances the initial Archean boundary is taken provisionally at 4000 Ma (4 Ga), it being reasonable to use a round number to bound Priscoan and Archean on the same principle as the round 2500 Ma terminal Archean boundary was chosen. This would not violate majority usage of the name.

Common use of 'the Archean' suggests that the notion of rock is in mind. This is especially evident where a diachron-

ous Archean–Proterozoic boundary has been proposed so as to accommodate 'truly' Archean rocks within an Archean division (e.g. Windley 1977, 1984, Tankard *et al.* 1982).

For the division of Archean time regional schemes have been proposed, some with the implication that the scheme should be adopted for the international scale. For example, the Geological Survey of Canada proposed divisions at 3.4 and 2.9 Ga (Douglas 1980). But the schemes vary from region to region.

At the fifth meeting of the ICS Precambrian Subcommission in Duluth, Minnesota, in 1979 it was proposed to divide Archean time at 3.5 Ga and 2.9 Ga in a three-fold classification 'to fit the Archean geologic record as now known in most parts of the world'. Little progress has been made subsequently by the Precambrian Subcommission, which has been preoccupied with the division of Proterozoic time. Until the matter is decided the manifestly artifactual scale adopted in GTS 82 is retained. It begins at 4.0, divides at 3.5 and 3.0, and ends at 2.5 Ga, without compromise towards a 'natural' scale.

2.4.3 The Proterozoic Eon

When Sedgwick in 1838 proposed his **Palaeozoic Series** he also named underlying **Primary Stratified Groups**, saying that if and when the older groups might prove to hold organic remains they should be termed the **Protozoic System**. Shortly afterwards in 1841, when Phillips extended the meaning of Palaeozoic to include Cambrian to Permian, he introduced the terms **Hypozoic** or **Prozoic** for the pre-Paleozoic rocks in order to match his new sequence with Mesozoic and Cainozoic. Murchison, De Verneuil & Keyserling in 1845 proposed **Azoic** instead of Hypozoic and Prozoic and then in 1875 Dana (pp.138–9) proposed an **Eozoic** Era to follow Azoic. Proterophytic, although more appropriate, has hardly been used. Uncertainty about the history of early life led Irving in 1887 to suggest that the rocks between Archean and Palaeozoic should be referred to as **Agnotozoic** (or alternatively in 1888 to **Eparchean**). There was a general tendency then to refer to metamorphosed rocks as Archean and to include the earliest unmetamorphosed rocks with Cambrian (Wilmarth 1925, Harland 1974).

Proterozoic was first named by Emmons in 1888 and became established largely through use by Van Hise in 1892 (pp.492–8) and Van Hise & Leith in 1909. In the United States Geological Survey (USGS) Proterozoic then became more or less synonymous with Precambrian until at least 1925, though Chamberlin & Salisbury (1906) had restricted it as a division following Archaeozoic or Archean and preceding Paleozoic. Proterozoic time was confirmed as beginning with the deposition of the earliest Huronian rocks (Archean prior to that time) by the National Committee on Stratigraphical Nomenclature (Alcock 1934). In this sense it persisted in the Geological Survey of Canada.

At about the same time **Algonkian** was conceived in the USGS by Powell in 1890 as a rock system, rather than a time interval, underlying Cambrian rocks; Keweenawan was regarded as its top series following Huronian. Chamberlin & Salisbury regarded Algonkian as practically synonymous with Proterozoic (Gill 1957, Harland 1974). These divisions were part of the developing chronostratic time scale analogous to Phanerozoic (see under Huronian below).

However, in 1964 the GSC (Stockwell) defined Proterozoic chronometrically as succeeding Archean at 2390 Ma. From then on chronometric thinking took over, but with the serious implication that a numerical scale should replace rather than calibrate and supplement the divisions being established on the basis of rock sequences. Proterozoic was to be defined with an initial numerical boundary and, after much debate between competing values, this was decided at 2500 Ma exactly (Sims 1980).

Proterozoic divisions

In parallel with attempts to divide Archean time by numbers, there have been as many attempts to do the same for Proterozoic and in much the same way. At first, national and regional geological communities made proposals until it was seen necessary to agree international standards through the ICS.

As outlined above, the GSC introduced the names **Hadrynian, Helikian** and **Aphebian** for tectonic divisions of the Canadian Shield with boundaries at 880, 1640 and 2390 Ma (Stockwell 1964). These were subsequently rounded off to 1000, 1800 and 2500 Ma (Douglas 1980). At the same time the USGS accepted the principle of artifactual boundaries in round numbers for Precambrian W, X, Y and Z (James 1972). **Weltian** (W) was pre–2500 Ma, **Xenian** (X) spanned 2500 to 1600 Ma, **Yovian** (Y) to 800 Ma and then **Zedian** (Z) to 600 Ma. Competing opinions favoured boundaries at 1500 Ma and/or use of five divisions (see also Figure 2.1).

The ICS Precambrian Subcommission has made progress at meetings approximately every four years with intervening circulation of papers. In 1973 (Adelaide) Archean and Proterozoic were adopted; in 1977 (Cape Town) their boundary was decided chronometrically at 2.5 Ga; in 1979 (Duluth) three-fold division at 2.5, 1.6 and 0.9 Ga was proposed; in 1982 (Tanta) tripartite division was sustained and referred to as Proterozoic I, II and III.

In 1985 (Canberra), proposals were made with the hope that firm recommendations could be made by 1988 in time for ratification at the IGC in Washington in 1989. Three chronometric divisions were still recommended, presumably still divided at 1.6 and 0.9 Ga, with four alternative nomenclatures for consideration:

(1) Proterozoic I, II and III
(2) Early, Middle and Late (not preferred)
(3) Yovian, Xenian and Weltian (USGS and GTS 1982)
(4) Proteran, Deuteran and Triteran (newly proposed within the Subcommission)

A further possibility was suggested at the 1985 meeting to divide these three eras into eight periods also chronometrically defined and to name them Alphan, Betan, Gamman, Deltan, Epsan, Zetan, Etan and Thetan.

As this work was about to go to press, the minutes of the Eighth Meeting of the IUGS Subcommission on Precambrian Stratigraphy at Tianjin, China, 18 to 21 September 1988, came to hand (chairman K. A. Plumb,

Figure 2.2. Proterozoic subdivisions as proposed by the Precambrian Subcommission of ICS 1988.

Eon	Era	Period
Proterozoic	(Base of Cambrian)	
	Neoproterozoic	"Neoproterozoic III"
		650 Ma
		Cryogenian
		850 Ma
		Tonian
	—1000 Ma—	
	Mesoproterozoic	Stenian
		1200 Ma
		Ectasian
		1400 Ma
		Calymmian
	—1600 Ma—	
	Palaeoproterozoic	Statherian
		1800 Ma
		Orosirian
		2050 Ma
		Rhyacian
		2300 Ma
		Siderian
	—2500 Ma—	
Archaean		

The Greek derivation of period names and interpretation was given as follows.

Cryogenian – Cryos = ice; Genesis = birth: *'Global glaciation'*
Tonian – Tonos = stretch
Stenian – Stenos = narrow: *'Narrow belts of intense metamorphism and deformation'*
Ectasian – Ectasis = extension: *'Continued expansion of platform covers'*
Calymmian – Calymma = cover: *'Platform covers'*
Statherian – Statheros = stable, firm: *'Stabilization of cratons, cratonization'*
Orosirian – Orosira = mountain range (from modern Greek): *'Global orogenic period'*
Rhyacian – Rhyax = stream of lava: *'Injection of layered complexes (e.g. Bushveld)'*
Siderian – Sideros = iron: *'Banded iron formations'*.

secretary R. B. Flint). Although the conclusions of that meeting were in the form of recommendations yet to be balloted by post and then for ratification or otherwise by the ICS, it seems that after so long a gestation the Subcommission has brought forth a Proterozoic chronometric classification which is likely, at least in its broad divisions, to be acceptable. The arguments were minuted, but are not repeated here. The conclusions as issued by the secretariat are conveniently set out here in Figure 2.2. There seems little doubt that the three Proterozoic eras (re-divided at 1000 and 1600 Ma) will stand with their distinctive nomenclature, so avoiding confusion with **Early, Middle** and **Late** in their varied application.

The further division of eras into periods was more contentious, but perhaps only so far as the names are concerned. Happily these were proposed so as not to conflict with the expected introduction of chronostratic names. Greek roots were emphasized with meanings as explained in the caption to Figure 2.2. Geographical names had been rejected

not only because of possible confusion with chronostratic
division but also to avoid international rivalry. The names
chosen 'reflect globally significant geological events'.
Anticipating that perception of these events may change, it
was 'emphasized that these names are not definitive of the
units they describe. The process described by a term shall play
no role in assigning a rock unit to a period. The names are not
diagnostic of the Periods to which they apply. The names are
just labels, whose meaning will soon be lost . . .' One might
comment for a start that **Cryogenian** might well precede and
so not contain the most widespread of all glacial events but,
according to the principles minuted above, that would not
matter. It is to be hoped, in the interests of an agreed
international standard, that some such artifactual numerical
scheme will be adopted without attempting to mimic a
chronostratic one.

Terminal Proterozoic

It will be noted that Proterozoic is generally used *either*
as terminating at the initial Cambrian boundary, which will
certainly be defined chronostratically, e.g. Cowie *et al*. 1984
(see Section 2.9.1 below), *or* at some notional numerical limit
guessed to be at about that time, such as 570 Ma. If the
boundary were defined numerically there would almost
certainly be a gap or an overlap with the Cambrian Period.
However, most geologists apply the former option of using the
initial Cambrian boundary point (when defined) as the terminal point for the Proterozoic Eon, even though Proterozoic
thereby becomes a hybrid division beginning chronometrically
and ending chronostratically. This is not a significant problem.

2.5 Geochronology – the term

The term **geochronology** has been used in two senses.
One use refers to numerical time estimates in geoscience and
is sometimes distinguished from non-numerical time studies
and especially by those engaged in isotopic age determinations. The other usage is that of the North American Code
and the *International Stratigraphic Guide* (Hedberg 1976),
where geochronology also relates to the non-numerical time
aspect of chronostratic divisions.

Geochronology (*sensu lato*) thus deals with any ways of
assessing the age of events, chronometric and chronostratic,
and also their mutual calibration in what is generally known as
a time scale. Accordingly, this book is an essay in geochronology.

Geochronology (*sensu stricto* = Geochronometry) is the
laboratory discipline that determines dates in standard time
units (a, ka, Ma, Ga).

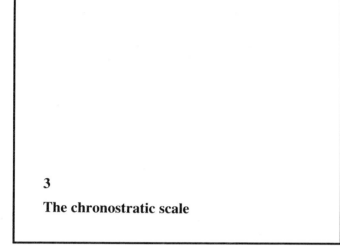

3

The chronostratic scale

3.1 Introduction

The principles distinguishing chronostratic (successional) from chronometric (numerical) scales were outlined in Chapter 1. Each is constructed by decisions on how to divide and name it. This chapter outlines currently accepted and newly proposed chronostratic scale divisions and conventions, with their historical origins, so as to provide a tool for immediate use. Perhaps the main interest in this book is the numerical calibrations (in Chapters 4, 5 and 6) of the Phanerozoic divisions set out in this chapter.

Some parts of the chronostratic scale are already well established by international decision and/or conventional usage; other parts are not. Where no standard usage has been agreed, one has been suggested. An interim scheme for the whole scale may be useful; at the same time it may provoke discussion and so hasten the day when a more complete scale is established.

Although the first steps towards such an international convention were taken at the International Geological Congress (IGC) in Bologna in 1881 (Anon. 1882), it was only by slow degree that the nature of the task was understood to

require the precise definition of boundaries. An attempt to agree a Pliocene–Pleistocene boundary in 1948 at the IGC in London was an important first step. A thorough application of the principle awaited the decision of the International Union of Geological Sciences (IUGS – set up in 1960) to establish the Silurian–Devonian boundary at Klonk in Bohemia – finally decided in 1972 at the IGC in Montreal (McLaren 1977). Under the IUGS International Commission of Stratigraphy (ICS) there are now active groups working on nearly every remaining period boundary, so that within another five or ten years we may expect the main points of the scale to be established.

The principal changes here since GTS 82 result both from publications in the meantime, and from correspondence with the chairmen of the various subcommissions and working groups of ICS. Developments concerning some geologic periods have been rapid and there is a consequent expansion of this chapter in addition to alteration and correction in response to suggestions and criticism of GTS 82.

3.1.1 Chart conventions in Chapter 3

Figures 3.1 to 3.17 depict the chronostratic divisions for each major interval such as a geologic period. In each case there is no vertical scale representing time numerically – such charts are plotted in Chapter 7. Here will be found the succession and approximate correlation. The vertical spacing is determined by the required lettering rather than by the duration of intervals. However, numerical ages are entered in a column in each chart as the current estimate from Chapters 5 and 6 where their derivation is discussed. Numerical ages are given here in million year units (Ma) without qualification.

A typographic convention for these charts is that names which designate geologic time (i.e. names in the chronostratic scale) are printed in lower case with initial capitals for proper names. On the right-hand side (of most charts) upper-case lettering is used throughout to indicate rock **successions**, which designation replaces **systems** of GTS 82; for example, Jurassic Period, Oxfordian Stage and JURASSIC SUCCESSIONS, OXFORD CLAY.

The boundaries of the rock units are not likely to coincide with those of the chronostratic scale. This work does not pretend to give the best possible correlations, a task which would be far greater than we have undertaken. The reason for including rock succession with only approximate time values is to allow the interpretation of the less familiar names from those better known. Where there is considerable uncertainty this is indicated by not extending the formational boundary lines to the edge of the column.

The biostratigraphic information has a similar indicative rather than definitive function. Many biozones are in effect potential chrons, so have an intermediate status until defined as chrons.

For time scale division a system of abbreviations is available (see Appendix 2). Abbreviations for eon, era and period names have been largely determined internationally. Account has also been taken of the symbols used in the *Geological World Atlas* (Choubert & Faure-Muret 1976). For the rest a three-letter system was introduced in GTS 82 and is extended here (two letters gave insufficient scope for recognition and uniqueness). This has already had wider use in databases and we have been urged not to change it. It has now been extended for all divisions. The numerical subscript is often preferred to indicate early, middle and late (as in the USSR). Two, three or more divisions can be so noted, and this usage avoids the unfortunate ambiguity of the abbreviation 'L' for Late or Lower. It is recommended generally to write early and late in full, and also to show by the initial if it is formal (with upper-case) or informal (with lower-case initial). If they must be shortened the use of Er- and Lt- for time terms is recommended, see Appendix 2.

Where necessary the transliteration of names has been revised:

Chinese in the contemporary Pinyin, and Russian according to the standard PCGN/BGN system (Permanent Committee on Geographic Names, US Board on Geographic Names, as used for example in *The Times Atlas*). As already observed we have accepted American spelling to avoid diphthongs in names such as Paleozoic. This has not found favour with some British colleagues, but the choice is made deliberately for an international time scale.

3.2   Requirements of a chronostratic scale

The essential requirement of such a scale is a sequence of reference points defined in (boundary) stratotype sections which have a good correlation potential. Cowie *et al.*(1986) used the expression GSSP for such 'Global Stratotype Sections and Points'. A sufficient spread of GSSP, named from the type locality (e.g. Klonk for Silurian–Devonian boundary) or otherwise labelled, would alone define a chronostratic scale (Hughes *et al.* 1967). When such a scale is established it does no more than provide a single agreed standard against which to correlate and for numerical calibration. It eliminates an element of uncertainty and ambiguity about divisions which may otherwise have alternative or vague definition. GSSP thus introduce a degree of rigour in the use of stratigraphic names.

In order to convert the traditional stratigraphic names (of TCSS) and classification to a single standard chronostratic scale (GCSS) for general use, three kinds of decision or agreement need to be made, and made by a single authority (i.e. ICS of IUGS): (1) a scheme of divisions with appropriate rank and classification together with (2) agreed names for each division that shall correspond to the time spans between the boundaries and (3) agreed standardization of the boundaries. Consideration of these three elements of a chronostratic scale follows.

3.2.1   Classification

Classification of time scale divisions has developed traditionally on a hierarchical basis with eons (e.g. Phanerozoic), eras (e.g. Mesozoic), periods (e.g. Jurassic), epochs (e.g. Late Jurassic or Malm), ages (e.g. Oxfordian) and chrons (e.g. Mariae). The number of ranks is not a matter of principle but of convenience and is an accident of history. There have been attempts to limit the ranks by standardization, but sub-eras, sub-periods, etc. have intervened. The rock that formed anywhere during such intervals has been commonly referred to respectively as the eonothem, erathem, system, series, stage and chronozone (but see Section 3.3 below).

It therefore matters little whether a rigid hierarchy be strictly maintained. Provided each span is properly defined, the usage will consolidate a suitable number of names. Higher ranks may have descriptive advantages for slow events or uncertain timing and they reduce the need to remember names of lower ranks.

Conterminosity. A convenient feature of the hierarchical classification is that boundaries of divisions of higher rank shall coincide with those of lower rank: this is the convention of conterminosity. For example, according to Figure 3.8 the initial GSSP for Triassic would also serve Mesozoic, Scythian, Griesbachian and the *Otoceras concavum* chron.

3.2.2   Nomenclature

The origins of the names used in the chronostratic scale are indicated where they occur in this chapter. Period names had varied origins as system names when the international application of biostratigraphy was discovered and appeared to reflect chapters in the history of the Earth early in the nineteenth century.

The **names** for stages mostly began as formational names, of which some were used more widely as parts of regional schemes. They thus mostly have original body-stratotype localities and sections. (See, for example, discussion of Jurassic stages introduced by d'Orbigny in Section 3.16.)

The standard scale on the other hand requires points and intervals between them, each of which needs a name. This has generally been selected from already familiar stage names. Their boundaries will be defined somewhere precisely, but almost never where the name was first used, or exactly in the original sense, because it was probably established from a good body stratotype (frequently with a disconformity below) whereas boundary stratotypes taken at a good through-succession are needed for definition. So in pressing the claims of a favoured name for international use, it is accepted that the boundaries, which newly define the named time span, might need to be fixed in more suitable localities elsewhere. Once the name has been clearly defined it should retain its meaning even when its rank in the hierarchy is changed.

For stages the suffix -ian is standard usage in English (-ien in French), but the majority of epochs do not conform and there is an advantage in preferring the shorter distinctive form where it is already familiar. The short form (Varanger, Ludlow, Zechstein, Malm, Eocene, etc.) is thus used.

Moreover, the introduction of defined names to replace early, (middle) and late when applied to major time divisions has two advantages. (i) It avoids ambiguity through conflicting usage in these terms. The division may comprise two, three or more parts; indeed, bipartite and tripartite schemes currently coexist. (ii) A named division then facilitates the further informal qualification; early, middle and late. It is also convenient to use **mid** instead of **middle** to indicate a time term.

Where alternative bipartite and tripartite schemes are in use, names for both alternatives are applied, in some case for clarification pending international decision. The name Dyfed is for mid-Ordovician, and Gallic refers provisionally to the middle of a tripartite scheme when Cretaceous is so divided. This is not to favour a tripartite rather than a bipartite classification but to enable any such use to be unambiguous.

A final point on the use of names in the time scale is that in English, at least, they are mostly in adjectival form and are best used to qualify rocks, events, time, configurations, sequences, meetings, literature, etc. Although colloquially common, it is better and generally shorter to avoid the substantive usage, e.g. 'events in the Cambrian' is better as 'Cambrian events'.

3.2.3   Definition and standardization

Definition and standardization is a two-fold process. Agreement is first necessary as to the approximate time in Earth history for a boundary to be convenient and acceptable. This may well be decided for Phanerozoic divisions on a bio-stratigraphic basis before the locality for the reference point in the boundary stratotype is decided. An official working group considers competing 'candidates' and sooner or later recommends to ICS a single reference point, the GSSP of Cowie *et al.* 1986. This may take many years. This chapter brings together the decisions made and the situation concerning others in the making. In the end the reference points (GSSP) define the chronostratic scale, and invite calibration in years – the subject of Chapters 4 and 5.

A further distinction made is between the terms division and unit. **Division** is appropriate for a component of a time scale. It is artifactual. All chronostratic names refer to divisions. **Unit** implies either a single entity (e.g. a year), or an object united by common characteristics or a common boundary such as a rock unit (e.g. a formation). A division is defined only by its boundary-stratotype points (GSSP). A formational unit may be defined by a body stratotype at the type locality.

3.3   Time and rock terminology

The terms system, series, etc. are commonly referred to as 'time–rock units'. It is argued here that they are now redundant and even confusing, being better replaced by reference to time divisions or to rock units as the meaning requires.

It is commonly assumed that the time–rock couplets era–erathem, period–system, epoch–series, age–stage, and chron–chronozone, as well as early–lower and late–upper, are precisely equivalent and should be selected only according to context in the sentence. It is simpler for the chronostratic scale to apply only the first term in each couplet. This is recommended as being convenient for both thought and expression. Some others adopt the opposite simplification, e.g. Lower and Upper for time as well as rock. On that basis Lower Cambrian time (for example) may be expanded to 'the time in which all Lower Cambrian rock formed (as well as intervening time not represented by rock) falling within the Early Cambrian time interval which is in turn defined by the two (initial and terminal) GSSP in rock, each point representing an event in time'. The use of time–rock terms (e.g. Lower Cambrian) predates the standardization of time terms, so it is an understandable perpetuation of an old habit that it is now nevertheless timely to replace. By referring to Early Cambrian rather than Lower Cambrian, the definition (and concept) is more direct. Early Cambrian rock means any rock formed in Early Cambrian time. The geologic period is defined by the initial and terminal events represented by the GSSP. The system is the rock estimated to have formed in that interval. It cannot define the period because the system boundaries are unknowable except at unconformities where the boundary rocks of uncertain age are missing.

This work eliminates the use of time–rock terms such as **system** without loss of meaning.

3.3.1   Age and stage

A further usage in this book is to depart from that which has been commonly accepted as proper even though commonly disregarded. We refer to the particular time–rock couplet: age–stage. Traditionally, **stage** is the rock formed in the corresponding age according to most national stratigraphic codes and to the ISSC *Guide* (Hedberg 1976).

The new departure here is to use the term **stage** for time as well as for rock. This has the advantage of liberating the term **age** for general use and so avoiding some ambiguity. In fact this is commonly done and few readers will notice this as an innovation. The hierarchy newly applied in this book is thus eon, era, period, epoch, stage, chron.

3.3.2   Chron and chronozone

International preoccupation has hardly arrived at the definition of GSSP for chrons except where they are conterminous with longer divisions. Biozones are commonly used somewhat ambiguously as referring either to rock characterized by certain fossils, or to rocks supposed to be of a certain age even if diagnostic fossils have not been found in the particular sample (= biochrons of Williams 1901).

Oppel, in 1856, was perhaps the first to use (Jurassic) zones systematically and effectively but he did not define a zone. 'He even says that although he has elected to name his zones after fossils, they could equally well have been named after places' (Arkell 1933, pp.16–17). It is clear that he conceived zones as time divisions. He tabulated his zones grouped in stages, and these into three divisions: Lias, Dogger and Malm for Lower, Middle and Upper Jurassic. D'Orbigny had in 1849–1852 employed zone as a division of a stage. Buckman used the term biozone in such a time sense in 1902 but happily in this sense the term biochron had been introduced by Williams in 1901 (p.22). Arkell (1933 and 1956) used zones in the sense of biochrons, and for this purpose he always arranged for the fossil named to be in Roman type. The practice is recommended especially if fossil names are used for a contemporary chron defined by GSSP, but many editors have not previously appreciated the reasons for this usage.

The conclusion here is to eliminate chronozone as the unnecessary part of the chron–chronozone couplet and to aim to standardize chron (formal) or biochron (informal) for time divisions.

3.4   The Precambrian chronostratic scale – introduction

A commonly expressed view is that while chronostratic divisions are suitable for Phanerozoic time, chronometric divisions alone should be used for Precambrian time. Chronostratic divisions were applied earlier to Precambrian rocks and are now usefully creeping back. There is no reason why they will not continue to do so. Each type of scale has its own use and limitations (Harland 1975). For example, were the initial Cambrian boundary to be now defined chronometrically we might not know whether Ediacara or olenellid faunas were Precambrian or Cambrian because of uncertain calibration; and if the initial Vendian boundary were similarly defined we should not know whether the Varanger tillites were Vendian or not, and so on. Therefore, while supporting the chronometric scale for Precambrian time (Chapter 2), there is need for a complementary distinct sequence of time divisions based on rock successions (Sections 3.5 to 3.7).

For Precambrian as well as for Phanerozoic chronostratic divisions rock successions are needed which have a global correlation potential by all methods (including numerical determinations) for the Global Stratotype Sections and Points (GSSP).

3.4.1   Displays of Precambrian stratigraphy in this volume

The Precambrian Chart 2.1 of GTS 82 is replaced in this edition by six displays.

Figure 2.1 in Chapter 2 shows some options for the classification of Precambrian time if defined chronostratically. The columns are explained in the text.

Figure 2.2 shows the Proterozoic chronometric scheme arising from the International Subcommission on Precambrian Stratigraphy meeting, 1988.

Table 3.1 here lists the chronostratic sequence of Precambrian successions that might be considered as the basis in due course for defined chronostratic divisions. The explanation is in Section 3.5.

Figure 3.1 is the only attempt at a Precambrian chart along the lines of the succeeding Phanerozoic ones. It treats the Vendian Period and is explained in the text of Section 3.7.

In Chapter 7 a linear scale plot of the above division for Precambrian time is presented in Figure 7.8 and a plot of Precambrian events (in addition to some Phanerozoic ones) is shown in Figure 7.9.

3.5 Precambrian chronostratic eras

This section expands the brief entries made in GTS 82 to illustrate the way forward to construct a Precambrian chronostratic scale. Today we share a similar state of ignorance with that of Phanerozoic stratigraphers perhaps 170 to 200 years ago and it is largely due to the relative lack of evident Precambrian characters for correlation. But the principles are identical and progress is proving to be no less rapid now than it was then. The names suggested here for eras have already been proposed for chronostratic (rather than chronometric) divisions in some form or other and are placed in a tentative sequence for a workable chronostratic scale (Table 3.1).

3.5.1 Hadean Era

The name **Hadean** was proposed by Cloud (1976) and elaborated by him in 1983 (Schopf) as a division of Earth history preceding Archean. Its initial boundary coincides with some event taken to signal the first step in the formation of the Earth (however that may be defined and however calibrated). Its terminal boundary might be taken in some terrestrial record as, for example, in the Isuan supracrustal succession; it being a chronostratic convention that the terminal boundary be defined by the initial boundary of the succeeding division (Cowie *et al.* 1986).

A pre-Hadean division to accommodate events prior to the Earth's formation may be considered; but not in this work. It may be simply referred to as pre-terrestrial time.

Meteorites (chondritic) appear to be the oldest accessible objects for determination. By Pb–Pb methods the following ages (in Ga) are noted: 4.52 to 4.57 (Tatsumoto *et al.* 1973); 4.635 (Tilton 1973); 4.565 (Chen & Tilton 1976); 4.559 (Chen & Wasserburg 1981); 4.57 to 4.66 (Ireland & Compston 1986); and 4.565 to 4.575 (Manhes *et al.* 1986). By Sm–Nd, 4.46 to 4.56 were obtained by Jacobsen & Wasserburg (1984).

Lacking a Hadean record of Earth history we look to the Moon, especially since the Apollo missions from 1969 to 1972. Before that time, however, Shoemaker & Hackman (1962) had postulated a 'stratigraphic basis for a lunar time scale'. Superposition of successive craters, based in part on studies of Lunar Orbiter photographs from Ranger craft in preparation for the Apollo missions, led them to propose the following consecutive stratigraphic sequence which was tentatively mapped throughout the Moon's surface.

'Present time
   Copernican period
   Eratosthenian period
   Procellarian period
   Imbrian period
   pre-Imbrian period
Beginning of lunar history'

Table 3.1. *Precambrian succession referred to in text listed as potential chronostatic divisions* (*guesstimated ages given without qualification or authority*)

			Approximate Age (Ga.)
Phanerozoic	Ph	Phc	
Paleozoic Era	Pz	Pzc	
Cambrian Period	E	Cbn	
Caerfai Epoch		Cae	
Tommotian Stage		Tom	
			——0.57
Precambrian			
Sinian Era	Z	Zin	
Vendian Period	V	Ven	
Ediacara Epoch		Edi	
[Kotlinian		Kot]	
Poundian Stage		Pou	
Wonokan Stage		Won	
			——0.59
Varanger Epoch			
Mortensnes Stage		Mor	
Smålfjord Stage		Sma	
			——0.61
Sturtia Period	U	Stu	
			——0.80
Riphean Era		Rif	
Karatau Period	R₃	Kar	——1.05
Yurmatin Period	R₂	Yur	——1.35
Burzian Period	R₁	Buz	——1.65
Animikean Era		Ani	
(e.g. Gunflint Period		Gun	
			——2.20
Huronian Era	H	Hur	
Cobalt		Cob	
Quirke Lake		QkL	
Hough Lake		Hou	
Elliot Lake		Etl	
			——2.45
Randian Era	W	Ran	
Ventersdorp		Vtp	
Central Rand		Ctd	
Turffontein		Ttn	
Johannesburg		Jbg	
Government		Gvt	
Hospital Hill		Hsp	
Dominion		Dom	
			——2.8
Swazian Era		Swz	
[Hammersley Basin]			——2.87
Pongola		Pgl	
			——3.31
Moodies		Moo	
Figtree		Fig	
[Warrawoona]			
Onverwacht		Ovt	
			——3.50
Isuan Era		Isu	
			——3.8
Hadean Era		Hde	
Imbrian Period (pars)		Imn	
Early Imbrian Epoch		Er-Imn	
			——3.80
Orientale		Ori	
Imbrium		Imm	——3.85
Nectarian Period		Nec	
Two epochs with 10 to 12 basins			
Nectarian Basin			
			——3.95
Pre-Nectarian		Pr-Nec	
Basin Groups 2 to 9		Bg2–Bg9	
Basin Group 1		Bg1	
South Pole-Aitken		Spa	——4.10
Procellarum		Pcl	——4.15
Cryptic division		Cry	
(with many episodes, some dated)			
Moon's origin			
Origin of Solar System			——4.56

Such a method was valid whatever the interpretation of lunar craters; and the sequence was largely confirmed by the Apollo missions. However, Procellarian was misplaced and Pre-Imbrian greatly extended. Mutch's *Geology of the Moon – A Stratigraphic View* (1972) developed these and other ideas but it was written before the main work on Apollo samples

Early isotopic age determinations of lunar samples showed the Copernican and Eratosthenian periods to be much younger than Isuan rocks. Indeed the **mare basalt** determinations yielded ages from 3.96 to about 3.16 Ga (Taylor 1975). Pre-Isuan lunar history (i.e. Hadean) would thus precede most of the mare basalt formations which are punctuated by a decreasing frequency of impact craters (Drury *et al*. 1976). Conversely, the record reflected a Hadean history, altogether more catastrophic, back to about 4.4 Ga. This was a time of major and repeated impacts recorded in the turmoil of shock-metamorphosed lunar highland rocks. According to the Charles Arms research group studies using U–Pb determinations (Drury *et al*. 1976) it was regarded as 'dark ages', possibly with fluctuating intensity of bombardment, the last phase of which is known as the 'Terminal lunar cataclysm' from about 4.0 to 3.8 Ga. A few determinations of rock fragments that survived these major impacts gave a clue to an earlier mantle and crust formation.

Taylor (1975, p.269) quoted a speculative attempt at a new chronology of early ringed basins thus:

> Orientale 3.85 Ga
> Imbrium 4.0
> Crisium 4.13
> Humorum 4.13–4.20
> Nectaris 4.2

Earlier events are recorded in fragments in **highland soils** with Pb ages at 4.6 Ga. Crustal formation was suggested at 4.43 and 4.47 Ga. Evidence from relict plagioclase cores suggested ages approaching 4.5 Ga.

A dunite sample with a Rb–Sr age of 4.6 Ga has been interpreted as a relict of a primordial frozen crust predating the Highland anorthosites. Relict lunar ages approach meteorite ages 'to suggest extensive melting, fractionation, and formation of the highland crust at a time close to the condensation of the Solar System at 4.6 aeons' (Taylor 1975, p.271).

From what is perhaps the ultimate work on lunar history by the USGS, which takes into account all available data (Wilhelms 1987), the following summary is abstracted. The lunar analytical work is of the highest possible contemporary standard comparing favourably with the best work on terrestrial samples (Taylor 1975, p.21).

Pre-Nectarian Period

'Lunar and pre-Nectarian history began with the formation of the Moon about 4.55 aeons ago', and similarly for accretion of the Earth, from a composition similar to that of chondritic meteorites. Melting and differentiation of feldspathic crust and ultramafic mantle ensued, ending between 4.4 and 4.2 Ga. Heavy impact rate with extensive brecciation and impact-melting left only a fragmentary record of pre-

Nectarian events. Isotopic ages are of plutonic-rock clasts (e.g. from crater ruins) with ages ranging from 4.54 to 4.17 Ga. Volcanic rock samples gave ages 3.98 Ga and younger.

Wilhelms (1987) suggested three major divisions [epochs] for the pre-Nectarian Period as follows.

(1) A **cryptic** division from the Moon's origin with differentiation, plutonism, volcanism, impact mixing and melting. Age estimates gave 4.56 to about 4.2 Ga. The epoch might even constitute another named period.

(2) **Basin Group 1** comprising the two very large basins Procellarum and South Pole–Aitken, at about 4.15 and 4.10 Ga respectively.

(3) **Basin Groups 2–9** comprising a further 28 basins (and 3400 smaller craters larger than 30 km diameter), ranked and ordered by crater diversity and superposition into eight successive age groups (stages). Their suggested ages are interpolated and extrapolated from about 4.1 to 3.92 Ga.

Nectarian Period

The initial boundary would be defined in rock from the impact that created the Nectaris Basin (about 3.92 Ga) and terminated with that from the Imbrium Basin impact at 3.85 Ga, i.e. a duration of about 70 Ma. From 10 to 12 basins formed in this period, and they were ranked into two age groups [epochs] averaging about 121 craters per basin group.

Possibly two other basins (Serenitatis and then Crisium) might be about the same age. The abundance of ages determined between 3.85 and 3.95 Ga and the paucity of earlier ages led to the suggestions of the 'terminal impact cataclysm'.

Imbrian Period

This period is based on the well-known Mare Imbrium, the third largest after Procellarum and South Pole–Aitken. It is divided into two quite distinct epochs.

Early Imbrian Epoch Two large impact craters (Imbrium and Orientale) dominate the lunar surface and ejecta reached at least 1000 km from the crater rim and formed secondary craters each with their own secondary ejecta. The average age of breccia is 3.85 Ga which conforms with ages of KREEP-rich volcanic basalt. The other large Early Imbrian basin, Orientale, formed about 50 Ma later, and ended the (Hadean) era of enormous impacts. The span of the Early Imbrian Epoch was postulated to be about 50 Ma, say 3.85 to 3.80 Ga.

Late Imbrian Epoch Henceforth the record is clearer with the preservation of widespread mare basalts, marked by relatively less-disturbing impacts. Three age groups were mapped (1) commonly at about 3.79 (say 3.80 to 3.75), (2) 3.75 to 3.50 Ga and (3) 3.50 to 3.20 Ga.

Altogether the Late Imbrian 'Epoch' lasted about 600 Ma, from about 3.8 to 3.2 Ga. However, from Late Imbrian time the chronostratic scale is based on terrestrial strata even though the lunar record is also clear from that time onwards.

Eratosthenian and Copernican Periods

To complete the lunar record the Eratosthenian Period (with mare basalts and craters) followed, lasting from about

2.1 Ga to 1.1 Ga. The Copernican Period (with minimal basalts and a record almost entirely of impact craters) began at about 1.1 Ga and has not yet been terminated (Wilhelms 1987).

3.5.2 Isuan Era

The name for this era is taken from the oldest known terrestrial rocks at **Isua** in southern West Greenland which for some time, at least, were thought to provide the earliest stratigraphic record on Earth (e.g. McGregor 1968, 1973). The era is thus intended to span the early recorded stages of Earth's crustal evolution.

An earlier crust must have provided source rocks for the Isua sediments which are associated with volcanics. The sediments include quartzites, siltstones, pelites, various calcareous and carbonate facies and ironstones (magnetite and quartz – i.e. early banded iron-formations (BIF)), as well as tuffs related mainly to basic and ultrabasic lavas. The BIF yielded a Pb–Pb whole-rock Isuan date of 3760 ± 70 Ma which probably represents a metamorphic event rather than a primary one. Later, or possibly contemporaneous with an upper acid volcanic part of the Isua strata, granites were intruded, and they were deformed and metamorphosed as were the Isua strata so becoming the Amitsoq gneisses. The age of the Amitsoq metamorphism was determined at about 3750 Ma. The Ameralik dyke swarm cuts the above and was succeeded by Malene supracrustals with sediments, basic and intermediate volcanics and intrusions. Anorthosites were then emplaced, followed by major tectogenesis and then further emplacement of ultrabasics. Intrusion of sheets of calc-alkaline rocks were deformed and became the Nuk gneisses, providing the next isotopic age determination at 3040 Ma (Bridgewater *et al.* 1976).

The initial boundary for the proposed Isuan Era might be selected within the metasedimentary sequence only the metamorphism of which has so far been dated. The remainder of the Isuan Era needs to be investigated but need not yet be defined. Its terminal boundary will be the initial GSSP of the succeeding Swazian Era. On that basis and with present data the Isuan interval might span 300 Ma, 3.8 to 3.5 Ga. It was a time of great magmatic activity. It seems unlikely that for such a time in Earth history an unmetamorphosed stratal record will be found. So for the time being the Isuan rocks provide an appropriate name even if a better sequence be found elsewhere for the initial GSSP.

Carbon isotope studies (e.g. Schidlowski 1988) lead to the near certainty that the Isuan rocks, even though metamorphosed, bear evidence that an oceanic biomass was already generating a significant proportion of organic carbon (rich in ^{12}C). The production was probably from photosynthesis by prokaryotic cyanobacteria.

Rocks probably as old as or older than the Isuan rocks are now known elsewhere. An example is the Limpopo belt of Southern Africa where the Sand River gneisses have yielded a minimum date of 3786 ± 61 Ma. It is possible that they were of sedimentary origin but there seems to be less supracrustal material to compete with the West Greenland rocks for a standard scale (e.g. Tankard *et al.* 1982, p.96). However, the oldest parts of the Zimbabwean sequence (Lower Greenstones) have been tentatively regarded as 2.9 Ga. Search for the oldest detrital zircons gave ages of 3.75 to 3.80 Ga with a possible source in the Tokwe–Zvishavane gneisses (Dodson *et al.* 1988).

Still older detrital zircons (4.2 Ga) have been reported from Western Australia (Froude *et al.* 1983, Compston & Pidgeon 1986). These may have derived from an earlier buoyant terrestrial crust of which we have no other record.

3.5.3 Swazian Era

A southern African sequence is suggested as providing the best international reference for the succeeding era. This follows Kent & Hugo (1978) in favouring the rich **Onverwacht–Figtree–Moodies** succession as potentially good for global correlation and they also include the Pongola Group in the Swazian interval.

Kent & Hugo (1978) were addressing a similar need for a chronostratic Precambrian scale, though entirely based on the wealth of material from southern Africa. Their case was strengthened by agreement on classification and nomenclature of the rocks by the South African Committee for Stratigraphy. The stratigraphic philosophy, following that of the IGC (Bologna) 1881, is similar to that in this volume in requiring two essentially different hierarchies; one for time (periods, etc.), and the other for rock (formations, etc.). The earlier South African 'systems' were converted to the latter category. In the course of their work Kent & Hugo proposed consecutive chronostratic divisions, the first two of which are followed here: Swazian and Randian. Their Swazian division is outlined as follows.

> Pongola Supergroup (Swaziland, etc.)
>   Mozoan Group
>   Nsuze Group
> Swaziland Supergroup (east Transvaal)
>   Moodies Group
>   Figtree Group
>   Onverwacht Group
>   Geluk Subgroup
>   Swatkoppie Fm
>   Kromberg Fm
>   Hoogenoeg Fm
>   Tjakastad Subgroup
>   Komati Fm
>   Theispruit Fm
>   Sandspruit Fm (oldest)

A fuller account of the granite–greenstone terrane (Kaapvaal Province) with detailed environmental interpretation is available in Chapter 2 of Tankard *et al.* (1982). Metamorphic dates for the **Onverwacht** Group cluster around 3.5 Ga with the Theispruit Formation at 3.54 Ga. The **Moodies** Group is older than 3.31 Ga and even the **Figtree** Group may be as old as 3.5 Ga (Schopf 1983). The **Swaziland** Supergroup is an apparently continuous succession of supracrustal rocks, originally volcanics and sediments (best seen in the Barberton Belt). The Pongola Supergroup with volcanics is bracketed between igneous dates: 2930 and 2870 Ma (Tankard *et al.* 1982, pp.80–81).

The **Sandspruit** Formation, dominantly ultrabasic, is

preserved as rafts in later intrusions and interpreted as primordial ocean crust. Metasedimentary rocks, mainly cherts, carbonates and siliceous schists, occur in the Theispruit Formation and in all three formations of the Geluk Subgroup. Oolitic and filamentous microstructures have been claimed (but not yet established) as biogenic. Oehler and others suggested that 'Carbon isotopes from the lower Onverwacht rocks are anomalously heavy compared with those from the upper Onverwacht, suggesting evolution from non-photosynthetic to photosynthetic organisms' (Tankard *et al.*, 1972, p.45). Whatever the interpretation of these phenomena there is a richness of material and variety of forms that merits consideration for a type section and possibly a GSSP low within it. Moreover, the Figtree and Moodies strata follow conformably with somewhat similar mixed facies.

In the **Pilbara Block**, Western Australia, the Warra-woona stromatolitic chert formation may be coeval with this earliest fossil record at about 3.5 Ga. It has one of the very few early authentic well-established Archean (prokaryotic) microbiotas. (Schopf & Packer 1987).

The **Pongola Supergroup** is separated from the Swaziland Supergroup in time (and space) by at least six further distinct igneous episodes. The Nsuze Group is dominantly volcanic with interrelated sandstones and the Mozoan Group is dominantly siltstones and shales, but perhaps it is best known for its red banded iron formations (BIF), probably related to volcanism, and stromatolites.

The name Swazian for an era spanning these events in Earth evolution is appropriate in that both supergroups have Swaziland connexions.

3.5.4 Randian Era

Randian is adopted from Kent & Hugo (1978) for and from the Witwatersrand Triad, which is rich in gold, uranium and many other features, not least in tillites. The name Wit-watersrand Triad was given by Hamilton & Cooke (1969) for the sequence:

> Ventersdorp System
> Witwatersrand System
> Dominion Reef System (oldest)

The Dominion Group has been included latterly in the Witwatersrand Supergroup. Kent & Hugo included the earlier Beitbridge rocks of the Limpopo Group in their Randian division, but the tectonic complexity and lack of a clear stratal sequence diminish the value of the group for a chronostratic type sequence. On the other hand, the Witwatersrand and Ventersdorp Supergroups are stratigraphically and sedimentologically very well known. The gold and uranium reefs and placers are of great economic importance. Their distribution was locally controlled whereas thin, irregular, but extensive diamictites are taken as time markers and some have long been interpreted as glacial (du Toit 1926, Harland 1981, 1982, Tankard *et al.* 1982). There is thus a potential for wider correlation. In the following summary of the 'Witwatersrand Triad' succession only a few selected formations are included.

> Ventersdorp Supergroup
> Priel Group (volcanics and sediments)
> Platberg Group (mainly sedimentary with stromatolites and ooids)
> Klipriviersberg Group (volcanics)

> Witwatersrand Supergroup
> Central Rand Group
> Turffontein Subgroup
> with Eldorado, Elsberg and Kimberley (Reefs) formations and one or more Kimberley tillites near base
> Johannesburg Subgroup with Buit volcanics and Main Leader at base

> Government Subgroup with Coronation Fm with tillites Welgegund Fm with (Reitfontein tillite) probably coeval and Bonanza formations with one tillite and Promise Fm with two tillites at least one being earlier than Reitfontein

> Hospital Hill Subgroup (sandstone and siltstones with Orange Grove Fm at base)

> Dominion Group
> Syferfontein Fm (and volcanics)
> Renosterhoek Fm (basic volcanics)
> Renosterspruit Fm (sandstone) (oldest)

Dominion lavas have yielded U–Pb dates of 2.83 to 2.80 and 2.75 Ga; similarly Ventersdorp lavas gave 2.40 to 2.30 Ga. However, the Rb–Sr ages of granite intrusions into Ventersdorp rocks were 2.6 and 2.64 Ga suggesting that the whole Randian sequence might be Archean (Harland 1981 and Tankard *et al.* 1982). In the Johannesburg Subgroup the Vaal Reef has gold and uranium-bearing kerogen-rich seams, estimated at 2.5 Ga (Schopf 1983). Kent & Hugo (1978) estimated their Swazian and Randian divisions as approximating 3750 to 2900 and 2800 to 2630 respectively. But they regarded the Randian rocks as of Proterozoic type and so, with some others, viewed the Archean–Proterozoic boundary as diachronous.

In Montana glacigenic rocks occur beneath the Stillwater Complex dated (by U–Pb in zircons from metasediments) at 2713 ± 3 Ma (Nunes 1981). A post-Stillwater igneous rock was dated at 2750 Ma (Page 1982). Whether or not this might be the earliest known glacial event, it might provide a candidate for the initial GSSP of the Randian Era in competition with some rock in the Dominion Group.

If the terminal boundary were defined at the initial GSSP in the Huronian sequence Randian could span to 2.4 Ga or less. Thus for the later part of the era the Hammersley Basin with its rich succession of BIF would be Randian. The Fortescue and Hammersley Group were estimated as forming between 2.6 and 2.2 Ga (Hickman 1983).

Two of the best authenticated Randian biotas occur in the Ventersdorp and Hammersley rocks.

3.5.5 Huronian Era

The name **Huronian** was intended by Logan (1863) to

include all the Precambrian sediments and eruptives above the Laurentian granites and gneisses, but below the Animikie Series. The rocks span the international boundary, and to clarify increasing confusion the International Committee on Geological Nomenclature, representing the Geological Surveys of the United States and Canada, proposed the following scheme for Precambrian divisions in the Lake Superior region (Van Hise 1905).

Keweenawan (Nipigon)

Huronian { Upper / Middle / Lower

Keewatin

Laurentian (oldest)

The Upper Huronian division comprised the Animikie rocks. Chamberlin & Salisbury followed this scheme in their influential textbook (1906) but grouped the lower two divisions as Archeozoic or Archean and the upper two as Algonkian or Proterozoic. They were ambiguous in separating the Animikie from Huronian rocks. Today the Huronian Supergroup does not include Animikie rocks and two eras are proposed here for these rocks (Huronian and Animikean).

The Huronian Supergroup now comprises four (instead of two) groups, three of which begin with glacigene formations, the latest of these being the Gowganda tillite of the Cobalt Group. Associated rocks have yielded a whole-rock Rb–Sr age of about 2.3 Ga, which could be too high. Otherwise the Supergroup is bracketed between rocks of the Superior Province altered by the Kenoran Orogeny at about 2.5 Ga and the later Nipissing dykes at 2219 ± 4 Ma (Corfu & Andrews 1986).

The succession is as follows.

Huronian Supergroup
 Cobalt Group
  Bar River Fm
  Gordon Lake Fm
  Lorrain Fm
  Gowganda Fm (with several tillite members representing two main glacial advances)

 Quirke Lake Group
  Serpent Fm
  Espanola Fm
  Bruce Fm (tillite)

 Hough Lake Group
  Mississagi Fm
  Pecors Fm
  Ramsay Lake Fm (tillite)

 Elliot Lake Group (oldest)
  with five formations including a lava resting unconformably on basement of the Superior Province

The initial GSSP of the Era might be selected in the Elliot Lake Group and it would terminate with a selected point in the Animikie rocks.

The Elliot Lake basal conglomerate contains placer deposits of uranium minerals indicating reducing conditions. These contrast in their drab colour (as also do the Hough and Quirk Lake Groups) with the Cobalt Group which has red beds. This could represent a significant increase in atmospheric oxygen.

The Griqualand West sequence in the Transvaal Basin in South Africa, with complex glacial episodes, is bracketed between Ventersdorp Supergroup lavas dated from 2300 to 2600 Ma and overlying lava at 2224 ± 21 Ma. It could well be the equivalent of one of the Huronian tillite formations. There is a potential for correlation.

3.5.6 Animikean Era

The **Animikean** Group was proposed by Hunt in 1873 for the Animike District, Port Arthur, Ontario, near Thunder Bay, Lake Superior (Van Hise 1892). In this type locality the Group comprises two formations resting on pre-Huronian basement rocks of the Canadian Shield Superior Province. The lower unit is the **Gunflint (Iron)** Formation with three members (Goodwin 1956, Moorhouse 1957). Overlying the Basal Conglomerate Member are the Lower and Upper Gunflint Members each representing a similar cycle beginning with algal chert passing through tuffaceous shales into taconite (a typical banded ironstone facies – BIF – with silica and siliceous red iron ores). The uppermost unit is the **Rove** Formation which is overlain by the (Keweenawan) Sibley sediments, followed by the classic, copper-bearing Keweenawan lavas.

The Gunflint cherts and coeval strata with visible algal mats have yielded the earliest known microbial structures in a rich biota of many species (e.g. Schopf 1983). Prior to Gunflint time only stromatolitic algal structures have been recorded (Schopf & Packer 1987).

The Animikean rocks are more extensive south of the Canadian border where deformation in the Penokean Orogeny, and some thermal metamorphism together with variation in these shallow-water facies make correlation difficult. Nevertheless, overlying late Huronian strata are ironstone formations outcropping in several distinct mining areas with well-known formations. These include **Bijiki** (Marquette district), **Vulcan** (near Menominee), **Ironwood** (at Gogebie), **Biwabik** (at Mesabi) and **Deerwood** (at Cuyuna). Altogether they are said to account for 95% of North American iron ore resources and production and were described in the classic account by Van Hise & Leith (1911).

The ironstone formations are overlain by the still more extensive **Michigamme Slate** Formation.

The Penokean (Hudsonian) Orogeny which affected these rocks has been dated from 1820 to 1650 Ma. The Animikie Group is probably partly equivalent to the Whitewater Formation which may rest unconformably on the Huronian Supergroup. It postdates the Sudbury eruptive dated at 1844 ± 2 Ma (Krogh & Davis 1974b and Krogh *et al.* 1982). The Animikie Group is commonly given an age from 2.1 to 1.8 Ga. It was given the status of Middle Precambrian by Goldich (1968).

Animikean is selected for the chronostratic era following Huronian with an initial definition near the base of the Gunflint Formation and extending in time to whatever GSSP be decided for the initial Riphean boundary.

3.5.7 Riphean Era

Riphean was proposed by Shatsky in 1945 from the Bashkir anticlinorium in the southern Ural Mountains with four major divisions often referred to as R_1, R_2, R_3 and R_4 (Shatsky 1960, Semikhatov 1966). The concept has been developed entirely in the USSR but is commonly used elsewhere by reference to Soviet sections and so Riphean is a serious contender for international use.

The classification of Riphean rocks has been proposed in a variety of ways which is not surprising considering the long time span involved. Semikhatov (1974), who had characterized the main Riphean divisions by their stromatolites, suggested dividing Proterozoic into Upper (Riphean) and Lower (Aphebian). As **Aphebian** is a Canadian division this suggests an international proposal.

There has been considerable debate as to how Vendian or Riphean 4 (R_4) (Kudashian) relate to Riphean. The All-Union Meeting on Subdivision of the USSR Precambrian Rock in 1977 proposed the following scheme (Keller 1979). It will be noted that Soviet authors use their divisions in a chronostratic sense adding numerical ages of boundaries as their current estimates and not as definitions. This principle was confirmed later (Cowie *et al.* 1986). The 'periods' are based on evolutionary stages of stromatolites and katagraphic microfossils.

Phanerozoic		
Cambrian (E or €)		
		570 ± 20 Ma
Upper Proterozoic (Ptz$_2$)		
Vendian (V)		
		650 ± 50 Ma
Riphean		
Kudash (R_4)	} Late Riphean	
Karatau (R_3)		
		1050 ± 50 Ma
Yurmatin (R_2)	Mid Riphean	
		1350 ± 50 Ma
Burzian (R_1)	Early Riphean	
		1650 ± 50
Lower Proterozoic (Ptz$_1$)		
= Karelian		
three divisions divided at		1900 ± 100 Ma
		2300 ± 100 Ma
		2600 ± 100 Ma

The Uralian stratotype of **Burzian(ian)** R_1 comprises the (oldest) Ai Formation with conglomerates, arkoses, basalt flows and then black shales; then the Satkai Formation with numerous stromatolites; and finally the Baikal Formation with iron ores, shales and conophytic dolostones. There would appear to be scope for correlation and for an initial Riphean Era boundary if it be decided in this sequence. The initial boundary for R_1 was estimated at 1650 ± 50 Ma (Keller 1979).

The proposal here is to adopt R_1, R_2 and R_3 for a Riphean Era in the global chronostratic scale; but so that the latest Riphean should give place to Sinian which has priority (see below). On this basis Kudash and possibly later Karatau

rocks might be Sinian. If Riphean were thus adopted for the international chronostratic scale its definition would not necessarily persist as hitherto in the USSR because the terminal boundary is determined here by the initial boundary of the succeeding Sinian division. On the other hand if Soviet geologists prefer to retain Riphean as a regional standard as defined in their country, some other name(s) for a global chronostratic era or eras would be sought elsewhere.

3.5.8 Sinian Era

The name **Sinian** was introduced for the latest Precambrian strata in the Yangtze Gorge where it was appreciated that strata extended downwards without apparent break from rocks of established Cambrian age (Willis, Blackwelder & Sargent 1907). Prior to that time it had been generally assumed that Cambrian strata rest unconformably on Precambrian rocks. The widespread realization that many Precambrian to Cambrian successions showed apparent continuity led to international use of the name Sinian for them.

Sinian and Vendian are contenders for the formal chronostratic divisions to precede Cambrian. Sinian has priority of publication (1907 compared with 1952) and was used internationally, not least in the USSR, as well as in China, for many years. Then, when Precambrian studies in the USSR advanced beyond those in China, the name Vendian was introduced and has been widely used. It seems almost to have been adopted internationally as is done here.

In the meantime research in China has surged and the claims of Sinian are being pressed again. Sinian had two widely differing meanings – including or excluding the thick northern succession at Jixian, representing a long time span which is probably altogether older than about 800 Ma. In the Jixian sense it was approximately equivalent to Riphean (Bates & Jackson 1980 *AGI Glossary of Geology*). But now that the older sequence has been officially excluded in China, Sinian is again restricted to rocks younger than about 800 Ma, as originally so named from the widespread platform successions typified in the Yangtze Gorge (as on Figure 3.1). A more complete geosynclinal sequence in Xinjiang may well provide a better standard, with four distinct tillite horizons that will probably facilitate international correlation elsewhere. These are the Bayishi, Altungol, Tereeken and Hangeerqiaoke of Wang *et al.* (1981).

Our recommendation, tabulated in Figure 3.1, as in GTS 82, is to establish a Sinian Era or Sub-era (Bayishi to Hangeerqiaoke) divided into two periods, **Sturtian** and **Vendian**.

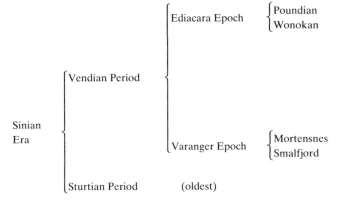

This matter is addressed with more urgency than that of the preceding proposed eras because a working group of ICS has already been set up for the as yet unnamed terminal Precambrian stratigraphic division.

We may add that the old name **Eocambrian** has been effectively ruled out because of ambiguity in usage. However, it may be useful as an informal name for rocks of uncertain age at about the Phanerozoic threshold.

The initial boundary of the Sinian Era would be as for that of the Sturtian Period selected at a GSSP in China, Australia or elsewhere. The Era thus spans the chronostratic interval between the Riphean (as defined here) and Paleozoic Eras.

3.6 The Sturtian Period

The **Sturt** tillite of South Australia was first recognized as glacial by Howchin (1901). Since then the Adelaide Geosyncline in which it occurs has been shown to contain one of the most complete, best preserved and best documented late Precambrian glacial sequences in the world (Preiss 1987). At the same time the rich variety of sedimentary facies in the successions includes many fossiliferous horizons. The Umberatana Group was defined to contain all the tillites. These have been separated into two sequences of **Sturtian** and **Marinoan** ages. The Marinoan tillites postdate Sturtian by about ten 'time slices' (Preiss 1987), and probably correlate with the early Vendian (Varanger) tillites. The 'time slices' are successive stratigraphic intervals each capable of interpretation with a distinctive paleogeologic configuration. The Sturtian sequence contains two main glacial episodes. It begins with the Pualco Tillite. Then after a break comes the Sturt Tillite with many correlative tillites suggesting at least two glacial phases (Hansborough and Appila).

Isotopic age determinations of Sturtian tillite horizons have large uncertainties about a mean of 0.8 Ga. Older rocks do not constrain their age significantly. The overlying Tapley Hill Formation yielded an Rb–Sr whole-rock isochron of 750 ± 53 Ma. Early Marinoan rocks (Willochra Subgroup) by similar Rb–Sr work were dated at 614 ± 98 Ma, the post-Marinoan tillite and pre-Ediacara Brachina Formation gave 601 ± 68 Ma and the slightly later Bunyeroo Formation gave 588 ± 35 Ma. All these and other determinations summarized by Preiss (1987) leave much to be desired, but nevertheless they seem to confirm earlier results and ideas that the Sturtian glacial episodes were much older than the Marinoan.

Current opinion in Australia (e.g. Preiss 1987) is to treat Sturtian, with the preceding Torrensian and the succeeding Marinoan, as regional time divisions. In view of its long history Sturtian was selected as the name for the first of two periods comprising the Sinian Era in GTS 82, whether or not the initial GSSP be taken in South Australia. The Changan and possibly the Nantuo tillites of China might even be Sturtian. In the list of abbreviations (Appendix 2) the symbol U (preceding V for Vendian) was selected partly for its recurrence in Umberatana, Sturt and Nantuo.

3.7 The Vendian Period
3.7.1 Introduction

Vendian will probably be chosen for the name of the period preceding Cambrian. It is commonly so used and the scheme as published in GTS 82 is retained (Cowie & Brasier 1989).

Vendian was used initially for the fourth and latest Late Precambrian division in the USSR following R_1, R_2 and R_3 (Sokolov 1952). There was, however, a paleontologic break earlier than his defined Vendian, and hence an alternative view to extend Vendian (*s.l.*) so as to begin at that change. Alternatively, the change was taken as the initial boundary of the fourth (Yudomian) assemblage or Terminal Riphean, also known as Vendomian (= Vendian *s.l.*). On this basis Riphean would equal Late Proterozoic in the USSR. The alternative point of view seems to have prevailed (Keller 1979, Sokolov & Fedonkin 1984), and is accepted here, in which Kudash was introduced as a division to equal early Vendian (*s.l.*) and to precede Vendian proper. So the sequence in the USSR is R_1, R_2, R_3, R_4 and V, spanning from 1650 ± 50 Ma to initial Cambrian. The initial Kudashian was estimated at 680 ± 20 Ma and the initial Vendian a little later (Chumakov & Semikhatov 1981). The main criterion for separating Vendian from Kudashian is by the Varanger glacial horizon as recommended by Harland & Herod (1975) and referred to in the USSR as Laplandian by Chumakov (1978). With one or other convention now commonly accepted (e.g. Sokolov & Fedonkin 1984, Cowie 1985) the definition of the Vendian initial boundary is likely to coincide with the initial Varanger boundary which has yet to be defined. So Vendian is retained from GTS 82 in the restricted sense commonly used and so as to include two epochs (e.g. Harland & Herod 1975), i.e. Varanger (Kulling 1951) and Ediacara (Cloud 1972). This would be appropriate if a Xinjiang section for Sinian correlation were used and it might avoid some difficult problems of correlation between Vendian and latest Riphean (e.g. Vidal 1979). Within this span there are relatively good palynological and climatological bases for correlation.

On the other hand, Vidal suggested a three-fold division of Vendian, namely Valdaian (latest), Varangerian and an early division. Certainly northern Europe has good successions for correlation in this interval, but Valdaian is not used here (the name is already established in Quaternary nomenclature). Vetternian was suggested by Vidal for the early division (personal communication 1981): it is related to a small and isolated outcrop in Sweden, but by the definition adopted here would probably be pre-Vendian.

A **Vendian Period** is therefore proposed with two epochs **Varanger** and **Ediacara**. The initial (Varanger) boundary was estimated at 650 ± 10 Ma (Sokolov & Fedonkin 1984).

3.7.2 The Varanger Epoch

The name is from Varanger Fjord in Finnmark, north Norway, where a tillite was one of the first two to be recognized as Precambrian. This proved to be the lower of two glacial horizons correlated in detail through north Norway, Spitsbergen and East Greenland (e.g. Hambrey 1983) and referred to by Kulling as the basis of the **Varanger Ice Age**

29

Figure 3.1. The Vendian Period and selected successions. Triangles indicate tillites.

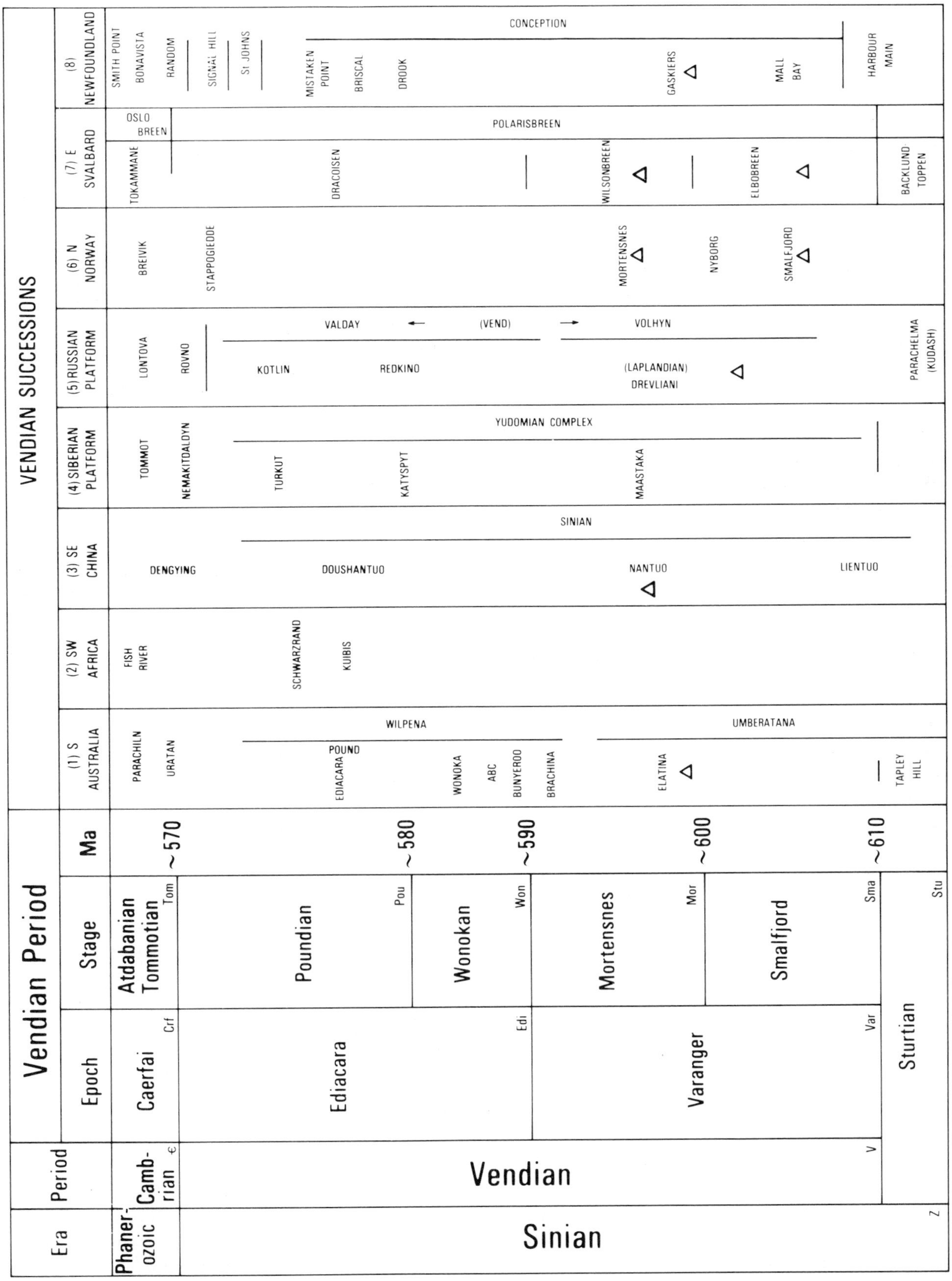

(1951). It was later argued that this ice age was global with evidence of low-latitude marine glacial deposits (Harland 1964), and the stratigraphic scheme was proposed (as here) with Vendian comprising Varangian and Ediacaran (Harland & Herod 1975).

The name **Varangerian** was later argued to be etymologically more correct than **Varangian**, but we now go back to the original name **Varanger**. This conforms with the majority of epoch names that omit the suffix -ian. A junior synonym is Laplandian in the USSR.

The two main glacial stages recognized in Finnmark are readily identified in the lands bounding the North Atlantic–Arctic region. Other intermediate or low-latitude paleomagnetic determinations confirm that the many apparently coeval tillites in most continents have the potential for more precise correlation. Climatic episodes distinguished by biostratigraphic, isotopic and magnetic determinations may well identify this epoch in many parts of the world. The initial Vendian (and Varanger) boundary would be defined by a GSSP somewhere near the lowest glacial facies.

3.7.3 The Ediacara Epoch

The **Ediacara** Member (Rawnsly Quartzite Formation, Pound Subgroup) of the Wilpena Group (overlying the Umberatana Group) yielded the famed Precambrian metazoan fossils. The first discovery in the Flinders Range (Sprigg 1947) was followed by records from many other localities in South Australia and of similar faunas in many other parts of the world. There is no question but that this named episode will be incorporated somehow into the international scale. Termier & Termier possibly first applied the name internationally in 1949. Cloud in 1973 used the name **Ediacarian** (Preiss 1987). Harland & Herod (1975) proposed **Ediacaran** as the second Vendian division (following Varangian). Jenkins (1981) made a detailed proposal for an **Ediacaran Period** beginning with a Wonokan division; whereas Cloud & Glaessner (1982) used Ediacarian for a period beginning earlier, immediately after the Marinoan glaciation. In all cases the division would terminate with the initial Cambrian boundary.

In this work the form **Ediacara** avoids the conflict of suffixes. As an epoch Ediacara does not compete with Vendian which is ranked as a period.

The rich fauna is the subject of innumerable papers conveniently summarized by Glaessner (1984) and Sokolov & Fedonkin (1984). GTS 82 followed Jenkins (1981) in distinguishing two Ediacara divisions, the Wonokan and Poundian ages. In this work these are stages. Sokolov & Fedonkin (1984) divided Vendian rocks of the USSR into three series or epochs: Laplandian (Varanger), Redkinian (Ediacara) and Kotlinian (Povaronian). The third division is bound up with problems of latest Precambrian trace fossils and small shelly fossils which are still awaiting resolution along with that of the initial Cambrian boundary. In this work the Ediacara Epoch terminates at the Cambrian boundary and so includes Kotlinian if that proves to be a useful post-Poundian stage. It is possible that there was a further weak glacial episode in latest Ediacara time (**Late Sinian** of Harland 1983a and Harland 1989).

Somewhere in the Lower Wonokan Formation (with the stromatolite *Tungussia* cf. *julia*) an initial Ediacara GSSP might be selected.

3.7.4 The Vendian chart (Figure 3.1)

On the left of Figure 3.1 the chronostratic scale adopted in this work (as in GTS 82) is tabulated. On the right are selected names of rock units or successions from eight areas (1) to (8) as follows:

(1) Names from South Australia are selected from a multitude of rock units described (e.g in Preiss 1987). From the Ediacara Formation, of the Pound Subgroup of the Wilpena Group, Jenkins (1981) proposed an Ediacaran system, ranked here as an epoch.

The Elatina and related tillites are taken as of Varanger age.

(2) The Schwarzrand and Kuibis formations contain Ediacara type fossils (e.g. Sokolov 1984).

(3) The Yangtze platform sequence as seen, for example, in the Yangtze Gorge region is the type Sinian locality (West Hubei) (Wang in Yang, Cheng & Wang 1986). The Nantuo Tillite is of unknown age; here it is tentatively plotted as of Varanger age.

(4) The Siberian Platform contains the distinctive Tommot fauna, a stage divisible locally into at least three chrons and younger than the Nemakit Daldyn units. The Yudomian complex is not well suited for a Vendian standard but the name has been widely used as an alternative to Vendian (e.g. Sokolov 1984).

(5) The Russian Platform sequence is in the type Vend area and the rock divisions in that column are as proposed by Sokolov (1984 and earlier) for Vendian classification. Laplandian is synonymous with Varanger.

(6) The North Norway sequence of two distinctive tillite formations occurs in the Varanger Fjord area of Finnmark in North Norway, hence the epoch name Varanger (e.g. Edwards & Føyn 1981).

(7) The succession in eastern Svalbard (Ny Friesland) is one of the best for Varanger rocks (e.g. Hambrey 1988).

(8) The Newfoundland succession is from Krogh *et al.* (1988).

3.8   The Phanerozoic Eon

3.8.1   The name Phanerozoic

'Post-Pre-Cambrian' was the alternative name for expressing Cambrian through Holocene time until Chadwick in 1930 proposed **Phanerozoic** for this span (Paleozoic + Mesozoic + Cenozoic) and **Cryptozoic** for Proterozoic + Archeozoic (e.g. Harland 1974).

All the above names carry a descriptive meaning related to the evolution of animal life (even when probably there was none), but they have also become conventional names in a stratigraphic hierarchy and this is how they are used here. In other words, initial boundaries are conterminous with successively lower ranks at the defined point. Thus the initial Phanerozoic boundary would be defined at the point that will define the initial Cambrian boundary and not according to successive fossil discoveries or opinions about animal evolution. This opinion was supported by the Precambrian Subcommission in 1988.

3.8.2   The Paleozoic Era

Palaeozoic Series was first proposed by Sedgwick in 1838 for the rocks overlying the **Primary Stratified Groups**. J. Phillips in 1840 applied Palaeozoic to those 'transitional' rocks up to and including the Old Red Sandstone, and in 1841 he extended it to include all rocks from Cambrian to Permian, when he also introduced the names **Mesozoic** and **Cenozoic** in the modern sense.

Paleozoic is commonly divided into two sub-eras with three periods in each.

3.8.3 The Mesozoic Era

In the latter part of the eighteenth century the word **Secondary** was used for stratified rocks younger than either Transition or Primary or Primitive rocks. It then came to refer to the sequence from Old Red Sandstone through 'Upper Fresh Water Beds' (Oligocene) in Britain (e.g. Challinor 1978). Arduino in Italy in 1759 had distinguished rocks as Primary, Secondary and Tertiary (Lyell 1832, p.56). Secondary rocks came to refer to rocks up to the top of the Chalk (Cretaceous) and in 1841 J. Phillips introduced **Mesozoic** (Middle life) to follow Paleozoic (including Permian). Thus Mesozoic was from its inception defined as it is now. Secondary became synonymous with Mesozoic and gradually lapsed.

3.8.4 The Cenozoic Era

While Tertiary (1759) is an older name than Cenozoic (1841) the two do not compete, referring as they do to different time spans. Cenozoic encompasses Tertiary and Quaternary, Quaternary having been introduced later (1854).

The spelling of the name in the form adopted here (**Cenozoic**) clearly worries some geologists as indeed other spellings worry others. The name **Cainozoic** (from the Greek *kainos*, recent) was coined in 1841 by J. Phillips in his table of British Strata. He did not use **Kainozoic** which might have spared us later confusion. Cainozoic became transformed into the Latinized-Greek hybrid form **Caenozoic**, the letter K not being Latin, the letter C being standard in English spelling (cf. Carnian for Karnian). It was thus commonly pronounced 'seenozoic' and so has become spelt in the 'barbaric' (*Open Earth* 1982 No. 18, p.42) form **Cenozoic** and **Cenozoique**.

Cenozoic has long been adopted in North American English and the Tertiary Sub-committee, and then the Stratigraphy Committee, of the Geological Society of London (in George *et al.* 1968) recommended the spelling Cenozoic partly for international conformity in the absence of a clearly agreed scholarly alternative. This usage was followed when that sub-committee eventually reported in full, and so Cenozoic was retained (footnote in Curry *et al.* 1978, p.2). Cenozoic is so spelled here for the same reason as Paleozoic (because language evolves and seems to have got so far).

As to the constituent parts of the Cenozoic Era, there has been a long history of varied opinion out of which Paleogene and Neogene have been accepted as periods. Tertiary, though it persists in North American usage, has become either informal or a sub-era.

It was only after the IUGS was set up around 1960 that there was a proper authority to settle these matters. So, in response to what seemed to be a prevailing opinion in Europe, George *et al.* (1968) in a submission to the International Commission on Stratigraphy (ICS) of the IUGS, recommended 'that the Cenozoic Era should be divided informally into Tertiary and Quaternary sub-eras, and that the formal division should include the Paleogene and Neogene

Periods/Systems. Paleogene was divided into Paleocene, Eocene and Oligocene Epochs/Series and Neogene into Miocene and Pliocene Epochs/Series' (Curry *et al.* 1978, p.2). There are now subcommissions of the ICS on both Paleogene and Neogene periods along with subcommissions for all other geologic periods. There is, however, a proposal under consideration to include Quaternary (Pleistocene and Holocene) also within the Neogene Period. This proposal is under consideration by the Neogene Subcommission and is often adopted but not yet ratified; INQUA is also concerned.

3.9 The Cambrian Period
3.9.1 History and classification

The history of the definition and classification of Cambrian rocks in the British Isles, especially in North Wales, from the initial publication of the name in 1835 by Adam Sedgwick to the ultimate resolution of the conflict about the Cambrian–Silurian boundary through establishment of the Ordovician System by Lapworth in 1879, has been outlined by Stubblefield (1956), Cowie, Rushton & Stubblefield (1972), Bassett (1985b) and Secord (1986). Cambria is a variant of Cumbria (ancient British kingdom, in present-day northwest England) latinized from the Welsh *Cymry* (= fellow country-man, compatriot against invading Anglo-Saxons). The Celtic *Cymru* survives only in Wales and Cambrian already pertained to Wales when Sedgwick adopted it.

Detailed stratigraphic information on the various Cambrian divisions and units proposed for the British Isles appear in the *Lexique stratigraphique international* Vol. 1, Europe Fasc. 3aIII by Stubblefield in 1959. This immense series of publications lists rock units of all ages and many regions. It is now mainly useful for the history of stratigraphic rock unit names, and will not be cited again in this volume.

Sedgwick had first used the group names Lower and Upper. His Lower Cambrian was divided into the Bangor Group (with Llanberis Slates and Harlech Grits) below, and the Ffestiniog Group (with Lingula Flags, Tremadoc Slates and Arenig Slates) above; his Upper Cambrian included Bala strata and extended up to the base of the Woolhope Limestone (now Silurian). Since 1879 the Cambrian System has comprised only Sedgwick's Lower Cambrian and not all of that. Arenig was the lowest series of the new Ordovician System and Tremadoc had long been a part of the Cambrian System, as argued by Whittington & Williams (1964) for example. However, the prevailing international practice is accepted here, whether or not based on a misunderstanding, to make Tremadoc the first Ordovician epoch.

In North America, Emmons in 1842 (Wilmarth 1925) proposed a New York Transitional System arranged in four groups: Erie (with Old Red Sandstone), Helderbergh, Ontario and Champlain. Beneath the New York System he proposed Taconic for the earliest system. The lowest Champlain Formation was Late Cambrian and the Taconic System was thus mainly Mid Cambrian and older (Wilmarth 1925). The difficulties of Cambrian classification and correlation arise partly from the paucity of appropriate fossils. In central Europe Barrande (in 1859) listed his oldest (primordial) fauna with *Paradoxides*, and then Brøgger showed the olenellid *Holmia* to be older than *Paradoxides* in central Norway (e.g. Harland 1974). These difficulties led in North America to the early placing of *Olenellus*-bearing rocks above, rather than below, those with the more abundant *Paradoxides*. The name Acadian, introduced by J. W. Dawson in 1868, was proposed for strata characterized by *Paradoxides*, etc. as 'the oldest member of the Paleozoic of America' by Walcott in 1891 when he also introduced the name Potsdam for Upper Cambrian. The order of strata was sorted out but not the nomenclature. These relationships were not understood until about 1900 (Cowie *et al.* 1972).

In 1903 Walcott replaced Potsdam by Saratogan and then in 1912 by St Croixian. He had used Georgian originally from 1882 (Hitchcock in 1861) for Lower Cambrian, and in 1912 Walcott recommended Waucoban as a provincial rock series and applied the name Taconian for general use for the Early Cambrian Epoch (Cowie *et al.* 1972).

In due course in North America these and other names have had regional significance often contrasted between Appalachian and Cordilleran usage, i.e. Georgian, Acadian and Potsdam (or Saratogan) in the east (e.g. Blackwelder 1912), and Waucoban, Albertan and Croixian somewhat later in the west. Usage in North America now tends to favour the western nomenclature.

This leaves us today with a general acceptance of three Cambrian epochs, but no international names recommended for them. Because of the widely different usages of the names Early, Middle and Late Cambrian, and also because of the need to refer informally to parts of epochs (e.g. early Early Cambrian or worse), a nomenclature defined by a stable classification and standardized in type areas, not necessarily near the eponymous localities, is to be preferred. Bassett (1984, pp.297–301) reviewed studies on Early Paleozoic Welsh rocks during the preceding 25 years. The choice made here is for names long used unambiguously for British Cambrian rocks and/or proposed by the Geological Society of London (GSL) Working Group (Cowie *et al.* 1972), i.e. Saint Davids, etc.

3.9.2 The initial Cambrian boundary

The initial Cambrian boundary has been discussed many times. In 1972 the Precambrian–Cambrian Boundary Working Group was formed and in 1974 this became Project 29 of the International Geological Correlation Programme (IGCP), leader J. W. Cowie (e.g. Cowie 1985). That group visited candidate stratotypes throughout the world and in 1983 the biostratigraphic level was decided as post-Ediacara and certainly predating the olenellid–fallotaspid faunas and probably pre- or early small shelly faunas.

It was also agreed by the Working Group to recommend that the initial Cambrian boundary, wherever defined, would also be the initial Paleozoic and initial Phanerozoic boundary. It would serve equally, as is customary, for the initial boundary of the earliest Cambrian epoch, stage and chron when their names have been agreed.

Boundary stratotypes for the reference point (GSSP) had been narrowed down to three: Newfoundland (Burin Peninsula), Yunnan (Meishucun) and Siberia (Ulakhan-Sulugur on the Aldan River). The Working Group voted decisively in favour of the Chinese section after they met in Bristol in 1983, but at the IGC in Moscow in 1984, because of the (perennial) 'need for more work' a decision by ICS was delayed. Without doubt, of all the Phanerozoic boundary standards, rocks of this age are the most difficult to correlate. There is also a commonly held view that it is an especially significant boundary to decide; but it is clear that the early evolution, as of all bio-evolution, of (preserved) metazoan life is rich in potential critical transitions at different times, all of them more or less significant (Cowie & Brasier 1989).

Although Newfoundland or Siberia may still have the best potential, the Yunnan section is summarized here (Cowie 1985) as having been almost accepted.

The Kunyang Quarry section (Meishucun, Jinning County, Yunnan Province, 24°44'N 102°34'E) is 75 km south of Kunming. It is conserved as a national monument and has good communications, facilities and climate. The exposure is almost continuous at Tuanshanding, Xiaowaitoushan and Badaowan and is being excavated throughout. The recommended reference point is placed at the base of unit 7 of the Yuhucun Formation in the section near Xiaowaitoushan in a succession as follows:

Qiongzhusi Formation	Yuanshan Member 72 m
	(Units 13–15)
	Badaowan Member 54 m
	(Units 9–12)
Yuhucun Formation	Dahai Member
	(Unit 8) 1.1 m
	(Unit 7) 0.3 m thick, grey, thin and moderately thick-bedded, quartzose, silty phosphatic and manganiferous dolomite, intercalated with striped cherts, yielding hyolithids, hyolithelminthids, tubetichitids, conodontomorphs, monoplacophorans, gastropods, brachiopods and trace fossils.
	Zhongyicun Member 11.3 m (proposed GSSP)
	(Unit 6) 5 m thick, blue grey moderately thick-bedded, pseudo-oolitic or oolitic, siliceous and dolomitic phosphorite intercalated with a thin layer (0.2 m) of phosphatic, intrapsammitic, lutaceous shale yielding hyolithids, hyolithelminthids, monoplacophorans and acritarchs. (Units 3–5) 6.3 m
	Xiaowaitonshan Member 8.2 m
	(Units 1–2)
	Bayanshao Member
	Jincheng Member

Faunas immediately above the proposed GSSP include small shelly fossils of the *Paragloborilus–Siphogonuchites* assemblage with *Ovalitheca*, *Tovellella*, *Paragloborilus*, *Lophotheca*, *Siphogonuchites*, *Zhijinites*, *Latouchella*, *Yangtzespira*, *Bemella*, *Ilsanella*, *Annella*, *Sunnaginia* and others, and the trace fossil assemblage *Didymanlichnus miettensis*.

Faunas immediately below the marker point include small shelly fossils of the *Anabarites–Circotheca* assemblage with *Circotheca*, *Tureatheca*, *Conotheca*, *Spinulitheca*, *Barbitositheca*, *Anabarites* and others, and the trace fossil assemblage *Cavaulichnus viatorus*.

Figure 3.2. The Cambrian Period and selected successions.

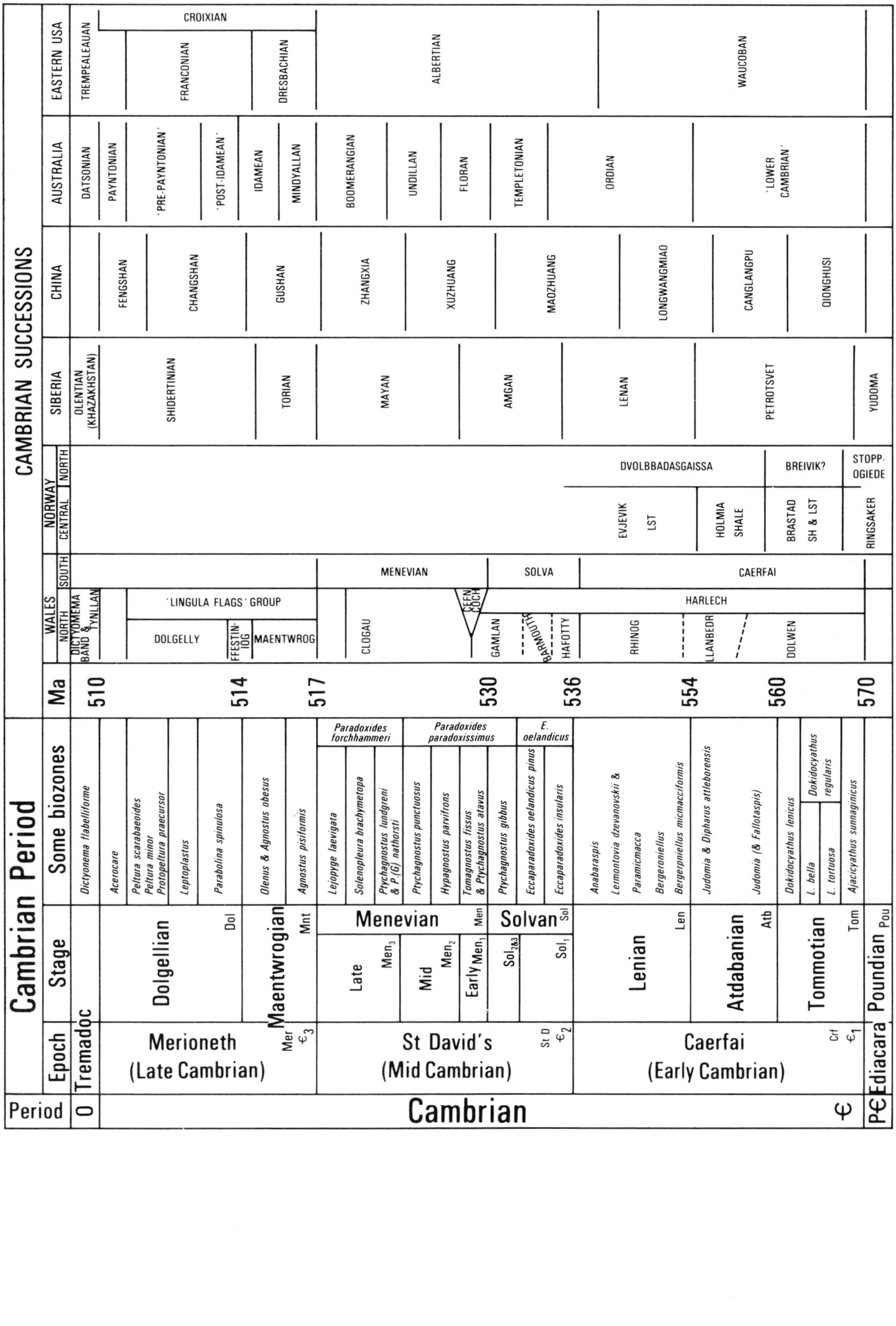

Further work planned includes especially an evaluation of the small shelly fossil faunas both taxonomically and for their global correlation potential.

3.9.3 The Caerfai Epoch

Caerfai is adopted in preference to Comley for the Early Cambrian Epoch because Comley, however well defined (Cowie *et al*. 1972), could mislead so long as the Upper Comley Group of Middle Cambrian age be so named. Although the sequence around Caerfai in South Wales cannot be said to be definitive, the name has, for example, been selected by the USGS (Cohee 1970) as the European standard name and it appears also on Van Eysinga's (1975) chart. It is already familiar internationally. Moreover, a Welsh name from the Principality where the Cambrian system was conceived is appropriate. It has, however, been suggested (Cowie personal communication) that if **Comley** were preferred the Cambrian rock unit might be renamed.

The Siberian sections have now long competed for recognition for the initial Cambrian boundary and wherever that boundary be standardized it would be appropriate to use Siberian names for the stages of the Early Cambrian Epoch.

The **Tommotian Stage** is based on that of Siberia (e.g. Raaben 1981). It could be expanded with chrons in the Tommotian Stage but they (with Atdabanian) seem not to be usable outside the USSR. Therefore, the informal proposal that Atdabanian and Tommotian, which are clearly distinguishable in the Aldan River area but not yet elsewhere, be combined for international correlation as sub-stages in a new **Aldanian stage** has been made (e.g. Cowie & Harland 1989/90).

Soviet geologists have suggested that still older rocks, yet probably still agreed as Cambrian, are found in the Nemakit-Daldyn area. **Rovnian** or **Kotlinian** have been in use in the USSR. Such a division would be pre-Tommotian and the name **Etcheminian** (Matthew 1899) from Newfoundland is already available. It seems that the succession in Newfoundland competes with those of Siberia and Yunnan and so it is an appropriate name for the earliest Cambrian stage to accommodate strata between approximately the first appearance of *Phycodes pedium* and trilobite-bearing rocks in Newfoundland. Etcheminian might compete with Aldanian or Meishucunian for this position. The matter may be resolved in the near future. In the meantime we leave the stated stages on Figure 3.2 as they were tabulated previously in GTS 82 but put forward the possible scheme shown below.

Lenian (or Toiotian)

Aldanian {
  Atdabanian
  Tommotian or Meishucunian
}

?Etcheminian

3.9.4 The Saint David's Epoch

Saint David's is the most appropriate name suggested for the Mid Cambrian Epoch (Cowie *et al*. 1972). Rocks with *Paradoxides* were first identified in the Saint David's area, in South Wales, and the divisions **Menevian** and **Solvan** from the same area in South Wales have a long history in Mid

Cambrian stratigraphic literature. For Mid Cambrian (Saint David's) stages, Cowie used three *Paradoxides* biozones.

3.9.5 The Merioneth Epoch

Merioneth (a former county in North Wales, now included in Gwynedd) is the name for the restricted Late Cambrian Epoch; the name makes it unambiguous in excluding Tremadoc.

In the Late Cambrian (Merioneth) Epoch, **Maentwrogian** and **Dolgellian** divisions are retained but the Ffestiniog stage that once separated them in a three-fold division (e.g. Cohee *et al*. 1967) has been dropped (Cowie *et al*. 1972). All are North Welsh names.

The sequence of stages listed here is taken from Cowie *et al*. (1972). Their identity is indicated in the column of correlative fossils listed also from that work.

3.9.6 The Cambrian chart (Figure 3.2)

The columns in Figure 3.2 are modified from Cowie *et al*. (1972) for the left-hand side of the chart. On the right the Norway column is from Martinsson (1974) and gives only Early Cambrian divisions because thereafter all the Scandinavian sequences were described in terms of biostratigraphic names or symbols rather than rock units. The China column is from Wang & Liu (1980). The other columns are from Cowie *et al*. (1972).

3.10 The Ordovician Period

3.10.1 History and classification

The Ordovician System was founded by Lapworth in 1879 and solved the Murchison–Sedgwick controversy of conflicting claims for their overlapping Silurian and Cambrian systems respectively. It resulted in the three periods of the Early Paleozoic Era.

Lapworth described 3600 m of volcanic ash and sedimentary rocks in the Arenig–Bala district, where the boundaries of the original system were clearly defined – above the Tremadoc (Amnodd Shales). Ordovician thus began with the Basal Grit of Sedgwick's Arenig Group and ended below the Llandovery (Cwm yr Aethnen Mudstones) at the top of the Foel y Ddinas Mudstone which was the youngest formation of Sedgwick's Bala Group.

Whittington & Williams (1964) recounted the complex history of error and confusion, a confusion that led to various departures from the above simple definitions and so gave some support for including Tremadoc rocks in the Ordovician System. In short, while priority and the logic of the initial definition would exclude Tremadoc rocks, the fact is that a majority of geologists throughout the world have allowed these mistakes to pass unchecked into general use and so established the convention that we now follow by beginning the Ordovician Period at the point that would make the Tremadoc Group also Ordovician. In fact, the typically Tremadoc graptoloid *Dictyonema flabelliforme* has been commonly taken as the index fossil in correlation of earliest Ordovician rocks. Furthermore, the International Working Group on the Cambrian–Ordovician Boundary has practically agreed that the boundary shall coincide with the initial Tremadoc boundary when a reference point in a stratotype has been decided (Derby 1986).

Figure 3.3. The Ordovician Period and selected successions.

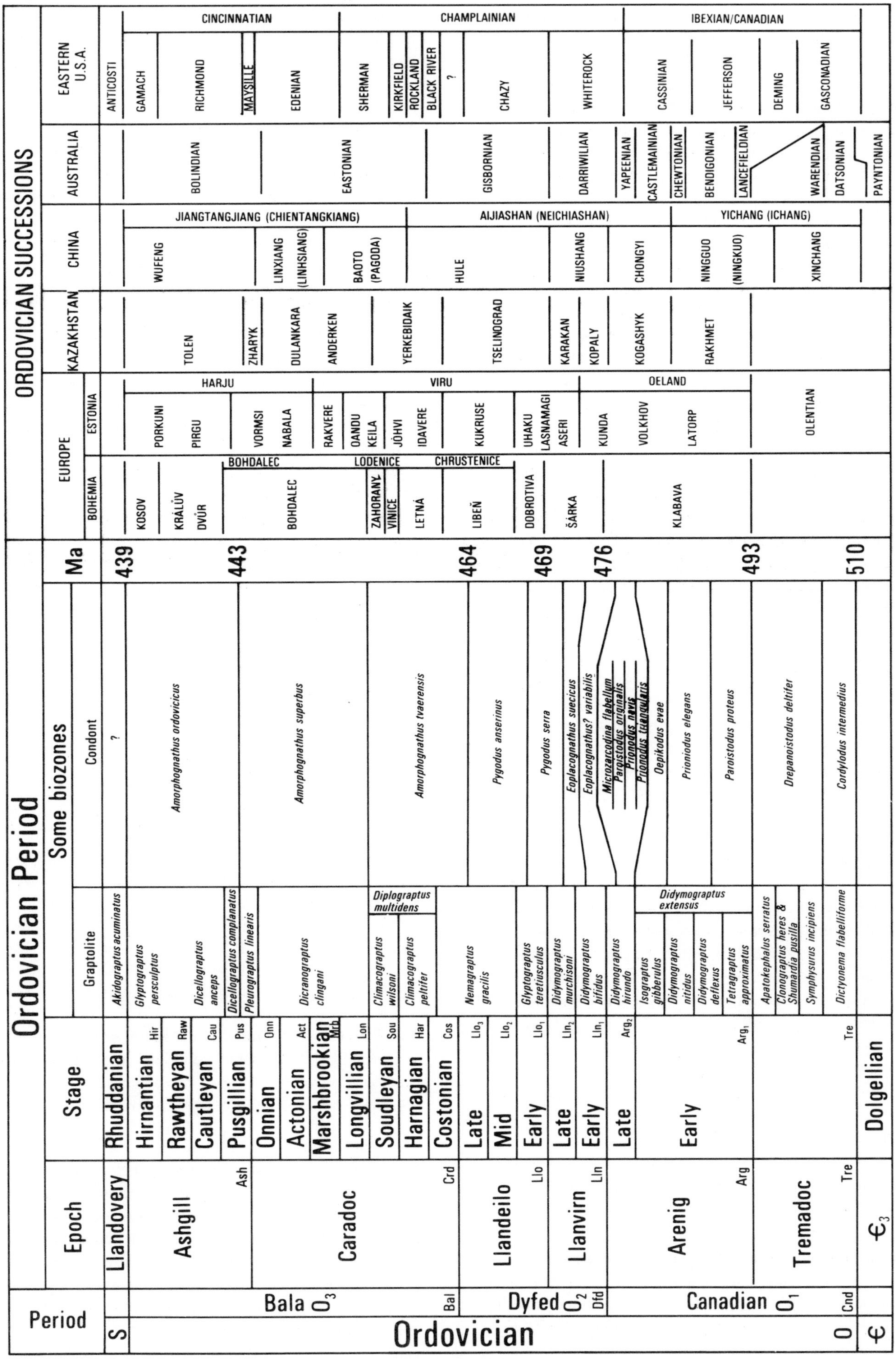

In North America the 'Original Champlain Group' of Emmon's New York System (see Section 3.9 above) ranged from the Potsdam rocks (late Cambrian) to Early Silurian and so encompassed the Ordovician Period. Clark & Schuchert in 1899 proposed the Champlainic era or system (exclusive of the Potsdam Sandstone) writing 'In face of Champlain, 1842, the term Ordovician has no standing'. However, the same author used Champlainian for Mid Ordovician only (Mohawkian and Chazy), as have many others, so the claim was ambiguous (Wilmarth 1925).

The divisions of the Ordovician Period at present are based on names and successions of rocks in Britain. These successions are, however, scattered and they do not comprise a single or even two or three standard sequences, so that no column of British rock units has been attempted in Figure 3.3. However, they are clearly depicted in the correlation charts in Williams *et al*. (1972).

Lapworth used two rather than three Ordovician divisions, which corresponded to Sedgwick's original Arenig and Bala. This two-fold division was reaffirmed by Whittington & Williams (1964). We adopt the same Late Ordovician Sub-period, bringing back the name **Bala** for it (Bala happens to coincide with Upper Cambrian as defined by Sedgwick in 1852 (Sedgwick & McCoy 1852). For the earlier part of Ordovician time there is some difficulty. Sedgwick's Arenig has since been divided into **Arenig**, **Llanvirn** and **Llandeilo** so cannot now be used to name a sub-period or epoch. A three-fold division of the Period is also commonly employed, especially in America, and, although Whittington & Williams argued against it, now that the Tremadoc division has been included with Ordovician this makes a three-fold classification more reasonable. The name **Canadian**, defined as **Tremadoc + Arenig**, is an unambiguous name for the Early Ordovician Sub-period in spite of Canadian facies being totally foreign to the classic British type successions.

For the time division between Canadian and Bala, Mid Ordovician has been ambiguous because in North America it has also encompassed most of Caradoc time, being mostly Champlainian. An unambiguous name for Mid Ordovician in the sense post-Canadian and pre-Bala, i.e. **Llanvirn + Llandeilo**, would complete the tripartite scheme. Pending an official recommendation, the name **Dyfed** is proposed here. It is the name of the Welsh county (formerly Pembroke, Carmarthen and Cardigan) where these rocks crop out.

There are thus six epochs which may conveniently be grouped in pairs. Pre-Bala epochs are not yet formally divided into named stages, whereas eleven Bala stages have been established for some years. All these divisions have been defined approximately in terms of graptolite zones as shown in Figure 3.3, but conodont divisions may be more discriminating.

An attempt was made to define the six epochs by Whittington *et al*. (1984) and this scholarly source is used below. These epoch names now appear to stand unchallenged.

3.10.2 Canadian (Early Ordovician) Sub-period
The name Canadian is applied to the two epochs Tremadoc and Arenig, thereby avoiding ambiguity in the use of Early Ordovician, as already suggested (GTS 82). Ross (1984) suggested that Ibex be used in this sense, Canadian having already changed in meaning from Dana's original type section. Use of Canadian would allow Ibex to retain its precise North American meaning in terms of the pre-Whiterock succession at Ibex Hills, Utah.

Tremadoc Epoch
Tremadoc is the first Ordovician epoch (Derby 1986). Rushton (in Bassett & Dean 1982 and in Whittington *et al*. 1984) outlined the history of this division in the literature. He recommended an initial Tremadoc boundary in the forestry road section at Bryn-llin-fawr, east of the Harlech Dome, North Wales, where, 184 m above an arbitrary datum, the reference point lies at the base of a mudstone bank yielding the lowest examples of *Dictyonema flabelliforme*, and is referred to the *sociale* sub-biozone. However, the International Working Group (Derby 1986) recommended not only that Tremadoc should be the earliest Ordovician division but that its initial boundary should be sought in a conodont-rich succession at or near the influx of (nema-bearing) dendroid graptolites with additional characters for non-biologic correlation. The base of the conodont biozones *Hirsutodontus simplex* and of *Cordylodus lindstromi* were considered. Boundary stratotype candidates in Utah, Kazakhstan and the Oslo region were rejected, so leaving the Cow Head or other regions of Western Newfoundland and Dayangcha, Jilin, China, for further consideration.

Arenig Epoch
Arenig was reduced to its present span by separation of Murchison's Llandeilo, e.g. by Ramsey and Salter in 1866 (Ramsay 1866). Fortey (in Whittington *et al*. 1984) discussed the Arenig division at length and in the light of Zalasiewicz's (1984) redescription of the type Arenig rocks as the Carnedd Iago Formation, with three members. He made a three-fold division, mainly on trilobite faunas, which could be made the basis of three stages. He gave faunal lists but the divisions are not named or defined.

It may be better to look outside North Wales and even outside Wales for good Arenig definitions. The *Tetragraptus approximatus* biozone has been widely taken as the initial Arenig boundary although it is not known in the type area, but Whittington *et al*. (1984) would have preferred a reference point near the base of the *extensus* biozone 'if it can be identified in a continuous sequence in Wales'.

3.10.3 Dyfed (Mid Ordovician) Sub-period
Hicks (1881) introduced Llanvirn together with Llandeilo between Arenig and Bala. These two epochs are grouped here as Mid Ordovician or Dyfed.

Llanvirn Epoch
The Llanvirn Group was named by Hicks for 'upper Arenig' and 'lower Llandeilo' and he referred to Llanvirn Farm in southwest Wales, Dyfed, where a quarry in prolific fossiliferous rocks was used as type. The epoch is conveniently formed of two faunal divisions, but no precise definitions were proposed except for the incoming of tuning-fork graptolites. Further work is in progress and *D. bifidus* may not be an appropriate name for the initial stage.

Llandeilo Epoch

The base of the type succession (Williams 1953) is unconformable. Above and to the northeast of this a distinctive graptolite fauna of the *Glyptograptus teretiusculus* biozone occurs and is commonly taken as characterizing the initial Llandeilo boundary pending precise standardization.

3.10.4 Bala (Late Ordovician) Sub-period

The Bala Sub-period comprises the Caradoc and Ashgill Epochs as understood today. Murchison in 1835 (1839) first used Caradoc for the whole division from which Marr (1905) separated Ashgill. The name Bala was introduced by Ramsay & Salter in 1866 (Ramsay 1866) and is in use today (e.g. British Geological Survey Regional Guides for North and South Wales).

The Caradoc Epoch

According to Dean (in Whittington *et al.* 1984) the name is from Murchison's Caradoc Sandstone near Church Stretton, Shropshire, and exposures in the River Onny near Horderley have been taken as typical. However, the base of the unit is unconformable and diachronous, and the oldest part of it is at Hoar Edge, 6 km northeast of Church Stretton.

The oldest of the seven Caradoc stages is **Costonian**, but there is no good section transitional between Llandeilo and Costonian yet described from the type region where an initial Caradoc boundary might be defined. The base of the *N. gracilis* biozone is not suitable to characterize it and a new standard is urgently needed.

A new stage, **Woolstonian**, was introduced in place of the 'Upper Longvillian sub-stage'. The redefined **Longvillian Stage** was the 'Lower Longvillian sub-stage'. Thus the **Marshbrookian Stage** follows the Woolstonian Stage (Hurst 1979).

The youngest Caradoc stage is **Onnian**.

The Ashgill Epoch

Marr (1905) named the 'Ashgill Series' after Ash Gill near Coniston in the southern Lake District. He did not designate a type locality in that paper but in 1914 he described Cautley in nearby northwest Yorkshire for this function and it remains a candidate for an international boundary stratotype.

Four stages are now recognized: **Pusgillian**, **Cautleyan**, **Rawtheyan** and **Hirnantian**.

3.10.5 Conclusion

Whittington *et al.* (1984), after exhaustive studies, appear to have found little in the way of good stratotypes for boundary reference points for Ordovician epochs in the British Isles and the search will proceed outside the British Isles for standards (probably from North America) to define them. It might be preferable to define divisions within the Canadian Sub-period from the Ibex Formation and for Llanvirn and Llandeilo from Whiterock and Chazy. Pertinent evidence from other parts of the world is also being drawn into the global discussion.

3.10.6 The Ordovician chart (Figure 3.3)

Figure 3.3 shows the main major British graptolite zones used for correlation, as in Williams *et al.* (1972); see also Bassett (1984).

The consolidation of two former biozones to *Diplograptus multidens* is based on common usage. Some biozones are of longer duration than the ages, whose finer division is nevertheless possible using fossil assemblages. The Tremadoc biostratigraphic correlation is based on Scandinavian biozones taken from Cowie *et al.*(1972). The post-Tremadoc divisions are from Williams *et al.* (1972), as are the columns for Bohemia, Estonia and Kazakhstan. The China section is from Sheng Shen-Fu (1980). The Australian stages are from Webby *et al.* (1981). The North American column is based on Barnes, Norford & Skevington (1981).

3.11 The Silurian Period
3.11.1 History and classification

The Silurian System was named by Murchison in 1835 (1839) for the Welsh Borderland tribe the Silures, and first included rocks that were claimed as Cambrian by Sedgwick. Silurian was subsequently used in two senses – to include or to exclude what are now recognized as Ordovician strata, so the less ambiguous name Gotlandian was proposed in 1893 by de Lapparent for the post-Ordovician period and competed with the name Silurian until, in 1960 at the IGC in Copenhagen, Silurian was officially adopted in its restricted sense, i.e. Murchison's Upper Silurian.

Murchison had included Llandovery, Wenlock and Ludlow in his Upper Silurian. Lapworth in 1879 and 1880 used Valentian (approximately = Llandovery) and Salopian (approximately = Wenlock + Ludlow), but this usage was generally discontinued (e.g. by O. T. Jones in 1929 in Evans & Stubblefield) when a fourth division (the Downton) came into use (including part of Murchison's Tilestones that he had included with what later became the Old Red Sandstone). Whittard (1961) and Cocks *et al.* (1971) summarized the confused nomenclatural history of the Period. However, Valentian persisted in the British Geological Survey, e.g. in earlier Regional Guides to North Wales (1961) and to South Wales (1948).

The four epochs (S_1 to S_4) were more recently recombined informally into two sub-periods (Early Silurian = $S_1 + S_2$ and Late Silurian = $S_3 + S_4$) when the divisions of the Wenlock (S_2) and Ludlow (S_3) epochs were recommended (as accepted here) by the ICS Subcommission on Silurian Stratigraphy (Holland 1980).

Early and Late Silurian sub-periods may confuse because of the ambiguity between the above and the original Valentian (S_1) and Salopian ($S_2+S_3+S_4$) divisions. It is therefore recommended where possible to use epoch names.

The Silurian Period is, perhaps, the first to have all of its (four) epochs and (seven) stages internationally agreed, defined and standardized at reference points in boundary stratotypes.

3.11.2 The Llandovery Epoch

'Llandovery' is from the type area in Dyfed, southern Wales. It has been agreed by the Commission of Stratigraphy (Bassett 1985b) that the Llandovery Epoch shall comprise three stages: **Rhuddanian**, **Aeronian** and **Telychian**. The names Idwian and Fronian (Cocks, Toghill & Zeigler 1970 and GTS 82) have been replaced by Aeronian: this stage is, however, not exactly equivalent to the two previous ones (see also Cocks *et al.* 1984 in Holland 1985).

Figure 3.4. The Silurian Period and selected successions.

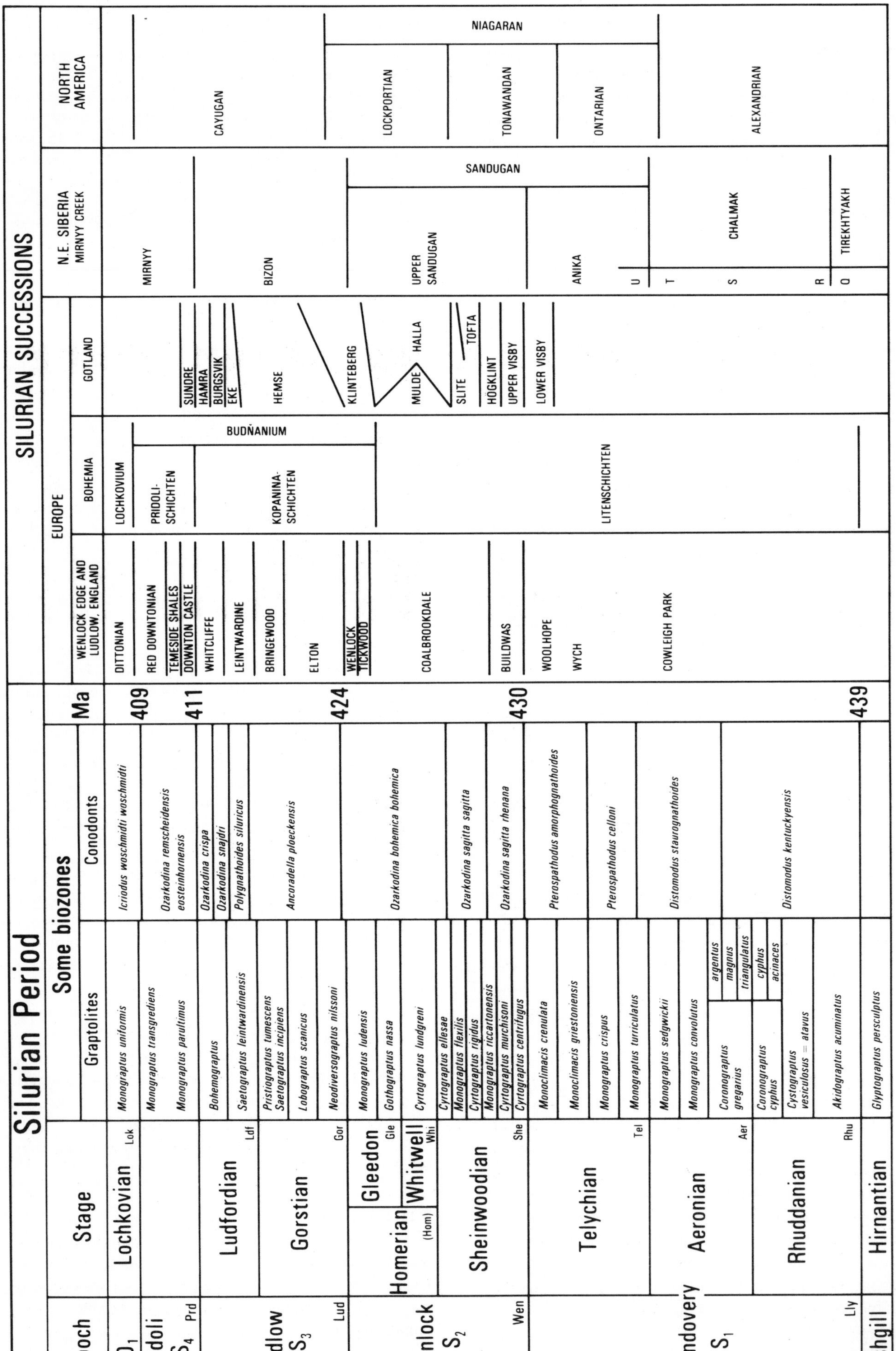

Silurian Period

Period	Epoch	Stage		Some biozones — Graptolites	Some biozones — Conodonts	Ma
Silurian (S)	Pridoli S₄ (D₁)	Lochkovian	Lok	Monograptus uniformis	Icriodus woschmidti woschmidti	409
		Ludfordian	Ldf	Monograptus transgrediens	Ozarkodina remscheidensis eosteinhornensis	411
	Ludlow S₃			Monograptus parultimus		
				Bohemograptus	Ozarkodina crispa	
					Ozarkodina snajdri	
				Saetograptus leintwardinensis	Polygnathoides siluricus	
	Gorstian	Gor	Lud	Pristiograptus tumescens / Saetograptus incipiens	Ancoradella ploeckensis	
				Lobograptus scanicus		
				Neodiversograptus nilssoni		424
	Wenlock S₂	Gleedon	Gle	Monograptus ludensis		
		Whitwell (Whi)		Gothograptus nassa	Ozarkodina bohemica bohemica	
		Homerian (Hom)		Cyrtograptus lundgreni		
				Cyrtograptus ellesae		
				Monograptus flexilis		
		Sheinwoodian	Wen	Cyrtograptus rigidus	Ozarkodina sagitta sagitta	
			She	Monograptus riccartonensis	Ozarkodina sagitta rhenana	430
				Cyrtograptus murchisoni		
				Cyrtograptus centrifugus		
				Monoclimacis crenulata	Pterospathodus amorphognathoides	
	Llandovery S₁			Monoclimacis griestoniensis		
		Telychian	Tel	Monograptus crispus	Pterospathodus celloni	
				Monograptus turriculatus		
				Monograptus sedgwickii	Distomodus staurognathoides	
		Aeronian	Aer	Monograptus convolutus		
				Coronograptus gregarius — argentus / magnus / triangulatus		
				Coronograptus cyphus — cyphus / acinaces		
		Rhuddanian	Rhu	Cystograptus vesiculosus = atavus	Distomodus kentuckyensis	
				Akidograptus acuminatus		439
	Ashgill (O)	Hirnantian		Glyptograptus persculptus		

SILURIAN SUCCESSIONS

	NORTH AMERICA	N.E. SIBERIA MIRNYY CREEK	EUROPE — GOTLAND	EUROPE — BOHEMIA	EUROPE — WENLOCK EDGE AND LUDLOW, ENGLAND
	NIAGARAN				DITTONIAN
	CAYUGAN	MIRNYY	SUNDRE / HAMRA / BURGSVIK / EKE	LOCHKOVIUM	RED DOWNTONIAN
				PRIDOLI-SCHICHTEN	TEMESIDE SHALES / DOWNTON CASTLE
			HEMSE		WHITCLIFFE
		BIZON		KOPANINA-SCHICHTEN	LEINTWARDINE
		SANDUGAN		BUDŇANIUM	BRINGEWOOD
	LOCKPORTIAN		KLINTEBERG		ELTON
		UPPER SANDUGAN	MULDE / HALLA		WENLOCK / TICKWOOD
	TONAWANDAN		SLITE / TOFTA		COALBROOKDALE
			HOGKLINT		BUILDWAS
	ONTARIAN	ANIKA	UPPER VISBY	LITENSCHICHTEN	WOOLHOPE
			LOWER VISBY		WYCH
		U / T			COWLEIGH PARK
	ALEXANDRIAN	CHALMAK (S)			
		TIREKHTYAKH (R / Q)			

The earliest Llandovery rocks are better for time correlation in the Southern Uplands of Scotland. There the Dobb's Linn section (where Lapworth conceived the Ordovician System) was described by Lapworth in 1878 and revised by Toghill in 1968, Ingham in 1979 and Williams in 1983 (Holland 1985).

The Rhuddanian Stage (Early Llandovery)

The stage is named from the Cefn-Rhuddan Farm in the type Llandovery area. It spans the *acuminatus* to *cyphus* biozones. The initial Rhuddanian boundary is, of course, defined at the same point as the initial Llandovery, and the Ordovician–Silurian boundary. The Ordovician–Silurian Boundary Working Group formally recommended the boundary stratotype to be at Dobb's Linn and the GSSP to coincide with the base of the *Parakidograptus acuminatus* biozone. This was after full consideration of the excellent competing reference section at Anticosti Island, eastern Canada, where conodonts and shelly faunas could be used for correlation. This recommendation was ratified by the ICS and by IUGS in 1984. The reference point is 1.6 m above the base of the Birkhill Shale Formation which, in its lower part, comprises the *persculptus* biozone. The latter was sometimes previously taken as the initial Silurian biozone.

Description of strata throughout the world with proposed correlations are available in a symposium volume (Cocks & Rickards 1988).

The Aeronian Stage (Mid Llandovery)

Named from Cwm-coed-Aeron Farm in the type Llandovery area, the Aeronian Stage spans the *triangulatus* to *sedgwickii* biozones, thus containing the earlier Idwian and about half the Fronian ages.

The initial boundary is defined in the Trefawr forestry road section 500 m north of the farm (National Grid SN 8380 3935) at a reference point in a continuous lithological section through part of the Trefawr Formation at the base of locality 72 in a transect that has been described by Cocks *et al.* (1984, p.165). It approximates to the base of the *Monograptus triangulatus* biozone (Bassett 1985a).

The Telychian Stage (Late Llandovery)

The name is from the Penlan-Telych Farm in the type Llandovery area. This age spans the *turriculatus* to *crenulata* biozones and so incorporates the later Fronian Stage (of GTS 82) with the Telychian Stage.

The initial boundary stratotype is at an old quarry on the west side of the Cefn Cerig Road (National Grid SN 7743 3232). The reference point is within a continuous lithological section through part of the Wormwood Formation at locality 162 in the transect of Cocks *et al.* (1984). This point correlates with the base of the *Monograptus turriculatus* biozone and is also marked by distinctive brachiopod faunal changes.

3.11.3   The Wenlock Epoch

The name Wenlock was first used by Murchison in 1839 from the type area of Wenlock Edge in the Welsh Borderlands, England. A scheme of formations and members was developed over the years. However, a Geological Society of London (GSL) initiative led to a reinvestigation of these rocks and to proposals for an international scale that could be correlated by the graptolite sequence (Bassett *et al.* 1975). This is adopted in the accompanying classification (without detailing the sequence of evolving classifications) and the definitions are abstracted below.

The initial **Sheinwoodian Stage** (Llandovery–Wenlock boundary) is seen in the standard section (National Grid SO 5688 9839) in Hughley Brook which lies 200 m southeast of Leasowes Farm and 500 m northeast of Hughley Church. The marker point for the boundary is taken in the left (north) bank of the stream, at the base of unit G, which is the base of the Buildwas Formation, and immediately above the Purple Shales in a measured section recorded by Bassett *et al.* (1975, p.13).

The **Homerian Stage** was divided into the two sub-stages – Whitwell and Gleedon – because of the possibility of international correlation by the graptolite zones and the likely ease of recognition of their bases elsewhere by correlation with the standard section. The spacing of these graptolite zones in the chart follows that of Bassett *et al.* (1975, p.2), presumably being the best time scale calibrated by their relative thicknesses in the 'Wenlock Shales' (mainly Coalbrookdale Formation).

The initial **Whitwell** boundary is located in the 'small side-stream (National Grid SO 6194 0204) to the tributary of Sheinton Brook which flows through Whitwell Coppice, 500 m north of Homer. The marker point . . . is within a more or less continuous section of olive to grey-green mudstones, blocky-fractured and thin-bedded, its exact position coinciding with the point at which the *ellesae/lundgreni* biozone boundary cuts the right (north) bank of the stream.' More biostratigraphic details were given for international correlation by Bassett *et al.* (1975).

The initial **Gleedon Sub-stage** boundary is defined (at National Grid SO 5016 8999) on the southeast side of the track 182 m east of Eaton Church and coincides with the point where the *lundgreni/nassa* biozone boundary cuts the tracks.

3.11.4   The Ludlow Epoch

The **Ludlow Epoch** (named for the town in the Welsh Borderlands, England, has been redivided (Holland 1980) by combining four stages used earlier (e.g. Cocks *et al.* 1971) into two that are now better established as follows:

Holland (1980)	Cocks *et al.* (1971)
Ludfordian	Whitcliffian Leintwardinian
Gorstian	Bringewoodian Eltonian

The initial **Gorstian Stage** (Wenlock–Ludlow) boundary was defined by Holland, Lawson & Walmsley (1963) at Pitch Coppice in the Ludlow anticline, in the standard section in the Old Quarry (National Grid SO 4726 7301) on the south side of the Ludlow/Wigmore road, about 2 km northeast of Aston Church. This definition is in younger strata than was sometimes earlier supposed.

The initial **Ludfordian** boundary was taken at the base of the beds with *Saetograptus leintwardinensis* (Holland 1980), but was defined precisely by Holland *et al.* (1963) in the section on the northwest face of Sunnyhill Quarry in Mary Knoll Valley, 2.8 km southsouthwest of Ludlow (National Grid SO 4953 7255).

3.11.5 The Pridoli Epoch

This epoch equals post-Ludlow, pre-Lochkovian time, and its name is now internationally established (Bassett 1985b). Downton is the name used in the Welsh Borderland sequence and follows the sequence Llandovery, Wenlock and Ludlow. Graptolites have not been found in this facies. Skala (from Podolia) was also proposed as a name for this epoch. However Pridoli, from Pridoli–Schichten (labelled E332 20–80 m in Barrande's section in Bohemia) was preferred partly because the classic Silurian–Devonian boundary is established there. It was used in GTS 82.

Stages within the Pridoli Epoch have not been internationally proposed. In the DNAG chart for northwestern Canada, Norford (1985) showed seven biozones (compared with only two for the Ludlow Epoch) as follows:

> *uniformis angustidens*
> *uniformis transgrediens*
> *uniformis bouceki*
> *uniformis chelmiensis*
> *uniformis bugensius*
> *uniformis formosus*
> *uniformis leintwardinensis* (oldest)

The Ludlow–Pridoli boundary stratotype is in the Pozary section near Reporyji, Prague. The marker point is at the base of bed 96 in the section, conforming with the base of the Pridoli Formation where it overlies the Copanina Formation. Czech stratigraphers have agreed to change the name of the Pridoli Formation so that the name shall have only one meaning. The point approximates in age to the base of the *Monograptus parultimus* biozone in the section.

3.11.6 The Silurian chart (Figure 3.4)

The epochs, ages and biostratigraphic correlation columns of Figure 3.4 are based on Cocks *et al.* (1971) with modifications from Bassett *et al.* (1975) and Holland (1980) as detailed above. R. B. Rickards kindly provided details of the zone fossils. The Wenlock Edge and Ludlow, Bohemia and Gotland columns are also from Cocks *et al.* (1971), the northeast Siberia column is based on Oradovskaya & Sobolevskaya (1979) and the North American column is from Norford *et al.* (1970, p.604). Alexandrian should be shown with an initial boundary probably later than Ashgill.

3.12 The Devonian Period

3.12.1 History and classification

The Devonian System was named from Devon, England, by Sedgwick & Murchison in 1839, in the course of a controversy which lasted from 1834 to 1842 and in which De la Beche, Greenough and Lonsdale also participated. The question was as to whether the rocks were Cambrian or Carboniferous. They were eventually shown to be coeval with the Old Red Sandstone. Possibly the most scholarly work available in the history of any geological problem expounds the above story in nearly 500 pages (Rudwick 1985).

The division of the System on the basis of the marine faunas came to be established through the work of Dumont, Beyrich, Roemer and many others in the Ardenne–Rhenish area, where more recent attempts to standardize the time divisions of the period have been based (Ziegler 1979, Ziegler & Klapper 1982). For a time the name Rhenian was preferred by some to Devonian.

The principal Devonian divisions – epochs and stages – were established by decision of the ICS in May 1984 (Ziegler & Klapper 1985).

Epochs	Stages
Late Devonian	Famennian
	Frasnian
Mid Devonian	Givetian
	Eifelian
	Emsian
Early Devonian	Pragian
	Lochkovian

Precise correlation is now largely dependent on a standard conodont zonation (e.g. Ziegler 1971).

3.12.2 The Early Devonian Epoch

The **Lochkovian Stage** has been officially adopted from the Bohemian Pridolian to Lochkovian sequence. It thus replaced Gedinnian as a name, but has the same meaning as Gedinnian in GTS 82 (Ziegler & Klapper 1985). Gedinnian had first been defined by Dumont in 1848 from Gédinne in Belgium (Appendix 1). The initial Lochkovian boundary is thus also the Silurian–Devonian boundary which was the first to be fully agreed by the ICS and IUGS (Montreal 1972). They defined the boundary-stratotype section at Klonk near Prague in the Barrandian area of Bohemia, Czechoslovakia. The reference GSSP selected is just below a bed with the first and abundant occurrence of *Monograptus uniformis* and *M. uniformis angustidens*, i.e. in bed No.20 (7–10 cm thick), described by Chlupac, Jaeger & Zacmundova in 1972. The base of this horizon was approximately of initial Gedinnian age so that the initial point of the Lochkovian Stage is now thereby standardized (McLaren 1977).

The **Pragian Stage** as a name was similarly adopted in place of Siegenian because of the better faunal sequence in Bohemia. It would probably refer to the same time interval as Siegenian but this still remains to be standardized. Siegenian was named by Kayser in 1881, from Siegen in Germany, and was formerly the Coblentzian (Coblenzian) of Dumont in 1848 and the lower Coblenzian of Gosselet in 1880–88 (Ziegler 1979).

The **Emsian Stage** was introduced by Dorlodot in 1900 from Bad Ems in Germany, to clarify confusion in nomenclature (discussed in detail by Ziegler 1979). The initial Emsian boundary might be taken, for example, at the base of the Gres de Vireaux (Ziegler 1979), but this has yet to be decided.

Figure 3.5. The Devonian Period and selected successions.

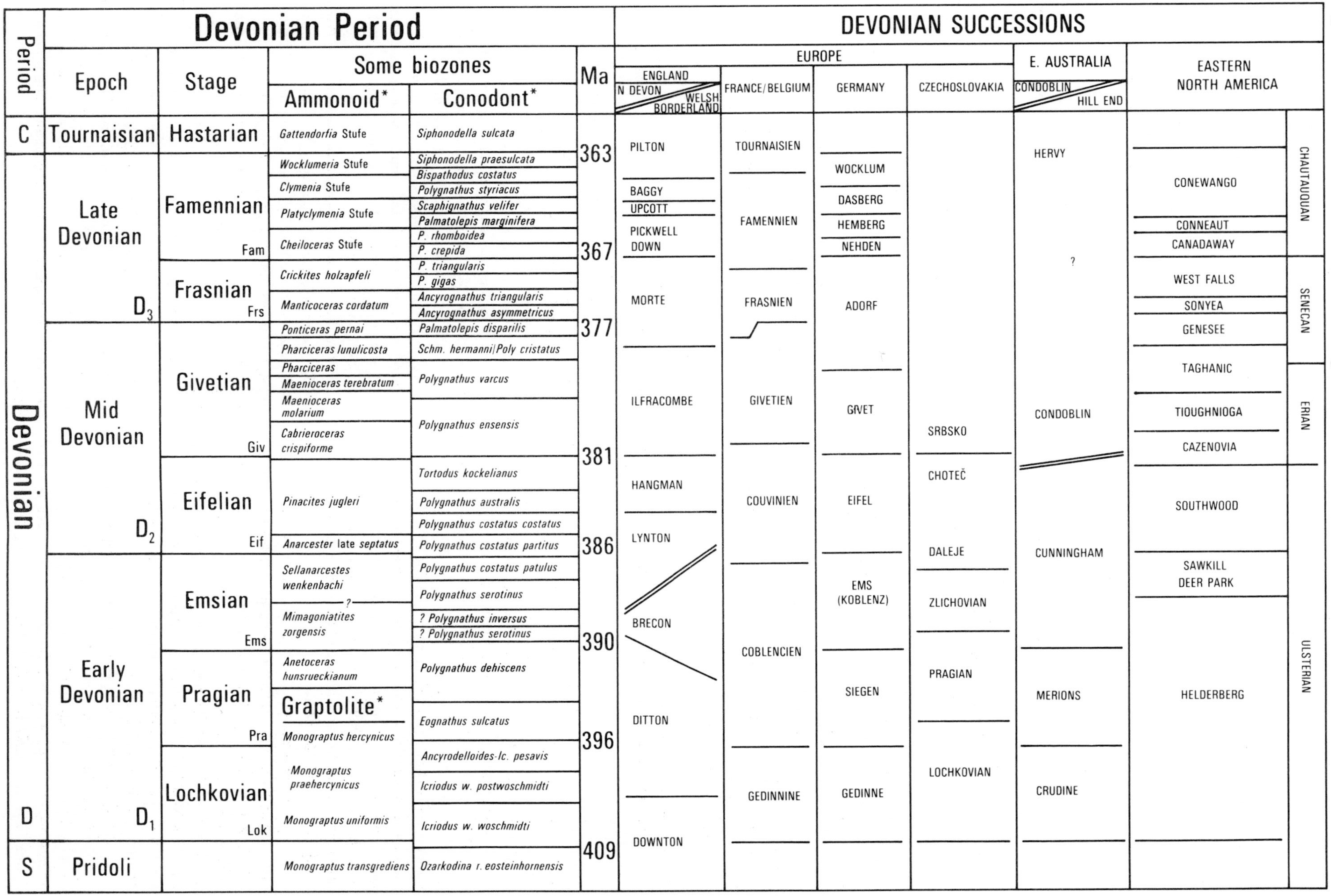

Devonian Period — Some biozones

Period	Epoch	Stage	Ammonoid*	Conodont*	Graptolite*	Ma
Devonian	Tournaisian	Hastarian	Gattendorfia Stufe	Siphonodella sulcata		
	Late Devonian (D₃)	Famennian (Fam)	Wocklumeria Stufe	Siphonodella praesulcata / Bispathodus costatus		363
			Clymenia Stufe	Polygnathus styriacus		
			Platyclymenia Stufe	Scaphignathus velifer / Palmatolepis marginifera		
			Cheiloceras Stufe	P. rhomboidea / P. crepida		367
		Frasnian (Frs)	Crickites holzapfeli	P. triangularis / P. gigas		
			Manticoceras cordatum	Ancyrognathus triangularis / Ancyrognathus asymmetricus		
			Ponticeras pernai	Palmatolepis disparilis		377
	Mid Devonian (D₂)	Givetian (Giv)	Pharciceras lunulicosta	Schm. hermanni / Poly cristatus		
			Pharciceras	Polygnathus varcus		
			Maenioceras terebratum			
			Maenioceras molarium	Polygnathus ensensis		381
			Cabrieroceras crispiforme			
		Eifelian (Eif)	Pinacites jugleri	Tortodus kockelianus		
				Polygnathus australis		
			Anarcester late septatus	Polygnathus costatus costatus		
				Polygnathus costatus partitus		386
	Early Devonian (D₁)	Emsian (Ems)	Sellanarcestes wenkenbachi	Polygnathus costatus patulus		
			Mimagoniatites zorgensis (?)	Polygnathus serotinus		
				? Polygnathus inversus		
				? Polygnathus serotinus		390
		Pragian (Pra)	Anetoceras hunsrueckianum	Polygnathus dehiscens		
		Lochkovian (Lok)		Eognathus sulcatus	Monograptus hercynicus	396
				Ancyrodelloides-Ic. pesavis	Monograptus praehercynicus	
				Icriodus w. postwoschmidti		
				Icriodus w. woschmidti	Monograptus uniformis	409
S	Pridoli			Ozarkodina r. eosteinhornensis	Monograptus transgrediens	

Devonian Successions

Ma	ENGLAND — N DEVON / WELSH BORDERLAND	EUROPE — FRANCE/BELGIUM	EUROPE — GERMANY	EUROPE — CZECHOSLOVAKIA	E. AUSTRALIA — CONDOBLIN / HILL END	EASTERN NORTH AMERICA
363	PILTON	TOURNAISIEN	WOCKLUM		HERVY	CHAUTAUQUAN: CONEWANGO
	BAGGY / UPCOTT	FAMENNIEN	DASBERG			CONNEAUT / CANADAWAY
367	PICKWELL DOWN		HEMBERG / NEHDEN			SENECAN: WEST FALLS / SONYEA / GENESEE
	MORTE	FRASNIEN	ADORF			TAGHANIC
377				SRBSKO		
381	ILFRACOMBE	GIVETIEN	GIVET	CHOTEČ	CONDOBLIN	ERIAN: TIOUGHNIOGA / CAZENOVIA
	HANGMAN		EIFEL			
386	LYNTON	COUVINIEN		DALEJE		SOUTHWOOD
390	BRECON	COBLENCIEN	EMS (KOBLENZ)	ZLICHOVIAN	CUNNINGHAM	ULSTERIAN: SAWKILL / DEER PARK
	DITTON		SIEGEN	PRAGIAN	MERIONS	HELDERBERG
396		GEDINNE	GEDINNE			
409	DOWNTON			LOCHKOVIAN	CRUDINE	

3.12.3 The Mid Devonian Epoch

The **Eifelian Stage** name is from Eifel in Germany. It was first applied by Dumont in 1848 to younger rocks but with a greater time span both older and younger than Eifelian as here. The standard stage as now understood was restricted to its present meaning in 1937 (see Richter 1942, Ziegler 1979). The name Couvinian (from Couvin in Belgium named by Dupont in 1885) (Appendix 1) has been widely used by French-speaking geologists; it corresponds to all present Eifelian and part of the upper Emsian strata and so it is not recommended here for the standard scale.

The initial Eifelian (and therefore initial Mid Devonian) boundary has been standardized in recommendations by the Subcommission of Devonian Stratigraphy (Ziegler & Klapper 1985). It is chosen to coincide with the first occurrence of the conodont *Polygnathus costatus partitus* (Klapper, Ziegler & Mashkova 1978), and is thus at the lower limit of the *partitus* zone. The boundary stratotype for the reference point is in a trench within the Heisdorf Formation first dug in 1937 and dug more deeply in 1982. It is traditional, accessible and protected; it lies southsoutheast of the town of Schönecken-Wetteldorf, Eifel Hills, Federal Republic of Germany. Full details with map and section were given by Ziegler & Klapper (1985).

The **Givetian Stage** (named from Calcaire de Givet, France, by Gosselet in 1879) has been used in approximately the present sense from its inception. The initial boundary of the stage is taken by modern German authors at or near the lower range of *Stringocephalus burtini* (Ziegler 1979). The boundary reference point remains to be formally defined.

3.12.4 The Late Devonian Epoch

The **Frasnian Stage**, from Frasnes, Belgium (Gosselet in 1880), is used approximately in its original sense. Adorfian and Manticoceras Stufe, as well as the North American Senecan have also been proposed as alternatives, but Frasnian still stands. The initial boundary was considered (House *et al.* 1977, p.8) to be 'at the base of the goniatite *Phaciceras lunulicosta* zone of the Manticoceras Stufe' which ammonoid workers have used throughout this century.

As a result of more precise conodont work it has now been recommended by Ziegler & Klapper (1985) to place the initial Frasnian boundary at a level to coincide with the base of the Lower *asymmetricus* conodont zone, defined by the first occurrence of *Ancyrodella rotundiloba* (Bryant) and referred to as the Lower *asymmetricus* Boundary. This boundary would probably precede the appearance of the goniatites *Koenenites*, *Probeloceras* and *Manticoceras*.

The **Famennian Stage**, from Famenne in Belgium, was first used by Dumont in 1885 for his earlier (1848) Systeme de Condroz. Gosselet in 1879 used it in the present sense. The initial boundary of the age would be near the base of the Schistes de Senzeilles. The terminal boundary is related to the top of the Wocklumeria Stufe (with *Wocklumeria sphaeroides*) and below the Gattendorfia Stufe (see discussion under Tournaisian in Section 3.13.2).

It had already been agreed (Ziegler & Klapper 1985) that the initial Famennian boundary reference point shall coincide with the lower limit of the middle *Palmatolepis triangularis* conodont zone as defined by the first occurrence of *Palmatolepis delicatula* Branson & Mehl. This is a subdivision

of the conodont standard zonation (Ziegler 1962). The boundary would thus be placed slightly earlier than the first occurrence of the goniatite *Cheiloceras* and later than the highest occurrence of *Crickites*. However, the matter is again in the melting pot and waited clarification at a meeting of the Devonian Subcommission at Rennes in July 1988 (J. W. Cowie personal communication).

It should be noted that the Frasnian–Famennian boundary approximates to McLaren's (1983) biotic crisis with mass extinction and was related by him to bolide impact. This event, if it occurred, would be restricted to the uppermost *gigas* zone and lower *triangularis* zone or parts thereof (Ziegler 1984), but the convention referred to above for the boundary was preferred to this possible event as a boundary marker.

3.12.5 The Devonian chart (Figure 3.5)

The time divisions and the two biostratigraphic correlation columns of Figure 3.5 are from House *et al.* (1977) upon which the four European columns have also been based. Details of eastern Australian rocks are taken from Hill (1967) and those of North America (Appalachian Basin) from Oliver *et al.* (1967).

3.13 The Carboniferous Period
3.13.1 History and classification

Because of their economic value and good outcrops in Britain, Carboniferous rocks were amongst the first to be classified. The name Coal Measures was proposed by Farey in 1807 and 1811, Millstone Grit by Whitehurst in 1778 and, with the Mountain or Carboniferous Limestone, these three major divisions (listed by William Phillips in 1818) constituted the Medial or Carboniferous Order (the latter including the Old Red Sandstone) as set out by Conybeare & Phillips (1822). In a detailed historical discussion Ramsbottom (1981) showed that Carboniferous was Conybeare's creation in that work (140 pages of stratal description in England and Wales). It was the first geological system to be established though it was referred to variously as an order, formation, group or series before system was applied by John Phillips in 1835 (Ramsbottom 1981). The three British units correspond to the three north European divisions. Green *et al.* in 1878 grouped the upper two units as Upper Carboniferous and a two-fold division was again used in Belgium by d'Omalius d'Halloy in 1808, the upper one being the Terrain Houiller (Zittel 1901, Ramsbottom *et al.* 1978).

The three divisions, **Namurian** (proposed by Purves in 1883), **Westphalian** and **Stephanian** (both by de Lapparent in 1893), were systematized by Jongmans in 1928 when Westphalian was divided into A, B and C based on goniatite biozones, and at the Heerlen Congress in 1935 Westphalian D was based on floras. Westphalian E was later referred to Stephanian.

The two-fold division of the Period was formalized by the Heerlen Congress in 1935 when **Dinantian** was introduced for Lower Carboniferous. The IUGS Subcommission on Carboniferous Stratigraphy proposed **Silesian** for the upper part in 1960. The same Subcommission in 1972 (George & Wagner) regarded Dinantian and Silesian as sub-systems, Namurian, Westphalian and Stephanian being ranked as series.

In North America the division of the Carboniferous (or Pennine) System into Coal Measures or **Pennsylvanian** (above) and Lower Carboniferous or **Mississippian** is attributed to H. S. Williams in 1891 (Wilmarth 1925). Williams also included in Pennsylvanian the 'Coal Measure Conglomerate' or Millstone Grit (Pottsville Formation). Mississippian had already been in use since Winchell proposed it in 1869 as the designation of the Carboniferous or Mountain Limestone of the United States. By 1891 it already encompassed three groups of several formations. Ulrich in 1911 divided the Mississippian into Waverleyan and Tennesseean 'systems'. Anthracolithic was also used for Carboniferous by the USGS, but not recently (Wilmarth 1925).

The Russian sequences have more to offer in the way of marine facies which are richly fossiliferous through to Permian rocks. Unlike those of western Europe and North America, Carboniferous rocks of the USSR were classified into Lower, Middle and Upper. This led to much ambiguity because Lower Carboniferous and Upper Carboniferous each spanned quite different times in each of the three regions.

Uralian (Ouralien) was proposed by de Lapparent in 1893 as the latest Carboniferous division, intended as a marine equivalent of Stephanian but discontinued when many supposedly Uralian rocks proved to be Permian. **Orenburgian** was also proposed as the latest Carboniferous division in the southern Urals, and even recently this was shown as a Carboniferous stage above the **Gzelian**, but it was later abandoned as being part Permian (Sherlock 1948, p.14).

The conclusion of a long story, in which three areas (western Europe, Russia and North America) seemed to be competing equally for the role of providing names and possible stratotypes for a global Carboniferous scale, seems now to be approaching, as in the proposal made here. Figure 3.6 combines the recent thinking of some authorities and it seems that the Carboniferous Subcommission of ICS may be formalizing some such scheme (e.g. Ramsbottom 1981). GTS 82 followed Bouroz and others in 1977 (Rotai 1979) in adopting Mississippian and Pennsylvanian as sub-periods for the international scale, partly because of priority over Dinantian and Silesian, but mainly because their mutual boundary coincided approximately with that of the Lower–Middle Carboniferous boundary in Russia. A largely West European scheme for Mississippian divisions was thus used and a largely Russian scheme for the major part of Pennsylvanian. Since 1982 this scheme seems to have been in the main confirmed internationally except for a proposal to move the Mississippian–Pennsylvanian boundary (see below).

Apart from acknowledging by names the three regional communities that have contributed, the scheme made use of the detailed preparatory work in defining stages as well as epochs in the two Special Reports of the Geological Society of London (George *et al.* 1976 and Ramsbottom *et al.* 1978) and in the symposium: *The Carboniferous of the USSR* (Wagner, Higgins & Meyen 1979). Its epochs and stages are outlined in order below.

3.13.2 The Mississippian Sub-period

The **Devonian–Carboniferous** boundary necessarily coincides with the initial Mississippian, Tournaisian and Hastarian boundaries and all will be defined by the one GSSP when it is decided. The Working Group on this boundary (Paproth & Streel 1985) agreed so far as possible to apply principles of priority and least disruption of conventions already in the literature. The initial Hastarian boundary has been described at the stratotype at Hastière (Belgium). Accordingly, what was written in GTS 82 is not materially altered and the situation can be updated as follows.

The operational definition 'is linked to a level near the base of the *Gattendorfia* goniatite zone, and the level with the oldest representative of the conodont species *Siphonodella sulcata* in an evolutionary lineage.' The desired lineage *S. praesulcata* – *S. praesulcata sulcata* – *S. sulcata* has yet to be found in one section.

Three candidates for the boundary stratotype for the reference points remain as Hasselbachtal (FRG), Berchogur (USSR) and Muhua (China) but submission of further candidates was invited as each of these seemed to contain some gaps. Thus the possible sections mentioned in GTS 82, namely in the Kinsale Formation (Courceyan) and at the base of the Hangenberg Kalk in Germany, may have been bettered.

The **Tournaisian Epoch** is divided into two stages based in Belgium, namely **Hastarian** and **Ivorian** (zonal scheme K + Z respectively). These correspond to the British scheme and were to be used by the British Geological Survey as well as internationally by the ICS Subcommission on Carboniferous Stratigraphy (Ramsbottom 1981). So it seems the Belgian names for these divisions are likely to be used. It does not, of course, follow that the stages will be defined at boundary stratotypes in the eponymous localities and it is not yet clear where the GSSP will be.

The **Visean Epoch** is defined approximately according to the Belgian biostratigraphic sequence (see Figure 3.6b). Attempts at precise definitions of the constituent stages made by George *et al.* (1976) are only briefly indicated here as follows.

The initial **Chadian Stage** boundary (after St Chad) lies within the Chatburn Limestone Group at the base of the Bankfield East Beds near Clitheroe, Lancashire, England (National Grid SD 7743 4442).

The initial **Arundian Stage** boundary lies at the base of the Pen y holt Limestone on the east side of Hobbyhorse Bay in south Pembrokeshire, Wales (*arundo*, Latin for hobby-horse, National Grid SR 8800 9563).

The initial **Holkerian Stage** boundary lies at the base of the Park Limestone, in sea cliffs near Holker Hall, Cumbria, England (National Grid SD 3330 7827).

The initial **Asbian Stage** boundary lies at the base of the Potts Beck Limestone at Little Asby Scar, Cumbria, England (National Grid NY 6988 0827).

The initial **Brigantian Stage** (named for the Celtic tribe of Brigantes) boundary lies at the base of the Peghorn Limestone in the east branch of River Eden, 5 km southsoutheast of Kirby Stephen, Cumbria, England (National Grid NY 7832 0375). The terminal Brigantian boundary would end the Dinantian regional sub-period.

The **Serpukhovian Epoch** is perhaps an unsatisfactory name for the late Mississippian epoch; it approximates to the

Figure 3.6a. The Carboniferous Period.

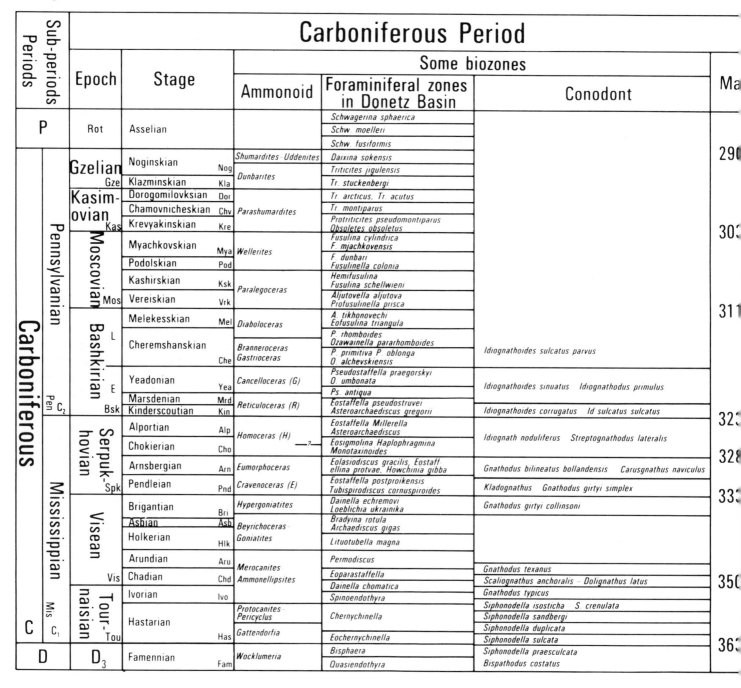

Periods	Sub-periods	Epoch	Stage	Ammonoid	Foraminiferal zones in Donetz Basin	Conodont	Ma
P		Rot	Asselian		Schwagerina sphaerica		
					Schw. moelleri		
					Schw. fusiformis		290
C (Carboniferous)	Pennsylvanian	Gzelian (Gze)	Noginskian (Nog)	Shumardites-Uddenites	Daixina sokensis		
			Klazminskian (Kla)	Dunbarites	Triticites jigulensis		
					Tr. stuckenbergi		
		Kasimovian (Kas)	Dorogomilovksian (Dor)		Tr. arcticus, Tr. acutus		
			Chamovnicheskian (Chv)	Parashumardites	Tr. montiparus		
			Krevyakinskian (Kre)		Protriticites pseudomontiparus / Obsoletes obsoletus		303
		Moscovian (Mos)	Myachkovskian (Mya)	Wellerites	Fusulina cylindrica / F. mjachkovensis		
			Podolskian (Pod)		F. dunbari / Fusulinella colonia		
			Kashirskian (Ksk)	Paralegoceras	Hemifusulina / Fusulina schellwieni		
			Vereiskian (Vrk)		Aljutovella aljutova / Profusulinella prisca		311
		Bashkirian (Bsk) L	Melekesskian (Mel)	Diaboloceras	A. tikhonovechi / Eofusulina triangula		
			Cheremshanskian (Che)	Branneroceras Gastrioceras	P. rhomboides / Ozawainella pararhomboides / P. primitiva P. oblonga / O. alchevskiensis	Idiognathoides sulcatus parvus	
		Bashkirian E	Yeadonian (Yea)	Cancelloceras (G)	Pseudostaffella praegorskyi / O. umbonata / Ps. antiqua	Idiognathoides sinuatus Idiognathodus primulus	
			Marsdenian (Mrd)	Reticuloceras (R)	Eostaffella pseudostruvei		
			Kinderscoutian (Kin)		Asteroarchaediscus gregorii	Idiognathoides corrugatus Id sulcatus sulcatus	323
	Mississippian	Serpukhovian (Spk)	Alportian (Alp)	Homoceras (H)	Eostaffella Millerella Asteroarchaediscus	Idiognath noduliferus Streptognathodus lateralis	
			Chokierian (Cho)	—?—	Eosigmolina Haplophragmina Monotaxinoides		328
			Arnsbergian (Arn)	Eumorphoceras	Eolasiodiscus gracilis, Eostaffellina protvae, Howchinia gibba	Gnathodus bilineatus bollandensis Carusgnathus naviculus	
			Pendleian (Pnd)	Cravenoceras (E)	Eostaffella postproikensis / Tubispirodiscus cornuspiroides	Kladognathus Gnathodus girtyi simplex	333
		Visean (Vis)	Brigantian (Bri)	Hypergoniatites	Dainella echremovi / Loeblichia ukrainika	Gnathodus girtyi collinsoni	
			Asbian (Asb)	Beyrichoceras Goniatites	Bradyina rotula / Archaediscus gigas		
			Holkerian (Hlk)		Lituotubella magna		
			Arundian (Aru)	Merocanites Ammonellipsites	Permodiscus		
			Chadian (Chd)		Eoparastaffella	Gnathodus texanus	
						Scaliognathus anchoralis – Dolignathus latus	350
					Dainella chomatica	Gnathodus typicus	
		Tournaisian (Tou)	Ivorian (Ivo)		Spinoendothyra	Siphonodella isosticha S. crenulata	
			Hastarian (Has)	Protocanites-Pericyclus	Chernychinella	Siphonodella sandbergi	
						Siphonodella duplicata	
				Gattendorfia	Eochernychinella	Siphonodella sulcata	363
D	D3		Famennian (Fam)	Wocklumeria	Bisphaera	Siphonodella praesculcata	
					Quasiendothyra	Bispathodus costatus	

old meaning of Namurian A. The four Serpukhovian stages are standardized in the British Isles (Ramsbottom *et al.* 1978, Ramsbottom 1981).

The initial **Pendleian** boundary (as well as initial Serpukhovian, Namurian A, and Silesian boundaries) was proposed by the Heerlen Congress in 1958 to be at the base of strata containing the 'earliest occurrence' of *Cravenoceras leion* Bisat. A stratotype marker point has been suggested at Little Mearley Clough, Pendle Hill, Lancashire, England (Ramsbottom 1981).

Pendleian, **Arnsbergian**, **Chokierian**, **Alportian** (i.e. to end of Namurian A and to the initial Pennsylvanian boundary) are each approximately equivalent to the goniatite zones used in Britain for many years, respectively E1, E2, H1 and H2,

initiated by Bisat in 1928 (Ramsbottom 1981). More precise definitions are not available.

3.13.3 The Pennsylvanian Sub-period

The **Bashkirian Epoch** as defined in GTS 82 began with the initial Kinderscoutian boundary which also defines the initial Pennsylvanian boundary. The whole epoch is usefully described with evidence for international correlation by Semichatova and others in Wagner *et al.* (1979).

However, a potential change in the scheme is reported. There is a proposal which appears to have gathered international momentum to lower the initial 'Mid-Carboniferous boundary' to allow the later division to begin with Chokierian. The case presented seems only to be on the

Figure 3.6b. Selected Carboniferous successions.

CARBONIFEROUS SUCCESSIONS

	NORTH WEST EUROPE			CHINA	USSR MOSCOW BASIN & URALS			USA	
	BRITISH ISLES		GERMANY / BELGIUM						
Ass			AUTUNIAN		ASSELIAN			WOLFCAMPIAN	
Gze			OTTWEILER U M L / STEPHANIAN C B A	MAPING	NOGINSKY / KLAZ'MINSKIY	GZELIAN (U CARB C₁)		WABAUNSEE / SHAWNEE / DOUGLAS / LANSING	VIRGILIAN (MISSOURIAN)
Kas	CANTABRIAN		Ctb		DOROGOMILOVSKY / KHAMOVNICHESKY / KREVYANINSKY	KASIMOVIAN		KANSAS CITY / PLEASANTON	
Mos	BOLSOVIAN (D,C,B) / DUCKMANTIAN / LANGSETTIAN (A) WESTPHALIAN	COAL MEASURES U L / WESTPHALIAN D C B A	SAARBRUCKER	DALA	MYACHKOVSKY / PODOL'SKY / KASHIRSKY / VEREISKIY	MOSCOVIAN (MIDDLE CARBONIFEROUS)		MARMATON / CHEROKEE	DESMOINESIAN / ATOKAN (PENNSYLVANIAN)
Bsk	YEADONIAN (G₁) / MARSDENIAN (R₂) / KINDERSCOUTIAN (R₁)	MILLSTONE GRIT / NAMURIAN	SILESIAN	HUASHIBAN	MELEKESSKY / CHEREMCHANSKIY / PRIKANSKY / SEVEWROKEL'TEN-SKIY / KRASNOPOLYANSKY	BASHKIRIAN (C)		WINSLOW / BLOYD / HALE	MORROWAN
Spk	ALPORTIAN (H₂) / CHOKIERIAN (H₁) / ARNSBERGIAN (E₂) / PENDLEIAN (E₁)		NAMURIAN	DEWU	VOSNESENSKY / ZAPALTYUBINSKY / PROTVINSKY / STESHEVSKY / TARUSSKY	SERPUKHOVIAN		ELVIRIAN / HOMBERGIAN	CHESTERIAN
Vis	BRIGANTIAN (P₂, D₂) / ASBIAN (D₁) / HOLKERIAN (S₂) / ARUNDIAN / CHADIAN (C₂ S₁)	CARBONIFEROUS LIMESTONE / V3c, V3b, V3a, V2b, V1b V2a, V1a	VISEAN / DINANTIAN	DATANG	VENEVSKY / MIKHAILOVSKIY / ALEKSINSKY / TUL'SKIY / ILYCHSKIY / PESTER'KOVSKIY	VISEAN (LOWER CARBONIFEROUS)		GASPERIAN / ST GENEVIEVE / ST LOUIS / SALEM / WARSAW / KEOKUK	MERAMECIAN / OSAGEAN (MISSISSIPPIAN)
Tou	COURCEYAN (C₁, Z, K)	Tn3, Tn2, Tn1b	TOURNAISIAN	YANGYUAN	KOS'VINSKIY / KIZELOVSKY / CHEREPETSKY / UPINSKY / MALEVSKY	TOURNAISIAN		BURLINGTON / FERN GLEN / MEPPEN / CHOUTEAU / HANNIBAL / GLEN PARK	KINDERHOOKIAN
Fam		Tn1a STRUNIAN		SHAODONG	KALINOVSKY / ZAHOLZHSKY	C		LUOISIANA	

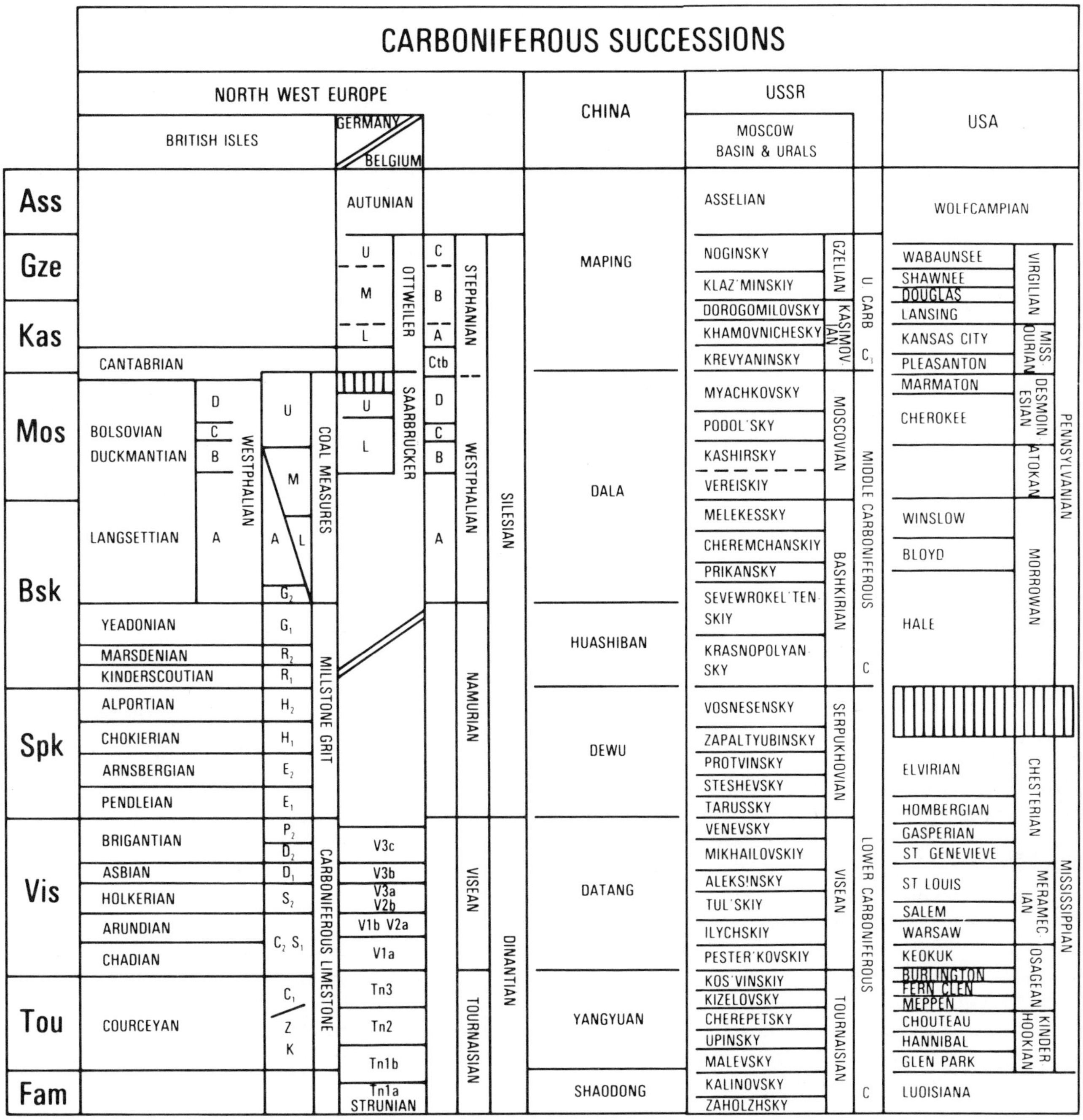

grounds that it would be paleontologically convenient for a division to be made at a point to correspond to appearance of the conodont *Declinognathodus noduliferus* and coinciding approximately with the *Eumorphoceras–Homoceras* ammonoid zonal transition' (Lane & Manger 1985). To make sense this would require that Pennsylvanian and Bashkirian be extended, and Mississippian and Serpukhovian be diminished. These changes have still to be decided and their intervention simply delays definition of a stable chronostratic scale. If the Arnsbergian–Chokierian boundary proves to be a good one for correlation, it is of value whether or not it coincides with a sub-period boundary.

The three stages **Kinderscoutian**, **Marsdenian** and **Yeadonian** were first defined in the British Isles, on goniatite zones of Namurian B and C (Ramsbottom *et al.* 1978). They may be referred collectively to Early Bashkirian. The Late Bashkirian stages **Cheremshanskian** and **Melekesskian** are part of the standard Russian sequence and, with all the later stages to the end of the Carboniferous Period, are based on successions in the Moscow Basin and the Urals. They are adopted for the standard scale but, as yet we think, without definition of boundary-stratotype points.

The **Moscovian Epoch** is divided into four stages which are described in detail with biostratigraphic evidence for international correlation by Ivanova and others in Wagner *et al.* (1979). The ages have been defined in biostratigraphic terms but not standardized at reference points.

The initial boundary of the **Kasimovian Epoch** (also the

base of the Russian Upper Carboniferous) is suggested by Rotai (1979) to correspond to the incoming of *Protriticites pseudomontiparus – Obsoletes obsoletus*, i.e. at the base of the Limestone N$_2$ of the Donetz succession. The three Kasimovian stages originated by reference to three foraminiferal zones.

The **Gzelian Epoch**, as defined by Rotai (1979), corresponds to three fusulinid zones and is divided into two stages. It terminates with the initial Asselian (Permian) boundary. As already indicated, the latest Carboniferous and earliest Permian stratigraphy has been confused, in part by difficulties arising from previous correlations. Until recently the Upper Carboniferous in Soviet usage comprised an upper Orenburgian and a lower Gzelian stage, so it may seem odd that Figure 3.6 shows an earlier Kasimovian and a later Gzelian stage to terminate the Carboniferous Period. But this arrangement appears now to be well established internationally.

3.13.4 The Carboniferous chart (Figure 3.6 a and b)

Figure 3.6 is constructed from the papers cited above. The biostratigraphic scheme has been taken from Rotai (1979, p.245), except that the biostratigraphic correlation of the Hastarian, Ivorian and Chadian stages has been modified. The principal change was to show *Gattendorfia* rather than *Wocklumeria* as the initial Carboniferous zone.

3.14 The Permian Period

3.14.1 History and classification

In 1841, after a tour of Imperial Russia, R. I. Murchison with the cooperation of Russian geologists named the **Permian System** to take in the 'vast series of beds of marls, schists, limestones, sandstones and conglomerates' that surmounted the Carboniferous System throughout a great arc stretching from the Volga eastwards to the Urals and from the Sea of Archangel to the southern steppes of Orenburg. He named it from the ancient kingdom of Permia in the centre of that territory, and the city of Perm which lies in the flanks of the Urals. In 1845 he included rocks now known as Kungurian to Tatarian in age and for a time the underlying strata (Artinskian, etc.) were known as Permo–Carboniferous, i.e. intermediate between Carboniferous and Permian (Dunbar 1940).

Already by 1822 (e.g. Conybeare & Phillips) the **Magnesian Limestone** and **New Red Sandstone** of England were well known as were the equivalent German **Rotliegendes** and **Zechstein** (a traditional miner's name) with its valuable Kupferschiefer. However, all these rocks lacked richly fossiliferous strata, were difficult to correlate, and so inadequate to justify the erection of a new system in western Europe. The lack of fossils had been noted by d'Omalius d'Halloy in 1808 who referred to the Kupferschiefer as **Terrain Penéen** including at first a part of the Triassic Bunter. Permian in due course displaced Penéen in general use.

In North America, J. Mancou in 1853 recognized Permian rocks in a large area from the Mississippi to the Rio Colorado and noted two divisions analogous to those in western Europe. He accordingly suggested the name **Dyassic** as more suitable than Permian and proposed a combined Dyas and Trias as a major period (Zittel 1901).

Karpinskiy in 1874 extended the Permian System in Russia downwards to include Artinskian and Sakmarian sediments.

The name **Thuringian** was introduced in 1874 as equivalent to Zechstein and in 1893 de Lapparent used a three-fold division with Thuringian as Upper Permian, **Saxonian** (= Upper Rotliegendes) as Middle Permian and **Autunian** (from Autun, in France) for Lower Permian.

The case for a three-fold division has been pressed intermittently ever since. For example Waterhouse (1978) suggested a middle epoch spanning Kungurian (Filippovian) to Djulfian or Dzhulfian (named after Dzhul'fa on the River Araks, Caucasus) and ?part Tatarian, and in 1982 he proposed two tripartite schemes for global use, preferring Lopingian (Late), Guadalupian (Middle) and Cisuralian (Early). His stage names also changed to where the best sequences seemed to be in each case.

The confusion has yet to be resolved at international level and stems partly from the fact that the Russian sections for later Permian, while better than those of western Europe, are still inadequate, with few clear standards for correlation. It has long been known that eastern Asia and especially South China contains richly fossiliferous successions for the later Permian record (e.g. Kayser 1883).

Within the classic Ural sequence fusulines are practically limited to Early Permian while the Chinese **Lopingian** is characterized by *Codonofusulina* and *Palaeofusulina* and has a better marine sequence than, say, the Tatarian succession of the USSR.

The Chinese name for Permian (*Erdie*) may be freely translated as 'two-fold' or 'two-cycled' followed by *Sandie* for the tripartite Triassic. These names have force because they are clearly based on Dyas and Trias. However, Chinese geologists, while using two Permian epochs, have defined their later epoch as limited to Lopingian (see below) and their earlier epoch spans all of Permian time before that.

The alternative of two- or three-fold classification (each with more than one variant) leaves the use of Early (Middle) and Late Permian as ambiguous.

Pending a decision by the ICS, this work retains two major sub-systems or epochs and uses Rotliegendes and Zechstein for them as is already commonly done. **Rotliegendes** is the original Permo-Carboniferous and **Zechstein** is approximately Murchison's Permian. In this usage the two names are detached from precise definition in western Europe, in favour of definition wheresoever it prove best. Rotliegendes is well identified in the Russian sequence although there is some doubt about Kungurian which might better be included in Zechstein.

The case for a tripartite division of Permian could still be met by dividing Zechstein into two sub-epochs. Following Waterhouse and encouraged by indications from Jin Yugan (personal communication) and from informal correspondence with J. M. Dickins, **Guadalupian** is used for Early and **Lopingian** for Late Zechstein.

Two American names to divide Guadalupian are introduced: **Wordian** and **Capitanian** (distinguished from the Word and Capitan formations by their suffix). Similarly two Chinese names divide Lopingian (**Longtanian** and **Changxingian**).

Further confusion is not risked by selecting from the very many suggestions which have been made. The Russian names are retained in Figure 3.7 to indicate their approximate equivalence. These were the stage names adopted by the Geological Society of London Working Group (Smith *et al.* 1974) which had been proposed by Likharev and others in 1966 and followed in GTS 82.

3.14.2 The Rotliegendes (Early Permian) Epoch

In this work as in GTS 82 the Rotliegendes Epoch comprises four stages: Asselian, Sakmarian, Artinskian and Kungurian.

The Asselian Stage

The earliest **Asselian** rocks in the Urals correlate with the earliest formed rocks containing *Triticites californicus* in North America, at the base of the Wolfcamp; the initial Permian boundary would therefore be defined between the zones of latest Pennsylvanian *Triticites coronadoensis* and the earliest *T. californicus*. It will be noted that the long uncertainty in both classification and correlation may well account for some rocks traditionally regarded as Carboniferous (e.g. later Stephanian C) being indeed Asselian. The boundary problem with respect to the USSR was reviewed by Rauser-Chernousova & Schegolev (1979).

The Sakmarian Stage

Rocks of **Sakmarian** age (from Sakmara, a tributary of the River Ural, in the southern Urals) originally included all deposits from the top of the Upper Carboniferous to the base of the Artinskian. They were distinguished by *Pseudoschwagerina*. In 1950 Ruzhentsev distinguished two sub-stages – **Asselian** (lower) and **Sakmarian** (upper). These were then raised in rank.

The Artinskian Stage

Murchison, De Verneuil & Von (1845) described the 'grits of Artinsk' which were the basis of Karpinskiy's 1874 Artinskian Stage (Appendix 1) then referred to the Carboniferous System. The Permian was thought to begin with the overlying Kungurian. This was generally accepted until 1889 when, from a study of ammonoids, Karpinskiy in 1889 concluded that Artinskian belongs to Permian. Although named from the town of Artinsk in the middle Urals, the type section was designated to be in the Orenburg district in the southern Urals. It applied to a narrow belt of detrital facies. This facies also characterized some older rocks in the geosynclinal belt. It was only later that their equivalence to and continuity with the carbonate facies further west was appreciated with zones of *Schwagerina* and *Parafusulina*.

The Kungurian Stage

Named after the town of Kungur (near Perm) is a series of evaporites with important salt deposits. They overlie Artinskian strata and are capped by red beds. This interval between Artinskian and Ufimian is generally classed with Early Permian (as here).

In North America Early Permian rocks have been classed in two main divisions: **Wolfcampian** and **Leonardian**.

Stevens, Wagner & Sumsion (1979) described and listed twelve fusuline zones for the Early Permian succession in central Cordilleran America. Because they were not correlated with the Russian stages they have not been fitted into Figure 3.7, but are listed below, to indicate the sequence of fusuline faunas.

Leonardian
Parafusulina spiculata
Parafusulina communis
Parafusulina allisonensis
Parafusulina leonardensis

Wolfcampian
Schwagerina aculeata
Pseudoschwagerina convexa
Schwagerina cf. *S. crebrisepta*
Eoparafusulina linearis
Pseudofusulina hueconensis
Schwagerina bellula
Pseudofusulina attenuata
Triticites californicus

3.14.3 The Zechstein (Late Permian) Epoch

In North Germany Zechstein is an old mining term. The Zechsteinkalk, following Kupferschiefer and being followed by other well-known formations including the Harptdolomit and the Stassfurt evaporites, have made Zechstein the widely used name for Late Permian in a bipartite classification. The Buntsandstein, following Zechstein, is the lowest of the three divisions of the German Trias and much of it is now thought to be Permian. Although Zechstein facies in northwest Europe are not suitable for an international Permian standard the name has widespread currency as well as priority and is chosen for the Late Permian Epoch in a two-fold classification.

Zechstein is divided here into two sub-epochs (Guadalupian and Lopingian) facilitating a three-fold division of Permian if desired. A schematic representation of the regional series or epoch and period boundaries follows, with the names selected for the classification in this work in bold letters. A somewhat more detailed scheme is shown in Figure 3.7.

Germany	Russia	Texas	Southern China
Bunter	Tatarian	Ochoan	**Lopingian** (2 stages)
Zechstein	Kazanian	**Guadalupian**	Maokouan (3 stages)
—	Ufimian Kungurian	—	
Rotliegendes	Artinskian Sakmarian Asselian	Leonardian Wolfcampian	Qixian etc.

In GTS 82 the Late Permian Epoch according to current practice composed three stages: Ufimian, Kazanian, Tatarian. It was the original Permian of the early writers and difficulties have always attended its application because of the

Figure 3.7. The Permian Period and selected successions.

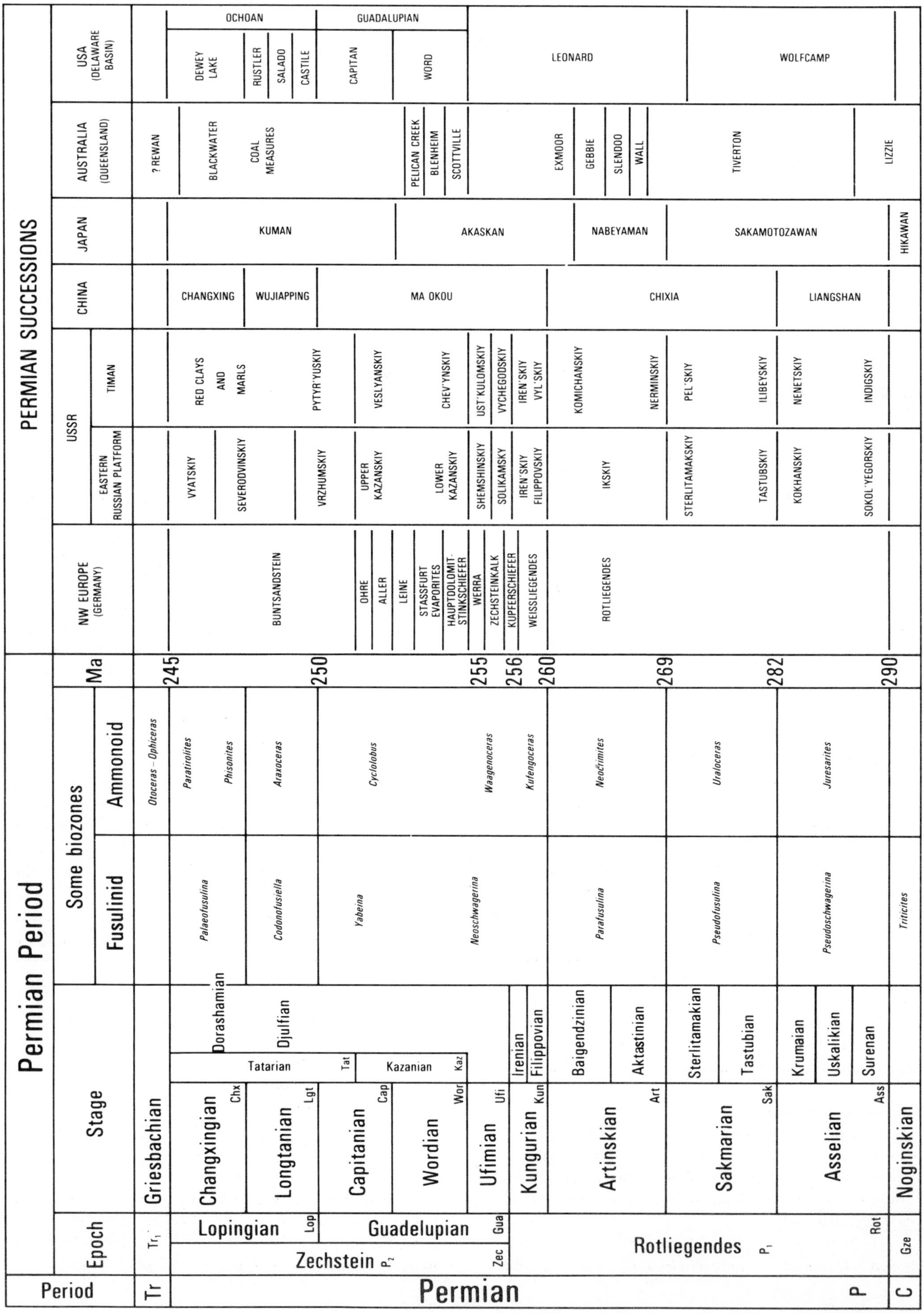

PERMIAN SUCCESSIONS

USA (DELAWARE BASIN) — OCHOAN: DEWEY LAKE, RUSTLER, SALADO, CASTILE; GUADALUPIAN: CAPITAN, WORD; LEONARD; WOLFCAMP

AUSTRALIA (QUEENSLAND) — ? REWAN, BLACKWATER, COAL MEASURES, PELICAN CREEK, BLENHEIM, SCOTTVILLE, EXMOOR, GEBBIE, SLENDOO, WALL, TIVERTON, LIZZIE

JAPAN — KUMAN, AKASKAN, NABEYAMAN, SAKAMOTOZAWAN, HIKAWAN

CHINA — CHANGXING, WUJIAPPING, MA OKOU, CHIXIA, LIANGSHAN

USSR (TIMAN) — RED CLAYS AND MARLS, PYTYR'YUSKIY, VESLYANSKIY, CHEV'YNSKIY, UST'-KULOMSKIY, VYCHEGODSKIY, IREN'SKIY, VYL'SKIY, KOMICHANSKIY, NERMINSKIY, PEL'SKIY, ILIBEYSKIY, NENETSKIY, INDIGSKIY

USSR (EASTERN RUSSIAN PLATFORM) — VYATSKIY, SEVERODVINSKIY, VRZHUMSKIY, UPPER KAZANSKIY, LOWER KAZANSKIY, SHEMSHINSKIY, SOLIKAMSKIY, IREN'SKIY, FILIPPOVSKIY, IKSKIY, STERLITAMAKSKIY, TASTUBSKIY, KOKHANSKIY, SOKOL'YEGORSKIY

NW EUROPE (GERMANY) — BUNTSANDSTEIN, OHRE, ALLER, LEINE, STASSFURT EVAPORITES, HAUPTDOLOMIT-STINKSCHIEFER, WERRA, ZECHSTEINKALK, KUPFERSCHIEFER, WEISSLIEGENDES, ROTLIEGENDES

Ma: 245, 250, 255, 256, 260, 269, 282, 290

Permian Period — Some biozones

Ammonoid: Otoceras – Ophiceras, Paratirolites, Phisonites, Araxoceras, Cyclolobus, Waagenoceras, Kufengoceras, Neocrimites, Uraloceras, Juresarites

Fusulinid: Palaeofusulina, Codonofusiella, Yabeina, Neoschwagerina, Parafusulina, Pseudofusulina, Pseudoschwagerina, Triticites

Stage: Griesbachian, Changxingian (Chx), Dorashamian, Longtanian (Lgt), Djulfian, Tatarian (Tat), Capitanian (Cap), Kazanian (Kaz), Wordian (Wor), Ufimian (Ufi), Irenian, Filippovian, Kungurian (Kun), Baigendzinian, Aktastinian, Artinskian (Art), Sterlitamakian, Tastubian, Sakmarian (Sak), Krumaian, Uskalikian, Surenan, Asselian (Ass), Noginskian

Epoch: Tr₁, Lopingian (Lop), Guadelupian (Gua), Zechstein (P₂) (Zec), Rotliegendes (P₁) (Rot), Gze

Period: Tr, Permian (P), C

predominance of continental facies in most West European and Russian successions especially for later Zechstein time.

The new scheme adopted here takes account of what appear to be good marine sequences in North America for Early Zechstein time and in South East Asia – especially southern China – for Late Zechstein time. Accordingly Zechstein is divided here into two sub-epochs: **Guadalupian** and **Lopingian**.

The Guadalupian Sub-epoch

Guadalupian is taken from its widespread use in North America for the division following Leonardian and preceding Ochoan and comprising the Word and Capitan formations. It was named for the Guadalupe Mountains which lie between the Glass Mountains in Texas and the Capitan range in New Mexico. Guadalupian has been a regional series name in the USA corresponding to Middle Permian in a commonly used tripartite classification. Waterhouse & Gupta (1982), in proposing the international use of this name, suggested the Guadalupian comprise Kungurian, Kazanian and Punjabian (p.218) or Roadian, Wordian and Capitan (p.219) to use American rather than Eurasian divisions. They argued for the inclusion of Kungurian and its approximate equivalent Roadian = basal Word of Glass Mountains, Texas (King 1931) as the initial Guadalupian stage on paleontological grounds. These were that primitive neo-schwagerines and cyclolobids, characteristic of Guadalupian, commenced in basal Word, and brachiopods were renewed in Kungurian time. Because of uncertainty of correlation of Kungurian and Ufimian rocks the traditional **Ufimian** stage is continued as initiating the Late Permian sequence followed by **Wordian** and **Capitanian**.

The Ufimian Stage

The name is from Ufa in Russia, and Ufian has been used as an alternative name. Ufimian rocks were previously known as the 'lower red unit' or the 'Lower division of the Permian system P' and the 'Lower Permian Red Group'. Ufimian was adopted as the initial Late Permian stage in 1960 (in the USSR) and probably replaces part of what was formerly known as Kungurian. Some authors have combined Ufimian with Kadanian as the Kana stage.

Ufimian is in part probably equivalent to the American Roadian. Definition and correlation at this time is much needed to clarify the position.

The Wordian Stage

Wordian is from the Word Formation of the Glass Mountains, Texas. It is distinguished by limestones with *Parafusulina* and is approximately equivalent to the Kazanian Stage (after the town of Kazan of the middle Volga).

The Capitanian Stage

Capitanian is from the Capitan Formation of Capitan Peak, New Mexico, and together with the Word Formation is best developed in the western Texas–New Mexico region.

The Lopingian Sub-epoch

Since the Loping fauna of south China was discovered on Richthofen's expedition and described (e.g. Kayser 1883) it has been known as one of the richest Permian faunas and

corresponds to relatively barren continental strata in North America, western Europe and western USSR in contrast to neighbouring areas in the eastern Pacific region of the eastern Tethys (Huang 1932a, 1932b, 1933, Tozer 1988).

In Russia the Tatarian rocks, named in 1887 from the Tartar people (Kotlya 1977), probably span much of the Capitanian and Lopingian time. In western Europe part of the Bunter is probably of this age and in America Ochoan has been used for a division between Capitanian and Triassic but with little potential for international correlation.

The Lopingian is commonly divided into two stages with alternative nomenclature: **Longtanian** or **Wujiapingian** and **Changxingian** or **Talungian** (Yang *et al.* 1986). Rocks of both ages are widespread in south China and beyond, represented by many formations. Two works may be referred to, namely the comprehensive memoir by Huang (1932) and the much later review in Yang *et al.* (1986).

The Longtanian Stage

The name is from Longtan, Nanjing, and the type section is at Tianboashan also near Nanjing. Rocks of this age are characterized by *Codonofusiella*. They are probably coeval with the proposed *Djhulfian* Stage of Waterhouse. Wujiaping is arguably a better succession biostratigraphically than Longtan. Several biozones have been distinguished. Further west in China basaltic and coal-bearing facies are widespread as at Emeishan (Mt Omei).

The Changxingian Stage

This stage is named from Changxing, Zhejiang Province, with a rich marine carbonate sequence as well as coal facies. Rocks of this age are characterized by *Palaeofusulina*, and several biozones have been distinguished.

The Lopingian division, so rich in fossiliferous strata in China and the eastern Tethys, is not well represented in the West. The introduction of this division, and the consequent reclassification of the Zechstein Epoch, is a novel feature of this work. A significant time interval has probably been tele-scoped in earlier time scales, and so may have in part reinforced the concept of the remarkable biostratigraphic changes at the Permian–Triassic boundary.

3.14.4 The Permian chart (Figure 3.7)

The fusulinid zones on Figure 3.7 are taken from a list from Japan, which is rich in fossils of the period (Takai, Matsumoto & Toriyama 1963). Ammonoids in the adjacent column are from the Canadian Arctic. The list and correlation are from Smith *et al.* (1974).

The other columns on Figure 3.7 are derived as follows: northwest Europe and USA from Smith *et al.* (1974); Japan from Takai *et al.* (1963); Australia from Waterhouse (1978); the USSR columns from the unnumbered table that plots a comparative scheme of Permian sections abstracted from the whole work by Likharev (1966). Details of these two USSR columns are summarized in that work in Tables 1, 2, 4 and 5 respectively.

3.15 The Triassic Period
3.15.1 Introduction

The Triassic System was established in Germany from the three-fold lithologic division into **Bunter**, **Muschelkalk** and **Keuper** by Alberti in 1834. The traditional stages (Scythian, Anisian, Ladinian, Carnian, Norian) were established in the marine successions of the Northern Calcareous Alps of Austria where, however, the ammonite zones first used later proved to be incomplete or not always in chronological sequence. For this reason, and because of the excellent Arctic successions and those from the Western Cordillera of North America, Tozer (1967) proposed 'a standard for Triassic time' in which ammonite zones were used to characterize the standard stages more precisely. New stages were proposed to divide Scythian or Early Triassic. For this epoch the name Scythian is useful and is retained. In that work Tozer went so far as to define boundary reference points in type sections; and because this is the essential method for defining the chronostratic scale his standard is adopted here almost in its entirety (Tozer 1967, table II, 1982, 1984). Silberling & Tozer (1968) elaborated this scheme for North America as a whole.

In 1984 Tozer modified his 1967 divisions in part, combining Dienerian and Smithian in Nammalian and also proposing to eliminate Rhaetian (but see below).

3.15.2 The Paleozoic–Mesozoic boundary

It is perhaps remarkable that since 1841 Phillips' proposal for the three eras should not only have been retained till today but still has force as a natural division in biologic history.

The apparent contrast between brachiopod-rich Permian faunas and predominantly molluscan Mesozoic faunas has been accentuated by the general absence in the West of the later Permian marine faunal record. Diener (1908, 1912) noted in the Himalaya that productids were dominant until Triassic time. It proves, as might be expected in transitional sequences, that productids do indeed overlap with the ranges of the early Triassic ammonoids *Otoceras* and *Ophiceras*. This led to a proposal that *Otoceras*-bearing strata be treated as late Permian (e.g. Waterhouse 1978). However, by international convention, it is now established that the Permian–Triassic boundary be taken so that *Otoceras* would characterize the earliest Triassic strata. Tozer's initial Griesbachian reference point (see below) is adopted here until an international convention be established. We note that Tozer (1984) proposed that the Permian–Triassic boundary should be in the Himalaya where the biozone *Otoceras woodwardi* is recognized, but competition from a section in Meishan in Zhejiang, South China, may be entertained. *Otoceras* is not common outside Arctic regions and a Tethyan standard could be more useful. 'It seems valid that the Permo–Triassic boundary is placed between the base of the *Otoceras* zone or its equivalent *Hypophiceras* zone, and the top of *Rotodiscoceras* and *Pleuronodoceras* zones or *Palaeofusulina sinensis* zone, representing the top of the youngest Permian . . .' (Tozer 1984). The problem of the initial Triassic boundary was further reviewed by Tozer (1988).

3.15.3 The Scythian (Early Triassic) Epoch

Some scientists may doubt the value of dividing the Scythian Epoch into stages, it having been regarded as a stage itself (e.g. Kummel in 1957). However, Spath in 1935 divided Early Triassic into Early and Late 'Eo-Trias' each with three subdivisions. His Early and Late divisions corresponded to the Induan and Olenekian stages of northern Siberia introduced by Kiparisova & Popov in 1956 (but changed in 1964).

Tozer's four stages (1967) were used in GTS 82, his Dienerian–Smithian boundary divided Scythian into two pairs of stages: Griesbachian, Dienerian and Smithian, Spathian. However in 1984 he adopted Nammalian to embrace the two sub-stages Dienerian and Smithian while the sub-stages of Griesbachian were given the names Gangetian and Ellesmerian. This scheme is adopted in Figure 3.8.

The **Griesbachian Stage** (from Griesbach Creek, Axel Heiberg Island) was proposed by Tozer in 1965 (1967) and divided into two sub-stages each with two ammonite zones:

The **Gangetian Sub-stage** is characterized by two *Otoceras* chrons.

The **Ellesmerian Sub-stage** begins with *Ophiceras* followed by a *Proptychites* chron.

The **Nammalian Stage** was proposed by Guex (1978) from the Himalaya and recommended by Tozer (1984). Its definition in terms of its two constituent sub-stages (Dienerian and Smithian) is continued here.

The initial **Dienerian** (from Diener Creek, Ellesmere Island) boundary is in the Blind Fjord Formation of northwestern Ellesmere Island (Tozer 1967, note 16) and is generally recognized by the appearance of Gyronitidae. It is correlated in the Himalaya at the boundary between *Otoceras* and *Meekoceras* beds (Diener in 1912, see Tozer (1967)), and in the Salt Range between the *Ophiceras connectens* bed and the Lower Ceratite Limestone (Kummel & Teichert 1966).

The initial **Smithian** (from Smith Creek, Ellesmere Island) is also in the Blind Fjord Formation of northwestern Ellesmere Island, corresponding to the boundary between the Induan and Olenekian stages of Kiparisova & Popov (1956, 1964).

The **Spathian Stage** is defined with its initial Spathian boundary (from Spath Creek, Ellesmere Island) in the lower shale member of the Blaa Mountain Formation. The type locality is at Spath Creek, Ellesmere Island (see Tozer 1967, note 11), with a useful additional sequence in Axel Heiberg Island.

3.15.4 The Mid Triassic Epoch

The original type locality for the **Anisian Stage** is in Austria, but it lacks ammonoids near its base. Tozer (1967) defined its initial boundary by the *caurus* zone in the type locality (east limb of anticline west of Mile Post 375, Alaska Highway, northeast British Columbia, note 28). Despite the North American definition of Anisian, Tozer's three sub-stages have the Mediterranean names **Aegean**, **Pelsonian** and **Illyrian**, of which Pelsonian has three Tethyan-based chrons and Illyrian two (Tozer 1984).

The initial **Ladinian Stage** (from Ladini, people of Tyrol) boundary is (for the same reason) not defined in the original Ladinian rocks of Italy but in the Humboldt Range, Nevada, USA (Tozer 1967). Five North American chrons compare with only two Tethyan ones (Tozer 1984).

3.15.5 The Late Triassic Epoch

The **Carnian Stage** is named from the Carnic Alps (the alternative spelling Karnian as used by Tozer 1967 is from German rather than from the Latin form preferred here). The stage was introduced by Mojsisovics in 1869. Its initial boundary is based in the type locality at Ewe Mountain, four

Figure 3.8. The Triassic Period and selected successions.

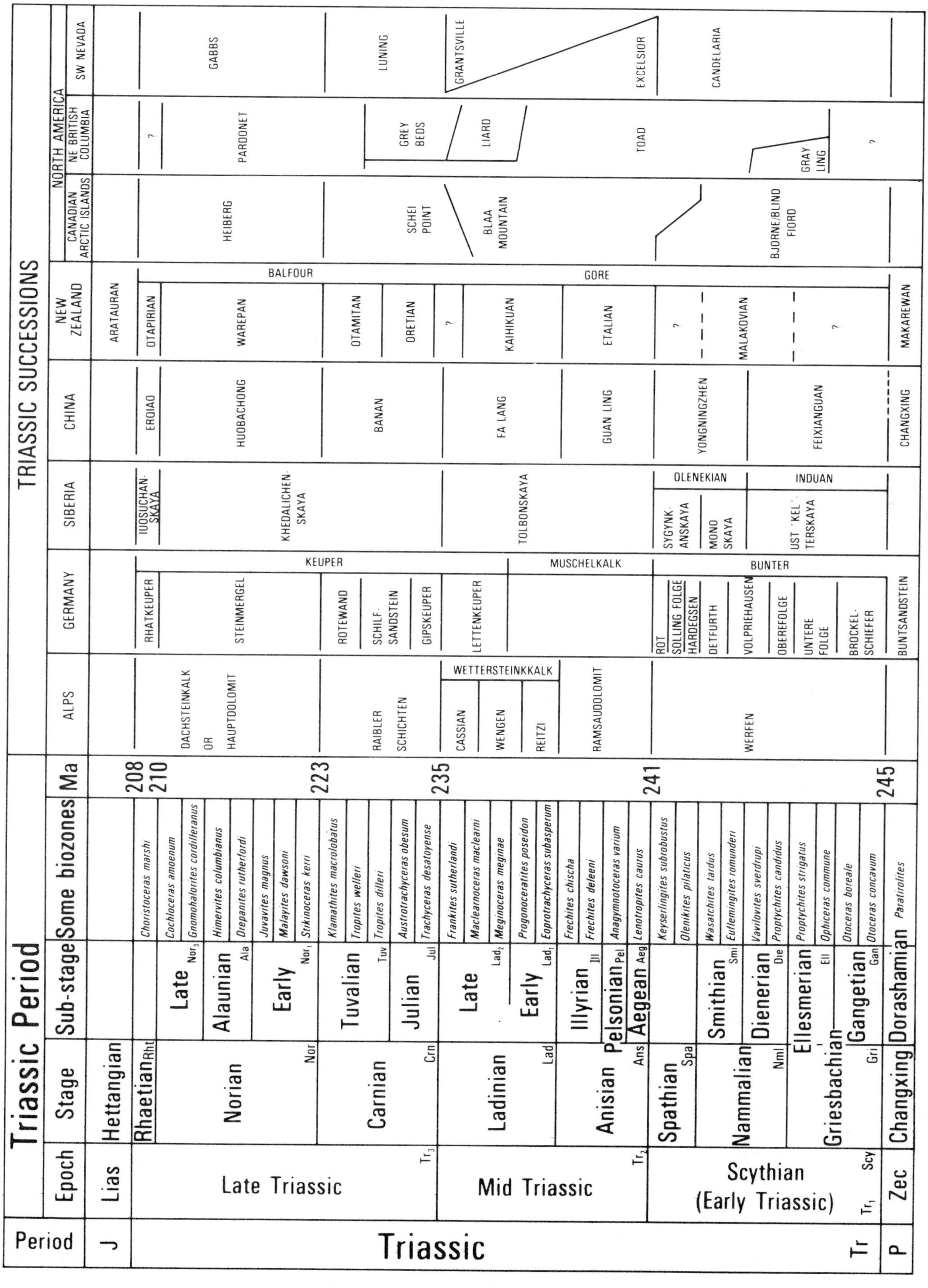

TRIASSIC SUCCESSIONS

Period	Epoch	Stage	Sub-stage	Some biozones	Ma	ALPS	GERMANY	SIBERIA	CHINA	NEW ZEALAND	CANADIAN ARCTIC ISLANDS	NE BRITISH COLUMBIA	SW NEVADA
J	Lias	Hettangian			208 / 210					ARATAURAN			GABBS
Triassic	Late Triassic (Tr₃)	Rhaetian (Rht)		*Choristoceras marshi*		DACHSTEINKALK OR HAUPTDOLOMIT	RHATKEUPER	IUOSUCHAN-SKAYA	EROIAO	OTAPIRIAN	HEIBERG	?	
		Norian (Nor)	Late (Nor₃)	*Cochloceras amoenum* / *Gnomohalorites cordilleranus*			KEUPER / STEINMERGEL			BALFOUR / WAREPAN		PARDONET	LUNING
			Alaunian (Ala)	*Himavites columbianus* / *Drepanites rutherfordi*				KHEDALICHEN-SKAYA	HUOBACHONG				
			Early (Nor₁)	*Juvavites magnus* / *Malayites dawsoni* / *Stikinoceras kerri*						OTAMITAN	SCHEI POINT	GREY BEDS / LIARD	GRANTSVILLE / EXCELSIOR
		Carnian (Crn)	Tuvalian (Tuv)	*Klamathites macrolobatus* / *Tropites welleri* / *Tropites dilleri*	223	RAIBLER SCHICHTEN	ROTEWAND / SCHILF-SANDSTEIN		BANAN	ORETIAN	BLAA MOUNTAIN		
			Julian (Jul)	*Austrotrachyceras obesum* / *Trachyceras desatoyense*			GIPSKEUPER			?			
	Mid Triassic (Tr₂)	Ladinian (Lad)	Late (Lad₂)	*Frankites sutherlandi* / *Maclearnoceras maclearni* / *Meginoceras meginae*	235	CASSIAN / WENGEN	LETTENKEUPER		FA LANG	KAIHIKUAN		TOAD	CANDELARIA
			Early (Lad₁)	*Progonoceratites poseidon* / *Eoprotrachyceras subasperum*		REITZI / WETTERSTEINKKALK	MUSCHELKALK	TOLBONSKAYA					
		Anisian (Ans)	Illyrian (Ill)	*Frechites chischa* / *Frechites deleeni*		RAMSAUDOLOMIT			GUAN LING	ETALIAN			
			Pelsonian (Pel)	*Anagymnotoceras varium*									
			Aegean (Aeg)	*Lenotropites caurus*	241		ROT / SOLLING FOLGE / HARDEGSEN	SYGYNK-ANSKAYA	YONGNINGZHEN	MALAKOVIAN		GRAY LING	
	Scythian (Early Triassic) (Tr₁)	Spathian (Spa)		*Keyserlingites subrobustus* / *Olenikites pilaticus*		WERFEN	DETFURTH / VOLPRIEHAUSEN	OLENEKIAN / MONO SKAYA			BJORNE/BLIND FIORD		
			Smithian (Smi)	*Wasatchites tardus* / *Euflemingites romunderi*			BUNTER / OBEREFOLGE	UST' KEL-TERSKAYA / INDUAN	FEIXIANGUAN	?			
		Nammalian (Nml)	Dienerian (Die)	*Vavilovites sverdrupi* / *Proptychites candidus*			UNTERE FOLGE						?
			Ellesmerian (Ell)	*Proptychites strigatus* / *Ophiceras commune*			BROCKEL-SCHIEFER						
		Griesbachian (Gri)	Gangetian (Gan)	*Otoceras boreale* / *Otoceras concavum*	245		BUNTSANDSTEIN						
P (Zec)		Changxing	Dorashamian (Dor)	*Paratirolites*					CHANGXING	MAKAREWAN			

miles eastnortheast of Triangulation Station 6536, Toad River area, northeast British Columbia (Tozer 1967, note 23).

The **Norian Stage** of the Eastern Alps (Mojsisovics in 1895, and Diener in 1926 and earlier referred to as Jovavic) was originally divided into three parts but doubt has been thrown on their sequential arrangement in that region. The Norian stage now appears as one of the longest time spans in the Triassic or even Phanerozoic scale. Three sub-stages have therefore been suggested as in Figure 3.8 here (Tozer 1984).

Rocks of the **Rhaetian Stage** (from the Rhaetic Alps) are not rich in pelagic fossils. They were earlier included within the Jurassic System (e.g. Arkell 1933 but not by Arkell 1956). The younger rocks contain the bivalve *Rhaetavicula contorta*.

Tozer's type locality of the initial Rhaetian boundary is at Brown Hill, Peace River, northeast British Columbia (1967, note 20). The type locality for the original stage is at Kendelbachgraben, St Wolfgang, Austria, and the Rhaetian division (being clearly older than Hettangian with *Psiloceras planorbis*) is now unambiguously Triassic. However, according to the chart it comprises only one chron in contrast to the six Norian zones. Tozer (1979) suggested abandoning the Rhaetian division as a stage and incorporating it in the late Norian. The *Rhabdoceras suessi* and *Choristoceras marshi* zones would be replaced by three zones of a revised Sevatian sub-stage, namely *Cordilleranus* zone, *Amoenum* zone (both approximately equivalent to the original Suessi zone), and a third *Crickmayi* zone to replace the *Marshi* zone. However, the ICS Subcommission on Triassic Stratigraphy decided (Moscow 1984) to retain Rhaetian as a useful stage at least for the present.

The initial Jurassic boundary is characterized by the arrival of *Psiloceras planorbis*. The implication is that strata such as the pre-Planorbis beds of the English Blue Lias are therefore mostly Triassic together with the Watchet Beds and the Penarth Group (see Section 3.16.2).

3.15.6 The Triassic chart (Figure 3.8)

The columns to the left of Figure 3.8 summarize the discussion above.

The columns to the right of Figure 3.8 were based as follows: Alps on Sherlock (1948); Germany on Warrington *et al.* (1980); Siberia on Kiparisova, Radchenko & Gorskiy (1973); China on Chen (1974); New Zealand on Suggate, Stevens & Te Punga (1978); Canadian Arctic Islands and northeast British Columbia on Tozer (1967); and southwest Nevada on Kummel (1961, p.574).

3.16 The Jurassic Period
3.16.1 History and classification

Between 1797 and 1815 William Smith produced successions and geologic maps of England and Wales in which detailed stratigraphy of successive strata (of Jurassic age) played a key part, the sequence in England being especially well displayed. Many of these were grouped as the Oolite Formation by Buckland in 1818, or Oolitic Series (divided into Lower, Middle and Upper Oolites) overlying the Lias by Conybeare & Phillips (1822). They were equated with the Jura-Kalkstein of Alexander von Humboldt who in 1795 so referred to the Calcaire de Jura. From this Alexandre Brongniart in 1829 first used the name Terrains Jurassiques but only for the Lower Oolitic Series of Conybeare and

Phillips. In Britain for many years the name Jurassic coexisted with the earlier named constituent parts: Lias and Oolites. The above history is surveyed by Zittel (1901), Wilmarth (1925), Arkell (1933, 1956) and Torrens in Cope *et al.* (1980a, 1980b).

Because of the immense wealth of Jurassic fossils, particularly ammonites, biostratigraphic zonation advanced more rapidly than for rocks of other periods.

D'Orbigny saw clearly the possibility of distinguishing divisions based on fossil sequences from formations characterized by lithology. He proposed ten stages each with two or three ammonite zones (1842–9). Although he conceived each as representing the remains of a distinct population which was swept away and replaced by the next, his stage names (ending in -ien) are in use today (except for Liasien). Other (stage) names then proliferated in France in this form but tended to be little more than formational names.

Biostratigraphy developed rapidly with many workers; the most notable was Oppel whose systematic studies resulted in 33 Jurassic zones for correlation between England, France and Germany (1856 *et seq.*). These were in effect chrons for, though named for ammonite species, he said they could equally well have been named after places.

Oppel grouped his zones into stages and these into three divisions: **Lias**, **Dogger** and **Malm**. The divisions correspond to those of Conybeare & Phillips (1822): Lias = Lias, Dogger = Lower Division of Oolites, and Malm = Middle and Upper Oolites (Arkell 1933). The scheme is followed approximately today except that *Ammonites macrocephalus* was his initial Malm zone whereas it is the intial Callovian chron, Callovian now being the youngest Dogger stage.

Arkell (1933) reviewed the Jurassic rocks of Britain, and then of the world (1956). The scheme here is the same as that developed by Arkell as a European standard from 1946 to 1956 except that Aalenian has been introduced by dividing Bajocian while Tithonian replaces Arkell's Portlandian and Purbeckian; the latter with freshwater facies had no ammonites and is in any case partly Cretaceous in age.

For the three standard Jurassic epochs (Early, Mid and Late), Lias, Dogger and Malm are applied here following European usage and in recognition of Oppel's work.

In the following list of stages the definition of the initial boundary points follows the UK recommendations to the Colloque Jurassique in Luxembourg (July 1967) and the UK Contribution to the International Geological Correlation Programme submitted in Prague in 1968 to the IUGS by the Royal Society (later reprinted in George *et al.* 1969). The boundary definitions are only indicated here and in any case they generally await a firm decision. These stages are summarized and approximately equivalent rock sequences in some other parts of the world are given in Figure 3.9. The classic (74) or so zones are listed in Figure 3.10 (Cope *et al.* 1980a, 1980b). It may be noted that each stage on average spans about five potential chrons and 10–15 subchrons. As now used they are conceived as standard time divisions or chrons. To make this clear chrons and subchrons, although named from the original zonal name, should in due course be defined by GSSP and then printed wholly in Roman rather than partly in italic type.

Each of the 11 Jurassic stages has been divided informally into two or three sub-stages.

3.16.2 The Lias (Early Jurassic) Epoch

The **Hettangian** initial boundary (name from Hettange, France) is recognized by the first appearance of the genus *Psiloceras*. The initial boundaries of the Planorbis Chron and of the Jurassic Period coincide with it. Oppel (in 1856, pp.24–8) described sections at Lyme Regis and in quarries near Uplyme, Dorset, England, as being characteristic of the *planorbis* zone; he also referred to the coastal section of Watchet, Somerset, England. Morton (1971, p.84) recommended that the coastal section between Blue Anchor and Quantox Head, in the Watchet area, be regarded as the type area of the *planorbis* zone. George *et al.* (1969, p.53) and then Cope *et al.* (1980b) regarded the *planorbis* subzone as being 'clearly and unequivocally acceptable as the basal subzone of the basal Jurassic' (p.22). The GSSP has yet to be defined, and possibly in South America (J. W. Cope personal communication).

The **Sinemurian** initial boundary (name from Semur, France) is also that of the Bucklandi Chron and the Conybeari Subchron. No type area for the *bucklandi* zone was given by Oppel in 1856. The *conybeari* subzone would be founded in the Keynsham area, Somerset, England, but lack of permanent sections raises questions about its use as a type area. Morton (1971, p. 85) recommended that the type area be designated 50 miles south-southwest of Keynsham, on the Dorset coast, southwest of Lyme Regis. Here the base of the *conybeari* subzone would be placed at the base of Lang's bed 21, which outcrops at Seven Rock Point and at Devonshire Head. An exact type locality remains to be decided and this may turn out to be in France (Cope personal communication).

The **Pliensbachian** initial boundary is also that of the Taylori Subchron and the Jamesoni Chron. According to Morton (1971, p.85) 'there is no explicit type section for the Taylori Subzone, but it was first used with reference to the Dorset coast section' (south England). There it is seen in Lang's (1928) bed 105, the base of the Belemnite Marls, separated from the underlying bed 104 by a non-sequence (Spath 1956, p.148). Bed 105 outcrops near Charmouth. At Pliensbach in southwest Germany, the upper two subzones of the Sinemurian Raricostatum Chron are absent (Geyer 1964, p.165). Morton (1971, p.85) regarded the Pliensbach section as being 'acceptable for defining the Pliensbachian Stage in terms of its basal subzone . . . Here, the Taylori subzone rests nonsequentially on the lower part of the Raricostatum Zone.' This is likely to be accepted for GSSP (Cope personal communication).

The **Toarcian** initial boundary (name from Thouars, France) is also that of the Tenuicostatum Chron. Morton (1971, p.85) designated the outcrop west of Kettleness, on the north Yorkshire coast, England, as the type section for the Tenuicostatum Chron, and therefore for the basal Toarcian. The marker point for the base of the zone is at the base of bed 29 of Howarth (1955).

3.16.3 The Dogger (Middle Jurassic) Epoch

The name **Aalenian** was proposed by Mayer-Eymar in 1864 for the lowest part of the 'Braunjura' in the vicinity of

Aalen, Germany, at the northern edge of the Swabian Alps. The initial Aalenian (and so initial Dogger) boundary is recognized at the base of the Opalinum Zone. However, only later Aalenian is represented near the present-day village of Aalen-Attenhofer so an initial boundary stratotype elsewhere is needed. Some workers have regarded Aalenian as Early Bajocian in a three-fold division (e.g. Morton and others at the Colloque Jurassique at Luxembourg, 1967). Aalenian is accepted as a distinct stage and so Bajocian divides into Early and Late only. Arkell's Scissum Zone inserted between Opalinum and Murchisonae zones is not so easy to apply in continental Europe where the index fossil is found in earlier and later beds.

Bajocian was introduced by d'Orbigny in 1852 for strata outcropping near Bayeux, France (hence the name). Its initial boundary would be taken somewhere at the base of the Discites Zone and probably in France (since the Aalenian is being treated here as a separate stage). On the northwestern periphery of the Anglo–Paris Basin, the Bajocian is represented by the Middle and Upper Inferior Oolites and the Aalenian by the Lower Inferior Oolite.

The **Bathonian** initial boundary (name from Bath, England) is that of the Zigzag Chron (and the Convergens Subchron) at the base of bed 23 of Sturani (1967) at the Bas Auran section, 4 km west of Barreme, Basses-Alpes, southeast France.

The initial **Callovian** boundary (name from Kellaway, England; Kellaways Rock) and that of the initial Macrocephalus Subchron was proposed with the Chippenham–Trowbridge area, Wiltshire, England, as the type area (of the Macrocephalus Subzone). No exact type section has yet been designated but this may be selected in Germany.

3.16.4 The Malm (Late Jurassic) Epoch

The initial **Oxfordian** boundary (name from Oxford, England; Oxford Clay Formation) is that of the Mariae Chron and Scarburgense Subchron, and has been selected defined in the cliff of Cornelian Bay (Cayton Bay, 3 km southeast of Scarborough, Yorkshire, England), rather than in the earlier standard on the shore at Auberville, Normandy, France. Fortunately, conflict is avoided since 'the Oxford Clay of the Yorkshire coast begins as far as one can tell with beds of the same age as those at the base of the Mariae Zone in its type-locality, Normandy; and the Mariae Zone is divided into subzones, the lowest of which has its type section on the Yorkshire coast' (Morton 1971, p.89), i.e. the Scarburgense Subzone.

The initial **Kimmeridgian** boundary (name from Kimmeridge, originally Kimeridge, Dorset, England; Kimmeridge Clay Formation; = Kimeridge of Arkell) and that of the Baylei Chron, is better defined between Osmington (Black Head) and Ringstead Bay, on the coast of Dorset, England, than on the coast of Normandy, France. Morton (1971, p.90) specified that the base of the Baylei Zone be defined at Ringstead, Dorset, and this has been accepted.

There is some difference of opinion as to the actual span of this stage. In this work Tithonian follows it (see below) so that the 'Kimmeridgian' employed here is that of French usage and restricted to Early Kimmeridgian of others, with the initial

Figure 3.9. The Jurassic Period and selected successions.

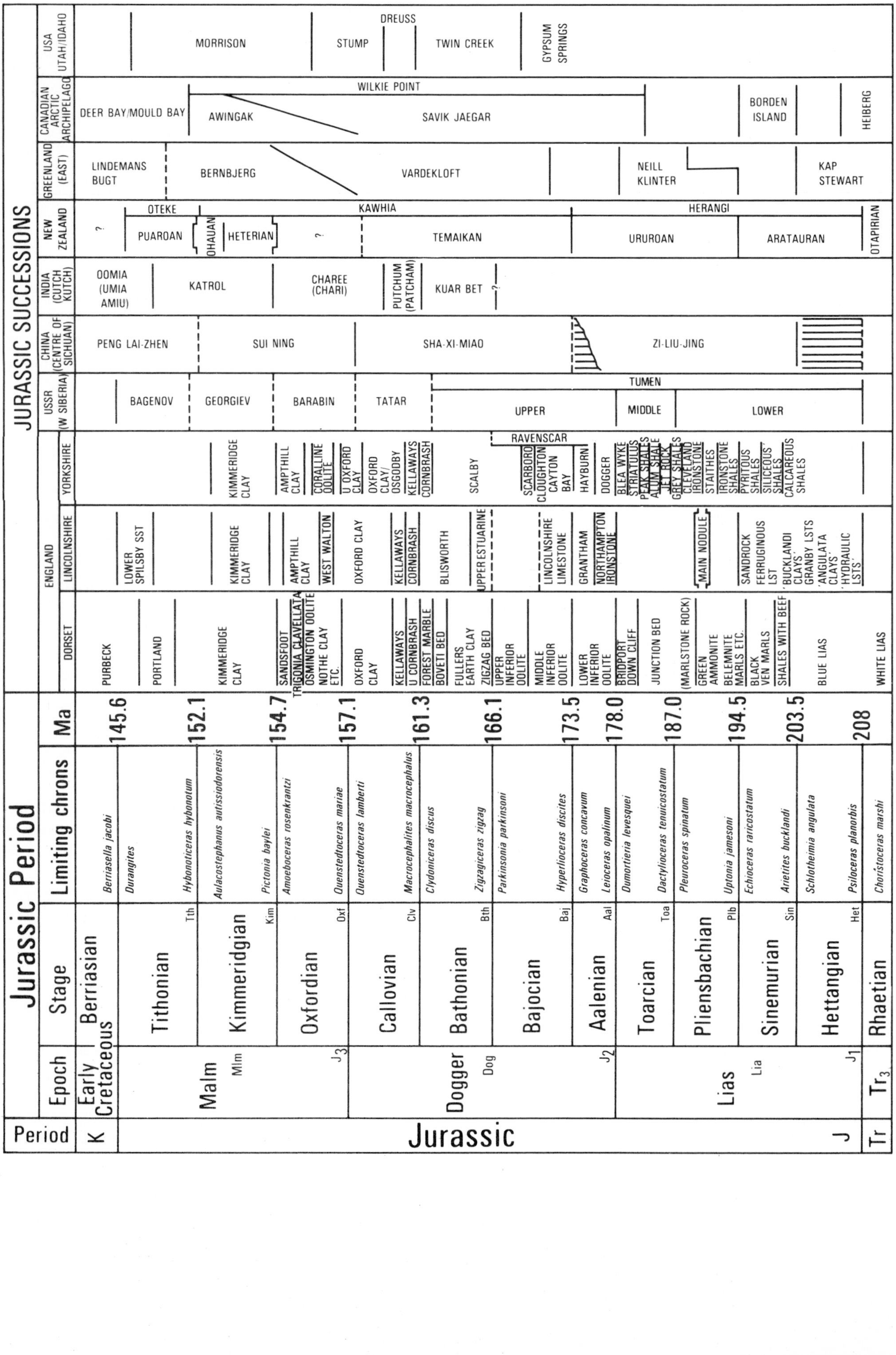

Period	Epoch	Stage	Limiting chrons	Ma	DORSET	ENGLAND LINCOLNSHIRE	YORKSHIRE	USSR (W SIBERIA)	CHINA (CENTRE OF SICHUAN)	INDIA (CUTCH KUTCH)	NEW ZEALAND	GREENLAND (EAST)	CANADIAN ARCTIC ARCHIPELAGO	USA UTAH/IDAHO
K	Early Cretaceous	Berriasian	Berriasella jacobi	145.6	PURBECK			BAGENOV	PENG LAI-ZHEN	OOMIA (UMIA AMIU)	OTEKE / PUAROAN	LINDEMANS BUGT	DEER BAY/MOULD BAY	MORRISON
Jurassic	Malm (Mlm)	Tithonian (Tth)	Durangites			LOWER SPILSBY SST					OHAUAN / HETERIAN	BERNBJERG	AWINGAK	
			Hybonoticeras hybonotum	152.1	PORTLAND			GEORGIEV	SUI NING	KATROL	KAWHIA ?		WILKIE POINT	STUMP / DREUSS
		Kimmeridgian (Kim)	Aulacostephanus autissiodorensis		KIMMERIDGE CLAY	KIMMERIDGE CLAY	KIMMERIDGE CLAY						SAVIK JAEGAR	TWIN CREEK
			Pictonia baylei	154.7	SANDSFOOT TRIGONIA CLAVELLATA OSMINGTON OOLITE NOTHE CLAY ETC.	AMPTHILL CLAY WEST WALTON	AMPTHILL CLAY CORALLINE OOLITE	BARABIN	SHA-XI-MIAO	CHAREE (CHARI)	TEMAIKAN	VARDEKLOFT		GYPSUM SPRINGS
	J3	Oxfordian (Oxf)	Amoeboceras rosenkrantzi		OXFORD CLAY	OXFORD CLAY	U OXFORD CLAY OXFORD CLAY/ OSGODBY							
			Quenstedtoceras mariae	157.1				TATAR		PUTCHUM (PATCHAM)				
		Callovian (Clv)	Quenstedtoceras lamberti		KELLAWAYS U CORNBRASH FOREST MARBLE BOVETI BED	KELLAWAYS CORNBRASH	KELLAWAYS CORNBRASH			KUAR BET	?	NEILL KLINTER	BORDEN ISLAND	
			Macrocephalites macrocephalus	161.3		BLISWORTH		TUMEN UPPER	ZI-LIU-JING		?			HEIBERG
	Dogger (Dog)	Bathonian (Bth)	Clydoniceras discus		FULLERS EARTH CLAY ZIGZAG BED UPPER INFERIOR OOLITE	UPPER ESTUARINE	SCALBY				URUROAN			
			Zigzagiceras zigzag	166.1			RAVENSCAR SCARBORO CLOUGHTON CAYTON BAY	MIDDLE			HERANGI	KAP STEWART		
		Bajocian (Baj)	Parkinsonia parkinsoni		MIDDLE INFERIOR OOLITE	LINCOLNSHIRE LIMESTONE	HAYBURN							
	J2	Aalenian (Aal)	Hyperlioceras discites	173.5	LOWER INFERIOR OOLITE	GRANTHAM NORTHAMPTON IRONSTONE	DOGGER BLEA WYKE STRIATULUS	LOWER			ARATAURAN			
			Graphoceras concavum				PEAK SHALES ALUM SHALE JET ROCK GREY SHALES							
			Leioceras opalinum	178.0	BRIDPORT DOWN CLIFF		CLEVELAND IRONSTONE STAITHES							
		Toarcian (Toa)	Dumortieria levesquei		JUNCTION BED		IRONSTONE SHALES PYRITOUS SHALES							
			Dactylioceras tenuicostatum	187.0	(MARLSTONE ROCK)	MAIN NODULE	SILICEOUS SHALES				OTAPIRIAN			
	Lias (Lia)	Pliensbachian (Plb)	Pleuroceras spinatum		GREEN AMMONITE BELEMNITE MARLS ETC. BLACK VEN MARLS	SANDROCK FERRUGINOUS LST	CALCAREOUS SHALES							
			Uptonia jamesoni	194.5		BUCKLANDI CLAYS 'GRANBY LSTS'								
		Sinemurian (Sin)	Echioceras raricostatum		SHALES WITH BEEF	'ANGULATA CLAYS' 'HYDRAULIC LSTS'								
			Arietites bucklandi	203.5	BLUE LIAS									
	J1	Hettangian (Het)	Schlotheimia angulata											
			Psiloceras planorbis	208	WHITE LIAS									
Tr	Tr3	Rhaetian	Choristoceras marshi											

Figure 3.10. Jurassic planktonic zonation.

JURASSIC PLANKTONIC ZONATIONS

Period	Epoch	Stage	Substage	AMMONITE ZONES (from Cope, *et al* 1980 a & b)	DINOFLAGELLATE ZONES (from Woollam & Riding 1983)	NANNOFOSSIL ZONES (from van Hinte, 1978b)
Jurassic	Malm	Tithonian	Volgian (Vol); Portlandian (Por)	*Subcraspedites lamplughi*	*Gochteodinia villosa*	*Nannoconus colomi*
				Subcraspedites preplicomphalus		
				Subcraspedites primitivus		
				?Titanites (Paracraspedites) oppressus		
				Titanites anguiformis	*Ctenidodinium cumulum/Ctenidodinium panneum*	
				Galbanites (Kerberites) kerberus		
				Galbanites okusensis		
				Glaucolithites glaucolithus		
				Progalbanites albani		
			Kimmeridgian Late (Kim)	*Virgatopavlovia fittoni*	*Glossodinium dimorphum/Dingodinium tuberosum*	*Parhabdolithus embergeri*
				Pavlovia rotunda		
				Pavlovia pallasioides		
				Pectinatites (Pectinatites) pectinatus		
				Pectinatites (Arkellites) hudlestoni		
				Pect (Virggatosphinctoides) wheatleyensis		*Watznaueria communis*
				Pectinatites (Virgato.) scitulus		
		Tth		*Pectinatites (Virgato.) elegans*		
		Kimmeridgian	Kimmeridgian Early	*Aulacostephanus autissiodorensis*		
				Aulacostephanus eudoxus	*Scriniodinium luridum*	
				Aulacostephanoides mutabilis		
				Rasenia cymodoce		
		Kim		*Pictonia baylei*	*Gonyaulacysta jurassica/Scriniodinium crystallinium*	
		Oxfordian	Late	*Amoeboceras rosenkrantzi*		*Vekshinella stradneri*
				Amoeboceras regulare		
				Amoeboceras serratum		
				Amoeboceras glosense		
			Mid	*Cardioceras tenuiserratum*		
				Cardioceras densiplicatum	*Acanthaulax senta*	
			Early	*Cardioceras cordatum*		*Actinozygus geometricus*
	Mlm	Oxf		*Quenstedtoceras mariae*	*Wanaea fimbriata*	*Diadozygus dorsetense*
	Dogger	Callovian		*Quenstedtoceras (Lamberticeras) lamberti*	*Wanaea thysanota*	*Discorhabdus jungi*
				Peltoceras athleta		
				Erymnoceras coronatum	*Ctenidodinium ornatum/Ctenidodinium continuum*	*Podorhabdus rahla*
				Kosmoceras (Gulielmites) jason		*Podorhabdus escaigi*
				Sigaloceras calloviense		*Stephanolithion bigoti*
		Clv		*Macrocephalites (M.) macrocephalus*		*Stephanolithion hexum*
		Bathonian	Late	*Clydoniceras (Clydoniceras) discus*	*Ctenidodinium combazii/Ctenidodinium sellwoodii*	*Stephanolithion speciosum var. octum*
				Oppelia (Oxycerites) aspidoides		
				Procerites hodsoni		*Diazomatolithus lehmani*
			Mid	*Morrisiceras (Morrisiceras) morrisi*		
				Tulites (Tulites) subcontractus		
				Procerites progracilis		
			Early	*Asphinctites tenuiplicatus*		
		Bth		*Zigzagiceras (Zigzagiceras) zigzag*		
		Bajocian	Late	*Parkinsonia parkinsoni*	*Nannoceratopsis gracilis* (range, down to *Pleuroceras spinatum*); *Acanthaulax crispa*	*Stephanolithion speciosum s.s.*
				Strenoceras (Garantiana) garantiana		
				Strenoceras subfurcatum	*Nannoceratopsis gracilis*	
			Early	*Stephanoceras humphriesianum*		
				Emileia (Otoites) souzei		
				Witchellia laeviuscula		
		Baj		*Hyperlioceras discites*		
		Aalenian		*Graphoceras concavum*	*Mancodinium semitabulatum*	
				Ludwigia murchisonae		
	Dog	Aal		*Leioceras opalinum*		*Discorhabdus tubus*
	Lias	Toarcian		*Dumortieria levesquei*		*Podorhabdus cylindratus*
				Grammoceras thouarsense		
				Haugia variabilis		
				Hildoceras bifrons		
				Harpoceras falciferum		
		Toa		*Dactylioceras tenuicostatum*		
		Pliensbachian		*Pleuroceras spinatum*	*Luehndea spinosa*	
				Amaltheus margaritatus		
				Prodactylioceras davoei	*Liasidium variabile*	
				Tragophylloceras ibex		*Crepidolithus crassus*
		Plb		*Uptonia jamesoni*		
		Sinemurian	Late	*Echioceras raricostatum*		
				Oxynoticeras oxynotum		*Palaeopontosphaera dubia*
				Asteroceras obtusum		
			Early	*Caenisites turneri*	*Dapcodinium priscum*	*Parhabdolithus liasicus*
				Arnioceras semicostatum		*Parhabdolithus marthae*
		Sin		*Arietites bucklandi*		*Crucirhabdus primulus*
		Hettangian		*Schlotheimia angulata*		*Annulithus arkelli*
				Alsatites liasicus		
J	Lia	Het		*Psiloceras planorbis*		

Tithonian boundary corresponding to that of the Hybonotum Chron. The middle and upper divisions of the original tripartite Kimmeridgian Stage are now described by correlation as of Tithonian age.

Tithonian, from Tithon, spouse of Eos (Aurora), Goddess of Dawn (of Cretaceous Period), was defined by Oppel in 1863, in the Mediterranean area, to include all deposits which lie between a restricted Kimmeridgian (as used here) and 'Valanginian', with its lower boundary coinciding with the base of the Gravesiana Zone (corresponding to Elegans–Hybonotum Zone of the zonation used here). No representative section was designated. Tithonian is preferred to Volgian (e.g. of Cope *et al.* 1980a) because it is based on Tethyan, rather than Boreal, faunas and so serves better as a standard for correlating northern and southern hemispheres. It is now regarded as being followed naturally by the earliest Cretaceous stage (Berriasian). The two versions of the initial boundary of the Volgian, shown on Figure 3.10, are from Gerasimov *et al.* (1975), with the earlier boundary being that used by Russian workers in the northern Urals, and the later one being that which has been used by Casey (1963, 1967) for the English succession. Arkell's (1933 and 1956) scheme of Jurassic stages used the full tripartite Kimmeridgian followed by Portlandian and Purbeckian. His Portlandian is thus contained entirely within the Tithonian age span, while his Purbeckian strata in England are not a satisfactory basis for international correlation, being largely of non-marine facies.

3.16.5 The Jurassic charts (Figures 3.9 and 3.10)

The three British columns in Figure 3.9 are based on Cope *et al.* (1980a, 1980b); Siberia is from Kryngol'ts (1972); Sichuan, China, from Wang & Liu (1980); Cutch, India, from Arkell (1956, p.386); New Zealand from Suggate *et al.* (1978); East Greenland from Surlyk (1977); Canadian Arctic Archipelago from Johnsen & Hills (1973); and Utah/Idaho, USA, from Imlay 1952.

Figure 3.10 summarizes Jurassic planktonic zonation.

3.17 The Cretaceous Period

3.17.1 History and classification

Chalk characterizes the Anglo–Paris–Belgian area and was the basis for the Cretaceous System as one of five major terrains set up by J. J. d'Omalius d'Halloy in 1822. In 1823 he defined as Cretaceous terrains those corresponding 'to the formation of the chalk, with its tufas, its sands and its clays'.

William Smith had already mapped four strata between the 'lower clay' (Eocene) and the 'Portland Stone' (Jurassic) namely: 'White Chalk, brown or grey chalk, Greensand and Micaceous clay or brick earth' (the last later referred to as Blue Marl and in 1788 as Gault). In 1822 Conybeare & Phillips listed these in two groups, Chalk and the formations below, so a two-fold division, adopted in England and France at an early stage, has persisted in the two epochs or sub-periods commonly used.

In California the Shasta Series was recognized in 1869 (by W. M. Gibb) as early Cretaceous, and in 1887 P. T. Hill showed that the Comanche Series in Texas was of similar age and older than the late Cretaceous Gulf Series. Chamberlin &

Salisbury in 1906 proposed a Comanchean System, but the USGS has used the name as a provincial series of lesser span. Comanche and Gulf have been suggested as suitable international names for the two major time divisions of the Cretaceous Period (Wilmarth 1925) and Gulf serves this purpose for the second epoch. Comanche could be released for international use if the USGS were to redefine its regional rock succession.

Alternative three-fold divisions have also been proposed, for example by Leymerie in 1841 who introduced Neocomian for the lower division and the lower white chalk as its upper division. D'Orbigny developed this into five stages: Neocomian, Aptian, Albian, Turonian and Senonian; he later added Urgonian (approximately = Barremian) and Cenomanian. The current complement of 12 stages into which this developed is now accepted internationally. The well-established names Neocomian and Senonian need to be clearly defined anew, because of different usages; for example, Haug included Albian to Turonian as Middle Cretaceous (Zittel 1901). Because of this three-fold tendency these divisions are included in Figures 3.11 and 3.12 and defined below.

Neocomian division

Ever since Thurmann, in 1836, coined the name Neocomian for strata in the vicinity of Neuchâtel, Switzerland, there has been confusion as to how much of the early Cretaceous Period should be embraced by the name. Some authors have used it to signify earliest Cretaceous up to and including the Valanginian Stage; others have extended it to include Barremian and even Aptian stages. The recommendation of Barbier, Debelmas and Thieuloy in the Colloque sur le Crétacé inférieur (Barbier & Thieuloy 1965) is adopted here, namely that Neocomian be regarded as an informal term embracing the lowest three Cretaceous stages, i.e. Berriasian, Valanginian and Hauterivian.

Gallic division

For those who prefer a three-fold Cretaceous division, and to avoid ambiguity with Early, Middle and Late, Gallic is introduced provisionally to complete the trio and comprises five stages: Barremian, Aptian, Albian, Cenomanian and Turonian. This follows from the definition of Neocomian adopted here.

The five stages were named for places throughout France so no lesser area is appropriate. Gaul, with its tripartite connotation (Caesar, Gallic Wars, Book I first sentence), comprehends these original type localities. Gallic would include Comanchean in its current usage.

Senonian division

Senonian, like Neocomian, is an informal name which has signified different things to different authors. The question was whether or not the Maastrichtian is included. According to the *Lexique stratigraphique international* (Vol.1, Europe, fasc. 4a, Crétacé, p.318) 'dans le sens de d'Orbigny (1842), createur du terme, le Sénonien correspond à l'ensemble des couches comprises entre le Turonien et le Danien, c'est-à-dire qu'il débute avec la craie de Villedieu et se termine avec la craie de Maestricht'. This is the sense in which Senonian is regarded in this work, i.e. comprising four stages: Coniacian, Santonian, Campanian and Maastrichtian (Rawson 1983).

Leaving aside further discussion of the complex history of Cretaceous stages, the sequence of time divisions that has come to be accepted internationally follows. The account is revised from GTS 82 in the light of comments from N. F. Hughes and from P. F. Rawson (1983) of the 'Pre-Albian Stages Working Group' of the ICS Cretaceous Subcommission.

3.17.2   Early Cretaceous Epoch

No suitable name for this division has been proposed. Comanchean is a possible contender but has a much more limited application in North America. Six stages are well established as follows.

The **Berriasian Stage** was proposed by H. Coquand in 1871. The body stratotype is near the village of Berrias, Ardeche, southeast France. This is the initial Cretaceous stage and is adopted here in preference to the Boreal Ryazanian division because it was defined in the Tethyan province, i.e. for the same reason that Tithonian was preferred to Portlandian or Volgian. Originally conceived as a subdivision of Valanginian, it was subsequently often referred to as 'Infra-Valanginian' until the name 'Berriasian' was eventually brought back into use. The initial Berriasian boundary thus equals the initial Cretaceous boundary; it would be placed at, or near, the introduction of *Berriasella grandis*, but there is not yet a defined marker point in the boundary stratotype.

A study of nannofossil zonation in relation to ammonite *Buchia*, calpionellid and dinoflagellate zones, and magnetic anomaly polarity, by Bralower, Monechi & Thiestein (1989), has greatly increased resolution, precision and reliability of correlation especially around the Jurassic–Cretaceous boundary. They showed *Durangites* as the latest Tithonian ammonite zone followed by *Berriasella jacobi*, then *B. grandis* as the earliest Berriasian. *Buchia* aff. *B. skansis* gives place at about the same level to *B. uncitoides* as does *Crassicolaria* to *Calpionellaalpina*. So characterized, this boundary would fall within the newly proposed nannofossil zone of *Microstaurus chiastius* and its subzone *Rotelapillus laffitei* and approximately at the transition to magnetic anomaly polarity chron CM18. The study showed that several nannofossil events fall approximately in each chron at about this time of rapid increase in nannofossil abundance and diversity.

The **Valanginian Stage** was proposed by E. Desor in 1853, with a type locality at the Seyon Gorge, near Valangin, Neuchâtel, Switzerland (Valendis in German). Desor defined Valanginian as comprising all post-Jurassic rocks to the base of the 'Marnes de Hauterive' (= approximately Neocomien inferieur of Compiche), and it was better defined by Desor & Greesly in 1859.

The original Swiss stratotype is incomplete and not suitable for wider correlation and most biostratigraphic work has been done across the French Vocontian trough. Valanginian has been divided into early and late sub-stages, each with three biozones. The initial Valanginian boundary has been suggested at approximately the base of the old *roubaudiana* zone equivalent to the base of the *Kilianella pertransens* or *Thurmanoceras otopeta* zone. The initial Late

Valanginian boundary would be around the sudden appearance of *Saynoceras verrucosum* (Rawson 1983). A reference point remains to be selected.

The **Hauterivian Stage** was proposed by Renevier in 1874, from Hauterive, near Neuchâtel, Switzerland. Two sub-stages have been proposed. Early Hauterivian (with four biozones) begins with *Acanthodiscus radiatus* which is approximately equivalent to *Endomoceras amblygonium*. The Late Hauterivian sub-stage (with three biozones) begins with *Subsaynella sayni*, and is characterized by *Simbirskites* (Rawson 1983). No reference points have yet been designated.

The **Barremian Stage** was named by Coquand in 1861 who mentioned the localities at Barrême, near Digne and Angles (Basses-Alpes, southeast France). Busnardo (1965) subsequently designated the Angles roadside section as the stratotype. There is probably scope for an initial boundary point in that section, at or near the introduction of *Raspailiceras* and *Barremites*.

The **Aptian Stage** was proposed by d'Orbigny in 1840 for strata containing an 'Upper Neocomian' fauna, and named after the village of Apt, Basses-Alpes, southeast France. Three sub-stages are based on the Bedoulian (Toucas 1888) with seven ammonite zones, Gargasian (Kilian 1887) and Clansayesian (Breistroffer 1947); these names are concerned principally with the Urgonian limestone facies (see Figure 3.11) and are not applied beyond that area. The Aptian initial boundary would be near the arrival of *Prodeshayesites fissicostatus* and *Deshayesites deshayesi*.

The **Albian Stage** was proposed by d'Orbigny in 1844 for the interval between Aptian and what is now Cenomanian. Its name was derived from Alba, the Roman name for Aube, France. It appears to be a stage of considerable duration, so a three-fold subdivison into Early, Mid and Late is generally adopted. Following Breistroffer (1947) its initial boundary is regarded as commencing with beds containing *Leymeriella tardefurcata*. A composite stratotype has been proposed by Larcher, Rat & Malapris (1965) based on several exposures in the Aube area. The latest part of the Late Albian chron was referred to as 'Vraconian' by Renévier in 1867; this term is now regarded as being superfluous, as it is merely another name for the time interval represented by Spath's (1923) *Stoliczkaia dispar* zone (Rawson 1983).

3.17.3   The Gulf (Late Cretaceous) Epoch

For a two-fold Cretaceous division Gulf is the later epoch. It avoids the ambiguity of 'Late Cretaceous'. Six stages are well established.

The **Cenomanian Stage** was proposed by d'Orbigny in 1847 with a type locality in the environs of Le Mans (Roman Cenomanum), in Sarthe, France; however, he did not designate a type section. More recently Marks (1967) proposed a composite section in the area of St Ulphace-Theligny-Moulin de l'Aunay as the stratotype for the Cenomanian stage. The Cenomanian initial boundary would be defined near the arrival of *Mantelliceras mantelli*.

The **Turonian Stage** was introduced by d'Orbigny in 1842 but in 1847 he separated the lower part as Cenomanian. The name Turonian is derived from Tours (Roman Turones) or Touraine (Roman Turonia). No stratotype was proposed,

Figure 3.11. The Cretaceous Period and selected successions. (The numerical ages are rounded to 0.5 Ma intervals.)

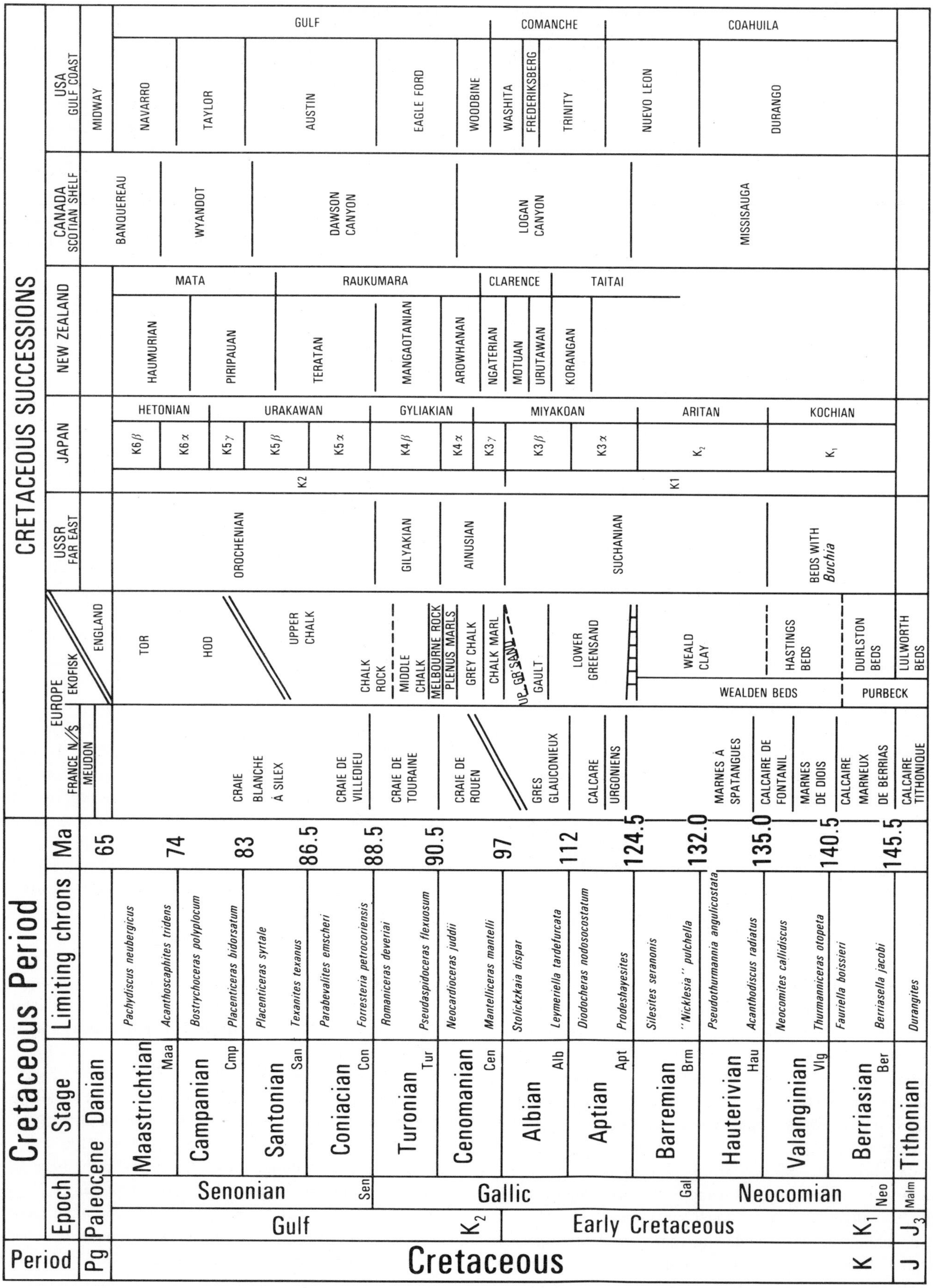

CRETACEOUS SUCCESSIONS

USA Gulf Coast

Major groups: GULF · COMANCHE · COAHUILA

MIDWAY, NAVARRO, TAYLOR, AUSTIN, EAGLE FORD, WOODBINE, WASHITA, FREDERIKSBERG, TRINITY, NUEVO LEON, DURANGO

Canada Scotian Shelf

BANQUEREAU, WYANDOT, DAWSON CANYON, LOGAN CANYON, MISSISAUGA

New Zealand

Groups: MATA · RAUKUMARA · CLARENCE · TAITAI

Stages: HAUMURIAN, PIRIPAUAN, TERATAN, MANGAOTANIAN, AROWHANAN, NGATERIAN, MOTUAN, URUTAWAN, KORANGAN

Japan

HETONIAN, URAKAWAN, GYLIAKIAN, MIYAKOAN, ARITAN, KOCHIAN

$K6\beta$, $K6\alpha$, $K5\gamma$, $K5\beta$, $K5\alpha$, $K4\beta$, $K4\alpha$, $K3\gamma$, $K3\beta$, $K3\alpha$, K_2, K_1 (K2 / K1)

USSR Far East

OROCHENIAN, GILYAKIAN, AINUSIAN, SUCHANIAN, BEDS WITH *Buchia*

Europe

Ekofisk / England: TOR, HOD, UPPER CHALK, CHALK ROCK, MIDDLE CHALK, MELBOURNE ROCK, PLENUS MARLS, GREY CHALK, CHALK MARL, UP. GRSAND, GAULT, LOWER GREENSAND, WEALD CLAY, HASTINGS BEDS, DURLSTON BEDS, LULWORTH BEDS, WEALDEN BEDS, PURBECK

France N/S (Meudon): CRAIE BLANCHE À SILEX, CRAIE DE VILLEDIEU, CRAIE DE TOURAINE, CRAIE DE ROUEN, GRES GLAUCONIEUX, CALCAIRE URGONIENS, MARNES A SPATANGUES, CALCAIRE DE FONTANIL, MARNES DE DIOIS, CALCAIRE MARNEUX DE BERRIAS, CALCAIRE TITHONIQUE

Cretaceous Period

Period	Epoch	Stage	Limiting chrons	Ma
Pg	Paleocene Danian	Danian		65
Cretaceous	Senonian (Sen) / Gulf (K_2)	Maastrichtian (Maa)	*Pachydiscus neubergicus*	
			Acanthoscaphites tridens	74
		Campanian (Cmp)	*Bostrychoceras polyplocum*	
			Placenticeras bidorsatum	83
		Santonian (San)	*Placenticeras syrtale*	
			Texanites texanus	86.5
		Coniacian (Con)	*Parabevahlites emscheri*	
			Forresteria petrocoriensis	88.5
		Turonian (Tur)	*Romaniceras deveriai*	
			Pseudaspidoceras flexuosum	90.5
			Neocardioceras juddii	
	Gallic (Gal) / Early Cretaceous	Cenomanian (Cen)	*Mantelliceras mantelli*	97
		Albian (Alb)	*Stolickzkaia dispar*	
			Leymeriella tardefurcata	112
		Aptian (Apt)	*Diadocheras nodosocostatum*	
			Prodeshayesites	124.5
		Barremian (Brm)	*Silesites seranonis*	
			"*Nicklesia*" *pulchella*	132.0
	Neocomian (Neo) / K_1	Hauterivian (Hau)	*Pseudothurmannia angulicostata*	
			Acanthodiscus radiatus	135.0
		Valanginian (Vlg)	*Neocomites callidiscus*	
			Thurmanniceras otopeta	140.5
		Berriasian (Ber)	*Fauriella boissieri*	
			Berriasella jacobi	145.5
J	J_3 (Malm)	Tithonian	*Durangites*	

Figure 3.12. Cretaceous biostratigraphic zonation. (The numerical ages are rounded to 0.5 Ma intervals.)

CRETACEOUS BIOSTRATIGRAPHIC ZONATIONS

Epoch	Stage	TETHYAN PELAGIC MACROFOSSIL ZONES (FROM VAN HINTE, 1978a)		FORAMINIFERAL ZONES (SIGAL, 1977)	AGE IN Ma	CALCAREOUS NANNOPLANKTON ZONES (SISSINGH, 1977)
Late	Maastrichtian L	Pachydiscus neubergicus		Glt. mayaroensis	65	26 Nephrolithus frequens
						25 Arkhangelskiella cymbiliformis
				Glt. gansseri		24 Reinhardtites levis
	E Maa	Acanthoscaphites tridens		Glt. stuarti/ Glt. falsostuarti	74	23 Tranolithus phacelosus
	Campanian L	Bostrychoceras polyplocum		Glt. calcarata		22 Tetralithus trifidus
		Hoplitoplacenticeras vari		Glt. elevata/Glt. stuartiformis		21 Tetralithus nitidus
		Delawarella delawarensis				20 Ceratolithoides aculeus
						19 Calculites ovalis
	E Cmp	Placenticeras bidorsatum			83	18 Aspidolithus parcus
						17 Calculites obscurus
	Santonian L	Placenticeras syrtale		Glt. concavata carinata		16 Lucianorhabdus cayeuxii
		Eupachydiscus isculensis				15 Reinhardtites anthrophorus
	E San	Texanites texanus (Inoceramus undulatoplicatus)			86.5	14 Micula staurophora
	Coniacian L	Parabevalites emscheri (Protexanites, Paratexanites) T. pseudotexanus)		Glt. concavata		13 Marthasterites furcatus
	E Con	Forresteria petrocoriensis				12 Lucianorhabdus maleformis
				Glt. sigali/Glt. schneegansi	88.5	
	Turonian L	Romaniceras deveriai				
	M	Romaniceras ornatissimum Romaniceras bizeti		Glt. helvetica		11 Tetralithus pyramidus
	Tur E	Pseudaspidoceras flexuosum (Kanabiceras septemseriatum, Metoicoceras whitei, Inoceramus labiatus)		Wh. arcchaeocretacea	90.5	
	Cenomanian L	Neocardioceras juddii		Rtl. cushmani		10 Microrhabdulus decoratus
	M	Acanthoceras rhotomagense				
	Cen E	Mantelliceras mantelli		Rtl. globotruncanoides Rtl. brotzeni		9 Eiffellithus turriseiffeli
Early	Albian L	Stoliczkaia dispar		Rot. appenninica/Pl. buxtorfi	97	
		Mortoniceras inflatum		Tic. breggiensis		
		Diploceras cristatum		Hed. rischi/Tic. primula		8 Prediscosphaera cretacea
	M	Hoplites lautus/H. nitidus				
		Hoplites dentatus		Hed. planispira		
	Alb E	Douvilleiceras mammilatum				
		Leymeriella tardefurcata		Tic. bejaouaensis	112	
	Aptian L	Diodochoceras nodosocostatum		Hed. trochoidea.		7 Chiastozygus litterarius
		Cheloniceras subndosocostatum		Gld. algeriana.		
	E	Aconoceras nisus		Gld. ferreolensis Schk. cabri		
	Apt	Prodeshayesites		Gld. maridelensis/Gld. blowi Gld. gottisi/Gld. duboisi	124.5	
	Barremian L	Silesites seranonis	Pulchella provincialis	Hed. similis Ctes. aptiensis (? Ctes. intercedens)		6 Micrantholithus obtusus/hoschulzii
	E	"Nicklesia" pulchella	Pulchella caicedi	Clh. eocretacea		
			Pulchella didayi			5 Lithraphidites bollii
	Brm		Pulchella pulchella	Hed. sigali	132	
	Hauterivian L	Pseudothurmannia angulicostata		Ctes. bartensteini Gav. gr. diaffaensis – sigmoicosta		
		Subsaynella sayni		Cauc. gr. hauterivica		4 Cretarhabdus loriei
	E	Crioceras duvali		Doroth. ouachensis		
	Hau	Acanthodiscus radiatus		Hapl. vocontianus		
				Lent ouachensis var bartensteini/Doroth. hauteriviana	135	
	Valanginian L	Neocomites callidiscus		Lent. gr. eichenbergi-meridiana		3 Calcicalathina oblongata
		Himantoceras trinodosum				
		Saynoceras verrucosum		Lent. busnardoi		
		Killianella campylotoxa				
	E	Killianella roubaudi		Lent. guttata		2 Cretarhabdus crenulatus
	Vlg	Thurmanniceras otopeta			140.5	
	Berriasian	Ryazanian	Fauriella boissieri			1 Nannoconus steinmanni
	Ber	Volgian	Berriasella jacobi		145.5	

The key to abbreviations of generic names in Sigal's foraminiferal zonation is as follows: *Glt.* – *Globotruncana*, *Wh.* – *Whiteinella*, *Rtl.* – *Rotaliporal*, *Pl.* – *Planomalina*, *Hed.* – *Hedbergella*, *Tic.* – *Ticinella*, *Gld.* – *Globigerinelloides*, *Schk.* – *Schackoina*, *Ctes.* – *Comorotalites*, *Clh.* – *Clavihedbergella*, *Gav.* – *Gavelinella*, *Cauc.* – *Caucasella*, *Doroth.* – *Dorothia*, *Lent.* – *Lenticulina*.

but in 1852 d'Orbigny designated the type area as lying between Saumur (on the river Loire) and Montrichard (on the river Cher), France.

The lowest formation of the type Turonian succession, within d'Orbigny's geographical limitations, contains *Mammites nodosoides*. The initial appearance of *M. nodosoides* is taken as the initial Turonian boundary; the *M. nodosoides* zone is recognizable over much of the world. Marks in 1977 proposed the first appearance of the planktonic foraminifer *Praeglobotruncana helvetica* as indicative of this same boundary. Unfortunately, *Praeglobotruncana helvetica*, widely regarded as an index form for the Turonian in Tethyan regions, has not yet been found in Touraine.

The **Coniacian Stage** was introduced in 1856, when Coquand divided the 'Upper Chalk' of the Saintogne (Charente-maritime, France) into three stages with the lowest being subdivided into three sub-stages. In 1857, he raised the lowest two sub-stages to stage rank, as the Coniacian, with the stratotype being the section described by Coquand in 1857 at Cognac, Charente (hence the name), in the northern part of the Aquitaine Basin, France. The stratotype appeared to contain no foraminifers or nannofossils. No fossils of correlative value have been reported from the basal sands at Cognac. However, the initial Coniacian is by definition also the initial Senonian boundary, the standard formation for which is the Craie de Villedieu, of Touraine, whose lowest member contains an assemblage containing *Barroisiceras haberfellneri*. The initial appearance of this *Barroisiceras* species is thus regarded as indicating the initial Coniacian boundary.

The **Santonian Stage** is the uppermost 'sub-stage' of Coquand (in 1856) and was raised to stage rank in 1857 (and named after the village of Saintes, France). In 1858, Coquand designated a section of the road between Javresac and Saintes as the stratotype. At present it is not possible to determine faunally the Santonian base in its stratotype area.

The **Campanian Stage** is the second stage of Coquand who introduced the name 'Campanian' as a stage in 1857. In 1858 Coquand made it clear that the stratotype was the hillside section at Aubeterre-sur-Dronne (La Grande Champagne – hence 'Campanian'). De Grossouvre, between 1895 and 1901, worked out possible ammonite zonations leading to the present one. The initial Campanian boundary approximates to the base of the *Placenticeras bidorsatum* zone. The stage itself does not seem to be well founded although many zones have been established. Recent work on the microfauna of the sections at Aubeterre has shown that the bulk of the type Campanian is actually Maastrichtian as now understood (Rawson *et al.* 1978). Seronie-Vivien designated parastratotype sections near Ne in the Charente valley, but since the macrofossils there are scarce and undistinctive, they do not provide a working standard for the Campanian base (Rawson *et al.* 1978).

The **Maastrichtian Stage** was introduced in 1849 by Dumont who separated the 'Calcaire de Maastricht' from the 'craie senonienne'. The stratotype has now been fixed by the Comité d'étude du Maastrichtian, as units Ma-Md of Uhlenbroek in 1911 – exposed in the ENCI quarry at St Pietersburg on the outskirts of Maastricht (in South Holland).

The type Maastrichtian corresponds only to the later Maastrichtian of current usage. A revised concept, based on the belemnite succession, regards the Maastrichtian as being coextensive with the range of *Scaphites (Hoploscaphites) constrictus*, with its initial boundary marked by the entry of *Belemnitella lanceolata* (Rawson *et al.* 1978).

3.17.4 The Cretaceous charts (Figures 3.11 and 3.12)

The Cretaceous ammonite zones shown in Figure 3.12 are taken from the compilation of Van Hinte (1978a, with sources cited). Of the many Cretaceous zonal schemes that have been erected on the basis of planktonic foraminifera, that of Sigal (1977) has been selected here for inclusion in Figure 3.12; this zonation is particularly applicable to the Mediterranean area in general. The calcareous nannoplankton zonation in Figure 3.12 is that of Sissingh (1977) and is calibrated, as far as is possible, with the planktonic foraminiferal zonation.

Figure 3.11 was compiled with the assistance of N. F. Hughes, for the European columns. The USSR column is from Nalivkin (1973, tables 55 and 56); the Japan column is from Takai, Matsumoto & Toriyama (1963); the New Zealand column is from Stevens (1980); the Canada column from Ascoli (1976); and the USA column from Murray (1961) and Postuma (1971).

3.18 The Tertiary Sub-era

The name '**Tertiary**' is a survivor from an early attempt at stratigraphic classification, i.e. Primary, Secondary, etc. (now otherwise obsolete). It was applied in Italy by Arduino in 1759 to rocks that are still recognized as Tertiary. Brogniart, in 1810, applied the term 'tertiaire' to strata which overlay the Cretaceous chalks in the Paris Basin. In its present usage the name is useful and unambiguous, referring to the post-Mesozoic–pre-Quaternary interval (Lyell 1832).

The word '**Neozoic**' was used by Lyell and his contemporaries to embrace all systems from Triassic onwards. The intention at that time was to complement Paleozoic. Then J. Phillips, in 1841, divided these systems between his Mesozoic and Cainozoic groups, and the name 'Neozoic' has fallen into disuse through being used variously as a synonym of 'Tertiary' and also of 'Cenozoic' (Tertiary + Quaternary); otherwise it would be a useful name and it might even return for post-Paleozoic.

Lyell, in 1833, divided European Tertiary into **Newer Pliocene**, **Older Pliocene**, **Miocene** and **Eocene**, based on the percentage of living amongst living plus extinct mollusc species. Further studies in the middle of the nineteenth century established the presence of extensive marine, brackish, freshwater and continental sediments in northern Europe, between Lyell's Eocene and Miocene; for these Beyrich, in 1854, proposed the name '**Oligocene**'. The Tertiary classification attained its modern aspect in 1874 when Schimper distinguished the early 'Eocene' as '**Paleocene**'. This was based on younger Paleocene strata in western Europe and was distinguished more by facies than by an evolutionary sequence; so separation from Eocene strata was hardly justified by the initial descriptions. But when the American Paleocene, with richly fossiliferous strata, could be considered, the division was well supported.

The two-fold classification of Tertiary time derives from

European stratigraphy where it is divided by the climax of the Alpine Orogeny. The older division was indeed Lyell's original Eocene and the younger was his Miocene and Pliocene. To combine the two latter, Hornes, in 1853 introduced the name **Neogene**. He also proposed **Paleogene** which was at first synonymous with Eocene. It had more use when Oligocene was distinguished, and it came into its own after 1874 when Paleogene usefully combined Paleocene, Eocene and Oligocene. These divisions became established in Europe: for example E. Suess established his five Mediterranean stages which divided the Neogene within Alpine Europe, and in France in 1902 E. Haug used the name **Nummulitique** (for Paleogene) and this usage continued there for 50 years. The division of Tertiary into the Paleogene and Neogene periods is now well established ICS practice and is followed here.

A comprehensive review of the historical development of the numerous Tertiary 'stages' was provided by Berggren (1971). See also the discussion of Cenozoic above (Section 3.8.4).

3.19   The Paleogene Period

3.19.1   Introduction

The definition of stages by boundary stratotypes is not easily carried out in the Paleogene basins of western Europe. A major reason is the lack of continuous sections at the boundaries; instead there are many well-known but relatively small outcrops scattered throughout these basins, mainly in France, England, Belgium, Germany and Denmark. The intercalation of marine, marginal-marine and non-marine sediments, the lateral facies changes and the nature of the fossil assemblages themselves, all combine to render inter-basin correlation difficult, and the possibility of world-wide correlation is even more remote. These factors have hindered the correlation and relative positioning of the various proposed stratotypes and have given rise to a bewildering number of stage names; 'most of the European Tertiary stage names in common use today were originally defined as lithostratigraphical (facies) units (i.e. formations within cycles of sedimentation' (Berggren 1971, p.696). In Appendix 1 the origins of 50 Paleogene stage names are listed.

The development, over the last three decades or so, of a biostratigraphic zonation based upon planktonic foraminifers has allowed detailed and long-ranging correlations. Such a zonation was initially restricted to low-to-middle latitudes; but with the addition of a calcareous nannoplankton zonation and, latterly, of a radiolarian zonation, the biostratigraphic scheme can be extended into higher latitudes and into the deep ocean (DSDP cores). The integration of these zonal schemes forms a biostratigraphic framework into which the various European Paleogene stratotypes may be placed in their most plausible relative positions. Examples of this are shown in Hardenbol & Berggren (1978, figure 4) and Aubry (1983). Until more precise standardization has been achieved some stages are redefined in terms of the biostratigraphic zonal schemes as tabulated in Figure 3.13. Figure 3.14 lists the nominate taxa upon which the P-zone scheme of Blow (1969) is based, together with the calcareous nannofossil taxa typifying the NP-zonal scheme (after Martini 1971). The zonation of Blow

(1969, 1979) is based on planktonic foraminifers. In addition to referring to the Cenozoic zones by their nominate taxon (or taxa), Blow numbered them in 'ascending' order in two groups. The prefix 'P' is applied to the Paleogene zones (P1 to P22) and the prefix 'N' to Neogene and Quaternary zones (N4 to N25). Martini (1971) similarly numbered his nannoplankton zones with the prefix 'NP' for the Paleogene zones (NP1 to NP25), and 'NN' for the Neogene (and Quaternary) zones (NN1 to NN25).

3.19.2   The initial Paleogene boundary

Along with much publicity for catastrophic events at around the Mesozoic–Cenozoic boundary, the official Working Group on the Cretaceous–Paleogene Boundary is still working on precise correlation standards before recommending a reference point in a stratotype. It seems that no formal recommendation has yet been made. Thomsen in 1981 suggested a boundary stratotype at Nye Klor (Jutland) rather than at Stevns Klint (Zealand).

The three epochs of the Paleogene Period are considered in sequence: Paleocene, Eocene and Oligocene.

3.19.3   The Paleocene Epoch

In Europe, this epoch is represented by two main lithostratigraphic units:

(1) a lower, carbonate unit, on which the Danian Stage is based, and

(2) an upper, clastic unit, on which the Thanetian Stage is based.

The **Danian Stage**, named by Desor in 1847, has type localities at Stevns Klint and Faxse (= Faxe = Faxie), Denmark. Danian (with typical chalk facies) was commonly considered as the latest Cretaceous division, to be succeeded by Montian (Belgium) as the earliest Tertiary stage. When it was shown that Montian was a different facies of Danian age the debate as to whether to assign it to the Mesozoic or Cenozoic Era was settled in favour of the latter (about 1960) probably because already the equivalent Midway rocks (with rich faunas) in the Gulf region of the USA were widely regarded as basal Tertiary. Danian has few equivalent stages in Europe (Montian in Belgium is in part equivalent to the type Danian), although sediments which have been correlated with it are widespread, especially in central and eastern Europe and the North Sea.

Hansen (1970, p.25) defined Danian as 'the time interval between the rocks found above the Maastrichtian White Chalk exposed on Stevns Klint up to and including the time of deposition of the rocks found below the Selandian basal conglomerate exposed at the locality Hvallose, in Jutland'. This definition would include part of zone P1a to P2. Martini (1971) and Perch-Nielsen (1972) have recognized zones NP1 to NP3 and an equivalent to NP4 (Perch-Nielsen 1971, 1979) from outcrops in the type area. Hardenbol & Berggren (1978) extended the definition of Danian to include all of P1 and P2 (= NP1 to NP3 part), see Figure 3.14.

The **Thanetian Stage**, proposed by Renévier in 1873, is named after the Thanet Sands, Isle of Thanet, Kent, England. Equivalents in Europe are the type **Selandian** (Denmark), type **Landenian** and type **Heersian** (Belgium), and the '**Sable de Bracheux**' (Paris Basin). Martini (1971) reported zone NP8

Paleogene Period

Period	Epoch		Stage		Ma	Some biozones — Foraminifera		Nanno-fossils	Radiolaria
Ng	Miocene	Mio	Aquitanian		23.3	N4	Globorotalia kugleri	NN1	
	Oligocene	Oli₂	Chattian	Cht		P22	Globigerina ciperoensis	NP25	12
						P21 b / a	Globorotalia opima opima	NP24	
					29.3	P20	Globigerina ampliapertura	NP23	13
		Oli₁	Rupelian	Rup		P19	Cassigerinella chipolensis		
						P18	Pseudohastigerina micra	NP22	
		Oli			35.4	P17		NP21	14
						P16	Globorotalia (T.) cerroazulensis	NP20	
	Eocene	Eoc₃	Priabonian	Prb		P15	Globigerinatheka semiinvoluta	NP19	
					38.6			NP18	
			Bartonian	Brt		P14	Truncorotaloides rohri	NP17	15 / 16 / 17
		Eoc₂			42.1	P13	Orbulinoides beckmanni	NP16	18
						P12	Morozovella lehneri		
			Lutetian	Lut		P11	Globigerinatheka subconglobata	NP15	19 / 20
					50.0	P10	Hantkenina aragonensis	NP14	21 / 22
						P9	Acarininia pentacamerata	NP13	23
		Eoc₁	Ypresian	Ypr		P8	Morozovella aragonensis	NP12	24
						P7	Morozovella formosa formosa		25
		Eoc			56.5	P6	Morozovella subbotinae	NP11 / NP10	26
						P5	Morozovella velascoensis	NP9	
	Paleocene	Pal₂	Thanetian	Tha		P4	Globorotalia pseudomenardii	NP8 / NP7	
						P3	Globorotalia pusilla pusilla	NP6 / NP5	
					60.5		Morozovella angulata	NP4	
						P2	Globorotalia (A.) praecursoria	NP3	Not zoned
		Pal₁	Danian	Dan		P1 d	Globorotalia (T.) inconstans		
						P1 c / b	Globorotalia (T.) pseudobulloides	NP2	
		Pal			65.0	P1 a	"Globigerina" eugubina	NP1	
K	Gulf	Gul	Maastrichtian				Abathomphalus mayaroensis		

PALEOGENE SUCCESSIONS

NEW ZEALAND		USA WEST COAST	U.S.S.R (a)	U.S.S.R (b)	USA GULF COAST
LANDON	WAITAKIAN	ZEMORRIAN		POLTAVA	HACKBERRYFRIO
	DUNTROONIAN				
	WHAINGAROAN			KHARKOVIAN	VICKSBURG
ARNOLD	RUNANGAN	REFUGIAN	ALMINIAN		JACKSON
	KAIATAN	NARIZIAN	BODRAKIAN (BELBEKIAN)		CLAIBORNE
	BORTONIAN				
	PORANGIAN	ULATIZIAN	SIMFEROPOLIAN	KANEV	WILCOX
	HERETAUNGAN				
	MANGAORAPAN	PENUTIAN	BAKHCH-ISARAIAN		
DANNEVIRKE	WAIPAWAN	BULITIAN	KACHIAN	KAMYSHINIAN	
				SARATOVAN	MIDWAY
	TEURIAMN	YNEZIAN	INKERMANIAN		
		'DANIAN'		SYZRANIAN	

Figure 3.14. Paleogene biostratigraphic zonation (from GTS 82 chart 3.14).

KEY TO PALEOGENE PLANKTONIC ZONATIONS		
FORAMINIFERA	**CALCAREOUS NANNOPLANKTON**	**RADIOLARIA**

FORAMINIFERA

N 3 / P 22	Globigerina angulisuturalis	P-R-Z
N 2 / P 21	Globigerina angulisuturalis / Globorotalia (T.) opima opima	Conc.-R-Z
N 1 / P 19 / P 20	Globigerina sellii / Globigerina ampliapertura	P-R-Z
P 18	Globigerina tapuriensis	P-R-Z
P 17	Globigerina gortanii gortanii / Globorotalia (T.) centralis	P-R-Z
P 16	Cribrohantkenina inflata	T-R-Z
P 15	Porticulasphaera semiinvoluta	P-R-Z
P 14	Globorotalia (M.) spinulosa spinulosa	P-R-Z
P 13	Globigerapsis beckmanni	T-R-Z
P 12	Globorotalia (M.) lehneri	P-R-Z
P 11	Globigerapsis kugleri / Subbotina frontosa boweri	Conc.-R-Z
P 10	Subbotina frontosa frontosa / Globorotalia (T.) pseudomayeri	Conc.-R-Z
P 9	Globorotalia (A.) aspensis / Globigerina lozanoi prolata	Conc.-R-Z
P 8 — 8b	Globorotalia aragonensis / Globorotalia (M.) formosa	Conc.-R-Z
P 8 — 8a	Globorotalia (M.) formosa / Globorotalia (M.) lensiformis	P-R-Z
P 7	Globorotalia (A.) wilcoxensis berggreni	P-R-Z
P 6	Globorotalia (M.) subbotinae subbotinae / Globorotalia (M.) velascoensis acuta	P-R-Z
P 5	Muricoglobigerina soldadoensis soldadoensis / Globorotalia (M) velascoensis pasionensis	
P 4	Globorotalia (G.) pseudomenardii	Conc-R-Z / P-R-Z
P 3	Globorotalia (M.) angulata angulata	P-R-Z
P 2	Globorotalia (A.) praecursoria praecursoria	P-R-Z
P 1 — 1b	Globorotalia (T.) compressa compressa / Eoglobigerina eobulloides simplicissima	Conc-R-Z
P 1 — 1a	Globorotalia (T.) pseudobulloides / Globorotalia (T.) archeocompressa	Conc-R-Z
Pα	Globorotalia (T.) longiapertura	P-R-Z

CALCAREOUS NANNOPLANKTON

NP 25	Sphenolithus ciperoensis
NP 24	Sphenolithus distentus
NP 23	Sphenolithus predistentus
NP 22	Helicopontosphaera reticulata
NP 21	Ericsonia ? subdisticha
NP 20	Sphenolithus pseudoradians
NP 19	Isthmolithus recurvus
NP 18	Chiasmolithus oamaruensis
NP 17	Discoaster saipanensis
NP 16	Discoaster tani nodifer
NP 15	Chiphragmalithus alatus
NP 14	Discoaster sublodoensis
NP 13	Discoaster lodoensis
NP 12	Marthasterites tribrachiatus
NP 11	Discoaster binodosus
NP 10	Marthasterites contortus
NP 9	Discoaster multiradiatus
NP 8	Heliolithus riedeli
NP 7	Discoaster gemmeus
NP 6	Heliolithus kleinpelli
NP 5	Fasciculithus tympaniformis
NP 4	Ellipsolithus macellus
NP 3	Chiasmolithus danicus
NP 2	Cruciplacolithus tenuis
NP 1	Markalius inversus

RADIOLARIA

Lyc.b.	Lychnocanium bipes
Dor.at.	Dorcadospyris ateuchus
The.t.	Theocyrtis tuberosa
Thy.b.	Theocyrtis bromia
Thy.t.	Thyrsocyrtis tetracantha
Pod.g.	Podocyrtis goetheana
Pod.c.	Podocyrtis chalara
Pod.m.	Podocyrtis mitra
Pod.a.	Podocyrtis ampla
Thy.tr.	Thyrsocyrtis triacantha
The.m.	Theocampe mongolfieri
The c.	Theocotyle cryptocephala
Pho.s.	Phormocyrtis striata striata
Bur.c.	Buryella clinata
Bek. b.	Bekoma bidartensis

FORAMINIFERA ; Zonation of Blow (1979)
P-R-Z = Partial Range Zone
Conc-R-Z = Concurrent Range Zone
T-R-Z = Total Range Zone

CALCAREOUS NANNOPLANKTON ;
Zonation of Martini (1971)

RADIOLARIA ;
Zonation from Hardenbol & Berggren (1978)
(In: Cohee et al., 1978)

as Thanetian, which can be correlated with zone P4 (Bramlette & Sullivan 1961). El-Naggar (1966a, 1966b) reported zone P3 in the type Heersian of Belgium. Hardenbol & Berggren (1978) extended the Thanetian to include zones P3 to lower P6, i.e. latest NP3 to NP9.

We note that Berggren et al. (1985a) used Selandian in place of Thanetian on the basis that Thanetian could be restricted in meaning so as to leave an unnamed division x between it and Danian. Thanetian and x would be their sub-stages. However, pending a decision of the ICS, no change is made here.

3.19.4 The Eocene Epoch

In western Europe Eocene consists traditionally of three chronostratic divisions, in which scheme we list four stages.

(i) Early Eocene

The **Ypresian Stage** (Dumont in 1849) is named from the Ypres Clay, south of Ypres (Ieper), Belgium. In the type area, shallow to moderately deep-water sands and clay are developed. Equivalents in other basins are London Clay (England), Rosnaes Clay (Denmark) and Eocene 3 (Germany). The Cuisian stage of the Paris Basin is represented by sandstones believed to be late Ypresian.

Zones NP11 and NP12 have been reported from the Ypresian type area (Hay & Mohler 1967, Martini 1971), and also from the Rosnaes Clay and the Eocene 3. Zone NP12 has been correlated with foraminiferal zone P8 by Hay and others in 1967. Hardenbol & Berggren (1978) extended the Ypresian chronostratic unit to include the biostratigraphically identified zone P6 (upper part) to P9, i.e. zones NP10 to NP13 (Berggren 1971, 1972).

(ii) Mid Eocene

(a) The **Lutetian Stage** (de Lapparent in 1883) has no precise body stratotype designated, but it was proposed for the 'Calcaire Grossier' of the Paris Basin. This is represented mainly by shallow-water carbonates; equivalents in other basins are part of the Bracklesham Beds (England) and the Brussels sands (Belgium). The Lutetian chronostratic division was traditionally regarded as the Middle Eocene stage and so is often taken to include all foraminiferal zones P10 to P14. Bouche (1962) and Hay & Mohler (1967) identified NP14 in the Lutetian type area, and Bukry & Kennedy (1969) made a tentative correlation of this with foraminiferal zone P10.

(b) The **Bartonian Stage** (Mayer-Eymar in 1858) was thought to be the northern equivalent of the Mediterranean Priabonian, but the latter contains Isthmolithus recurvus, i.e. zone NP19, while the Bartonian does not and is, therefore, presumably older. However, zone NP17 has been reported from the Barton Beds, Hampshire, England (Martini 1971), and this has been correlated with foraminiferal zone P14 (Berggren 1972) or even with P13 (Roth, Baumann & Bertolino 1971). These datings would place the Bartonian within the traditional Lutetian chronostratic division. Since there seems to be no suggestion of an overlap of Lutetian *sensu stricto* and Bartonian *sensu stricto*, Hardenbol & Berggren (1978) considered that confusion in the world-wide correlation framework would be minimized by restricting Lutetian to foraminiferal zone P10 to P12 (= NP14 to lower NP16) and regarding Bartonian as P13 and P14 (= upper NP16 to NP17), all being Middle Eocene.

(iii) Late Eocene

The **Priabonian Stage** (Munier-Chalmas and de Lapparent in 1893) is named from Priabona, Vicenza province, northern Italy, where the stratotype was restricted by Roveda (1961) to the section from Boro to Grenella. The type section has been found to contain zone P16 and part of P17 (Hardenbol 1968); also the nannofossil zones NP19 (Martini 1971) and NP19 and NP20 (Roth et al. 1971). At the Eocene Colloquium of 1968, Priabonian was considered to include foraminiferal zones P15 to P17 (part), i.e. nannofossil zones NP18 to NP21 (part).

3.19.5 The Oligocene Epoch

The complex history of nomenclature of the Oligocene Epoch was reviewed by Berggren (1971). The scheme followed here is that of Hardenbol & Berggren (1978), who recognized two distinct lithostratic units in northwest Europe: (i) a lower, moderately deep-marine, clayey unit, which includes the typical Rupelian rocks, and (ii) an upper, predominantly shallow-marine, sandy unit, which contains the Chattian type section.

IGCP Project 174 working on the Eocene–Oligocene boundary narrowed candidates for a standard to sections at Contessa (Italy), Bath Cliff (Barbados) and in the Betic Cordillera (Spain).

The **Rupelian Stage** (Dumont in 1849) was named from the 'argile de Rupelmonde', which is a strict junior synonym of 'Argile de Boom' (Koninck in 1837), at Boom (Anvers), Belgium. According to Banner & Blow (1965) the type Rupelian exhibits foraminiferal zones from the upper part of P18 to the upper part of P19. The nannofossil zone NP23 has been recognized in Boom Clay samples (Martini 1971). Hardenbol & Berggren (1978) 'arbitrarily placed the boundary between the Chattian and Rupelian Stages at the top of P19, which corresponds approximately to the middle of NP23'. The boundary position was thought to have the best chance of falling between the Rupelian and Chattian stages as defined in their respective type areas.

The **Chattian Stage** (Fuchs in 1894) was named from the latinized 'Kasseler Meeressand', near Kassel, northwest Germany, where the holostratotype was designated in 1958. Nannofossil zones NP24 and NP25 have been recognized tentatively in the Chattian (Neochattian) type section at Doberg, near Munde, Germany (Martini 1971, Benedek & Müller 1974, Martini & Müller 1975). Zone NP25 correlated with the foraminiferal zone P22 and the lower part of N4 (of

Blow 1969). The *Globigerinoides* Datum (first evolutionary appearance of *G. primordius*), which defines the base of zone N4, has been shown to occur in Upper Oligocene pre-Aquitanian levels (Anglada 1971a, 1971b, Scott 1972, Alvinerie et al. 1973, Theyer & Hammond 1974, Lamb & Stainforth 1976, Van Couvering & Berggren 1977). The Oligocene–Miocene boundary therefore occurs within zone N4, i.e. top of zone NP25.

3.19.6 The Paleogene charts (Figures 3.13 and 3.14)

The left-hand columns in Figure 3.13 tabulate the above conclusions. The named foraminiferal zones are from Stainforth et al. (1975). The New Zealand column is from Stevens (1980); USA (West Coast) from Berggren & Van Couvering (1974, figure 1); USA (Gulf Coast) from Murray (1961, figure 6.3); USSR (a) from Berggren (1971, table 52.19) and USSR (b) from Nalivkin (1973, table 1).

Figure 3.14 plots some Paleogene biostratigraphic zonation schemes. It was reproduced wthout change from GTS 82 and is in some respects superseded by Figure 3.13.

3.20 The Neogene Period

3.20.1 Introduction

In 1833, the Austrian geologist Hörnes proposed the name Neogene as follows (translated): 'The common occurrence of the Vienna mollusca not only in the type Miocene but in the type Pliocene induces me to join the nomenclature of these suites in a single category, the two suites to be known under the name Neogene . . .'

During the nineteenth century many Neogene stage names were proposed for Europe; these were mainly lithostratigraphic and were merely a reflection of facies variations. Most of them are no longer in common use, but an historical review of their development was given by Berggren (1971). Stratigraphic nomenclature has now almost stabilized and the following Neogene stages are now widely accepted as standard. In current usage Miocene and Pliocene are both epochs.

Moves have been made to include Pleistocene or Quaternary in Neogene as though stratigraphic classification should reflect some tidy notions of Earth history. Against this is the advantage of stability of an established convention which is followed here until ICS decide otherwise. But as noted earlier many recent papers extend Neogene to include Pleistocene.

3.20.2 The Miocene Epoch

Lyell, in 1833, named the second oldest of his four Tertiary divisions Miocene and considered the outcrops in the Touraine (southwest Paris Basin) as representing the type Miocene area.

(i) Early Miocene

The **Aquitanian Stage** was introduced by Mayer-Eymar in 1858 for the largely lagoonal, mollusc-rich beds in the Aquitaine Basin, Gironde, France. He noted that the best outcrops occurred in a 10 km section along the stream Saint Jean d'Etampes which runs through Saucats (Moulin de l'Eglise) and La Brede (Moulin Bernachon). This section, designated as the stratotype by Dollfus in 1909, was accepted as such by the Committee on Mediterranean Neogene Stratigraphy (1958 meeting at Aix-en-Provence).

The Neogene (Aquitanian) initial boundary might be defined near the base of this section if it is well enough exposed (George *et al.* 1969). It occurs within planktonic foraminiferal zone N4. The terminal boundary is close to the top of the succeeding zone, N5.

In discussing the stratal sequence in northern Europe Mayer-Eymar in 1858 included in his Aquitanian stage strata which had been explicitly included by Beyrich in 1854 and in 1856 in his (Upper) Oligocene. 'This has led, in no small way, to confusion regarding the affinities of the Aquitanian and assignment of Aquitanian to the Oligocene by some and to the Miocene by others' (Berggren 1971, p.747).

The **Burdigalian Stage** was introduced by Deperet in 1892 for marine strata overlying Aquitanian in the Aquitaine Basin, France (the name is from the richly fossiliferous 'Faluns de Bordeaux'). No single stratotype was designated in 1892, but *faluns* (crags = shelly deposits) of Saucats and Leognan were cited as being Early Burdigalian in age. Dollfus in 1909 designated Le Coquillat, near Leognan, as the type Burdigalian locality. He also referred to Mayer-Eymar's beds 8–10 in the Saint Jean d'Etampes section as Burdigalian. The designation of Le Coquillat was confirmed at the Congrés du Néogène Mediterreanéen in Vienna in 1960.

Since the type locality is a small isolated outcrop (a ditch), and difficult of access, E. Szöts (Berggren 1971, p.752) suggested designation as a 'neostratotype' the section along the Saint Jean d'Etampes stream between Moulin de l'Eglise and Pont-Pourquey, near Saucats; this section has the merit of being continuous with the Aquitanian stratotype. But the one boundary point has yet to be decided.

Examination of material from the Aquitanian–Burdigalian type area, in southwest France, led Berggren (1971, p.753) to the conclusion that the Burdigalian appears to be approximately equivalent to the *Catapsydrax dissimilis* and *C. stainforthi* zones of Bolli (N5 and N6 of Blow, 1969) and possibly to the lower part of the *Globigerinatella insueta* zone (N7) of Stainforth *et al.* (1975). Anglada (1971b) reported zone N5 from near the lower boundary of the Burdigalian neostratotype.

(ii) Mid Miocene

The **Langhian Stage** was introduced by Pareto in 1864 for outcrops in the heart of the Langhe, north of Ceva, northern Italy. Cita & Premoli Silva (1960) designated the Bricco della Croce section as the stratotype section, from near Cessole northwards to Case dei Rossi. The stratotype includes most of the 'Pteropod marls'. It appears that Burdigalian is earlier than the *Globigerinoides sicanus* zone (of Bolli), i.e. pre-N8 of Blow, and earlier than the *Praeorbulina glomerosa* zone of Stainforth *et al.* (1975, and Figure 3.15). From the work of Cita & Elter (1960) it appeared that '. . . the base of the Langhian may satisfactorily be drawn at the base of the *G. sicanus* zone' (Berggren 1971, p.753 and Figure 3.16), i.e. at the base of N8 and the base of the *P. glomerosa* zone in Figure 3.15. To avoid the Langhian stage straddling the *Orbulina* Datum (base of N9) and so be partly Early Miocene and partly Middle Miocene in age, the Langhian Stage is currently taken to begin at the first appearance datum (FAD) of *Praeorbulina glomerosa* (N8) and this is also taken to coincide with the initial Middle Miocene boundary.

The **Serravallian Stage** was named by Pareto in 1865 for outcrops in the vicinity of the village of Serravalle Scrivia (Alessandria, Italy), which is the type area (not Serravalle, Firenze, Italy). It was proposed as a stage intermediate between Langhian and Tortonian as then understood. The Serravallian Stage has, historically, been 'correlated' with various parts of the 'Helvetian' stage. Drooger (1964) suggested that Serravallian should be used to replace 'Helvetian'. A stratotype has been designated and described by Vervloet (1966), i.e. the Serravalle Formation in the Scrivia valley.

Serravallian embraces more than the average number of Neogene planktonic zones – i.e. zone N10 to within N15 of B low (1979) (*Globorotalia (T.) peripheroacuta* to *G. (T.) continuosa*) – and from within calcareous nannoplankton zone NN5 to NN9 (*Sphenolithus heteromorphus* to *Discoaster hamatus*) of Martini (1971). The relative fineness of these zonations attests to the general diversity 'explosion' of Middle Miocene time, following the relatively low diversities of Oligocene and, to a large extent, of Early Miocene time.

(iii) Late Miocene

The **Tortonian Stage** was named by Mayer-Eymar in 1858 for the 'Blaue Mergel mit *Conus canaliculatus* und *Ancellaria glandiformis* von Tortona'. The stratotype, designated by Gianotti in 1953, is exposed in the valley of Rio Mazzapiedi–Castellania (Tortona, Alessandria Province, Italy); it has been described by Gino *et al.* (1953).

In 1868, Mayer-Eymar restricted his original concept of Tortonian, distinguishing the regressive, increasingly brackish-water strata (Cerithium marls) above marine Tortonian (his lower Messinian) which is a part of the Paratethys regressive sequence. However, he did also refer a series of marine units near Messina, Sicily, to his Messinian, thus providing a more objective basis for comparison and correlation. Banner & Blow (1965) recognized late N15 and N16 zones in the type Tortonian. Berggren & Van Couvering (1974) considered that the early part of zone N17 was also represented.

The **Messinian Stage** is from marine strata near Messina, Sicily, which were referred to by Mayer-Eymar in 1868. Owing to sedimentological and tectonic complications in the vicinity of Messina itself, Selli (1960) selected and described a neostratotype, exposed between Mt Capodarso and Mt Pasquasia, on either side of the Morello River between Caltanaisetta and Enna, Sicily. The section is bounded below by Tortonian marls and above by basal Pliocene ('trubi'), and its microfauna has been described by d'Onofrio (1964).

The original concept of Messinian included the diatomaceous marls at the base ('tripoli'), evaporites ('gesso'), and the overlying 'trubi' (deep-water lutites with rich planktonic faunas). However, Seguenza in 1868 and 1879 restricted Messinian to the 'tripoli' and the 'gesso', considering the 'trubi' to be basal Pliocene.

Planktonic foraminiferal zone N17 (*Globorotalia (G.) tumida plesiotumida*) has been recognized in the type Messinian section by d'Onofrio (1964), Colalonga (1970) and Blow (1969). Berggren & Van Couvering (1974) concluded that Messinian is equivalent to only the later part of zone N17, i.e. it was thus of particularly short duration.

3.20.3 The Pliocene Epoch

Lyell in 1833 proposed the name Pliocene for the youngest Tertiary deposits which he recognized at the time. He divided his Pliocene (more than 50% of molluscan species still living) into an 'Older Pliocene', which would correspond to the Astian–Piacenzian as subsequently recognized in Italy, and a 'Newer Pliocene' (90–95% of molluscan species still living), for which he subsequently (in 1839) introduced the name Pleistocene. The 'Subapennine' strata of northern Italy were regarded by Lyell in 1833 as typical for his 'Older Pliocene' (= Pliocene *sensu stricto*).

Although some Italian geologists favour a three-fold division of the Pliocene Epoch, traditionally a two-fold division is used (e.g. Berggren & Van Couvering 1974). The

Figure 3.15. The Neogene Period and selected successions.

Figure 3.16. Neogene planktonic zonation.

KEY TO NEOGENE PLANKTONIC ZONATIONS

FORAMINIFERA			CALCAREOUS NANNOPLANKTON		RADIOLARIA	
N 21	Globorotalia (T.) tosaensis tenuitheca	Cons-R-Z	NN18	Discoaster brouweri	Pte. p.	Pterocanium prismatium
N 20	Globorotalia (G.) mutticamerata / Pulleniatina obliquiloculata obliquiloculata	P-R-Z	NN17	Discoaster pentaradiatus	Spo. p.	Spongaster pentas
N 19	Sphaeroidinella dehiscens dehiscens / Globoquadrina altispira altispira	P-R-Z	NN16	Discoaster surculus	Sti. p.	Stichocorys peregrina
N 18	Sphaeroidinellopsis subdehiscens paenedehiscens / Globorotalia (G) tumida tumida	P-R-Z	NN15	Reticulofenestra pseudoumbilica	Omm. p.	Ommatartus penultimus
N 17	Globorotalia (G) tumida plesiotumida	Cons-R-Z	NN14	Discoaster asymmetricus	Omm.a.	Ommatartus antepenultimus
N 16	Globorotalia (T) acostaensis / Globorotalia (G) merotumida	P-R-Z	NN13	Ceratolithus rugosus	Can. p.	Cannartus (?) pettersoni
N 15	Globorotalia (T) continuosa	Cons-R-Z	NN12	Ceratolithus tricorniculatus	Dor. al.	Dorcadospyris alata
N 14	Globigerina nepenthes / Globorotalia (T) siakensis	Conc-R-Z	NN11	Discoaster quinqueramus	Cal. c.	Calocycletta costata
N 13	Sphaeroidinellopsis subdehiscens subdehiscens / Globigerina druryi	P-R-Z	NN10	Discoaster calcaris	Cal. v.	Calocycletta virginis
N 12	Globorotalia (G) fohsi	P-R-Z	NN 9	Discoaster hamatus	Lyc. b.	Lychnocanium bipes
N 11	Globorotalia (G) praefohsi	Cons-R-Z	NN 8	Catinaster coalitus		
N 10	Globorotalia (T) peripheroacuta	Cons-R-Z	NN 7	Discoaster kugleri		
N 9	Orbulina suturalis / Globorotalia (T) peripheroronda	P-R-Z	NN 6	Discoaster exilis		
N 8	Globigerinoides sicanus / Globigerinatella insueta	P-R-Z	NN 5	Sphenolithus heteromorphus		
N 7	Globigerinoides trilobus / Globigerinatella insueta	P-R-Z	NN 4	Helicopontosphaera ampliaperta		
N 6	Globigerinatella insueta / Globigerinita dissimilis	Conc-R-Z	NN 3	Sphenolithus belemnos		
N 5	Globoquadrina dehiscens praedehiscens / Globoquadrina dehiscens dehiscens	P-R-Z	NN 2	Discoaster druggi		
N 4	Globigerinoides primordius / Globorotalia (T) kugleri	Conc-R-Z	NN 1	Triquetrorhabdulus carinatus		

FORAMINIFERA : Zonation of Blow (1969, 1979)
P-R-Z = Partial Range Zone
Cons-R-Z = Consecutive Range Zone
Conc-R-Z = Concurrent Range

CALCAREOUS NANNOPLANKTON : Zonation of Martini (1971)

RADIOLARIA : Zonation of Riedel & Sanfilippo (1971)

planktonic foraminiferal zones N18 to N21 of Blow (1969, 1979) span the Pliocene interval. Berggren (1973) constructed an alternative, more detailed, six-fold Pliocene subdivision based on planktonic foraminifera – his zones P11 to Pl6.

The **Zanclian Stage** was proposed by Seguenza in 1868 from Zancla (the pre-Roman name for Messina) for foraminiferal white marl (the 'trubi' of Baldacci in 1886) followed by coral limestones and yellowish sandy marls, in the vicinity of Messina, Sicily. No stratotype was designated until Cita in 1974 proposed one at Capo Rosello, near Agrigento, on the southern coast of Sicily. The base of the stratotype section signifies the beginning of permanent open-marine conditions in the Mediterranean following the restricted deposition and desiccation (evaporites) of the latest Miocene (Messinian). The name Zanclian is preferred to the more northerly Tabianian, since the cooler-water fauna of the latter is less favourable for correlation. Banner & Blow (1965) recognized planktonic foraminiferal zone N18 in the type Zanclian, and its base (the Miocene–Pliocene boundary) has been variously placed in the middle of this zone or at its base; the latter course is followed here, to be consistent with Berggren & Van Couvering (1974).

The **Piacenzian Stage** was introduced by Mayer-Eymar in 1858 (as the 'Piacenzische Stufe') for the argillaceous facies of the Lower Pliocene with *Nassa semistriata* in northern Italy. It was originally distinguished as an Astian sub-stage (de Rouville in 1853) and subsequently erected as a stage by Renévier (1897). The name Astian now refers to the sandy facies which interfingers with the Piacenzian clays and marls in northern and central Italy. Pareto (1865) adopted the name,

using the French equivalent Plaisancian, and specified that the typical development of the stage was to be seen in the hills around Castell' Arquato in northern Italy (Berggren & Van Couvering 1974).

The name derives from the city of Piacenza, which lies approximately midway between Parma and Milan in Italy. The name Plaisancian, as being the French translation of Piacenzian, does not refer to the small village of Plaisance situated between Castell' Arquato and Lugagnana, as erroneously attributed by Movius in 1949.

No specific type section has been designated, but the vicinity of Castell' Arquato appears to meet general approval as a type area, and the Castell' Arquato section was referred to as 'classical' by di Napoli-Alliata in 1954.

3.20.4 The Neogene charts (Figures 3.15 and 3.16)

The divisions of the period described above are illustrated in Figure 3.15. The named foraminiferal zones are from Stainforth *et al.* (1975); the New Zealand divisions are from Suggate *et al.* 1978; the USSR South Russian Platform from Nalivkin (1973); and the Californian west coast from Berggren & Van Couvering (1974). The foraminiferal, calcareous nannoplankton and radiolarian zone abbreviations given on Figure 3.15 are expanded on Figure 3.16 where references are given.

3.21 The Quaternary Sub-era (Pleistogene Period)

This latest period in Earth history has been variously named. In the early years of the nineteenth century **Alluvium** was established and in 1823 Buckland regarded the somewhat older **Diluvium** as being a product of the Biblical Flood.

In the years leading up to 1840 the widespread erratics in Alpine and northern Europe were interpreted as deposits of former extensive ice. In 1837 the term **Drift** came to be used for the widespread sands, gravels and boulder clays newly thought to have been deposited from floating ice. Lyell proposed the **Pleistocene Period** in 1839 to accommodate this ice age and to post-date his already established Pliocene. Pliocene was the last period in the Tertiary Era, thus there was room for a post-Pliocene **Quaternary** division – so named by Morlot in 1854. Quaternary was different from Pleistocene in that it also contained Lyell's original 'Recent', later named **Holocene** by the IGC in 1885. So the classification determined by historical priority and long usage is:

	Quaternary Sub-era	(Anthropogene or Pleistogene Period)	Holocene Epoch
			Pleistocene Epoch
Cenozoic Era		Neogene Period	Pliocene Epoch
			Miocene Epoch
	Tertiary Sub-era		Oligocene Epoch
		Paleogene Period	Eocene Epoch
			Paleocene Epoch

Quaternary is not a satisfactory name in the scheme; Primary and Secondary have been replaced and Tertiary has been replaced by Paleogene and Neogene as formal period names, so alternatives Anthropogene (often in use in the USSR) and Pleistogene have been proposed, the latter better fitting the overall nomenclature. However, tradition may well prevail with Quaternary and is acknowledged here. Alternative schemes have been proposed (e.g. to include Pleistocene in Neogene). An analogous proposal has been made to include Holocene as a Pleistocene stage. These would run counter to history and to an immense literature and would serve no great purpose. However, the ICS has the matter under consideration.

3.21.1 The Pliocene–Pleistocene boundary

An attempt to standardize the initial Pleistocene boundary was made at the IGC in London in 1948 when it was realized that an objective reference in a stratotype near or at the base of **Calabrian** strata (a stage introduced by Gignoux in 1910) in Italy would be appropriate. The Pliocene–Pleistocene boundary stratotype should be one of the marine sections preserved on land in southern Italy.

Estimates of its age have ranged from 0.6 to 4 Ma (Haq, Berggren & Van Couvering 1977). Correlation using paleontological criteria proved difficult. The initial Calabrian boundary was thought to be marked by the first appearance of both *Arctica islandica* and *Hyalinea baltica* (Sibrava 1978) but Ruggieri (INQUA publication in 1979) showed that the latter appears slightly later. Futhermore, Ruggieri argued for the suppression of Calabrian and replacement by Santernian along with a revision of the rest of the sequence.

Various sections in southern Italy have competed for the position of stratotype. On the basis of calcareous plankton at Le Castella and Santa Maria di Catanzaro, as well as of deep-sea sediments, Haq *et al.* (1977) correlated the boundary with the top of or slightly above the Olduvai magnetostratigraphic event at 1.6 Ma.

However, the matter was at last resolved by the decision of ICS on a recommendation of the Pliocene–Pleistocene Boundary Working Group in 1984 (Bassett 1985a). The initial boundary stratotype for the Pleistocene Epoch is within Subsection B of the Vrica Section, approximately 4 km south of Crotone in the Marchesato Peninsula, Calabria, southern Italy, 39°32′18.61″N, 17°08′05.79″E. The marker point is at the base of the claystone conformably overlying the sapropelic marker bed E in the section.

The boundary lies between the last appearance datum (LAD) of *Discoaster broweri* (below) and the LAD of *Globigerinoides obliquus extremus* and *Cyclococcolithus macintyri* and the FAD of *Geophyrocapsa oceanica* and *Globigerinoides tenellus* (above).

It also lies between the top of the Olduvai normal polarity zone and a zone with dominantly left-coiled specimens of the foraminifer *Neogloboquadrina pachyderma*. The boundary is some 3 to 6 m (representing an interval of 10–20 thousand years) above the top of the Olduvai normal polarity subchron (Bassett 1985a). Details of this boundary in its geologic setting were provided by Aguirre & Pasini (1985). They estimated the chronometric age of this boundary to be about 1.64 Ma. Magnetostratigraphy was reviewed by Tauxe *et al.* (1983).

It is thus fitting that the first boundary to which the golden-spike philosophy was proposed in 1948 by W. B. R. King, has been decided accordingly by the ICS of IUGS which was not then in existence.

3.21.2 Division of the Pleistocene Epoch

It seems that no formal decisions have yet been made, but also that an official scheme of stages may not be far off.

Competing bipartite and tripartite schemes have been in informal use and it now seems that a tripartite scheme is likely to be favoured. A 'note' to test opinion in the USGS (kindly supplied by G. M. Richmond in December 1984), on the basis of proposals by the INQUA/ICS Working Group on Major Subdivision of the Pleistocene is as follows.

'It is proposed that the initial **Middle Pleistocene** boundary be placed at the Matuyama–Brunhes magnetic polarity reversal. The reversal has not been dated directly by radiometric controls. It is significantly older than the Bishop Tuff (revised K–Ar age 738 ka; Izett, 1982a), and the estimated K–Ar age of 730 ka assigned to the reversal by Mankinen & Dalrymple (1979) is too young. In Utah, the Bishop volcanic ash bed overlies a major paleosol developed in sediments that record the Matuyama–Brunhes reversal (Eardley *et al.*, 1973). The terrestrial geologic record is compatible with the astronomical age of 788 ka assigned to the reversal by Johnson (1982).'

'The initial **Late Pleistocene** boundary is placed arbitrarily at the beginning of marine oxygen isotope substage 5e (at Termination II or the stage 6/5 transition). That boundary also is not dated directly. It was assigned provisional ages of 127 ka by CLIMAP Project members (1984) and 128 ka by SPECMAP Project members (Ruddiman & McIntyre 1984), based on Uranium-series ages of the substage 5e high eustatic sea level stand. A siderial age of 132 ka is derived by projection of the boundary on to the astronomical time scale of Johnson (1982).'

3.21.3 Pleistocene terrestrial sequences

Attempts to make a stratigraphic sequence out of successive cycles of continental glaciation have proved to be totally inadequate in distinguishing particular fluctuations of climate amongst so many fluctuations. Furthermore, continental classifications have been shown to be oversimplified and incomplete, being based to a large extent on erroneous concepts.

In 1973 a Geological Society of London Working Group revised the Quaternary stratigraphy of the British Isles (Mitchell *et al.* 1973). The new terminology for the British stages was partly intended to discourage older terrestrial correlations and to face up to the recognition of a large number of glacial stages in the Quaternary Sub-era. However, this scheme is far from satisfactory, being essentially a climatic rather than a rock sequence. Lack of knowledge in key areas has led to the unfounded assignment of ages in places far removed from the stratotypes (Bowen 1978). Continental sequences of glacial tills, terraces, etc. were worked out first in the Alps and northern Europe long before there was any means of correlation between them or an adequate time scale by which to order them. They are still difficult to correlate and so have not been included in the chart. Two such sequences are given in Evans (1971) and Berggren & Van Couvering (1974), who each fitted the divisions differently on their time scales, and they are combined in the list below. Evans compressed all the glacial and interglacial periods of Europe into 1 Ma, although he recognized that some terrestrial sequences contained a gap representing 1 Ma and that the Pliocene-Pleistocene boundary occurred at 2.1 Ma.

For the Alps the sequence in increasing age is:

Wurm Glacial
    Riss/Wurm Interglacial
Riss Glacial
    Mindel/Riss Interglacial
Mindel Glacial
    Günz/Mindel Interglacial
Günz Glacial
(Berggren & Van Couvering gave two glacial Günz sub-stages)
    Donau/Günz Interglacial
Donau Glacial
(Evans listed four glacials under Donau)
?Biber Glacial
(Berggren & Van Couvering place Biber glacial at around 3 Ma)

For northern Europe the sequence (increasing age) is:

Flandrian (Holocene)
Weichselian Glacial
    Eemian Interglacial

Saalian Glacial
    Holsteinian Interglacial
Elsterian Glacial
    Cromerian Interglacial
Menapian Glacial
    Waalian Interglacial
Eburonian Glacial
    Tiglian Interglacial
    (Berggren & Van Couvering indicate this as extending back into pre-Quaternary time)

In the USSR the following sequence (increasing age) was given by Flint (1971) with additions from Sachs & Strelkov (1961):

Valdayan/Zyryanka Glacial
    Mikulino Interglacial
Moscovian Glacial
    Odintsovo Interglacial
Dnepr/Samarovo Glacial
    Likhvin Interglacial
Oka/Demyanka Glacial

Since GTS 82 there have been many developments which render much of that work out of date. From the Chairman of the INQUA/ICS Working Group on Pleistocene Subdivision (G. M. Richmond) a private communication reads: 'With respect to the USA . . . the terms Nebraskan and Kansan have been abandoned . . . Afton is essentially useless except at the type locality and Yarmouth and Sangamon can range very greatly depending on that region. Current knowledge would be published in a series of 18 regional papers and correlation charts that would appear as a section of the International Geological Correlation Programme (IGCP) volume on Quaternary glaciation of the Northern Hemisphere. Suffice it to say here that we have stratigraphic, tephrochronologic, radiometric dating and paleomagnetic evidence for 14 major glaciations in the Pleistocene and at least two in the late Pliocene. This includes early and late Wisconsin as two glaciations and also the two glaciations equivalent to marine stages 5a and 5b . . .'

An independent record of Late Pleistocene and Holocene climatic changes has been derived from $^{18}O/^{16}O$ ratios in cores through the Greenland and Antarctic ice sheets (Johnsen *et al.* 1972) and from other areas.

3.21.4 Pleistocene marine sequences

Because the span of Quaternary time is so near our own, a different order of discrimination is possible and different methods are being rapidly developed. The principal development in a new Pleistocene time scale depends on the regularity of the climatic cycle that was discovered around 1875 by Croll and developed especially by Milankovitch. This approach was not taken too seriously by Quaternary geologists until there were quantitative ways of testing it. Zeuner (1945), Emiliani (1965) and Evans (1971) were amongst those to recalculate and relate the astronomical parameters, testing, for example, 40 000 and 100 000 year cycles against other phenomena, such as the newly established oxygen-isotope

Figure 3.17.

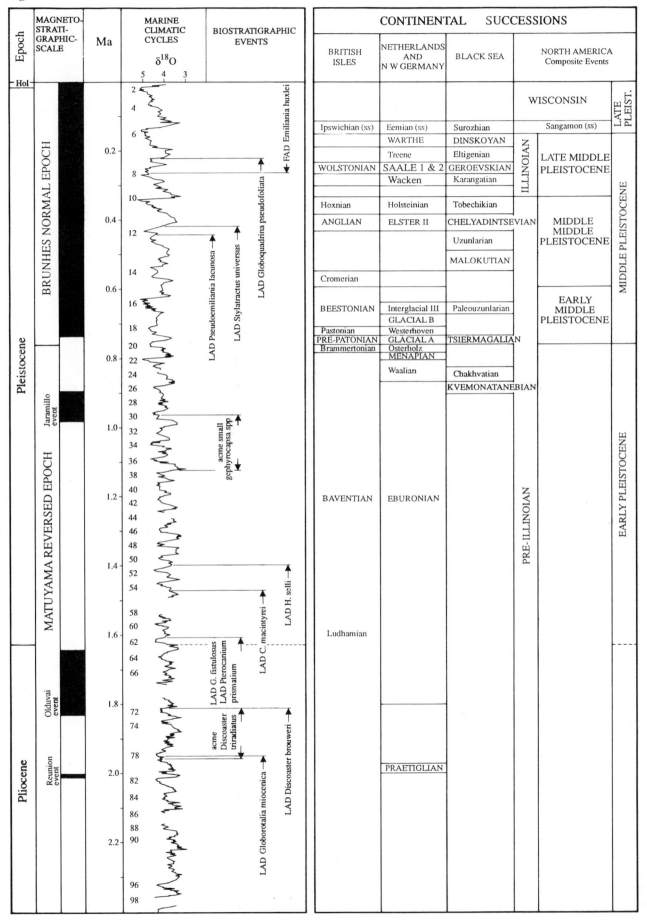

curve from the oceans. The first rigorous treatment using wide ranging techniques was by Hays, Imbrie & Shackleton (1976). Isotope studies from the bottom sediments of the Atlantic and Pacific Oceans since then have indicated as many as 21 Quaternary glacial ages and, as with Cenozoic studies, the continental evidence is so incomplete as compared with the oceanic sequences that terrestrial glacial–interglacial stratigraphy in future must depend on the ocean record for interpretation.

The discussion that follows is a brief review of the situation as outlined in GTS 82. Isotopic ages within the last 40 000 years or so depend on ^{14}C and, for the main part of Pleistocene time, on K–Ar, ^{230}Th–^{234}U and ^{231}Pb–^{230}Th determinations. Berggren et al. (1980) outlined the chronologies of different classes of events which need to be calibrated by chronometric methods. The principal scales used are: magnetostratigraphic; the oxygen-isotope scale, which extends from the present-day to the Jaramillo magnetic reversal event (0.91–0.97 Ma), i.e. stages 1 to 26; calcium carbonate content curves for the equatorial Pacific, back to the same reversal event; microfossil-assemblage changes; and planktonic-foraminiferal coiling scales in certain parts of the oceans. Correlation between the various scales is established by simultaneous analysis and comparison with magnetic reversal stratigraphy in individual cores, and by many other methods, including comparison of amino acid content and thermo-luminescence.

Datum levels relate to the apparent first and last appearance of taxa in the evolution of oceanic biota (FADs and LADs respectively): they are mainly determined from the appearances, acmes and extinctions of calcareous nannoplankton, diatoms, planktonic foraminifera and radiolaria.

The correlation of Pleistocene marine sediments is based either directly or indirectly on the magnetic polarity time scale, which in turn rests upon K–Ar dating of terrestrial lavas. K–Ar-derived ages are available for the chron boundaries. The ages of events are often interpolated from sea-floor magnetic anomalies and marine sediments, assuming constant rates of sea-floor spreading and deposition respectively.

The oxygen-isotope scale makes use of the fact that, when continental ice builds up as a result of global cooling and sea level is lowered, the ice is depleted in ^{18}O relative to the ocean water, leaving the ocean water enriched in ^{18}O. The oxygen-isotope composition of calcareous foraminifera and coccoliths, and of siliceous diatoms, varies in direct proportion to that of the water (see Shackleton & Opdyke (1976) for discussion of the limitations of isotope stratigraphy). The 16 stages of Emiliani (1966) obtained from Caribbean and Atlantic sediment cores were extended to 22 by Shackleton &

Opdyke (1973) after analysis of an equatorial Pacific core. These cores did not extend to the Pliocene–Pleistocene boundary, but subsequently another equatorial Pacific core (Shackleton & Opdyke 1976) and an Atlantic core (Van Donk 1976) extended through the boundary to provide a complete record of Quaternary climatic change. The Brunhes–Matuyama boundary, and the Jaramillo and Olduvai events were all recognized, thus enabling an average accumulation rate of 1 cm per 1000 years to be determined. However, Shackleton & Opdyke regarded their earlier core as showing steadier accumulation rates during the Brunhes Chron; thus they regarded their estimates of ages determined by interpolation in this earlier core as being the best available. They have suggested that this core be used as the Late Pleistocene stratotype. Van Donk recognized a total of 21 isotopically determined interglacial stages and an equal number of glacial or near-glacial stages. However, his numbering scheme (which extended beyond 23 stages) has not been formally adopted.

A biostratigraphic framework for the Pleistocene Epoch, based on datum levels of calcareous nannoplankton, was established by Gartner from several cores in equatorial and temperate regions (Berggren et al. 1980). His revised zonation (Gartner 1977) as given by Berggren et al. (1980) was shown in GTS 82; it had been tied in to the geomagnetic polarity scale and ages have been assigned to the interpolation after correlation between cores. Eight diatom and two silicoflagellate datum levels have been correlated with the paleomagnetic time scale and indirectly to the oxygen-isotope record. These are ranked into first and second order according to their reliability (Berggren et al. 1980). Planktonic foraminiferal datum levels (not shown here) have also been determined.

3.21.5 The Pleistocene–Holocene boundary

Corresponding to a climatic event around 10 000 radiocarbon years BP, this boundary is likely to be standardized in a lacustrine varved sequence in Sweden. It was proposed at the Eighth INQUA Congress in Paris in 1969 and has since been accepted in principle. The change is well established in a variety of sediments, notably in Scandinavia, and corresponds to the following boundaries: European Pollen Zones III/IV, the Younger Dryas/Preboreal and Late Glacial/Postglacial (Mörner 1976a). However, this boundary definition has yet to be decided by ICS on a recommendation being prepared by INQUA. If it were to be defined precisely at 10 000 years as some favour, it would be the first stratigraphic boundary later than Proterozoic to be defined chronometrically.

3.21.6 The Holocene Epoch

Holocene is the preferred equivalent to Recent as a name for the interval for the last 10 000 years or so for Earth history.

In prehistoric times as well as later, climatic events have largely served to identify the divisions therein, calibrated by ^{14}C and dendrochronology as well as successively by archeology and human history.

Those referred to above, who seek to reclassify

Figure 3.17. Quaternary (Pleistocene) chronostratic scale and paleomagnetic time scale with ^{18}O/^{16}O marine climatic stages compiled for this publication by N. J. Shackleton. In the continental successions names in upper case lettering refer to glacials and lower case to interglacials. See Section 3.21.7. and Chapter 5.

Cenozoic divisions into two periods only (Paleogene and Neogene), not only propose to include Pleistocene in Neogene but to include Holocene in Pleistocene. On this basis Quaternary as well as Tertiary might continue to be used informally. It is planned that the matter be discussed at the Washington IGC in July 1989 (J. W. Cowie personal communication).

3.21.7 The Quaternary chart (Figure 3.17)

This figure was designed by LEC and drawn by R. Khan (CASP) from the curve of ^{18}O climatic stages plotted against linear time and the magnetostratigraphic time scale prepared for us by N. J. Shackleton from oceanographic data. The Quaternary continental successions in four columns (three European, one North American) were selected from the 19 columns in figure 2 of Jenkins *et al.* 1985 (pp.202–3). Those successions were correlated against Shackleton's new climatic curve rather than against the linear time scale in their figure 2. However, it must be pointed out that the execution of the drawing did not accurately correspond to other data in this work. Therefore, the matching of the curve to the scale must be regarded as indicative rather than precise. When these discrepancies were noted it was too late to redraw the figure without seriously delaying publication.

4

Isotopic methods, dates, precision and database

4.1 Introduction

The major differences between the discussion of isotopic dates and the chronometric calibration in GTS 89 and the corresponding chapter in GTS 82 are:

(1) division into two chapters (4 and 5)

(2) addition of a Cenozoic database (Table 4.2)

(3) updating of the Mesozoic and Paleozoic databases (Table 4.2)

(4) estimates of chronostratic correlation errors based on the duration of chronostratic chrons (Section 5.3 and Table 5.1)

(5) use of relative numbers of chrons for interpolation between tie-points (Sections 5.7.2 and 5.8)

(6) further classification of chronostratic divisions (as outlined in Chapter 3)

(7) taking glauconites as minimum ages for the age calibration of stratigraphic boundaries older than 115 Ma (Sections 4.3 and 5.5)

(8) a review of dating methods (Sections 4.3 to 4.12 and Appendix 3)

In GTS 82 Berggren & Van Couvering's (1974) and Hardenbol & Berggren's (1978) ages were adopted for Cenozoic stratigraphic boundaries, as modified by Ness, Levi & Couch (1980) to take into account the newly recommended decay constants (Steiger & Jäger 1977). The ages of several sub-epoch boundaries that were poorly defined on the basis of isotopic dating were adjusted by using magnetostratigraphy

(Chapter 4 in GTS 82). Although we considered in GTS 82 that the chronogram method (Section 5.4) would not produce results that were notably different from these time scales, more recent reviews (Odin 1982b, Berggren, Kent & Van Couvering 1985) present time scales that are significantly different from each other. We have therefore compiled a Cenozoic database and attempted to evaluate independently the ages of the stratigraphic boundaries.

The Mesozoic and Paleozoic age determinations in GTS 82 were largely from RLA's pre-Cenozoic compilation in Cohee, Glaessner & Hedberg (1978), most of which was presented at the International Geological Congress in Sydney in 1976. GTS 82 is therefore essentially a 1976 time scale. The new database has been compiled and assessed by RLA and includes ages published up to early 1988 (Table 4.2). Some late additions that have not been used in the chronograms are also listed and included in the discussion.

The majority of the isotopic dates presented here are determined by the K–Ar method and our discussion will be centred around them. A few are determined by other techniques. All methods of isotopic dating deserve comment as to their relative reliability and use in time scale calibration. It is not our goal here to repeat the basic theory and technical details of the various dating techniques. These are covered in a number of books and reviews (Schaeffer & Zahringer 1966, Hamilton & Farquhar 1968, Dalrymple & Lanphere 1969, Faure & Powell 1972, York & Farquhar 1972, Fleischer, Price & Walker 1975, Faure 1977, 1986, Jäger & Hunziker 1979, Odin 1982a,b).

For purposes of time scale calibration, we assume that common sense prevails and that analyses are only made of unweathered material. Intense weathering has been proven to disturb all isotope systems; analyses of weathered samples produce discordant, generally low, dates (Goldich & Gast 1966, Stern, Goldich & Newell 1966, Bottino & Fullagar 1968, Dasch 1969, Worden & Compston 1973, Gleadow & Lovering 1974, Brass 1975, Fullagar & Ragland 1975, Clauer 1981).

4.2 Closure temperatures

Any dating method involves the analysis of daughter products which may diffuse through solids, or the measurement of radiation damage which may anneal with time. Diffusion and annealing rates are highly temperature dependent. The temperature at which daughter re-equilibration or loss becomes geologically significant is slightly below the 'closure temperature' cited in geochronometry papers. Strictly speaking, the 'closure temperature' is the temperature of the material at the calculated isotopic date, assuming the daughter product is all retained (Dodson 1973, 1976). The geologic closure temperature is dependent on cooling rate. It lies between the temperature where daughter products are effectively retained and the temperature at which the loss of daughter products is rapid on a geologic time scale. The difference between these two temperatures is usually a few tens of degrees. When cooling rates are low the quoted closure temperature is a few degrees lower than in the case of rapid cooling, but for a general appreciation of isotopic dates this difference is not important.

The closure temperatures listed in Table 4.1 give a first

Table 4.1. *Compilation of closure temperatures for intermediate cooling rates (1 to 10 °C/Ma) and of relative retentivities for a variety of dating methods (after Odin et al. 1982b, Dewitt et al. 1984, Van Breemen & Dallmeyer 1984, Parrish & Roddick 1984, Watson, Harrison & Ryerson 1985)*

Method	Material (in diminishing order of retentivity)	Closure temperature where estimates are available (°C)
Sm–Nd	Whole rock	650 – >800
Rb–Sr	Crystalline whole rock	650 – >800
U–Th–Pb	Zircon – concordia upper intercept	>800
Rb–Sr	10 cm slabs	~600
Rb–Sr	Amphibole	
U–Th–Pb	Monazite	500 – 600
U–Th–Pb	Amphibole	
Pb/Pb	Whole rock	
Ar/Ar	Amphibole plateau	530 ± 40
K–Ar	Amphibole	500 – 550
U–Th–Pb	Epidote	400 – 600
U–Pb	Sphene	500
Rb–Sr	Sphene	~550
Rb–Sr	Muscovite	>500
U–Pb	Apatite	~500
Rb–Sr	Plagioclase	~350
Rb–Sr	K-feldspar	~350
K–Ar	Muscovite	~350
Th–Pb	Sphene	
Th–Pb	Apatite	
Rb–Sr	Biotite	320 ± 40
Fission track	Epidote	
Fission track	Sphene	290 ± 40
Ar/Ar	Biotite	280 ± 40
K–Ar	Biotite	280 ± 40
K–Ar	Plagioclase	
U–Pb	Feldspars	
Rb–Sr	Glauconite	~230
Fission track	Zircon	200 ± 50
K–Ar	Glauconite	~200
Ar/Ar	K-feldspar plateau	130 ± 15
Ar/Ar	K-feldspar low-T release	~110
Fission track	Apatite	100 – 110
Fission track	Glass	<90

order ranking of different phases to thermal resetting of isotopic dates. For the same phase the loss of radiation damage usually occurs at lower temperatures than diffusive loss of Ar, and that loss occurs below the temperature at which diffusion equilibration of large cations is important. The size of the system analysed is another major factor in loss or re-equilibration of daughter products. Larger grains tend to be more resistant to resetting than smaller grains. Several dating techniques take advantage of whole-rock analyses, representing volumes many orders of magnitude larger than mineral grains and thus much more resistant to resetting

(Compston & Jeffery 1959, Gray & Compston 1978, Hofmann 1979, Köhler & Muller-Sohnius 1980, Van Breemen & Dallmeyer 1984).

Several studies have emphasized the importance of fluids and deformation in resetting isotopic dates (Arnold & Jäger 1965, Armstrong 1966, Råheim & Compston 1977, Black et al. 1979, Field & Råheim 1979, Hickman & Glassley 1984). These effects are less easy to model or quantify than temperature effects but it is clear that dry, undeformed rocks may preserve isotope systematics that would be totally obliterated in the presence of fluids or during deformation. Such behaviour contrast has been exploited in dating structural and metamorphic episodes.

Some authors have questioned the applicability of closure temperature concepts in the interpretation of relict isotopic dates in metamorphosed rocks (Chopin & Maluski 1980, Verschure et al. 1980, Del Moro et al. 1982). Such exceptions are rare and should not detract from the general utility of the ranking of dating methods given in Table 4.1. Some of these exceptions may be explained by such factors as the lack of chemical potential gradients to drive diffusion at temperatures where it may normally occur, rather than requiring a rejection of the concept of closure temperature. For time scale work it is best to avoid metamorphic rocks and stick to phases in environments that may be reasonably expected to be closed systems following rapid cooling through their closure temperatures.

4.3 Interpretation of K–Ar dates

Fresh volcanic rocks represent nearly ideal materials for isotopic dating and time scale calibration (McDougall 1966). Under normal circumstances, they cool rapidly and their phases are initially free of radiogenic Ar at the ± 0.3 Ma level (Dalrymple 1969, Krummenacher 1970). Decades of experience have shown that concordant and stratigraphically consistent dates can be obtained using hornblende, biotite, and high-temperature feldspar of lavas and high-temperature pyroclastic rocks (Baadsgaard, Lipson & Folinsbee 1961, Fechtig, Gentner & Kalbitzer 1961). Low K minerals are less suitable because of larger analytical errors and greater vulnerability to contamination, or to problems with excess Ar.

Fine-grained but holocrystalline whole-rock samples of composition ranging from basalt to rhyolite have also been shown to give satisfactory dates. Glass is prone to Ar loss and usually gives only minimum dates. The only exception is young, anhydrous glass which may be quantitatively retentive (Marvin, Mehnert & Noble 1970, Kaneoka 1972). Rocks with interstitial glass or very fine-grained groundmass or altered groundmass give low whole-rock dates (Mankinen & Dalrymple 1972). Visual screening of whole-rock samples reduces the number giving discrepant dates but the best whole-rock samples can never be assigned the same confidence given to concordant mineral dates. Rocks more than a few million years old are invariably altered when compared with modern lavas so that scepticism is justified for any whole-rock date older than Late Cenozoic. As a rule the amount of atmospheric Ar in a whole-rock sample increases with age and amount of alteration (Armstrong 1978b). This effect is notable even before the sample is visually suspect. Porous volcanic rocks, especially tuffaceous ones, are much more susceptible to isotopic modification than are massive rocks; flow interiors are often preferable to rims because they are better crystallized. Devitrified rhyolite is probably more retentive than basalt.

Some of the problems of Ar loss, particularly partial loss, may be overcome using the ^{40}Ar/^{39}Ar technique (Merrihue & Turner 1966, Mitchell 1968, Dalrymple & Lanphere 1971). Stepwise Ar extraction may allow a more correct high-temperature date to be measured when the total Ar, conventional date is low due to a thermal or alteration overprint (Lanphere & Dalrymple 1971, Dallmeyer 1979, Harrison 1983). This is a particularly powerful method for the dating of hornblende, feldspar and whole-rock samples.

There are circumstances, fortunately less common, where isotopic dates for volcanic rocks are too old. The cause may be simple contamination – a problem of particular concern with volcanic ash, even moderately welded ash flows. There is a proven problem with xenoliths and xenocrysts in lavas (Dalrymple 1964, Funkhouser, Barnes & Naughton 1966, Armstrong 1978a), and exotic detrital minerals in epiclastic volcanic deposits (Curtis 1966). Excess Ar may occur in submarine basalts, quenched at high pressure (Dalrymple & Moore 1968, Funkhouser, Fisher & Bonatti 1968, Noble & Naughton 1968), or in ice-contact lavas (Rutford et al. 1972, Armstrong 1978a), but seems rare in shallow-water pillow lavas (this may be an illusion resulting from their usual high degree of alteration and consequent Ar loss or reduced retentivity). Excess Ar has been observed in both phenocrysts and lavas – most known extreme examples being alkalic rocks (Dalrymple 1969, McDougall, Polach & Stipp 1969, Krummenacher 1970, Fisher 1971, Grant, Freeth & Rex 1972, Mohr, Mitchell & Raynolds 1980). This problem is significant at levels in excess of 1 Ma in only a few per cent of all the volcanic suites that have been investigated. The ^{40}Ar/^{39}Ar technique can recognize excess Ar in some cases, but not all (Lanphere & Dalrymple 1976b, Dallmeyer & Rivers 1983).

Altered lavas with high atmospheric Ar corrections sometimes give anomalously old dates (Baksi 1974, Cassignol & Gillot 1982, Souther, Armstrong & Harakal 1984). The likely explanation is fractionation of absorbed atmospheric Ar by diffusion from the sample during the pumpdown and bakeout that precedes analysis (preferential removal of ^{36}Ar – leading to an incorrect calculation of the atmospheric Ar correction and thus an overestimate of the amount of radiogenic Ar).

Intrusive igneous rocks may also be suitable for K–Ar dating of hornblende, biotite and muscovite. Feldspars, because of their complex inversions and exsolution during slow cooling, are less used because practical experience has shown that they often lose a fraction of their radiogenic Ar (Wetherill, Aldrich & Davis 1955, Goldich, Baadsgaard & Nier 1957). For this reason whole-rock dates of coarse intrusive rocks are suspect in toto. The stratigraphic age of intrusive rocks is normally more loosely bracketed than the age of volcanic rocks but intrusive rocks nevertheless provide many useful time scale points. The reasons for this are that intrusive rocks are usually massive and nearly anhydrous, and thus more resistent to resetting than associated volcanic rocks,

or the critical volcanic rocks may be absent due to erosion. In many magmatic episodes the ages of intrusive and volcanic rocks are essentially identical and it is merely a matter of choosing the best material for dating.

Intrusive rocks are prone to slow cooling after emplacement, and may undergo later thermal metamorphism yet show little evidence of that overprint. Incipient weathering may go unrecognized. Intrusive bodies buried by younger fossiliferous sediments may undergo heating and alteration after burial. Any of these reasons may limit the usefulness of intrusive rocks in providing maximum dates for fossil zones. The most confidence can be given when dates for several high closure temperature phases and for different methods are concordant. Intrusive rock dates are more confidently treated as minimum numbers for biostratigraphic horizons that predate their emplacement. It is in this way that they are most useful for time scale work.

Excess Ar in intrusive rocks is rare – the most notable examples being in hornblende from hypabyssal porphyry bodies in and near the Colorado Plateau (Armstrong 1969, Banks et al. 1972, Banks & Stuckless 1973, Bromfield et al. 1977), in alkalic intrusives (McDougall 1971), and in low K minerals of pegmatites (Damon & Kulp 1958, Laughlin 1969).

Metamorphic rock dates provide few time scale calibration points. In a few situations, where metamorphism followed closely upon deposition, K–Ar minerals or whole-rock dates are useful minimum values for the age of fossils in the rocks affected. The examples that come to mind are of subduction zone metamorphism and mica dates on slates from the peripheral parts of metamorphic belts (Dodson 1963, Harper 1964). But whole-rock dates for slates must be viewed with caution because detrital material may be incompletely degassed or the ambient metamorphic fluids may contain radiogenic Ar (Schamel, unpublished Ph.D. thesis, Yale University, Connecticut, USA, 1973, Harper & Schamel 1971). Metamorphic dates as maximum values for the age of overlying fossiliferous strata are rarely of use in time scale calibration for much the same reasons as just outlined for intrusive rocks. Such dates are only minimum values for a maximum age. One can usually do better.

Excess Ar in relict or new growth minerals, especially biotite, of igneous and metamorphic rocks – a phenomenon observed in areas where old rocks are being rejuvenated (Banks et al. 1972, Wilson 1972, Roddick, Cliff & Rex 1980, Dallmeyer & Rivers 1983) – has not yet been a significant issue in time scale work.

Ar in authigenic and detrital sedimentary minerals has been extensively investigated. Most important is the mineral family glaucony, which provides an important fraction of the time scale calibration points younger than 100 Ma. Little can be added to the definitive discussions in Odin's papers on that substance and the book he edited (Odin et al. 1982b). To paraphrase: with care in selection of well-formed, high K glauconies, the K–Ar dates obtained are in good agreement with high-temperature mineral dates. But the material is very susceptible to Ar loss due to thermal or tectonic overprints (Evernden et al. 1960, Hurley et al. 1960, McRae 1972, Thompson & Hower 1973) and there are cases where a detrital component leads to apparent Ar excess, especially in young samples (Curtis & Reynolds 1958, Obradovich unpublished

Ph.D. thesis, University of California, Berkeley, USA, 1964). Much early work was less critical of sample character and thus many published glaucony dates are suspect, and many proven to be discordant – usually too low when compared with other results. Our analysis shows no systematic deviation of the glaucony dates listed in Odin (1982b) for the last 20 to 115 Ma but significant discrepancies for older samples. Given the many reservations expressed about this material in the past, its relatively weak Ar retentivity, and the fluid-rich chemical environment of sedimentary rocks, this observation is no surprise.

K–Ar dating of illite gives few time scale points, and those probably only by coincidence – poor retentivity being compensated for by detrital contamination (Hower *et al.* 1962, Hurley *et al.* 1963, Hofmann 1971). Authigenic feldspar can only provide minimum dates for enclosing rocks and is likewise unimportant in time scale work.

4.4 Analytical precision of K–Ar dates

The analysis of interlaboratory standards has shown that K–Ar dates can be routinely measured with an interlaboratory precision (1 sigma) of 2.6 to 1.1% (Lanphere & Dalrymple 1967, 1976a, Odin *et al.* 1982b). Tabor, Mark & Wilson (1985) show a within-laboratory precision of 2.9 to 1.0% for replicate Ar analyses of hundreds of samples with more than 50% radiogenic Ar. At low percentages of radiogenic Ar the errors increase dramatically – reaching ± 15% at 10% radiogenic Ar. More precise intercomparison of dates, to 0.5 to 0.7%, may be achieved within a single laboratory (Baadsgaard & Lerbekmo 1982, Odin *et al.* 1982b), and as good as 0.3% where carefully controlled $^{40}Ar/^{39}Ar$ measurements are made (Roddick 1983, Kunk & Sutter 1984, Lippolt, Hess & Burger 1984). None of these figures include the uncertainty of decay constants which is of the order of 0.5% (Gale 1982). Furthermore, in actual practice there may also be a significant geologic variance (in addition to questions of systematic Ar loss or excess). Kistler (1968) and Armstrong (1970) observed a precision of 3 to 4% for mineral separates from multiple sample localities and of different mineral phases of the same ash flow layers in Nevada and Utah. This was greater than within-laboratory precision, based on replicate analysis of standards, by approximately a factor of two. Dalrymple & Hirooka (1965) observed a precision of 1.9% for K–Ar dates of one basalt hand specimen and 2.1% for dates of the same basalt flow at different localities. Cox & Dalrymple (1967) found the error in dating 2.5 Ma old volcanic rocks to be 3.6%.

An individual K–Ar date can only rarely be considered to have a gross precision greater than about 3% when it is compared to dates by other methods and from other laboratories. The analytical methods and precision of conventional K–Ar dates have been only slightly refined since the mid 1950s, so the vintage of a date is rarely of concern in its use for time scale calibration.

4.5 Interpretation of Rb–Sr dates

Fresh volcanic or intrusive igneous rocks are ideal for Rb–Sr dating, using minerals or whole-rock specimens of varying Rb/Sr ratio. In practice the highest Rb/Sr ratios are provided by micas. Mica Rb–Sr dates are of equivalent significance to mica K–Ar dates, even preferable in the case of muscovite because of its higher Rb–Sr closure temperature. Incipient weathering can affect Rb–Sr dates of biotite – resetting them to meaningless lower values (Satir 1974, 1975, Clauer 1981) – so that concordance of both Rb–Sr and K–Ar is important. There is no case where K-mica and whole-rock pairs have given Rb–Sr dates greater than their true crystallization age; Rb–Sr dates of micas are of value in confirming the presence of excess Ar.

The closure temperature for Rb–Sr in feldspars is in the same range as for micas. Feldspar isochrons and feldspar-mica isochrons are likewise interpretable like K–Ar mica dates. When these minerals do not fall within the error limits of an isochron, partial resetting is indicated and the date would not be likely to be useful for time scale calibration. Hornblende and sphene have been observed to have relatively high retentivities for Rb–Sr. They are thus particularly useful for demonstrating discordance in a rock but they only rarely give precise Rb–Sr dates.

Whole-rock Rb–Sr dating is extensively used to date times of differentiation, because of its resistance to resetting, but it is not without numerous pitfalls. Identical $^{87}Sr/^{86}Sr$ initial ratios for all samples of a cogenetic suite are assumed but this is rarely true in detail and sometimes spectacularly false. Studies of chemically heterogeneous young magmatic suites show an intrinsic initial variability of $^{87}Sr/^{86}Sr$ of the order of ±0.0003 (Brooks *et al.* 1976, Basaltic Volcanism Study Project 1981, Armstrong 1986, Armstrong *et al.* 1986). The initial ratio may vary with Rb/Sr ratio, giving a systematic error for the time of differentiation and emplacement (mantle isochrons of Brooks 1976, Brooks & Hart 1978; magma chamber residence time of Souther *et al.* 1984), or be random. Usually it is a bit of both. Initial ratios in S-type magmas are spectacularly variable and may totally defeat any attempts at whole-rock Rb–Sr dating (Armstrong, Taubenneck & Hales 1977, Pigage & Anderson 1985, Armstrong 1986).

Open system behaviour and daughter re-equilibration on a variety of scales may occur in whole-rock suites, and this has generated considerable disagreement. Volcanic rocks, with their initial porosity and unstable mineralogy, are particularly susceptible to resetting. Intrusive rocks are not immune and petrographic criteria alone may be inadequate to screen out all altered rocks. The common symptoms are droopy isochrons – high Rb–Sr ratio points being more discordant – and discordance with other types of dates, especially U–Pb of zircon, hornblende K–Ar, or whole-rock Sm–Nd (Page 1978, Compston, McDougall & Wyborn 1982, Cooper, James & Rutland 1982). This problem bedevils the interpretation of several time scale points. The best isochrons, where initial ratio uncertainties are negligible and a line is well defined, are only minimum values for the age of their rocks. Poor whole-rock isochrons are useful only to test the consistency of any time scale calibration. Lack of scatter of data points about an isochron is not an adequate criterion for dismissing a small degree of resetting. In numerical experiments we have done on isochron resetting, a date reduction of several per cent, and often a perceptible increase in apparent initial ratio, precedes an increase in the scatter of points to give recognizable errorchrons (i.e. those with more scatter than is compatible with reported analytical errors). The creation of false isochrons by special cases of open system behaviour have been discussed by Riley & Compston (1962) and Matsuda (1974).

As a general rule, intrusive igneous rocks are more resistant to whole-rock Rb–Sr date resetting than volcanic rocks. In the plutonic realm thin aplite dikes are the most vulnerable to loss of radiogenic Sr; and they may also have anomalous initial ratios so that their Rb–Sr dates must be viewed cautiously.

Glaucony may be dated by Rb–Sr but rarely with the precision available using K–Ar methods (Odin 1982b). Rb–Sr dates for glauconies provide a consistency check, but are usually less definitive, and are about as easily reset as K–Ar dates. Initial ratio uncertainty and recycled glaucony are problems that must be addressed.

Rb–Sr dating of authigenic clay of shales has been extensively investigated (Cordani, Kawashita & Filho 1978, Clauer 1979a, Bonhomme 1982). Its usefulness is largely in dating Precambrian strata where there is no alternative. Rb–Sr dating of shales suffers from uncertainties in initial ratio and initial ratio uniformity from sample to sample and mineral fraction to mineral fraction (Dasch 1969, Clauer 1979b), variable times or long duration of diagenesis (Perry & Turekian 1974), susceptibility to resetting even at subgreenschist to greenschist metamorphic grade (Peterman 1966, Gebauer & Grünenfelder 1974, Bernard-Griffiths 1976), and systematic error due to detrital components (Whitney & Hurley 1964, Dasch 1969, Spanglet, Brueckner & Senechal 1978, Armstrong & Ramaekers 1985). It is usually a default, last choice, method for dating sedimentary rocks. Considering its problems it has had remarkable success.

4.6   Analytical precision of Rb–Sr dates

The precision of a Rb–Sr date is highly case-dependent so that assignment of an error is impossible, unless complete documentation of laboratory techniques and actual analyses are reported. The gross precision incorporates (1) uncertainty of the decay constant, now about 1% (Gale 1982); (2) analytical uncertainties in Rb/Sr ratios, which vary from less than 1% in carefully controlled mixed-spike isotope dilution work to 2% or more for X-ray fluorescence analyses; (3) uncertainties in $^{87}Sr/^{86}Sr$ ratios (which have varied from ± 0.003 in some early work to ± 0.0001 for most recent work to near ± 0.00001 for the most precise numbers now achievable) and (4) the scatter of points about regression lines or uncertainty in model assumptions.

There is a variety of regression techniques that have been used, each employing different residual minimization criteria. Several are discussed by Brooks, Hart & Wendt (1972). The most widespread practice today is the least squared cubic fit of York (1967). This procedure minimizes deviations from the regression line along diagonal lines of slope equal to the negative inverse of the regression line slope times the ratio of the weights of the two coordinates of individual data points. The individual point coordinate weights are inversely proportional to the squares of the point coordinate errors. This emphasizes data points close to the origin on Rb–Sr isochron plots. With good linear data sets the differences arising from choice of regression technique or model are not significant, but with data that scatter clearly beyond analytical error (scatterchrons or errorchrons where the mean square of weighted deviates, MSWD, exceeds 2.0), or where the range in measured ratios is small compared to analytical uncertainty, the true uncertainty of a reported date is probably much greater than that calculated objectively.

Only Rb–Sr dates that can justify a precision estimate of 5% or better are of any real use in time scale work. Each isochron has to be evaluated on its own merits. In the case of Rb–Sr dates for mica, with a large Rb/Sr ratio and very radiogenic Sr, Lanphere & Dalrymple (1976a) found the between-laboratory precision to be 3.0%. For a standard glaucony the same precision was 3.3% (Odin *et al.* 1982). Baadsgaard & Lerbekmo (1982, 1983) give 1 sigma analytical precision estimates as low as 0.2 to 0.4% for Rb–Sr dates of biotite.

4.7   Interpretation of U–Pb dates

The U–Pb dates used for time scale calibration were at one time mostly for minerals from pegmatites and veins. These can be dated with high precision but rarely have precise stratigraphic brackets. Much of the original work was by wet chemical techniques; these results are now largely of historical significance. U–Pb dating of U-rich black shale or sedimentary U deposits has contributed a few time scale calibration points or limits but these are uncertain due to parent and daughter mobility in the sedimentary rock environment (Cobb & Kulp 1961). Modern U–Pb dating is almost exclusively of zircon, monazite, sphene and apatite from igneous or metamorphic rocks. All four minerals are relatively 'refractory' when pristine. Metamict or impure zircon will show lead loss, even at low grades of metamorphism (Koppel & Sommerauer 1973, Gebauer & Grünenfelder 1976), so no *a priori* closure temperature can be used for the interpretation of zircon U–Pb dates from metamorphic rocks.

The dual isotopes of the parent U give rise to special graphical (concordia of Wetherill (1956) being the most familiar) and mathematical systematics in interpretation of discordant U–Pb dates. This approach can resolve the effects of *one* Pb loss event in the history of a U-bearing mineral. This is a great advantage, especially in dating old rocks with recent disturbance of U–Pb systems. The real-world problem is that zircon suites with only an age of crystallization and one later event are rare and that the analyses are tedious and expensive, so that only a few points of time scale significance have been published. There is often Pb loss in several events subsequent to crystallization, the last being recent weathering and leaching in the laboratory. For old and near-concordant minerals this does not significantly affect upper intercept dates. But for samples up to a few 100 Ma old the concordia curve is only slightly curved and very close to any possible Pb loss trajectories – leading to a low angle of intersection of concordia and multipoint discordia lines – and the consequence is large uncertainties in the upper intercept. Even more insidious is the effect of even traces of old Pb or xenocrystic material (e.g. Gulson & Krogh 1973, Higgins *et al.* 1977, Seiders & Zartman 1978, Williams 1978). This will produce upper intercepts that are older than the crystallization age. In this case the normal U–Pb dates will be minimum values for the time of crystallization. This effect is so common that often the U–Pb dates for Mesozoic-Cenozoic rocks are emphasized rather than Pb/Pb dates (Stern *et al*. 1981, Chen & Moore 1982). In practice the true crystallization age of a nearly concordant U mineral will be bracketed between its U–Pb and Pb/Pb dates. With a large inherited component even the U–Pb date may be excessive. Neither U–Pb or Pb/Pb date alone is definitive; the most confidence can be given on the rare occasion where both are essentially the same, i.e. internal concordance prevails. This is sometimes observed with zircon, but is more common with monazite (Gulson & Krogh 1973, Köppel 1974, Köppel & Grünenfelder 1975, Gebauer & Grünenfelder 1979), sphene (Tilton & Grünenfelder 1968, Gulson & Krogh 1973, Mattinsson & Echeverria 1980), and apatite (Oosthuyzen & Burger 1973) (which may be less often present, and is analytically more troublesome, or more easily completely reset). Various tricks (fractionation using physical property differences, hand picking different shape, colour, or inclusion types, acid leaching, sample abrasion, ion probe analysis – Krogh & Davis 1971, 1974a, Hinton & Long 1979, Davis *et al*. 1982, Krogh 1982a, 1982b) have been invented to obtain the most concordant zircon fraction and thus minimize uncertainties in concordia interpretations.

Whole-rock and multiple-mineral isochron U–Pb dates do not in general have the precision required to be of use in time scale work and they are often reset due to the mobility of both U and Pb (Farquharson & Richards 1975).

4.8 Analytical precision of U–Pb dates

The precision of U–Pb dates has never been evaluated by a multi-laboratory study of a standard mineral, but exchange of standard solutions and spike calibrations and analyses of chemical standards indicate that the analytical errors are small compared to uncertainties in interpretation of points on concordia plots. Most calibrations of mixed U–Pb spikes are probably good to better than 0.3% for Pb/U ratio, thanks to the US National Bureau of Standards certified reference U oxide and Pb metal standards. The half lives are known to better than 0.1% (Gale 1982). Pb isotopic analyses have improved over 30 years of changing technology from >0.1 to <0.01% absolute precision of isotopic ratios. Thus some concordia intercept dates (whether calculated following Ludwig 1980 or Davis 1982) for nearly concordant Precambrian zircons (in which case the dates only depend on measurement of the $^{206}Pb/^{207}Pb$ ratio) have a reported analytical precision better than 0.1% (e.g. Krogh, McNutt & Davis 1982). For the Phanerozoic the analytical precision of U–Pb dates is closer to the limit in measuring Pb/U ratios, typically about 1% but potentially better than 0.5% (Baadsgaard & Lerbekmo 1982, 1983); this is greater than the half life uncertainty but much less than the uncertainty in intepretation of most such dates.

4.9 Interpretation of fission-track dates

A recent development is the application of precisely determined fission-track dates to time scale calibration. The method is applicable to both igneous and metamorphic minerals, only the former being of interest here. Any igneous mineral, for example zircon or apatite, that has remained well below its closure temperature should, in theory, be datable (Naeser 1979). In practice the track density must be neither too low nor too high and that is a function of both U content and age. Apatite has been used to date Cenozoic volcanic rocks but in older material is usually at least partially reset because of its low closure temperature. Zircon dating is done on individual crystals, one at a time, so that xenocrystic or detrital contamination can be recognized and excluded. Zircon is more retentive and its dates by the detector method are proven to be reproducible. The fission-track method is most at an advantage in dating bentonite layers where other techniques are impractical or impossible. Its disadvantage is limited precision, so that it is largely a supplemental time scale calibration, a consistency check. It has had only a minor effect on time scale calibration, even in the Ordovician Period where it makes its greatest contribution (Ross *et al*. 1982).

4.10 Precision of fission-track dates

The precision and accuracy of fission-track dates is a thoroughly debated topic (Johnson, McGee & Naeser 1979, Hurford & Green 1982, 1983, Gale & Beckinsale 1983). The accuracy is dependent on calibration using fission-track standards separated from rocks that have been dated by other techniques. Neutron flux and decay constant uncertainties thus cancel out. The accuracy and precision of calibration and analysis depend on the number of tracks counted in mineral standards, flux monitors, and unknowns, and on the correct recognition of tracks. The errors reported for fission-track dates in recent literature appear to be correct and are taken at face value for time scale calibration.

4.11 Interpretation of Sm–Nd dates

Sm–Nd dating is a relatively new technique that has much in common with Rb–Sr dating. The mathematics and error analysis are identical: whole-rock isochrons and minerals are used to date differentiation, crystallization and metamorphic events (O'Nions *et al*. 1979). The advantage here is that Sm–Nd is more resistant to resetting than most other systems and the method can be applied to mafic rocks (e.g. feldspar–whole-rock or feldspar–mafic mineral isochrons, Ashwal *et al*. 1985) and mafic rock suites (e.g. basalt–komatiite) that would be unsuitable for other techniques. The mineral garnet, often important in achieving a range in Sm/Nd ratio because of its affinity for Sm, is relatively refractory so that garnet–feldspar or garnet–whole-rock mineral isochron dates may be preserved where many other mineral dates are reset (Humphries & Cliff 1982). In general rare earth elements are considered 'immobile' and this characteristic makes it possible for Sm–Nd dates to be preserved where Rb–Sr and U–Pb dates are reset. The difficulties of this technique include initial ratio variability (Cattell, Krogh & Arndt 1984, Chauvel, Dupre & Jenner 1985), rare earth element mobility (Wood, Gibson & Thompson 1976, Hellman, Smith & Henderson 1979, Windrim *et al*. 1984) and limited precision attainable.

4.12 Analytical precision of Sm–Nd dates

Sm–Nd dates are subject to the same types of uncertainties and error analysis as Rb–Sr dates. Each case must be considered on its own merits. The half life uncertainty is about 0.6% (Gale 1982). In a comparative study Sm–Nd dates agreed with U–Pb dates to better than 1% (Nunes 1981). The fairly limited range of Sm/Nd ratios in natural materials and small differences in Nd isotopic composition that must be measured place practical limitations on Sm–Nd dates. Few have precisions as good as ± 20 Ma and thus their usefulness for time scale calibration is largely confined to the early Paleozoic.

4.13 The isotopic database

Isotopic dates that are closely bracketed stratigraphically or provide precise control on one stratigraphic boundary are listed in Table 4.2. These have been gleaned from several generations of such compilations, several recent reviews, and from the geologic literature generally. The database covers the entire Phanerozoic era and not just Paleozoic and Mesozoic time as in GTS 82. This change alone nearly doubles the size of the list. We have depended heavily on a few references. First are the items in the two Phanerozoic Time Scale (PTS) publications (Harland *et al.* 1964 and Harland & Francis 1971) numbered 1–337 and 338–404. Of even greater importance, and generally treated as definitive for the interpretation of any item, is the listing of items in *Numerical Dating in Stratigraphy* (Odin 1982b) which totals 251. Some items have been retained from the listing of Armstrong (1978c) in so far as they are not duplicated in the previous sources or later revised. Other items come from reviews in Snelling (1985b). Beyond these major sources are a number of papers, mostly recent, that add useful time scale calibration data. These are cited, together with their file codes, in Table 4.3.

A large number of items that have at one time or another been proposed as time scale constraints have been excluded from the list. The criteria for exclusion include rejection by the original authors, excessive uncertainty in date or stratigraphic position (generally those that exceed 3 to 4% 1 sigma errors or that lie more than two to five time scale subdivisions away from any likely time scale). Items that are clearly anomalous with respect to the main body of data have also been excluded. In most of these cases we have acted in agreement with previous assessments of the excluded items as anomalous. For example, a number of NDS items are explicitly evaluated as unsuitable for time scale calibration and we are generally in agreement.

If we have erred it is in the direction of including items that more critical reviews have excluded as unsuitable. Our approach is a somewhat 'democratic' one. As many time scale items as possible are allowed to influence the calibration. No single point is given exceptional weight. The result is a compromise that accommodates the maximum number of reported facts, tempered by common sense and experience. Some PTS items may have been retained more out of tradition than current importance, but never when they failed to meet acceptance criteria or were clearly superseded.

4.14 Chronometric errors

In Table 4.2 we have tried to list consistently 1 sigma errors. Some authors have insisted on the use of 2 sigma,

95%, confidence limits, but that is a matter of fashion and taste. Either expression of uncertainty is correct and equally meaningful. Neither is likely to be more precise than two significant figures. The problem is sometimes figuring out just what the original authors intended. Some do not even report errors, others fail to state explicitly which error they are using – assuming everyone understands exactly what they do. The reported errors often represent the internal analytical precision of a laboratory but do not include allowance for interlaboratory variation. Odin recognized this distinction and commented on it several times in his book (Odin 1982b). The analysis of standard mica and glauconite has demonstrated that K–Ar and Rb–Sr interlaboratory precision is of the order of 2 to 3% (Lanphere & Dalrymple 1967, 1976a, Odin 1982b).

We have usually followed original sources or compilations in assigning analytical errors to the dates in Table 4.2. If extremely precise within-laboratory analytical precision is evidently reported we add 1% for interlaboratory comparisons. In many cases we have had to re-create errors or assign them arbitrarily. This is most easily done with K–Ar dates.

The coefficient of variation of K–Ar dates is largely dependent on the fraction of radiogenic ^{40}Ar (**r**) observed in the analysis (Dalrymple & Lanphere 1969, Tabor *et al.* 1985). The compilation by Tabor, Mark & Wilson of a large number of analyses done in one laboratory showed that the average coefficients of variation for Ar analyses fit a regression equal to $-2.43 + 3.38 \times r^{-0.65}$. The curve they show is very close to the hyperbola $100/(\mathbf{r} \times 100)$, or simply 100/(per cent radiogenic Ar). Where a K–Ar date is given without errors but with per cent radiogenic or per cent atmospheric Ar reported in the original analytical data, an error has been assigned equal to that quantity plus 2% (for K analysis and Ar calibration uncertainties). The results of this simple procedure are error estimates that agree well with errors when they are reported.

Where no information is given from which to estimate a K–Ar error, the value assigned is based on a plot of reported errors versus data for the compiled data set. In other words, it is assigned an error similar to that of nearby analyses. This arbitrary error varies from about 5% in Early Cenozoic to 3% in Early Paleozoic. This is consistent with the errors assigned by many authors. For example at one time the USGS assigned a blanket error of approximately 3% in papers on Mesozoic geochronometry (e.g. Dalrymple & Lanphere 1969), and Cowie & Cribb (1978) gave an uncertainty of 3% to Cambrian dates for which errors were not reported.

Rb–Sr dates for minerals with large Rb/Sr ratios and errors for U–Pb dates have likewise been approximated using errors which are typical for similar samples. This may be quite incorrect where the enrichment in radiogenic Sr is small, but that would affect few items. For whole-rock isochron dates an original error estimate is mandatory for inclusion in the database. As already discussed, there is an unknown systematic error, unstated or unmeasured geologic variance, interlaboratory variance, and uncertainty in decay constants which are not always included in published error estimates. The gross uncertainties are thus somewhat larger than the errors used and the time scale cannot be expected to pass through all error rectangles.

Key to Table 4.2

Column (1) Item codes

The letters give the identity of the reference as given below; the number gives the item number in the publication cited. The letter R after a number indicates that the date has been revised.

A	Armstrong 1978c
AB	Aubry *et al.* 1988
AP	Abele & Page 1974
B	Bouroz in Cohee *et al.* 1978
BAN	Bandet *et al.* 1984
BD	Bryan & Duncan 1983
BDJK	Bianchi *et al.* 1985
BERG	Berggren *et al.* 1985a
BFSB	Bellon *et al.* 1986
BL	Boles & Landis 1984
BLM	Baadsgaard *et al.* 1988
B/MH	Backman *et al.* 1984; Macintyre & Hamilton 1984
CC	Cowie & Cribb in Cohee *et al.* 1978
CJ	Cowie & Johnson in Snelling 1985
CPR	Culver, Pojeta & Repetski 1988
DB	Dallmeyer & Van Breeman 1981
DKTS	Dunning *et al.* 1987
FB	Fabre & Bellon 1985
FC	Fyffe & Cormier 1979
FHMB	Fitch *et al.* 1978
FLN	Flynn 1986
FW	Forster & Warrington in Snelling 1985
G	Gale, Beckinsale & Wadge 1979
GA	Gale in Snelling 1985
GBVK	Gonovin *et al.* 1985
GBW	Gale, Beckinsale & Wadge 1980
HAL	Hall *et al.* 1984
HB	Hardenbol & Berggren in Cohee *et al.* 1978
HL	Hess & Lippolt 1986
HLB	Hess, Lippolt & Borsuk 1987
HLHP	Hess *et al.* 1985
HMFT	Hamamoto *et al.* 1980
HMP	Hopson, Mattinson & Pessagno 1981
HS	Harper 1984
HUL	Hubacher & Lux 1987
J	Jones, Carr & Wright 1981
JE	Jeans *et al.* 1982
JEN	Jenkins *et al.* 1985
K	Kunk *et al.* in Snelling 1985
KFH	Kokelaar, Fitch & Hooker 1982
KK	Keller & Krasnobaev 1983
KM	Kimbrough & Mattinson 1984
LAN	Lanphere 1981
LBM	Lang & Mimran 1985
LI	Lippolt, Hess & Burger 1984; Hess & Lippolt 1986
LJ	Lanphere & Jones in Cohee *et al.* 1978
LT	Lanphere & Tailleur 1983
MAT	Mattison 1975
MB	Mussett & Barker 1983
MCK	McKerrow, Lambert & Chamberlain 1980
MLC	McKerrow, Lambert & Cocks in Snelling 1985
MON	Montanari *et al.* 1985
MP	McDougall & Page 1975
NDS	Odin *et al.* 1982b
OB	Obradovitch 1984
OGA	Odin *et al.* 1983
OGD	Odin, Gale & Dore in Snelling 1985
OHJS	Odin, Hunziker, Jeppsson & Spjeldnaes 1986
OHMT	Odin, Hurford, Morgan & Toghill 1986
ONI	Obradovich *et al.* 1982
OSK	Obradovich, Sutter & Kunk 1986
PPC	Peucot, Paris & Chalet 1986
PRO	Prothero & Armentrout 1985
PTS	GSL Phanerozoic Time-Scale 1964 (Items 1 to 337) and GSL Phanerozoic Time-Scale Supplement 1971 (Items 338 to 366)
R	Ross *et al.* 1982
RN	Ross & Naeser 1984
ROD	Rodda *et al.* 1985
RUN	Rundle 1986
RZM	Reynolds, Zentilli & Muecke 1981
S	Shibata 1986
SCH	Schweickert *et al.* 1984
SEID	Seidmann *et al.* 1984
SHE	Shepard 1986
SM	Shibata & Miyata 1978
SMYH	Shibata *et al.* in Cohee *et al.* 1978
SR	Styles & Rundle 1981
THW	Tilton, Hopson & Wright 1981
W	Wang 1983
WEST	Westphal *et al.* 1979
WP	Wilson & Pedersen 1981
WTCM	Williams *et al.* 1982
YQHT	Xing Yusheng *et al.* 1984
ZGH	Zhang, Ma & Lee 1984

Columns (2) and (4) Dating method and coding

Numbers in [ ] are abbreviated decay constants.

1 = Old Western K–Ar [4.72, 0.584, 1.19]
2 = Old USSR K–Ar [4.72, 0.557, 1.19]
3 = New K–Ar and Ar–Ar [4.962, 0.581, 1.167]
4 = Old Rb–Sr [1.39]
5 = Old Rb–Sr [1.47]
6 = New Rb–Sr [1.42]
7 = Old U–Pb [1.54, 0.971]
8 = New or corrected U–Pb [1.55125, 0.98485]
9 = (Not used)
10 = Fission track [7.03]
11 = Sm–Nd [6.54]

Reference Steiger & Jäger 1977

Column (3) Material dated

Bi = biotite; Fl = feldspar; Gl = glauconite; Gt = garnet; Hb = hornblende; Ill = illite; Misc = miscellaneous;

Mu = muscovite; Mz = Monazite; Ph = phlogopite; Sd = sanidine; Sp = sphene; Ur = uraninite; WR = whole rock; Zr = zircon.

Column (5) Published date in Ma

Column (6) Standardized date in Ma

Column (7) Error

Positive values are the 1 sigma errors as originally reported, corrected to 1 sigma, or estimated from data given for radiogenic content. Negative values are errors estimated from the standardized date and regression lines fitting the published errors to published dates of similar type. Errors for multiple determinations are reduced by division by the square root of the number of analyses. No errors are allowed to be less than 0.5%, even if better precision is claimed.

Columns (8) and (9) 'Stage' codes

The date is bracketed by the stages represented by the two numbers listed in Table 4.3. For example, a code of 16   16 means the rock dated lies wholly in the Lutetian or its stratigraphic equivalents; a code of 5   21 means that the oldest rocks known to post-date the dated rocks are partly Messinian in age and that the youngest rocks known to pre-date the dated rocks are partly Campanian in age.

Column (10) Geographic/geologic name and item cross-reference codes

Each item has been assigned a name for ease of reference. Letter/number codes refer to other discussions or listings of the same item. Reference codes are as in column 1.

Order in which items appear

Items are ordered by 'stage' codes. Where an item is confined to a single stage the item appears under the appropriate stage list. For example, MDMR3 is a feldspar, dated at 1.88 Ma, from a unit within the Piacenzian, coded as '3 3', and heads the Piacenzian list. All other Piacenzian items follow it, in order of increasing standardized date.

Where **adjacent** stages are involved the item is placed in the listing under the younger stage. Only where **large** time ranges are involved are the items placed in the most appropriate stage listing – the one for which the date provides a useful bracket. Younger brackets are listed first, e.g. 1 28 before 28 28 items, and older (e.g. 28 128) items are listed last under the relevant stage.

Table 4.2. *Isotopic database*

Item	Dating method and mineral			Pub. date	Stand. date	1 sigma error	'Stage' codes		Item name
(1)	(2)	(3)	(4)	(5)	(6)	(7)	(8)	(9)	(10)

Pleistocene

Pleistocene chronograms are not used. For Pleistocene ages, see Sections 3.21 (pp. 67–72) and Figure 6.4.

Piacenzian

Item	(2)	(3)	(4)	(5)	(6)	(7)	(8)	(9)	(10)
PTS245	KAr	Fl	1	1.75	1.79	0.05	3	3	Olduvai
MDMR3	KAr	Fl	3	1.88	1.88	0.02	3	3	KBS Tuff
ONI1	KAr	Hb	3	2.22	2.22	0.03	3	3	Vrica
ROD1	KAr	Bi	3	3.20	3.20	0.05	3	3	Suva
BDJK	KAr	Bi	3	3.20	3.20	0.03	3	3	Crete
PTS248	KAr	Bi	1	2.25	2.31	0.07	3	4	Coso
ROD2	KAr	Bi	3	3.50	3.50	0.05	3	4	Suva
PTS251	KAr	WR	1	4.10	4.20	0.42	3	4	Bidahochi
PTS250R	KAr	Fl	3	4.50	4.50	0.50	3	4	Glenns Ferry

Zanclian

Item	(2)	(3)	(4)	(5)	(6)	(7)	(8)	(9)	(10)
ROD3	KAr	Bi	3	3.85	3.85	0.05	4	4	Suva
HAL1	UPb	Zr	8	3.93	3.93	0.28	4	4	Cindery
HAL2	ArAr	Fl	3	3.94	3.94	0.05	4	4	Cindery
NDS245	KAr	HbBi	3	3.98	3.98	0.10	4	4	Aegina Island
ROD4	KAr	Bi	3	4.80	4.80	0.05	4	4	Suva
MP2	KAr	Bi	3	5.49	5.49	0.10	4	4	Koroimavua
MP1	KAr	WR	3	4.42	4.42	0.30	4	5	Mohole
MP3	KAr	Hb	3	5.85	5.85	0.10	4	5	Namosi
PTS252	KAr	Fl	1	5.20	5.33	0.22	4	6	Pinhole

Messinian

Item	(2)	(3)	(4)	(5)	(6)	(7)	(8)	(9)	(10)
JEN4	KAr	WR	3	6.40	6.40	0.60	5	5	Fort de France
MP4	KAr	Bi	3	6.93	6.93	0.75	5	5	Elba
MP5	KAr	Bi	3	7.44	7.44	0.75	5	5	Melilla
PTS253R	KAr	Sd	3	6.69	6.69	0.09	5	6	Rattlesnake
PTS254	KAr	Fl	1	8.10	8.30	0.74	5	6	Alturas
PTS255	KAr	Sd	1	8.90	9.12	0.26	5	6	Drewsey
PTS256	KAr	WR	1	9.20	9.43	0.31	5	6	Teewinot
PTS110	KAr	Gl	1	9.60	9.84	0.43	5	6	Kadenberge

Tortonian

Item	(2)	(3)	(4)	(5)	(6)	(7)	(8)	(9)	(10)
PTS28	KAr	Bi	1	9.30	9.53	0.50	6	6	Smith Valley
PTS258	KAr	Fl	1	9.89	10.14	0.78	6	6	Siesta
PTS257	KAr	WR	1	10.00	10.25	0.41	6	6	Hole in the Wall
PTS271	KAr	WR	1	9.80	10.04	1.28	6	7	Grizzly Peak
PTS272	KAr	WR	1	9.80	10.04	0.89	6	7	Grizzly Peak
PTS259	KAr	Bi	1	9.90	10.15	0.54	6	7	Ricardo
PTS260	KAr	Sd	1	10.70	10.97	0.56	6	7	Avawatz
PTS261	KAr	Sd	1	11.00	11.27	0.33	6	7	Avawatz
PTS274	KAr	WR	1	11.40	11.68	0.84	6	7	Grizzly Peak
MP6	KAr	WR	3	11.70	11.70	0.60	6	7	Mohole
PTS273	KAr	WR	1	11.70	11.99	0.62	6	7	Grizzly Peak
MP8	KAr	WR	3	12.06	12.06	1.15	6	7	Cronese Basin
PTS275	KAr	BiFl	1	11.82	12.12	0.50	6	7	Czechoslovakia
MP9	KAr	FlWR	3	14.06	14.06	0.20	6	7	Dunedin

Table 4.2 *continued*

Item	Dating method and mineral			Pub. date	Stand. date	1 sigma error	'Stage' codes		Item name
(1)	(2)	(3)	(4)	(5)	(6)	(7)	(8)	(9)	(10)
Serravallian									
JEN3	KAr	WR	3	12.80	12.80	0.60	1	7	Marin
PTS264	KAr	WR	1	10.60	10.86	0.53	7	7	Aldrich Station
PTS266	KAr	Bi	1	10.70	10.97	0.51	7	7	Esmeralda
PTS265	KAr	Bi	1	10.80	11.07	0.38	7	7	Coal Valley
PTS262	KAr	Bi	1	10.85	11.12	1.30	7	7	Aldrich Station
PTS263	KAr	Bi	1	11.00	11.27	1.44	7	7	Aldrich Station
PTS268	KAr	Bi	1	11.10	11.37	0.40	7	7	Esmeralda
PTS27	KAr	Bi	1	11.20	11.48	0.50	7	7	Coal Valley
PTS269	KAr	Bi	1	11.40	11.68	0.39	7	7	Esmeralda
PTS267	KAr	Sd	1	11.50	11.78	0.40	7	7	Esmeralda
MP7	KAr	WR	3	12.63	12.63	0.40	7	8	Mohole
MP11	KAr	Fl	3	14.47	14.47	0.40	7	8	California
PTS276	KAr	WR	1	14.50	14.86	0.65	7	9	Steens
JEN2	KAr	WR	3	15.00	15.00	0.30	7	9	Brassignac
PTS277	KAr	Fl	1	14.70	15.06	0.55	7	9	Steens
PTS38	KAr	Bi	1	15.20	15.58	0.50	7	9	Barstow
PTS39	KAr	Bi	1	15.20	15.58	1.72	7	9	Caliente
PTS281	KAr	Sd	1	15.40	15.78	0.47	7	9	Jarbridge
MP10	KAr	WRHb	3	14.11	14.11	1.25	7	10	New Guinea
Langhian-Late									
PTS279	KAr	Sd	1	15.05	15.42	0.47	8	9	Payette
MP12	KAr	Fl	3	15.71	15.71	−0.78	8	9	California
PTS282	KAr	Sd	1	15.60	15.98	0.49	8	9	Vya
PTS280	KAr	WR	1	15.40	15.78	2.23	8	10	Mascall
Langhian-Early									
BAN	KAr	WR	1	15.30	15.30	0.30	9	9	Fangario
JEN1	KAr	WR	3	16.85	16.85	0.82	9	10	Sainte Anne
MP15	KAr	WRFl	3	17.24	17.24	0.40	9	10	Manokou
PTS79	KAr	Bi	1	17.20	17.62	0.60	9	10	Kinnick
Burdigalian									
MP13	KAr	WR	3	16.06	16.06	0.35	10	10	Gergovie
NDS216	KAr	WR	3	18.50	18.50	0.25	10	10	Chateau de Beaulieu
NDS88	KAr	Gl	3	19.30	19.30	0.50	10	10	Bad Hall PTS82
NDS48	KAr	Gl	3	20.50	20.50	0.45	10	10	Soustons
PTS310	KAr	Fl	1	21.70	22.23	2.30	10	10	Santa Cruz
NDS45	KAr	Gl	3	18.20	18.20	0.82	10	11	Briglez
NDS46	KAr	Gl	3	18.55	18.55	0.27	10	11	Ponzone
PTS80	KAr	Bi	1	21.45	21.98	1.81	10	12	Harrison
PTS283	KAr	Fl	1	24.90	25.51	1.36	10	12	John Day
PTS284	KAr	WR	1	25.30	25.92	1.14	10	12	John Day
Aquitanian									
NDS156	KAr	WR	3	22.00	22.00	0.30	11	11	Maude MP16
NDS155	KAr	BiFl	3	22.50	22.50	0.17	11	11	San Emigdio
NDS47	KAr	Gl	3	22.60	22.60	0.60	11	11	Corleone
NDS52	KAr	WR	3	25.97	25.97	0.58	11	11	Monti Lessini
MP17	KAr	Fl	3	23.09	23.09	−1.15	11	12	California

continued

Table 4.2 *continued*

Item	Dating method and mineral			Pub. date	Stand. date	1 sigma error	'Stage' codes		Item name
(1)	(2)	(3)	(4)	(5)	(6)	(7)	(8)	(9)	(10)
Chattian									
NDS42	KAr	Gl	3	22.00	22.00	0.55	12	12	Eger
PTS81	KAr	Gl	1	22.50	23.05	1.14	12	12	Santos
NDS154	KAr	Fl	3	23.80	23.80	0.27	12	12	Iversen HB96
NDS124	KAr	Gl	3	26.20	26.20	0.50	12	12	Astrup HB7
PTS285	KAr	WR	1	25.60	26.23	1.30	12	12	Gering
NDS138	KAr	WR	3	26.50	26.50	0.50	12	12	Patagonia
AP	KAr	WR	1	26.75	27.40	0.30	12	12	Aireys
MON5b	RbSr	Bi	6	27.80	27.80	0.10	12	12	Gubbio
MON5a	KAr	Bi	3	28.10	28.10	0.15	12	12	Gubbio
PTS286	KAr	WR	1	29.50	30.22	1.08	12	12	Gering
NDS218	KAr	Bi	3	31.30	31.30	0.30	12	12	Rhodes-Dali
PRO6	KAr	WR	3	32.30	32.30	−1.13	12	12	Vieja, upper
PTS296	KAr	Bi	3	31.60	32.37	1.10	12	12	Natrona PRO4
NDS123	KAr	Gl	3	24.55	24.55	0.22	12	13	Lr Saxony
PTS83	KAr	Sd	1	25.70	26.33	0.80	12	13	John Day
NDS52	KAr	WR	3	29.40	29.40	0.74	12	13	Marostica
NDS51	KAr	WR	3	29.90	29.90	0.55	12	13	Marostica (altered)
PTS288	KAr	Sd	1	31.10	31.86	1.33	12	13	John Day
PTS289	KAr	WR	1	31.50	32.27	1.36	12	13	John Day
PTS295	KAr	Bi	3	32.60	33.39	1.15	12	13	Lone Tree Gulch PR03
Rupelian									
NDS58	KAr	Gl	3	28.60	28.60	0.62	13	13	Vicksburg
NDS217	KAr	Gl	3	30.10	30.10	0.30	13	13	Kassel-Ihringshausen
NDS20	KAr	Gl	3	30.90	30.90	0.85	13	13	Pellenberg (weathered)
MON4	KAr	Bi	3	32.00	32.00	0.40	13	13	Gubbio
PTS311	KAr	WR	1	31.65	32.42	1.13	13	13	Hsanda Gol
NDS218	KAr	Bi	3	32.60	32.60	0.30	13	13	Rhodes-Karakia
BERG3	KAr	WR	3	33.20	33.20	1.10	13	13	DSDP 448A
PTS291	KAr	SdBi	3	35.09	35.09	1.00	13	13	Lone Tree Gulch B PRO1
MON3	KAr	Bi	3	35.40	35.40	0.20	13	13	Gubbio
PTS293	KAr	BiSd	3	35.55	35.55	0.85	13	13	Natrona PRO2 (average PTS293+294)
HB168	KAr	Gl	1	35.50	36.36	0.70	13	13	Lehrte
NDS58	KAr	WR	3	36.70	36.70	1.81	13	13	Vicksburg HB66
HB171	KAr	Gl	1	36.61	37.56	0.40	13	13	Helmstedt Gehlberg (av HB171–173)
PTS84	KAr	Sd	3	33.91	33.91	1.11	13	14	Vieja
NDS40	KAr	Gl	3	37.50	37.50	0.35	13	14	Silberberg
NDS58	KAr	Gl	3	38.10	38.10	0.70	13	14	Jackson Shubuta HB65
Priabonian									
MON2b	RbSr	Bi	6	35.40	35.40	0.10	14	14	Gubbio
MON1	KAr	Bi	3	36.40	36.40	0.15	14	14	Gubbio
PTS23	KAr	Gl	1	35.60	36.49	4.00	14	14	Umantsevo
MON2a	KAr	Bi	3	36.90	36.90	0.65	14	14	Gubbio
PTS299	KAr	Sd	1	36.50	37.39	1.18	14	14	Mitchell
PRO7	KAr	WR	3	37.40	37.40	−1.13	14	14	Vieja, lower
NDS58	KAr	Gl	3	37.70	37.70	0.70	14	14	Pachuta HB64
PTS290	KAr	Fl	3	37.70	37.70	1.15	14	14	Natrona PRO5
PTS298	KAr	WR	1	37.50	38.41	1.60	14	14	Clarno
PRO1	KAr	WR	3	38.50	38.50	1.60	14	14	Lincoln Creek

Table 4.2 *continued*

Item	Dating method and mineral			Pub. date	Stand. date	1 sigma error	'Stage' codes		Item name
(1)	(2)	(3)	(4)	(5)	(6)	(7)	(8)	(9)	(10)
NDS26	KAr	WR	3	38.65	38.65	0.71	14	14	DSDP Leg 31
HB169	KAr	Gl	1	38.60	39.54	0.53	14	14	Lehrte (av HB169+170)
HB175	KAr	Gl	3	39.79	39.79	0.53	14	14	Helmstedt (av HB174+175)
BERG2	Fi	Zr	10	40.75	40.75	1.17	14	14	Polanyi
PTS300	KAr	WRBi	1	37.00	37.90	2.02	14	15	Norwood
PTS125	KAr	Gl	1	37.00	37.90	1.92	14	15	Kaiatan
NDS58	KAr	Gl	1	39.10	39.10	−1.56	14	15	Jackson Yazoo HB63
NDS58	KAr	Bi	3	39.50	39.50	−1.58	14	15	Jackson Yazoo HB62
NDS41	KAr	G	13	37.80	37.80	0.35	14	16	Helmstedt Gehlberg
Bartonian									
NDS58	KAr	Gl	3	38.30	38.30	1.53	15	15	Moodys Branch HB61
NDS1	KAr	Gl	3	38.90	38.90	0.90	15	15	Isle of Wight
NDS2	KAr	Gl	3	39.60	39.60	0.90	15	15	Hampshire
PTS114	KAr	Gl	1	39.40	40.35	1.47	15	15	Moodys Branch
NDS57	KAr	Gl	3	40.70	40.70	1.03	15	15	Claiborne Gosport (av HB59+HB60)
OSK	ArAr	Sd	3	42.00	42.00	−1.00	15	15	Cook Mtn
NDS5	KAr	Gl	3	42.40	42.40	0.90	15	15	Brockenhurst
NDS3	KAr	Gl	3	39.10	39.10	0.75	15	16	Studley Wood
NDS21	KAr	Gl	3	40.00	40.00	0.80	15	16	Assche
NDS23	KAr	Gl	3	41.00	41.00	0.90	15	16	Wemmel
NDS84	KAr	Gl	3	41.08	41.08	0.25	15	16	Bande Noire
NDS57	KAr	BiSd	3	42.80	42.80	1.11	15	16	Claiborne Cook (av HB52+HB57)
NDS84	RbSr	Gl	6	43.09	43.09	0.12	15	16	Bande Noire
PTS301	KAr	Fl	1	42.70	43.73	1.38	15	16	Alamo Creek
PTS86	KAr	Gl	1	43.20	44.24	2.00	15	16	Hernandez
PTS302	KAr	Bi	1	45.00	46.09	2.46	15	16	Badwater
PTS21	KAr	Gl	2	51.00	49.84	−2.49	15	16	Turgay
NDS50	KAr	WR	3	39.90	39.90	0.77	15	128	Priabona
Lutetian									
NDS4	KAr	Gl	3	40.40	40.40	0.53	16	16	Brook (low?)
NDS212b	RbSr	Gl	6	41.40	41.40	0.35	16	16	Brook
NDS19	KAr	Gl	3	41.70	41.70	1.05	16	16	Luna de Sus
NDS49	KAr	WR	3	42.60	42.60	1.06	16	16	East Lessini
PTS115	KAr	Gl	1	41.80	42.81	1.38	16	16	Domengine
NDS30	KAr	Gl	3	42.90	42.90	0.60	16	16	Trou des Halles
NDS31	KAr	Gl	3	43.20	43.20	0.65	16	16	English Channel
PTS124	KAr	Gl	1	42.40	43.42	1.43	16	16	Stave Creek
NDS6	KAr	Gl	3	43.60	43.60	1.40	16	16	Bracklesham IX
NDS33	KAr	Gl	3	43.70	43.70	1.05	16	16	St Leu d'Esserent
NDS7	KAr	Gl	3	44.00	44.00	0.35	16	16	Bracklesham VI
NDS9	KAr	Gl	3	44.20	44.20	0.65	16	16	Yateley
NDS29	KAr	Gl	3	44.40	44.40	1.15	16	16	Fosses
NDS90	KAr	Gl	3	44.43	44.43	0.39	16	16	Fosses PTS87
NDS91	KAr	Gl	3	44.50	44.50	0.78	16	16	Knap-Hazeley
NDS57	KAr	Gl	3	44.80	44.80	1.72	16	16	Claiborne Zilpha HB48
NDS24	KAr	Gl	3	45.00	45.00	1.10	16	16	Hougaarde
NDS160	RbSr	Gl	6	45.10	45.10	0.45	16	16	Bracklesham 2
NDS32	KAr	Gl	3	46.20	46.20	0.80	16	16	Verneuil en Halatte

continued

Table 4.2 *continued*

Item	Dating method and mineral			Pub. date	Stand. date	1 sigma error	'Stage' codes		Item name
(1)	(2)	(3)	(4)	(5)	(6)	(7)	(8)	(9)	(10)
Lutetian *continued*									
BD	KAr	Bi	3	46.30	46.30	0.30	16	16	DSDP 72–516
NDS10	KAr	Gl	3	46.40	46.40	0.75	16	16	Bracklesham 2
PTS303	KAr	Bi	1	45.40	46.49	1.42	16	16	Wagonbed Springs
NDS57	KAr	Gl	3	46.50	46.50	1.20	16	16	Claiborne Weches HB50 HB51
NDS57	KAr	Gl	3	46.80	46.80	1.74	16	16	Claiborne Lisbon HB49
HB71	KAr	Gl	1	47.00	48.13	3.00	16	16	Lutetian Type
FLN	KAr	BiSd	3	49.30	49.30	0.50	16	16	Bridger/Uintan
NDS57	KAr	Gl	1	49.00	50.18	2.00	16	16	Claiborne Weches PTS102
NDS25	KAr	Gl	3	46.30	46.30	0.50	16	17	Aeltre
NDS57	KAr	Gl	3	48.40	48.40	1.66	16	17	Claiborne Tallahata HB47
PTS304	KAr	Bi	1	49.00	50.18	1.65	16	17	Green Cove
PTS305	KAr	Bi	1	49.20	50.38	2.15	16	17	Wind River
PTS88	KAr	Gl	1	51.00	52.22	2.00	16	18	Ober Teisendorf
Ypresian									
GBVK1	KAr	Gl	3	46.00	46.00	1.00	17	17	Syria
GBVK2	RbSr	Gl	6	46.00	46.00	4.00	17	17	Syria
NDS11	KAr	Gl	3	46.10	46.10	1.05	17	17	Bracklesham IV
NDS35	KAr	Gl	3	47.30	47.30	0.70	17	17	Montataire
NDS15	RbSr	Gl	6	47.70	47.70	0.45	17	17	Nanjemoy
NDS35b	KAr	Gl	3	47.80	47.80	1.55	17	17	Alzy-Jouy
NDS13	KAr	Gl	3	47.95	47.95	0.80	17	17	Whitecliff Bay
NDS34	KAr	Gl	3	48.30	48.30	1.35	17	17	Villers St-Paul
NDS12	KAr	Gl	3	50.90	50.90	1.45	17	17	London
NDS37	KAr	Gl	3	52.80	52.80	1.03	17	17	Varengeville
NDS112	KAr	Gl	3	54.90	54.90	0.90	17	17	Manasquan
NDS56	KAr	Gl	3	51.65	51.65	0.90	17	18	Hatchetigbee Bashi HB44 HB45 HB46 HB76 PTS101
NDS14	KAr	Gl	3	52.10	52.10	1.50	17	18	Studland
BERG1	KAr	WR	3	56.50	56.50	0.60	17	18	Kap Brewster
Thanetian									
NDS27	KAr	Gl	3	52.60	52.60	1.20	18	18	Gelinden
NDS17	KAr	Gl	3	53.10	53.10	1.65	18	18	Bishopstone Glen
B/MH	ArAr	WR	3	53.50	53.50	1.90	18	18	DSDP 81–555 (dubious material)
NDS28	KAr	Gl	3	54.10	54.10	1.00	18	18	Lincent
NDS38	KAr	Gl	3	54.40	54.40	0.97	18	18	Guiscard
NDS16a	KAr	Gl	3	54.80	54.80	1.75	18	18	Herne Bay
NDS39	KAr	Gl	3	56.00	56.00	0.95	18	18	Cuise-la-Motte
NDS22	RbSr	Gl	6	56.60	56.60	3.40	18	18	Beaufort
NDS113	KAr	Gl	3	57.50	57.50	0.67	18	18	Vincetown
NDS16b	RbSr	Gl	6	57.50	57.50	1.50	18	18	Herne Bay
FIT77	KAr	Gl	1	56.80	58.25	0.31	18	18	Reculver
PTS105	KAr	Gl	1	57.50	58.87	2.01	18	18	Thanet
PTS113	KAr	Gl	1	58.50	59.90	2.53	18	18	Lodo
PTS106	KAr	Bi	1	59.50	60.92	3.00	18	18	Martinez
FHMB1	KAr	Gl	1	59.50	61.02	0.45	18	18	Fanet, Pegwell
NDS55	KAr	Gl	3	59.44	59.44	0.70	18	19	Midway (av HB41–43)
NDS247	KAr	Gl	3	60.50	60.50	0.35	18	19	Boryszew

Table 4.2 *continued*

Item	Dating method and mineral			Pub. date	Stand. date	1 sigma error	'Stage' codes		Item name
(1)	(2)	(3)	(4)	(5)	(6)	(7)	(8)	(9)	(10)
Danian									
PTS306	KAr	WR	1	58.70	60.10	1.82	1	19	South Table Mtn
NDS92	KAr	Gl	3	61.25	61.25	1.08	19	19	Hornerstown A467
PTS362	KAr	BiSd	1	60.70	62.06	1.50	19	19	Paskapoo
NDS114	KAr	Gl	3	62.10	62.10	1.55	19	19	Hornerstown
NDS126a	KAr	Sd	3	63.00	63.00	0.32	19	19	Red Deer Valley
NDS120	KAr	Gl	3	63.03	63.03	0.63	19	19	Salamanca
NDS127a	RbSr	Bi	6	63.70	63.70	0.32	19	19	Z-Coal (Baadsgaard)
NDS127b	UPb	Zr	8	63.90	63.90	0.35	19	19	Z-Coal (Baadsgaard)
BLM	KAr	Bi	3	65.40	65.40	1.10	19	19	KT Boundary, N. Am (McDougall)
OSK	ArAr	Sd	3	65.50	65.50	−1.00	19	19	Red Desert
NDS103	KAr	Bi	3	65.80	65.80	0.70	19	19	Denver
OB2	ArAr	Bi	3	65.80	65.80	0.80	19	19	Denver
OB1	ArAr	Sd	3	66.00	66.00	0.90	19	19	Z-Coal (Obradovitch)
PTS307	KAr	Fl	1	64.80	66.34	2.41	19	19	South Table Mtn
PTS329	KAr	WR	1	65.50	67.14	−3.02	19	19	Sikhote-Alin
Maastrichtian									
PTS198b	RbSr	Bi	4	64.00	62.65	5.00	16	20	Jamaica
PTS198c	UPb	Sp	7	64.00	63.36	5.00	16	20	Jamaica
PTS198a	KAr	Bi	1	65.50	67.17	5.00	16	20	Jamaica
NDS126b	KAr	Sd	3	63.33	63.33	0.32	20	20	Red Deer Valley PTS200 PTS363 PTS365
NDS115a	KAr	Gl	3	63.53	63.53	0.70	20	20	Monmouth, Red Bank
PTS364	KAr	Fl	1	63.78	65.41	1.00	20	20	Triceratops PTS199 PTS365
NDS36	RbSr	Gl	6	66.70	66.70	1.00	20	20	Peedee
S3	KAr	WR	3	66.80	66.80	−2.00	20	20	Hamanaka
PTS365	KAr	BiSd	1	65.50	67.00	0.70	20	20	Kneehills PTS200 PTS363 NDS126
A449	RbSr	Gl	4	68.50	67.05	−3.02	20	20	Navarro
NDS104	KAr	Bi	1	67.40	69.00	0.70	20	20	Fox Hills A450
S8	RbSr	Bi	6	69.80	69.80	−2.10	20	20	Oborogawa
S7	KAr	WR	3	69.80	69.80	−2.10	20	20	Oborogawa
NDS104	KAr	Bi	1	68.50	70.10	0.70	20	20	Pierre NDS104
S6	KAr	Bi	3	70.40	70.40	−2.11	20	20	Oborogawa
PTS54	KAr	Gl	1	69.00	70.63	1.93	20	20	Ripley
NDS139	KAr	Gl	3	71.50	71.50	0.74	20	20	Gulpen A511–A514
S9	RbSr	Bi	6	69.50	69.50	−2.10	20	21	Otamura
S4	KAr	Bi	3	70.30	70.30	−2.11	20	21	Otamura
PTS365	KAr	BiSd	1	72.34	74.18	0.77	20	21	Bearpaw PTS201
Campanian									
NDS116	KAr	Gl	3	71.60	71.60	1.35	21	21	Marshalltown
NDS105	KAr	Fl	1	71.30	72.98	0.70	21	21	Mancos A454
NDS105	KAr	Bi	1	71.50	73.19	0.70	21	21	Bearpaw A452
OSK	ArAr	Sd	3	73.40	73.40	−1.00	21	21	San Juan
NDS105	KAr	Bi	1	72.20	73.90	0.70	21	21	Pierre, DeGrey A455
S5	KAr	WR	3	74.00	74.00	−2.22	21	21	Nokkamappu
NDS105	KAr	Bi	1	72.60	74.31	0.70	21	21	Mancos A453
NDS105	KAr	Bi	1	72.80	74.52	0.70	21	21	Pierre, DeGrey A456
NDS116	KAr	Gl	3	74.60	74.60	1.35	21	21	Wenonah PTS12
NDS140	KAr	Gl	3	75.33	75.33	0.66	21	21	Limburg A516 A515

continued

Table 4.2 *continued*

Item	Dating method and mineral			Pub. date	Stand. date	1 sigma error	'Stage' codes		Item name
(1)	(2)	(3)	(4)	(5)	(6)	(7)	(8)	(9)	(10)
Campanian *continued*									
AB	ArAr	Sd	3	75.50	75.50	0.60	21	21	Anonna
NDS106	KAr	Bi	1	77.30	79.12	0.80	21	21	Claggett A458
NDS117	KAr	Gl	3	79.30	79.30	0.67	21	21	Merchantille A470
NDS106	KAr	Bi	1	77.90	79.73	0.80	21	21	Claggett A457
A410	KAr	Hb	1	78.00	79.83	2.00	21	21	Elkhorn
NDS106	KAr	Bi	1	78.20	80.04	0.80	21	21	Claggett A459
NDS163	KAr	WR	3	81.50	81.50	1.50	21	21	Pilot Knob
A508	KAr	SdBi	1	80.00	81.88	3.00	21	21	Vermilion, Riding Mtn
PTS62	KAr	Gl	1	81.00	82.90	2.84	21	21	Hanover
A417	KAr	Bi	2	86.00	84.02	5.00	21	22	Azerbaijan
Santonian									
HMFT	RbSr	WRBi	6	82.38	82.38	1.50	22	22	Hokkaido
OSK	ArAr	Sd	3	84.40	84.40	−1.00	22	22	Desmoscaphites
NDS107	KAr	Bi	1	82.50	84.43	0.80	22	22	Telegraph Ck A460
PTS229	KAr	Gl	1	83.00	84.94	2.99	22	22	Salzgitter
A509	KAr	Sd	1	87.00	89.03	3.00	22	22	Vermilion, Sask.
Coniacian									
NDS83a	KAr	Gl	3	85.87	85.87	0.89	23	23	Sainghin
MB	ArAr	WR	3	86.00	86.00	4.00	23	23	DSDP 72–516F (dubious material)
NDS86	KAr	Gl	3	86.80	86.80	1.65	23	23	Klement
NDS83b	RbSr	Gl	6	87.02	87.02	0.65	23	23	Sainghin
PTS57	KAr	Gl	1	86.50	88.67	3.07	23	23	Emscher, Hanover
NDS60	KAr	Gl	3	90.50	90.50	1.05	23	23	Maisieres
NDS108	KAr	Bi	1	86.80	88.82	0.90	23	24	Colorado A461
Turonian									
LAN1	ArAr	Hb	3	92.20	92.20	2.00	1	24	Samail
NDS94b	KAr	Gl	3	87.30	87.30	2.05	24	24	Munsterland PTS58 PTS59
NDS82b	RbSr	Gl	6	87.60	87.60	1.30	24	24	Thieu
NDS227	KAr	Gl	3	88.10	88.10	0.75	24	24	Rauen, Mulheim-Broich
NDS82a	KAr	Gl	3	88.10	88.10	1.50	24	24	Thieu
NDS164	KAr	Gl	3	88.70	88.70	1.10	24	24	Thieu
NDS226a	KAr	Gl	3	89.50	89.50	0.75	24	24	Rauen
SM1	RbSr	Bi	6	90.10	90.10	3.60	24	24	Obira
NDS109	KAr	Bi	1	88.90	90.97	0.90	24	24	Marias River A462
SM2	KAr	Bi	3	93.69	93.69	2.40	24	24	Yezo Group
NDS95	KAr	Gl	3	88.75	88.75	1.26	24	25	Essener PTS61
A510	KAr	Sd	1	90.00	92.09	3.00	24	25	Favel
NDS118	KAr	Bi	3	92.76	92.76	0.43	24	25	Seabee LT1
Cenomanian									
PTS335	KAr	WR	2	94.00	91.83	−3.67	1	25	Sukhaya River
NDS69	KAr	Gl	3	87.60	87.60	1.60	25	25	Portmuck
A418	KAr	Gl	2	90.00	87.93	4.00	25	25	Belorussia
NDS81	KAr	Gl	3	89.50	89.50	1.65	25	25	Bellignies
NDS59	KAr	Gl	3	89.80	89.80	1.80	25	25	Bettrechies
NDS62	RbSr	Gl	6	90.50	90.50	3.00	25	25	Cauville
NDS226b	KAr	Gl	3	90.60	90.60	0.75	25	25	Rauen

Table 4.2 *continued*

Item	Dating method and mineral			Pub. date	Stand. date	1 sigma error	'Stage' codes		Item name
(1)	(2)	(3)	(4)	(5)	(6)	(7)	(8)	(9)	(10)
PTS209	KAr	Gl	2	93.00	90.86	−3.63	25	25	Koberzhik
NDS81	RbSr	Gl	6	91.20	91.20	0.73	25	25	Bellignies
NDS85	KAr	Gl	3	92.20	92.20	1.55	25	25	St.Paul
NDS119a	KAr	Gl	3	92.40	92.40	0.82	25	25	Tourtia
NDS80a	KAr	Gl	3	93.30	93.30	0.78	25	25	Strouanne
NDS110	KAr	Bi	1	91.30	93.42	0.90	25	25	Frontier A463
NDS211	KAr	Gl	3	94.20	94.20	0.35	25	25	Salgzitter
NDS110	KAr	Bi	1	92.10	94.24	0.90	25	25	Greenhorn A464
NDS62	KAr	Gl	3	94.70	94.70	0.55	25	25	Cauville standard (GL-O)
NDS64	KAr	Gl	3	95.00	95.00	1.00	25	25	Octeville
NDS96	KAr	Gl	3	95.00	95.00	1.00	25	25	Haldon PTS227
NDS80b	RbSr	Gl	6	95.32	95.32	0.66	25	25	Strouanne
NDS67	KAr	Gl	3	95.40	95.40	0.75	25	25	Gouvix
NDS119b	RbSr	Gl	6	96.53	96.53	1.07	25	25	Tourtia
NDS68	RbSr	Gl	6	97.40	97.40	1.25	25	25	Nogent-le-Rotrou
PTS211	KAr	Gl	2	100.00	97.69	−3.91	25	25	Bukanskoye, Kaluga
PTS226	KAr	Bi	1	96.00	98.22	3.04	25	25	Cache Creek, Calif.
TWH1	UPb	Zr	8	96.50	96.50	0.60	25	128	Samail
Albian									
PTS217	UPb	Zr	7	115.00	113.85	2.00	20	26	La Grulla
PTS202	KAr	BiSd	1	100.00	102.31	3.00	25	26	Crowsnest
NDS96	KAr	Gl	3	95.00	95.00	1.00	26	26	Haldon PTS227
PTS230	KAr	Gl	1	94.00	96.18	3.11	26	26	Salzgitter
PTS51	KAr	Gl	1	94.00	96.18	3.14	26	26	Folkestone
NDS145	KAr	Gl	3	96.50	96.50	1.35	26	26	Frankenmühle
NDS111	KAr	Bi	3	97.50	97.50	1.00	26	26	Colorado A465
NDS144b	KAr	Gl	3	97.60	97.60	0.48	26	26	Morgenstern, upper
NDS111	KAr	Sd	3	97.60	97.60	1.00	26	26	Colorado A466
PTS204	KAr	BiSd	1	96.00	98.22	2.00	26	26	Mowry
PTS56	KAr	Gl	1	96.00	98.22	3.22	26	26	Lyme Regis
NDS65	KAr	Gl	3	98.35	98.35	1.16	26	26	Octeville
NDS61	KAr	Gl	3	98.70	98.70	2.50	26	26	Orphan Knoll A518
NDS79a	KAr	Gl	3	98.90	98.90	1.23	26	26	St Po
NDS63	KAr	Gl	3	99.00	99.00	1.12	26	26	Upper Gaize
PTS228	KAr	Gl	1	97.00	99.24	3.38	26	26	Salzgitter
NDS66	KAr	Gl	3	99.25	99.25	1.39	26	26	Pont L'Eveque
NDS157a	KAr	Bi	3	99.40	99.40	0.65	26	26	Newcastle
NDS67	KAr	Gl	3	99.60	99.60	2.50	26	26	Gouvix
NDS144d	KAr	Gl	3	99.70	99.70	1.10	26	26	Minimus
NDS79b	RbSr	Gl	6	99.72	99.72	0.76	26	26	St Po
NDS78a	KAr	Gl	3	99.77	99.77	0.98	26	26	Gardes
NDS144a	KAr	Gl	3	100.00	100.00	0.80	26	26	Finkenkuhle
PTS242	KAr	Gl	1	98.00	100.27	3.00	26	26	Lower Gault
NDS144c	KAr	Gl	3	100.60	100.60	0.50	26	26	Morgenstern, lower
NDS97	KAr	Gl	3	100.60	100.60	2.50	26	26	Salzgitter
PTS212	KAr	Gl	2	103.00	100.62	−4.02	26	26	Skryleyeva Ravine
PTS220	KAr	Gl	2	103.00	100.62	4.00	26	26	Caucasus
PTS336	KAr	WR	2	105.00	102.57	−4.10	26	26	River Badzhal
NDS144e	KAr	Gl	3	103.10	103.10	0.95	26	26	Gitterer
A428	KAr	Gl	2	106.00	103.55	4.00	26	26	Swinetch Mtn
NDS70	KAr	Gl	3	103.58	103.58	0.72	26	26	Bois De Perchois

continued

Table 4.2 *continued*

Item	Dating method and mineral			Pub. date	Stand. date	1 sigma error	'Stage' codes		Item name
(1)	(2)	(3)	(4)	(5)	(6)	(7)	(8)	(9)	(10)
Albian *continued*									
NDS157b	KAr	Bi	3	104.40	104.40	0.75	26	26	Skull Ck
NDS78b	RbSr	Gl	6	105.36	105.36	0.91	26	26	Gardes
NDS143	KAr	Gl	3	106.00	106.00	0.50	26	26	Brunswick
A429	KAr	Gl	2	110.00	107.45	5.00	26	26	N Caucasus
PTS219	KAr	Gl	1	108.00	110.48	−3.87	26	26	Manville
PTS233	KAr	Gl	1	112.00	114.76	−4.01	26	26	Manville
NDS98a	KAr	Gl	3	116.05	116.05	1.24	26	26	Folkestone PTS49
PTS60	KAr	Gl	3	104.53	104.53	3.89	26	27	Salzgitter
Aptian									
NDS77a	KAr	Gl	3	105.80	105.80	1.31	27	27	Verlinthun
NDS71	KAr	Gl	3	107.30	107.30	1.95	27	27	Compton Bay
NDS146	KAr	Gl	3	108.20	108.20	0.95	27	27	Borgers
NDS188	KAr	Fl	3	109.30	109.30	3.00	27	27	Atlas
NDS98b	KAr	Gl	3	112.05	112.05	1.17	27	27	Bargate PTS49
JE1	ArAr	Sd	3	113.00	113.00	1.00	27	27	Fuller's Earth
OSK	ArAr	Sd	3	113.00	113.00	0.70	27	27	Parahoplites
NDS77b	RbSr	Gl	6	115.80	115.80	1.24	27	27	Verlinthun
PTS50	KAr	Gl	1	115.00	117.62	3.88	27	27	Hythe
SMYH1	RbSr	WR	5	121.00	125.26	6.00	27	28	Miyako
SMYH2	RbSr	WR	5	128.00	132.51	12.00	27	28	Taro
Barremian									
NDS162b	KAr	Gl	3	126.00	126.00	1.50	28	28	Georgsdorf
PTS215	KAr	Gl	2	136.00	132.83	−5.31	28	30	Lenin Hills
Hauterivian									
LBM1	KAr	WR	3	133.50	133.50	2.84	1	29	Tayasir Volc
NDS162a	KAr	Gl	3	119.30	119.30	0.45	29	29	Moorberg
NDS162b	KAr	Gl	3	126.00	126.00	1.50	29	29	Georgsdorf
PTS75R	KAr	BiHb	3	133.60	133.60	2.00	29	33	Shasta Bally LJ3
LJ3	UPb	Zr	8	136.00	136.00	2.00	29	33	Shasta Bally
Valanginian									
A406	KAr	WR	1	124.00	126.81	−3.80	1	30	Hellhole-Williams
PTS322	KAr	Gl	2	125.00	122.09	−4.88	30	30	Egorev PTS72
A430	KAr	Gl	2	128.00	125.02	5.00	30	30	N Caucasus
LJ4	KAr	Bi	3	136.50	136.50	2.70	30	30	Hughes Quad
Berriasian									
PTS328	KAr	Bi	2	134.00	130.88	−5.24	1	31	Magadan
LBM2	KAr	WR	3	146.00	146.00	4.26	30	31	Tayasir
PTS177	KAr	Gl	1	131.00	133.95	4.00	31	31	Sandringham
W	Misc		0	138.00	138.00	4.00	31	32	E China
Tithonian									
NDS75	KAr	Gl	3	129.40	129.40	2.40	32	32	Gaillefontaine
NDS76	KAr	Gl	3	131.10	131.10	1.50	32	32	Wimereux
PTS322	KAr	Gl	2	136.00	132.83	−5.31	32	32	Egorev
PTS178	KAr	Gl	1	132.00	134.97	4.00	32	32	Sandringham
NDS99	KAr	Gl	3	136.33	136.33	1.20	32	32	Okus

Table 4.2 *continued*

Item	Dating method and mineral			Pub. date	Stand. date	1 sigma error	'Stage' codes		Item name
(1)	(2)	(3)	(4)	(5)	(6)	(7)	(8)	(9)	(10)
NDS228	KAr	Gl	1	139.00	141.40	1.75	32	32	Hanover PTS73
HMP2	UPb	Zr	8	152.50	152.50	2.00	32	128	Cuesta Ridge
HMP3	UPb	Zr	8	154.00	154.00	2.00	32	128	Del Puerto
A445	KAr	Hb	1	155.00	158.42	3.57	32	128	Coast Range
Kimmeridgian									
PTS76	KAr	Bi	2	143.00	139.66	4.00	1	33	Horseshoe Bar
HS2	UPb	Zr	8	150.50	150.50	2.00	1	33	Galice
SCH1	UPb	Zr	8	153.00	153.00	1.00	1	33	Nevadan
A480	KAr	Bi	1	150.00	153.32	−4.60	1	33	Fanos
SBG	UPb	Zr	8	155.00	155.00	3.00	1	33	Nevadan
A481	RbSr	Bi	5	150.00	155.28	−4.66	1	33	Fanos
NDS142	KAr	Gl	1	138.45	138.45	−1.06	33	33	Aargau A485
Oxfordian									
NDS141	KAr	Gl	3	148.00	148.00	0.87	34	34	Randen
HS1	UPb	Zr	8	157.00	157.00	2.00	34	128	Josephine
HMP1	UPb	Zr	8	161.00	161.00	2.00	34	128	Point Sal
HMP4	UPb	Zr	8	162.00	162.00	2.00	34	128	Paskenta/Elder Creek
Callovian									
BL1	KAr	Hb	3	159.00	159.00	5.00	35	35	Coloradito
Bathonian									
A419	KAr	WR	2	164.00	160.14	6.00	36	37	Abkhazia
BFSB6R	KAr	WR	3	168.50	168.50	4.00	36	37	Les Vignes
Bajocian									
WEST2	KAr	Fl	3	173.00	173.00	4.00	27	37	Maroc
PTS90	KAr	Bi	2	173.00	168.92	2.00	32	37	Kelasury
NDS102a	KAr	Bi	3	161.20	161.20	1.80	37	37	Carmel
NDS102b	RbSr	Bi	6	162.00	162.00	6.50	37	37	Carmel
NDS182	KAr	Hb	3	171.00	171.00	3.50	37	37	British Columbia
Aalenian									
A479	KAr	Bi	1	168.00	171.67	−5.15	34	38	Talkeetna
Toarcian									
NDS183a	KAr	Hb	3	182.50	182.50	2.84	39	40	Mt Sister Mary FW43
HLB1	ArAr	BiFl	3	185.00	185.00	3.00	39	40	N Caucasus
NDS183b	RbSr	Hb	6	190.00	190.00	7.00	39	40	Mt Sister Mary FW43
NDS184R	KAr	BiHb	3	198.86	198.86	2.65	39	40	Toodoggone
NDS184R	ArAr	Bi	3	197.60	197.60	2.50	39	41	Toodoggone
FW40	KAr	WRFl	3	191.25	191.25	3.38	39	45	Liberia
FW42	KAr	WR	3	193.00	193.00	1.50	39	45	Karroo
FW41	RbSr	WR	6	193.00	193.00	1.50	39	45	Freetown
Pliensbachian									
None									

continued

Table 4.2 *continued*

Item	Dating method and mineral			Pub. date	Stand. date	1 sigma error	'Stage' codes		Item name
(1)	(2)	(3)	(4)	(5)	(6)	(7)	(8)	(9)	(10)
Sinemurian									
NDS181	KAr	Hb	3	203.00	203.00	1.95	1	41	Topley & Talkwa A504 A506 A507
SEID	KAr	WR	3	190.00	190.00	4.00	41	42	Hartford
NDS202	ArAr	WR	3	198.50	198.50	5.78	41	42	Newark FW38 FW39
NDS203	KAr	WR	3	200.00	200.00	5.00	41	42	Holyoke FW40
NDS248	KAr	Bi	3	202.40	202.40	2.13	41	42	Brooks Range A409
WEST1	KAr	Fl	3	206.00	206.00	6.00	41	43	Maroc
Hettangian									
A478	KAr	Hb	1	206.00	210.36	6.00	1	42	Cuddy Mtn
NDS177	KAr	Bi	1	205.00	205.00	2.50	42	44	Guichon Creek A366
Rhaetian									
NDS137a	KAr	BiHb	3	216.00	216.00	2.00	43	44	Bangka A446 A447 FW35
NDS137b	RbSr	WRBi	6	216.00	216.00	3.00	43	44	Bangka A446 A447 FW35
NDS199	KAr	WR	3	213.00	213.00	3.50	43	128	North Arm
Norian									
NDS174	KAr	WR	3	204.00	204.00	4.50	1	44	Kemess
FW36	RbSr	WR	6	207.00	207.00	1.03	1	44	Kulim
NDS178b	RbSr	Bi	6	207.00	207.00	4.00	1	44	Coldwater
NDS179	KAr	Hb	3	208.00	208.00	2.00	1	44	Tulameen
NDS204	KAr	Bi	3	210.00	210.00	−1.05	1	44	Mae Sariang
NDS178a	KAr	HbBi	3	210.00	210.00	4.00	1	44	Coldwater
NDS178R	ArAr	HbBi	3	215.70	215.70	2.70	1	44	Coldwater SHE2
NDS176	KAr	Hb	3	225.00	225.00	8.00	1	44	Wrede
KM1	UPb	Zr	8	211.00	211.00	−4.22	44	44	Munhiku
NDS170	KAr	Hb	3	221.00	221.00	6.00	44	45	Hotailuh
Carnian									
PTS160	UPb	Ur	7	218.00	215.82	5.00	1	45	Chinle
NDS171	KAr	HbBi	3	220.00	220.00	5.00	1	45	Kaketsa
NDS201a	KAr	Ph	3	231.00	231.00	2.48	1	45	Mt Monzoni FW31
NDS201b	RbSr	Bi	6	236.85	236.85	1.51	1	45	Mt Monzoni FW31 A476
NDS187	KAr	WR	3	229.00	229.00	2.50	45	45	Ischigualasto-Ischichuca A519
NDS193	KAr	WR	3	230.50	230.50	2.48	45	46	Sugars FW32
NDS197	KAr	Hb	3	235.00	235.00	4.00	45	128	Djuan
Ladinian									
PTS358a	KAr	BiHb	1	218.00	222.56	2.00	1	46	Maryborough FW34
FW33	KAr	Sd	3	225.00	225.00	2.00	1	46	Val Serrata
PTS358b	RbSr	WR	5	218.00	225.68	8.00	1	46	Maryborough FW34
NDS194	KAr	Bi	3	233.50	233.50	2.48	1	46	Station Creek
PTS361	KAr	Bi	1	231.00	235.78	−7.07	1	46	Predazzo
PTS361	RbSr	WR	5	229.00	237.06	3.00	1	46	Predazzo
NDS196	KAr	Sd	3	233.00	233.00	2.50	46	47	Mt San Giorgio FW30
NDS194	KAr	WRHb	3	239.00	239.00	1.64	46	47	Neara FW27
NDS186	KAr	WR	3	237.00	237.00	2.20	46	48	Puesto Viego FW28 A520

Table 4.2 *continued*

Item (1)	Dating method and mineral (2)	(3)	(4)	Pub. date (5)	Stand. date (6)	1 sigma error (7)	'Stage' codes (8)	(9)	Item name (10)
Anisian									
KM2	UPb	Zr	8	240.00	240.00	−4.80	47	47	Murikiku
NDS194	KAr	BiHb	3	243.00	243.00	2.12	47	128	Eskdale
Spathian									
NDS158	KAr	Bi	3	248.00	248.00	7.00	48	52	Joblanica
PTS357	KAr	Bi	1	239.00	243.92	−7.32	48	53	Gyranda A357
PTS338	KAr	WR	2	255.00	248.85	15.00	48	54	Semeitau
Nammalian (Smithian)									
None									
(Dienerian)									
None									
Griesbachian									
PTS346	KAr	Bi	2	250.00	243.98	−4.94	51	84	Yatyrgvarta
Lopingian (Tatarian)									
None									
Guadalupian (Kazanian)									
A434	KAr	WR	2	255.00	248.85	−8.71	53	53	Aksahnt River
NDS205	KAr	WRFl	3	250.00	250.00	1.79	53	53	Gerringong PTS68
FW22	KAr	WR	3	253.00	253.00	2.50	53	53	Sydney Basin
(Ufimian)									
FW23	KAr	HbBi	3	250.00	250.00	2.50	1	54	Marlborough
Kungurian									
FW21	UPb	Zr	8	257.00	257.00	1.50	1	55	Quillabama
PTS68	KAr	Fl	1	252.00	257.54	−7.73	1	55	Berkley
A435	KAr	Hb	2	270.00	263.47	−9.22	55	57	N Caucasus
NDS231a	KAr	WR	3	283.00	283.00	2.50	55	58	Roubayre
NDS169	KAr	WR	3	286.00	286.00	3.50	55	58	Mauchline FW15
Artiniskian									
None									
Sakmarian									
PTS120	KAr	Gl	2	274.00	267.40	−9.36	57	57	Indiga (low K)
FW18	KAr	HbBi	3	270.00	270.00	1.00	57	57	Lizzie Ck Granites
FW17	KAr	Fl	3	275.00	275.00	3.00	57	57	Lizzie Ck Volcanics
A421	UPb	Zr	7	287.00	284.13	10.00	57	63	Akchatau
A420	KAr	Bi	2	293.00	285.87	15.00	57	63	Akchatau
Asselian									
NDS240	RbSr	WRBi	6	270.00	270.00	2.00	1	58	Oslo A503 FW20
NDS240	RbSr	WR	6	278.00	278.00	3.50	1	58	Oslo A503 FW19
PTS341	KAr	Bi	2	289.00	281.97	10.00	1	58	Kyzylkiy
PTS192	KAr	Bi	1	284.00	290.08	−8.70	1	58	Oslo
FW16	ArAr	Misc	3	282.50	282.50	4.29	51	58	Great Serp Belt

continued

Table 4.2 *continued*

Item	Dating method and mineral			Pub. date	Stand. date	1 sigma error	'Stage' codes		Item name
(1)	(2)	(3)	(4)	(5)	(6)	(7)	(8)	(9)	(10)
Asselian *continued*									
PTS122	RbSr	Bi	5	282.00	291.93	4.00	58	60	Castro Daire
PTS8b	RbSr	Bi	5	275.00	284.68	6.00	58	62	Dartmoor
PTS8a	KAr	Bi	1	295.00	300.78	5.00	58	62	Dartmoor
FW11	KAr	WR	3	301.00	301.00	2.50	58	65	Whin PTS176 PTS359
FW10	KAr	WR	3	303.00	303.00	2.50	58	65	Midland Valley
Noginskian									
PTS63	RbSr	Bi	5	288.00	298.14	8.00	59	60	Brassac
LI1	ArAr	Sd	3	300.30	300.30	3.70	59	62	Baden Baden
PTS340	KAr	WR	2	310.00	302.42	15.00	59	63	Taskuduk
PTS65	KAr	Bi	1	298.00	303.82	−9.11	59	63	Paterson PTS30
NDS231b	KAr	WR	3	300.00	300.00	1.77	59	64	Lassouts, Ossos
Klazminskian									
None									
Dorogomilovskian									
None									
Chamovnicheskian									
NDS231c	KAr	WR	3	295.00	295.00	3.55	62	63	Plan de la Tour
LI2	ArAr	Sd	3	302.90	302.90	3.70	62	63	Saarrevier
Krevyakinskian									
None									
Myachkovskian									
B2	RbSr	WR	5	300.00	310.56	5.00	64	64	St Jean
Podolskian									
PTS360a	KAr	WR	1	308.00	313.97	10.00	1	65	Barrow Hill
HL2	ArAr	Sd	3	309.30	309.30	4.60	65	65	Krenov
NDS232	KAr	Sd	3	303.00	303.00	3.00	65	66	Brassert bei Marl
FW9	KAr	Sd	3	304.00	304.00	3.00	65	66	Ruhr Tonstein
HL1	ArAr	Sd	3	309.00	309.00	3.70	65	66	Radnice
LI3	ArAr	Sd	3	310.10	310.10	3.80	65	66	Ruhrrevier
Kashirskian									
LI4	ArAr	Sd	3	310.70	310.70	3.90	66	67	Ruhrrevier
Vereiskian									
None									
Melekesskian									
None									
Cheremshanskian									
None									
Yeadonian									
None									

Table 4.2 *continued*

Item (1)	Dating method and mineral (2)	(3)	(4)	Pub. date (5)	Stand. date (6)	1 sigma error (7)	'Stage' codes (8)	(9)	Item name (10)
Marsdenian									
None									
Kinderscoutian									
PTS339	KAr	Bi	2	330.00	321.89	10.00	60	72	Keregetass
Alportian									
LI5	ArAr	Sd	3	319.50	319.50	3.90	73	74	Ostrauer
LI6	ArAr	Sd	3	324.60	324.60	4.00	73	76	Revier, CSSR
PTS191	KAr	WR	1	322.00	328.16	12.00	73	76	Hillhouse
Chokierian									
None									
Arnsbergian									
None									
Pendlian									
None									
Brigantian									
PTS31	KAr	Bi	1	327.00	333.23	10.29	64	77	Harzburger
PTS172	RbSr	Bi	5	328.00	339.55	3.00	77	77	Malavaux
PTS360c	KAr	WR	1	334.00	340.32	17.00	77	78	Little Wenlock
NDS167	KAr	Fl	3	327.00	327.00	2.50	77	81	Clyde Plateau FW5
PTS173	RbSr	Bi	5	334.00	345.76	7.00	77	81	Gien-sur-Cure
NDS133a	RbSr	WR	6	331.00	331.00	3.00	77	84	Chateaulin Basin
Asbian									
PTS66	KAr	Hb	1	327.00	333.23	−7.09	70	78	Lower Kuttung
PTS360d	KAr	WR	1	338.00	344.37	4.00	78	78	Burntisland
Holkerian									
NDS166	KAr	Sd	3	353.00	353.00	3.50	79	81	Garleton Hills FW4
Arundian									
PTS171a	KAr	Bi	1	317.50	324.10	−4.86	64	80	Vosges PTS356
PTS171b	RbSr	Bi	5	322.50	325.06	4.00	64	80	Vosges PTS356
PTS360e	KAr	WR	1	347.00	353.49	7.00	80	81	Arthur's Seat
Chadian									
NDS153	RbSr	Gl	6	354.00	354.00	1.50	81	82	Barnett
A413	RbSr	Bi	5	347.00	359.22	−10.77	81	128	Massif Central
Ivorian									
NDS152	RbSr	Gl	6	349.00	349.00	1.50	82	83	Houy
PTS360f	KAr	WR	1	359.00	365.64	6.00	82	83	Campsie Fells
NDS133b	RbSr	WR	6	342.00	342.00	6.50	82	84	Laval Basin
NDS165	KAr	WR	3	361.00	361.00	3.50	82	84	Birrenswark & Kelso FW3 MLC28
A424	KAr	Bi	2	355.00	346.22	10.00	82	85	Kazakhstan

continued

Table 4.2 *continued*

Item	Dating method and mineral			Pub. date	Stand. date	1 sigma error	'Stage' codes		Item name
(1)	(2)	(3)	(4)	(5)	(6)	(7)	(8)	(9)	(10)
Hastarian									
PTS174	KAr	Bi	2	350.00	341.36	7.00	58	83	Erzgebirge
PTS347	RbSr	WR	4	379.00	370.99	16.00	83	84	Fisset Brook
Famennian									
NDS235a	KAr	HbBi	3	362.00	362.00	−1.81	84	84	S Australia GA2
NDS235b	RbSr	WR	6	366.00	366.00	4.61	84	84	S Australia
SR1	RbSr	WR	6	369.00	369.00	12.00	84	85	Kennack
Frasnian									
A425	KAr	Gl	2	366.00	356.93	10.00	85	85	Kurskiy
NDS234	KAr	Bi	3	367.00	367.00	−1.84	85	85	Cerberean FW2 MLC27 GA3 PTS95 PTS354
NDS234	RbSr	BiWR	6	370.00	370.00	2.00	85	85	Cerberean PTS354 FW1 MLC27
MLC26	KAr	Hb	3	369.00	369.00	2.50	85	86	Mt Morgan GA34
Givetian									
NDS244	ArAr	WR	3	379.00	379.00	5.00	1	86	Hoy FW1
NDS233	KAr	BiHb	3	381.00	381.00	2.00	86	87	Mirimbah & Mt Stirling GA4
Eifelian									
NDS151	RbSr	Gl	6	386.00	386.00	1.93	87	87	Stribling J50
Emsian									
PTS98a	KAr	Bi	3	348.98	348.98	10.00	82	88	Nova Scotia
RZM	KAr	Bi	3	366.70	366.70	1.83	82	88	Nova Scotia
PTS98b	RbSr	Bi	6	376.81	376.81	14.00	82	88	Nova Scotia
J29b	RbSr	Bi	6	385.00	385.00	3.00	88	90	Vinalhaven
NDS241c	UPb	Zr	8	390.00	390.00	3.00	88	90	Shap PTS6
NDS237	RbSr	WR	6	393.00	393.00	3.50	88	90	Pembroke GA7 J20
NDS241b	RbSr	BiWR	6	394.00	394.00	1.50	88	90	Shap PTS6 MLC24
NDS241a	KAr	Bi	3	397.00	397.00	3.50	88	90	Shap PTS6 NDS192
J29a	KAr	Bi	3	407.00	407.00	−12.21	88	90	Vinalhaven
PTS6	RbSr	Bi	3	410.98	410.98	11.00	88	90	Shap NDS241
Pragian (Siegenian)									
PTS3	KAr	Sd	1	385.00	391.96	15.00	89	89	Shiphead, Gaspe G14 MLC14
PTS97	KAr	Bi	1	394.00	401.06	−12.03	89	92	Snowy River J22
Lochkovian (Gedinnian)									
DB1	RbSr	WR	6	394.00	394.00	4.00	1	90	Three Mile Pond
NDS236	KAr	HbBi	3	401.33	401.33	5.20	1	90	St George A472
HUL	ArAr	BiHb	3	402.00	402.00	3.00	1	90	NE Maine
NDS236	RbSr	WR	6	407.00	407.00	21.00	1	90	St George A471
FC2	RbSr	WR	6	409.00	409.00	12.50	1	90	Gulquac Lake
PTS5	KAr	Bi	1	404.00	411.17	8.00	84	90	Calais
NDS236b	RbSr	WRBi	6	397.64	397.64	3.00	85	90	Red Beach A488
NDS236a	KAr	Bi	3	412.50	412.50	−12.38	85	90	Red Beach
MLC22	RbSr	WR	6	400.00	400.00	2.00	90	90	Lorne GA8
NDS222	RbSr	WR	6	401.00	401.00	6.00	90	90	Eastport PTS355
NDS223	RbSr	WR	6	406.00	406.00	5.00	90	90	Hedgehog PTS355GA9

Table 4.2 *continued*

Item	Dating method and mineral			Pub. date	Stand. date	1 sigma error	'Stage' codes		Item name
(1)	(2)	(3)	(4)	(5)	(6)	(7)	(8)	(9)	(10)
FC1	RbSr	WR	6	409.00	409.00	17.50	90	90	Gulquac Lake
NDS210b	RbSr	WR	6	402.00	402.00	2.01	90	91	Gocup GA10 MLC20
PPC9	UPb	Zr	8	405.00	405.00	5.00	90	91	Mareuil-sur-Lay
NDS210a	KAr	Misc	3	409.00	409.00	2.02	90	91	Gocup GA10 MLC20
WP1	RbSr	WR	6	405.00	405.00	4.50	90	97	Fongen-Hillingen
Pridoli									
NDS192	KAr	Bi	3	399.00	399.00	4.26	1	91	Skiddaw MLC23
MLC21	RbSr	Bi	6	407.30	407.30	2.80	91	93	Wormit Bay
Ludfordian									
OHMT86	KAr	Bi	3	417.00	417.00	5.00	92	92	Woodbury
OHMT86	KAr	Bi	3	423.00	423.00	4.50	92	93	Woodbury
Gorstian									
MLC19	Fi	Zr	10	407.00	407.00	7.00	93	93	Bringewood R13
R12b	Fi	Zr	10	407.00	407.00	9.00	93	93	Middle Elton GA12 MLC18
GA11a	KAr	BiSd	3	414.50	414.50	5.00	93	93	Laidlaw MLC17
R12a	KAr	Bi	3	419.00	419.00	3.57	93	93	Middle Elton GA12 MLC18
GA11b	RbSr	WRBi	6	422.00	422.00	3.00	93	93	Laidlaw MLC17
K2	ArAr	Bi	3	423.70	423.70	−2.12	93	93	Hopdale, Shropshire
OHMT86	KAr	Bi	3	424.00	424.00	4.00	93	93	Millichope
A522	RbSr	WR	4	438.00	428.75	4.00	93	93	New South Wales
Gleedonian									
PTS93b	RbSr	Bi	5	399.00	413.05	16.00	1	94	Creetown
R10	Fi	Zr	10	412.00	412.00	12.00	94	94	Coalbrookdale
R11	Fi	Zr	10	416.00	416.00	9.00	94	94	Much Wenlock
OHJS	KAr	Bi	3	426.60	426.60	3.00	94	94	Gotland
Whitwellian									
None									
Sheinwoodian									
R9	Fi	Zr	10	422.00	422.00	7.00	96	96	Buildwas MLC16
OHJS	KAr	Bi	3	430.50	430.50	3.00	96	97	Gotland
Telychian									
A521	RbSr	WR	4	455.00	445.39	15.00	97	98	State Circle
Aeronian									
NDS128	ArAr	Hb	3	436.20	436.20	2.50	98	99	Equisbel A415 MLC15 GA14 J2 K1
Rhuddanian									
R8	Fi	Zr	10	437.00	437.00	11.00	99	99	Birkhill
A490	KAr	Bi	1	437.00	444.52	13.00	99	104	Beemerville
A491	RbSr	Bi	5	430.00	445.14	20.00	99	104	Beemerville
Hirnantian									
None									

continued

Table 4.2 *continued*

Item	Dating method and mineral			Pub. date	Stand. date	1 sigma error	'Stage' codes		Item name
(1)	(2)	(3)	(4)	(5)	(6)	(7)	(8)	(9)	(10)
Rawtheyan									
R7	Fi	Zr	10	434.00	434.00	12.00	101	101	Hartfell
Cautleyan									
None									
Pusgillian									
RN3	Fi	Zr	10	447.00	447.00	12.00	100	103	Wufeng
Onnian									
A487	RbSr	WR	4	460.00	450.28	10.00	1	104	Penobscot
PTS351	KAr	Bi	1	445.00	452.60	5.00	104	110	Bail Hill GA23 MLC11
Actonian									
R6	Fi	Zr	10	466.00	466.00	12.00	105	109	Acton Scott G23
Marshbrookian									
RN5	Fi	Zr	10	453.00	453.00	3.00	106	110	Carters
NDS161R	KAr	Bi	3	454.10	454.10	−2.27	106	110	Tyrone GA19 GA20
K3	KAr	Bi	3	454.10	454.10	−2.27	106	110	Carters
NDS129a	KAr	BiSd	3	455.00	455.00	5.00	106	110	Carters PTS156 GA18
RN4	Fi	Zr	10	455.00	455.00	15.00	106	110	Tyrone
Longvillian									
R5	Fi	Zr	10	465.00	465.00	9.29	107	107	Pont-y-ceunant MLC13
PTS157	RbSr	BiFl	6	445.00	445.00	2.80	107	110	Kinnekulle Chasmops GA21 MLC12b WTCM1
MLC12c	RbSr	BiSd	6	447.00	447.00	−2.24	107	110	Kinnekulle Chasmops
PTS157	KAr	BiSd	3	454.50	454.50	3.00	107	110	Kinnekulle Chasmops GA21 MLC12a WTCM2
Soudleyan									
RUN1a	KAr	WR	3	443.00	443.00	3.50	108	108	Shelve
R4	Fi	Zr	10	450.00	450.00	12.00	108	108	Frondderw
RUN1b	RbSr	WR	6	454.00	454.00	7.00	108	108	Shelve
Harnagian									
None									
Costonian									
MLC10	SmNd	Gt	11	457.00	457.00	2.00	110	113	Borrowdale
A411R	RbSr	WR	6	480.62	480.62	2.00	110	116	Caledonian
Llandeilo–3									
MLC7	ArAr	Hb	3	468.46	468.46	2.50	111	116	Bay of Islands A416 GA29
NDS135	RbSr	WR	6	469.00	469.00	2.50	111	128	Benan Clasts MLC9 GA34
NDS135	UPb	Zr	8	475.00	475.00	2.37	111	128	Benan Clasts MLC9
Llandeilo–2									
None									

Table 4.2 *continued*

Item	Dating method and mineral			Pub. date	Stand. date	1 sigma error	'Stage' codes		Item name
(1)	(2)	(3)	(4)	(5)	(6)	(7)	(8)	(9)	(10)
Llandeilo–1									
R3	Fi	Zr	10	477.00	477.00	7.50	113	113	Bach-y-graig MLC8 GBW29
DKTS1	UPb	Zr	8	462.00	462.00	2.31	113	114	Victoria Lake Gp
Llanvirn–2									
MLC6	KAr	Bi	3	484.00	484.00	10.00	114	116	Colmonell, Ballantrae GA25
Llanvirn–1									
NDS191	KAr	HbBi	3	468.00	468.00	5.00	115	115	Great Cockup GA26
R2	Fi	Zr	10	487.00	487.00	13.00	115	115	Stapley
DKTS2	UPb	Zr	8	473.00	473.00	2.37	115	116	Buchans
DKTS3	UPb	Zr	8	473.00	473.00	2.37	115	116	Roberts Arm
MLC5	ArAr	Hb	3	480.00	480.00	5.00	115	116	Hare Bay GA27
A412	UPb	Zr	7	510.00	504.90	10.00	115	128	Cashel-Lough-Wheelan MCK23
Arenig									
NDS125	KAr	Gl	3	473.00	473.00	8.50	116	116	Sukrumagi
NDS134	KAr	Hb	3	481.00	481.00	4.26	116	116	Ballantrae MLC4a PTS350
NDS134	UPb	Zr	8	483.50	483.50	2.42	116	116	Ballantrae MLC4b
NDS132	KAr	Gl	3	488.00	488.00	14.00	116	116	Stora Stoland
R1	Fi	Zr	10	493.00	493.00	11.00	116	116	Llynfnant GBW33
MLC3	SmNd	WR	11	490.00	490.00	7.00	116	128	Southern Uplands GA33
MLC2	RbSr	WR	6	498.00	498.00	3.50	116	128	Twt Hill GA30
Tremadoc									
A414	RbSr	WR	4	515.00	504.12	7.00	1	117	Stappogiede
NDS122	KAr	Hb	3	508.00	508.00	5.50	117	117	Rhobell MLC1 GA31 KFH1
NDS130a	RbSr	WR	6	491.00	491.00	7.00	117	119	Krivoklat-Rokycany GA32 A407 MCK31
Dolgellian									
NDS150	RbSr	Gl	6	516.00	516.00	7.50	118	119	Morgan Creek
Maentwrogian									
PTS186	UPb	Zr	7	523.00	517.77	7.00	119	128	Wichita Mtns
Menevian–3									
RN1	Fi	Zr	10	535.00	535.00	12.00	120	122	Mauv
A426	KAr	Gl	2	530.00	516.35	20.00	120	124	Salt Range
A473	RbSr	Gl	4	542.50	531.04	20.00	120	124	Flathead
A474	RbSr	WR	4	555.00	543.27	18.00	120	124	Flathead
Menevian–2									
None									
Menevian–1									
RN2	Fi	Zr	10	563.00	563.00	12.00	122	123	Bright Angel
Solvan–2									
None									

continued

Table 4.2 *continued*

Item (1)	Dating method and mineral (2)	(3)	(4)	Pub. date (5)	Stand. date (6)	1 sigma error (7)	'Stage' codes (8)	(9)	Item name (10)
Solvan–1									
None									
Lenian									
PTS185	KAr	Gl	2	550.00	535.77	−16.07	125	126	Olenek
PTS183	RbSr	Gl	4	584.00	571.66	30.00	125	126	Murray
CC11	KAr	Gl	3	545.00	545.00	16.00	125	127	Chaya
CC8	KAr	Gl	3	549.00	549.00	16.00	125	127	Asha
CC12	KAr	Gl	3	558.00	558.00	17.00	125	127	Olenek
CC10	KAr	Gl	3	569.00	569.00	17.00	125	127	Chikanda
CC29	KAr	Gl	3	582.00	582.00	17.00	125	127	Olenek
Atdabanian									
NDS242	UPb	Zr	8	531.00	531.00	2.65	1	126	Ercall CJ15
NDS242	RbSr	WRBi	6	533.00	533.00	6.00	1	126	Ercall CJ14
CPR	ArAr	Mu	3	555.00	555.00	5.00	120	126	Walidiala
ZGH2	UPb	WR	8	568.00	568.00	6.00	126	126	Shuijintuo CJ2
ZGH3	UPb	WR	8	573.00	573.00	16.00	126	126	Shuijintuo CJ3
NDS251	RbSr	WR	6	529.00	529.00	4.00	126	128	Mandar CJ20
NDS136	UPb	Zr	8	534.00	534.00	5.00	126	128	Morocco OGD3 CJ13
NDS121	UPb	Mz	8	540.00	540.00	5.00	126	128	Vire-Carolles CJ12 A486 OGD2
NDS121	RbSr	WR	6	542.00	542.00	31.00	126	128	Vire-Carolles CJ12 OGD1
NDS249	RbSr	WR	6	547.00	547.00	6.00	126	128	Lezardrieux OGA2
NDS251	RbSr	WR	6	548.00	548.00	2.50	126	128	Amram, Sinai CJ19
NDS121	KAr	Bi	1	553.00	561.50	−16.85	126	128	Vire-Carolles CJ12 PTS42 A486 OGD1
NDS250	UPb	Zr	8	563.00	563.00	2.82	126	128	Ouarzazate
PTS352	RbSr	WR	5	574.00	594.21	11.00	126	128	Holyrood
NDS121	RbSr	WRFl	6	605.00	605.00	9.57	126	128	Vire-Carolles
Tommotian									
CC31	KAr	Gl	3	553.00	553.00	17.00	127	127	Aldan
CC5	KAr	Gl	3	578.00	578.00	17.00	127	127	Aldan
CC1	KAr	Gl	3	587.00	587.00	10.00	127	127	Yudoma
CC30	KAr	WR	3	587.00	587.00	18.00	127	127	Yudoma
KK1	KAr	Gl	3	590.00	590.00	−17.70	127	128	Blue Clay
KK2	RbSr	WR	6	592.00	592.00	16.00	127	128	Lontova
Precambrian									
NDS249	RbSr	WR	6	554.00	554.00	9.50	1	128	Port-Scarff/Loquivy OGA1
PTS352	UPb	Zr	8	620.50	620.50	3.00	126	128	Holyrood
CJ18	RbSr	WR	6	558.00	558.00	8.00	128	128	Uriconian
PTS116	KAr	Gl	2	573.00	558.10	−16.74	128	128	Ashinsk
A427	KAr	Gl	2	590.00	574.61	15.00	128	128	Dnieper
PTS117	KAr	Gl	2	595.00	579.46	−17.38	128	128	Laminarites
CJ22R	UPb	Zr	8	586.00	586.00	3.00	128	128	Burin Peninsula
PTS118	KAr	Gl	2	615.00	598.87	−17.97	128	128	Karatau C3?
CJ21R	UPb	Zr	8	606.00	606.00	3.00	128	128	Burin Peninsula
PTS55	UPb	Ur	7	620.00	613.80	20.00	128	128	Shinkolobwe
CJ21	UPb	Zr	8	623.00	623.00	3.10	128	128	Burin

Table 4.2 *continued*

Item	Dating method and mineral			Pub. date	Stand. date	1 sigma error	'Stage' codes		Item name
(1)	(2)	(3)	(4)	(5)	(6)	(7)	(8)	(9)	(10)

Items not used in time scale construction (all illites)

J19	KAr	Ill	2	404.00	394.02	6.00	91	91	Kirovgrad
NDS130b	RbSr	Ill	6	510.00	510.00	10.00	118	119	Krivoklat-Rokycany
ZGH4	RbSr	Ill	6	572.00	572.00	7.00	126	126	Shuijintuo CJ4
ZGH1	RbSr	Ill	6	573.00	573.00	3.50	126	126	Shuijintuo CJ1
ZGH9	RbSr	Ill	6	569.00	569.00	6.00	127	127	Niutitang CJ9
YQHT	RbSr	Ill	6	573.00	573.00	7.50	127	127	Shijingtuo
YQHT10	RbSr	Ill	6	579.70	579.70	4.10	127	127	Badaowan CJ10
YQHT7	RbSr	Ill	6	584.70	584.70	7.60	127	127	Badaowan CJ7
ZGH6	RbSr	Ill	6	587.00	587.00	8.50	127	127	Badaowan CJ6
ZGH5	RbSr	Ill	6	588.00	588.00	6.50	127	127	Badaowan CJ5
YQHT8	RbSr	Ill	6	602.00	602.00	7.50	127	127	Tientzushan ZGH8 CJ8

Table 4.3. *'Stage' abbreviations and code numbers (see also Appendix 2)*

CENOZOIC

Quaternary

Hol	1	Holocene
Ple	2	Pleistocene

Pliocene

Pia	3	Piacenzian
Zan	4	Zanclian

Miocene

Mes	5	Messinian
Tor	6	Tortonian
Srv	7	Serravallian
Ln2	8	Late Langhian
Ln1	9	Early Langhian

Note: Late Langhian and Early Langhian used in database but now combined as Langhian (Lan)

Bur	10	Burdigalian
Aqt	11	Aquitanian

Oligocene

Cht	12	Chattian
Rup	13	Rupelian

Eocene

Prb	14	Priabonian
Brt	15	Bartonian
Lut	16	Lutetian
Ypr	17	Ypresian

Paleocene

Tha	18	Thanetian
Dan	19	Danian

MESOZOIC

Cretaceous

Maa	20	Maastrichtian
Cmp	21	Campanian
San	22	Santonian
Con	23	Coniacian
Tur	24	Turonian
Cen	25	Cenomanian
Alb	26	Albian
Apt	27	Aptian
Brm	28	Barremian
Hau	29	Hauterivian
Vlg	30	Valanginian
Ber	31	Berriasian

Jurassic

Tth	32	Tithonian
Kim	33	Kimmeridgian
Clv	35	Callovian
Bth	36	Bathonian
Baj	37	Bajocian
Aal	38	Aalenian
Toa	39	Toarcian
Plb	40	Pliensbachian
Sin	41	Sinemurian
Het	42	Hettangian

Triassic

Rht	43	Rhaetian
Nor	44	Norian
Crn	45	Carnian
Lad	46	Ladinian
Ans	47	Anisian
Spa	48	Spathian
Smi	49	Smithian
Die	50	Dienerian

Note: Smithian and Dienerian used in database but now combined as Nammalian (Nml)

Gri	51	Griesbachian

PALEOZOIC

Permian

Tat	52	Tatarian
Kaz	53	Kazanian

Note: Tatarian and Kazanian used in database but now replaced by Changxingian (Chx), Longtanian (Lgt), Capitanian (Cap) and Wordian (Wor)

Ufi	54	Ufimian
Kun	55	Kungurian
Art	56	Artinskian
Sak	57	Sakmarian
Ass	58	Asselian

Carboniferous

Nog	59	Noginskian
Kla	60	Klazminskian
Dor	61	Dorogomilovskian
Chv	62	Chamovnicheskian
Kre	63	Krevyakinskian
Mya	64	Myachkovskian
Pod	65	Podolskian
Ksk	66	Kashirskian
Vrk	67	Vereiskian
Mel	68	Melekesskian
Che	69	Cheremshanskian
Yea	70	Yeadonian
Mrd	71	Marsdenian
Kin	72	Kinderscoutian
Alp	73	Alportian
Cho	74	Chokierian
Arn	75	Arnsbergian
Pnd	76	Pendleian
Bri	77	Brigantian
Asb	78	Asbian
Hlk	79	Holkerian
Aru	80	Arundian

Table 4.3 *continued*

Carboniferous *continued*

Chd	81	Chadian
Ivo	82	Ivorian
Has	83	Hastarian

Devonian

Fam	84	Famennian
Frs	85	Frasnian
Giv	86	Givetian
Eif	87	Eifelian
Ems	88	Emsian
Sig	89	Siegenian
		(now replaced by Pragian Pra)
Ged	90	Gedinnian
		(now replaced by Lochkovian Lok)

Silurian

Prd	91	Pridoli
Ldf	92	Ludfordian
Gor	93	Gorstian
Gle	94	Gleedonian
Whi	95	Whitwellian
She	96	Sheinwoodian
Tel	97	Telychian
Aer	98	Aeronian

Note: Fronian (Fro) and Idwian (Idw) used in GTS 82 now combined into Aeronian

| Rhu | 99 | Rhuddanian |

Ordovician

Hir	100	Hirnantian
Raw	101	Rawtheyan
Cau	102	Cautleyan
Pus	103	Pusgillian
Onn	104	Onnian
Act	105	Actonian
Mrb	106	Marshbrookian
Lon	107	Longvillian
Sou	108	Soudleyan
Har	109	Harnagian
Cos	110	Costonian
Llo3	111	Late Llandeilo
Llo2	112	Middle Llandeilo
Llo1	113	Early Llandeilo

Note: shown as Lo3, Lo2 and Lo1 in chronograms; above abbreviations preferred

| Lln2 | 114 | Late Llanvirn |
| Lln1 | 115 | Early Llanvirn |

Note: shown as Ln2 and Ln1 in chronograms; above abbreviations preferred

| Arg | 116 | Arenig |
| Tre | 117 | Tremadoc |

Cambrian

Dol	118	Dolgellian
Mnt	119	Maentwrogian
Men3	120	Late Menevian
Men2	121	Middle Menevian
Men1	122	Early Menevian

Note: shown as Mn3, Mn2 and Mn1 in chronograms; above abbreviations preferred

| Sol2 | 123 | Late Solvan |
| Sol1 | 124 | Early Solvan |

Note: shown as Sl2 and Sl1 in chronograms; above abbreviations preferred

Len	125	Lenian
Atb	126	Atdabanian
Tom	127	Tommotian
	128	**Precambrian**

4.15 Concluding remarks

In Table 4.2 we have used all generally accepted isotopic ages whatever their geologic setting, but have done this in a way that will enable the interested reader to recognize and, should he wish, reject some of the data. The use of a large database that does not exclude any generally accepted data introduces considerable stability into the time scale.

5

Chronometric calibration of stage boundaries

5.1 Introduction

Our major reasons for revising the chronometric calibration of stage boundaries presented in GTS 82 are to update the Mesozoic and Paleozoic calibrations in the light of recently published isotopic dates and to apply the chronogram method of assessing age boundaries (Section 5.4) to the Cenozoic Era for the first time.

Given a reasonably well-distributed set of stratigraphically controlled isotopic dates, the setting up of a time scale would be straightforward. Standard statistical methods could be applied to the data to give the age of each stratigraphic boundary together with estimates of the errors involved. The statistical problem is similar to that of finding the ages of magnetic reversals from isotopic data (Cox & Dalrymple 1967, Mankinen & Dalrymple 1979).

Unfortunately, this straightforward procedure cannot be applied to the entire Phanerozoic record for several reasons. The first is that, except for Cenozoic and later Cretaceous stratigraphic boundaries, the data controlling the ages of chronostratigraphic boundaries are sparse and unevenly distributed.

A second problem is the heterogeneity of the available data set, which includes dates determined by several different methods on a range of minerals from a variety of geologic settings. For example, glauconite dates are used for dating throughout the Phanerozoic Eon, particularly in the Cenozoic

Era. However, opinion about their reliability varies (Sections 4.3 and 5.5; cf. Berggren et al. 1985b, Odin 1982b).

A third problem is that the biostratigraphic stage of a dated mineral may be difficult to establish locally and the correlation of the local biostratigraphic stage with the global stages may be imperfect. We have attempted to meet this problem by assigning a chronostratic error to each date (Section 5.3).

A final problem is that many of the available values do not date a stratigraphic division directly but serve only to bracket its age within broad limits.

We illustrate the problem by the following example. Suppose we have a set of isotopic dates with estimated total errors, i.e. chronometric and chronostratic errors. The dates are from samples in two adjacent stratigraphic units. In an ideal case the dates would form a numerical progression, with all the dates from the younger unit being smaller than all the dates from the older unit, except for dates on the boundary, which would be identical. Provided the data were reasonably closely spaced, the age of the boundary could be read directly by inspection. Here the errors are unimportant in the determination of the boundary age.

In practice, the dates on either side of the boundary are relatively sparse, have different errors and include overlapping dates. The problem of estimating the age of a boundary is thus more difficult. One method of choosing a boundary age is to assess the dates individually, selecting a few as the most reliable and basing a boundary age on these alone (e.g. Berggren et al. 1985b). A second is to draw up a numerical sequence showing all the data, from which an age for the boundary is estimated (e.g. Gale 1982). A third method is to display the dates on a plot of time against stratigraphic ages and to assess visually the best fit of the boundary (e.g. Armstrong 1978c, Gale 1985).

The major criticism of all these methods is that they are not reproducible, i.e. the time scale resulting from the same set of dates will differ from geologist to geologist. The dates available for calibration are so variable in quality and distribution (Chapter 4) that there is bound to be considerable discussion for the foreseeable future about where to place a particular boundary. However, we believe these discussions should be based on reasoned modifications to reproducible estimates of the boundary position, rather than on purely subjective assessments of the data. In GTS 82 we estimated the ages of stratigraphic boundaries by a modification of the method used to estimate the ages of the magnetic reversals from isotopic data (Cox & Dalrymple 1967). In this method **chronograms** are constructed to express the statistical suitability of any age value for the particular stratigraphic boundary being evaluated.

In some cases the recognized stratigraphic divisions have no available isotopic dates, as in several Carboniferous stages. In such cases chronograms will be obtained for the older and the younger stratigraphic boundaries of the stages, but they will be very poorly controlled.

In other cases the stratigraphic divisions have been modified after the isotopic database had been set up. Some of these modifications have been incorporated into the database;

others have been noted but not modified; still others cannot be modified because the stratigraphic correlations necessary to make the modifications are highly uncertain, as in the later Permian. All such modifications are noted in Table 4.3.

5.2 Estimating the boundary age of adjacent stratigraphic units

5.2.1 Essentials of the method

We start with the simplest situation: a date, t, from a mineral lying wholly within a stratigraphic unit B, say, which is overlain by a younger unit A and itself overlies an older unit C. The B/C boundary must be older than t and the A/B boundary must be younger than t.

Assume that we are trying to estimate the age of one boundary, B/C, where B is younger than but adjacent to C. Assume we have a set of dates, from unit B or from unit C, each with an estimated error s. The value of s includes the chronometric and chronostratic errors. How can we estimate the boundary age? Let B_i and C_i be dates of units B and C respectively. Let each date have an error s_{Bi} and s_{Ci}.

Suppose we estimate t as the age of the B/C boundary. We can ignore all dates from B that are younger than t and all dates from C (the older unit) that are older. For example, suppose we have dates of 0, 2, 4 and 6 from B and 4, 6, 8 and 10 from C. In isolation the date of 0 from B tells us that the B/C boundary is at 0 or a still older time. Similarly, the date of 2 from B tells us that the B/C boundary is at 2 or a still older time. When 0 and 2 are taken together, both must be considered if we start with an estimate of 0 as the B/C boundary, but only 2 is relevant when the estimate is greater than 0, e.g. 1, 2, 3, etc. Similarly, if we consider the data from C in isolation, then the date of 4 from C tells us that the B/C boundary is at 4 or a still younger age. When 4 and 6 are taken together, both must be considered if we start with an estimate of 6 as the B/C boundary, but only 4 is relevant from C if we start with an estimate of 4.

In other words, if we scan the dates the following are relevant to each scan:

Scanning age	Relevant dates	
	From B	From C
0	0, 2, 4, 6	None
2	2, 4, 6	None
4	4, 6	4
6	6	4, 6
8	None	4, 6, 8
10	None	4, 6, 8, 10

We now need to estimate the influence of each date on the position of the boundary. If the scanning age is 0 all the relevant dates (0, 2, 4 and 6) should influence the position of the boundary, with the most influence being exerted by the oldest date (6), since it is the oldest date from the younger stratigraphic unit. Thus for each scanning age we can construct an 'error function', E, given by the sum of the squares of the (relevant date – scanning age):

$$E = Sum\ (B_i - t_e)^2, \text{ where } t_e \text{ is now the scanning age.}$$

In the case considered E would be:

Scanning age	Contribution to error function		
	From B	From C	Total E
0	0+(4)+(16)+(36)	0	56
2	0+(4)+(16)	0	20
4	0+(4)	0	4
6	0	0+(4)	4
8	0	0+(4)+(16)	20
10	0	0+(4)+(16)+(36)	56

If all the errors associated with each date were identical, the minimum value of the above function would be the best estimate of the calibration age of the B/C boundary. The function is clearly symmetric; its minimum value must lie at 5. If the scanning age is 5 then the contribution of the date of 6 from B to the error function is (6–5) squared, or 1, and that of the date of 4 from C is (4–5) squared, or 1. The least value of the error function is therefore 2.

A plot of this error function against scanning age will look like a parabola, with a minimum of 2 at scanning age 5 (Figure 5.1). If the dates did not overlap, i.e. if the date of 6 from B and 4 from C did not exist, then the error function would be zero in the scanning range 4–6 (Figure 5.2): it would be flat-bottomed. If the dates from B did not exceed 5 and

Figure 5.1. Model chronogram in which dates from B are 0, 2, 4 and 6 Ma; from C 4, 6, 8 and 10 Ma. The error function is symmetric but not zero at its minimum as a result of overlapping dates.

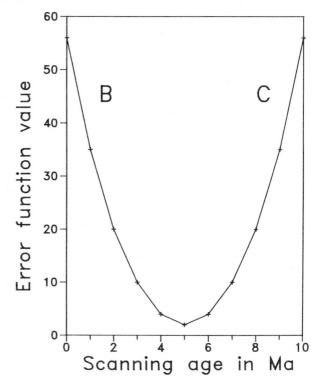

Figure 5.2. Model chronogram in which dates from B are 0, 2 and 4 Ma; from C 6, 8 and 10 Ma. The error function is symmetric, flat-bottomed and zero in the age range 4–6 Ma, as a result of non-overlapping dates.

Figure 5.3. Model chronogram in which dates from B are 0, 2, 4 and 5 Ma; from C 5, 6, 8 and 10 Ma. The error function is symmetric and zero at its minimum as a result of perfect data.

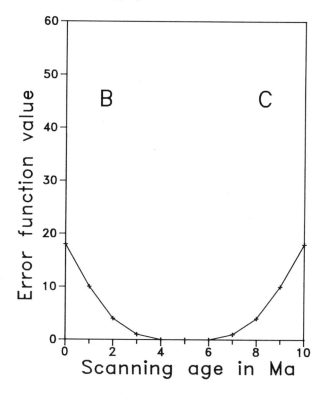

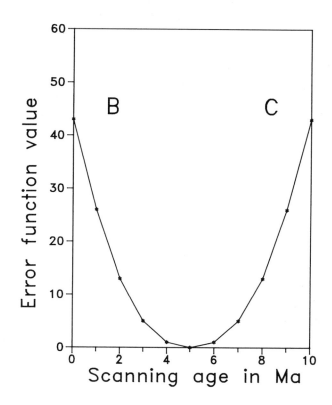

those from **C** were never less than 5, then the error function would just touch the scanning age axis (Figure 5.3).

Unfortunately, real dates have different errors associated with them. We need to weight each date accordingly. We have chosen to divide each of the above contributions to the error function by the square of the one sigma error. For example, suppose the date of 0 from **B** had an error of 2, we would divide the error contribution for that date by 4 before adding it to the error function.

Thus the full expression for the error function is:

$$E = (B_i-t_e)^2/s_{Bi}^2 + (C_i-t_e)^2/s_{Ci}^2 \qquad (5.1)$$

This error function is simply a mathematical attempt to express the contribution of each date to the **B/C** boundary by taking into account its date and error. We have experimented numerically with other error functions. For example, instead of squaring each term, it would be possible to use the modulus (i.e. the numerical value) of each term. However, we consider the above function is a good representation of the contributions of each date to the boundary. (In GTS 82 the function was described as E^2, rather than E, and E^2 was plotted against scanning age.)

5.2.2 Minimum dates

Sometimes a date from a mineral will be a minimum value, rather than an actual value. For example, we believe that glauconites are best treated as minimum dates for all stage boundaries older than 115 Ma (Section 5.5) rather than as true dates. How is this to be treated on a chronogram? Suppose the date is **x**, from unit **B**. Then what this date tells us

is that the **B/C** boundary is older than **x**. **x** contributes to the error function of the **B/C** and older boundaries for which $t_e <$ **x**. The date does not contribute at all to any estimates of the younger boundary of **B**, or to any other still younger boundaries.

If **C** has a glauconite dated at **y**, then this tells us nothing about the age of the **B/C** boundary unless we have some estimate of how much lower the dated glauconite is below the actual age. In the absence of such information the date **y** from **C** cannot contribute at all to the error function of the **B/C** or any younger boundaries, but will contribute to the error function for all older boundaries for which $t_e <$ **y**; i.e. it will control the position of the older boundary of **C** for which $t_e <$ **y**. For example, suppose the glauconite from **C** is dated at 50 Ma. Then this says nothing about the age of the **B/C** boundary – it could be virtually any age. However, it tells us that the **C/D** boundary – assuming **D** is the next unit older than **C** – is older than 50 Ma.

5.2.3 'Bracketed' dates

What happens if a date is 'bracketed' by units **A** and **D**? For example, consider a date from a granite which intrudes Kimmeridgian rocks and is overlain unconformably by Valanginian sediments. Here there are two missing units: Berriasian (younger) and Tithonian (older). The Valanginian/Berriasian boundary is younger than the date found and the Tithonian/Kimmeridgian boundary is older. The date does not contribute to a knowledge of the Tithonian/Berriasian boundary, i.e. we cannot say whether this boundary is younger or older than the date.

We can generalize then by saying that if **A** (youngest) and **D** (oldest) units form a sequence **A–B–C–D** then we can regard any date from **B** or **C** as lying within the composite unit **B–C**. Let a date be **x**. Then the **A/B** boundary $<$ **x** $<$ **C/D** boundary. If t_e is a trial estimate, then **x** contributes to the error function of the **A/B** (or younger) unit boundaries only if $t_e <$ **x**. It is used in the estimate of the **C/D** (or older) boundaries if $t_e >$ **x**. **x** does not contribute to the error function of the internal unit boundary, **B/C**.

What happens if units **E** and **K** 'bracket' a date, where **A–B–...J–K–L ...** form the sequence of units? This case is simply an extension of a bracketed date, where the composite unit is now **F–G–H–I–J**. The bracketed date contributes to the error function of unit boundaries **A/B, B/C, C/D, E/F** if $t_e <$ **x** and to that for unit boundaries **J/K, L/M ...** if $t_e >$ **x**. As in the previous case, a bracketed date does not provide any information about any unit boundary within the composite unit, such as **F/G, G/H, H/I** or **I/J**.

5.3 Chronostratic errors

In its original application to time scale calibration in GTS 82, the misfit error function took into account only the chronometric error, i.e. the estimate of the experimental error associated with each isotopic date. The contribution of an individual date to the error function in a chronogram is proportional to the inverse of the error squared. Thus the contribution of a date with a small reported error is much higher, as it should be, than a date with a larger error. Where data are sparse this method may give unduly heavy weighting to precise analytical values whose paleontologically determined chronostratic age is much less certain. For example, the error involved in correlating dates from rocks outside the standard sections to the standard sections might be several times the quoted analytical error. Clearly, it is misleading to ignore this chronostratic error when estimating the age of the boundary.

We know of no paleontologic procedure that will allow us objectively to estimate the chronostratic error. When the age of a sample is reported in the literature the biostratigraphic precision given to it is commonly a chronostratigraphic time interval, usually a 'stage', e.g. Maastrichtian. Sometimes the precision is much better, e.g. it may be assigned to a chronostratic chron such as *Pachydiscus neubergicus*; sometimes it is much worse, e.g. it may be assigned to an epoch, the Senonian. However, such an item would have little significance for the calibration of the time scale. We assume that in general the biostratigraphic precision is related to the stage assigned. All dates lying within a particular period are then assigned an average chronostratic error equal to half the average duration of each stage. For example, the average length of stages in Miocene and younger rocks is estimated at about 2 Ma, giving a chronostratic error of 1 Ma. The chronostratic errors are listed in Table 5.1.

The effect of including chronostratic errors is to reduce the influence of dates with small analytical errors on the calibration of the stage boundary. In most cases the inclusion of these chronostratic errors has little effect on the position of the minimum of the error function for a stage boundary, or on the estimated uncertainty in the age of the minimum value of

Table 5.1 *Chronostratic errors**

Strat. interval	'Stages'	Duration	Duration/ 'stages'	Age adopted	Chrono- stratic error (= age/2)
Q	3	Chronograms not used for Quaternary			
Ng	9	23	2.5	2.5	1.0
Pg	8	40	5.0	5.0	2.5
K	12	79	6.6	5.0	2.5
J	11	69	6.3	5.0	2.5
Tr	8	35	4.4	5.0	2.5
P	9	38	4.2	5.0	2.5
Pen	14	34	2.4	5.0	2.5
Mis	11	40	3.6	5.0	2.5
D	7	48	6.9	5.0	2.5
S	9	30	3.3	3.4	1.7
O	20	67	3.3	3.4	1.7
€	7	85	12	10	5†

*The GTS 82 data appear in this table because they were used for the chronograms that gave rise to the GTS 89 definitive scale and it makes little difference to the error values.

†12 Ma was considered an artificially high value for the age duration, so a chronostratic error of 5 Ma was adopted. The Cambrian data are so poor that its adoption would not greatly affect results.

the chronogram. We believe that progress in the precision with which stage boundaries can be determined may well depend on devising some numerical method for estimating how far from the top or the bottom of a stage boundary a sample lies: only in this way can the chronostratic error be reduced. A similar problem exists in specifying the location of stratigraphic boundaries within a magnetic anomaly reversal sequence (Chapter 6).

5.4 Chronograms

A chronogram is a plot of the error function against scanning age. The best estimate of the calibration age is then taken as the scanning age for which this function is minimum. Obviously, where no data exist in a particular scanning interval there is no unique minimum but rather a range in which the minimum must lie.

The ideal chronogram is symmetric, has steep sides and falls to zero at its mid-point (Figure 5.3). A chronogram whose minimum value is significantly different from zero shows the presence of a significant number of overlapping dates (Figure 5.1). A gap in the data is shown in Figure 5.2. Marked asymmetry is generally caused by a scarcity of data on one side of the minimum.

The chronogram program used for GTS 82 was rewritten and extended. Finding a chronogram is now completely automatic, with self-adjusting scaling for the scanning interval and for the error function. While this improves the appearance of the chronograms, it has the effect that scales may vary vertically and horizontally in successive chronograms: comparisons are less simple than they were in the original volume. The program has also been adjusted so that where permitted by the data the minimum value lies at an integral number of millions of years, or at some readily decimalized fraction such as 0.5, 0.25, etc.

Autoscaling has improved the precision with which chronogram ages can be estimated. For example, in GTS 82 the duration of short stages in the Cretaceous Period could be in error by at least 50%. This error arose because the minimum scanning interval for some of the Late Cretaceous stage boundaries was set at 0.5 Ma. Such a coarse scanning interval meant that Cretaceous ages could vary by as much as 1 Ma, irrespective of any errors in the data. The duration of the Coniacian was assessed at 1 Ma, but it could have been as little as 0.5 Ma or as much as 1.5 Ma – a range of a factor of 3. In the Paleozoic where the scanning interval was even coarser, stages as short as the Coniacian might well be missed altogether in the scan for the minimum error.

The inclusion of a chronostratic error in the data has increased some of the estimated errors in the chronogram ages compared with those given in GTS 82. Strictly speaking, the errors given in GTS 82 are the chronometric errors in the analytical data for each chronogram age. If a boundary section has been studied in detail in one area the chronometric error may be quite small, but unless it is part of the internationally recognized standard section there will be an unavoidable and generally significant chronostratic error which was ignored in GTS 82.

The available data are too sparse for rigorous statistical methods to be applied. We have attempted to give a quantitative estimate of the error in the chronogram age by taking the total error as the age range for which the error function did not exceed its minimum value by more than 1.0 (Table 5.4). In general the error is symmetric about the chronogram age, but where the chronogram is asymmetric, the error range may be asymmetric as seen in the Santonian/ Coniacian chronogram (Figure A4.20).

The plausibility of this choice for the error range is readily seen where only two identical dates determine a boundary, one of them being from the younger unit at the boundary, the other from the older unit. From equation 5.1, E is zero at the boundary. It rises to 1.0 on both sides of the boundary when the scanning age differs from the date by the error. We call this difference t. If there are N identical dates, with identical errors, e, then E rises to 1.0 when t is reduced in the ratio $1/N^{1/2}$, which is a reasonable relationship between the number of observations and the error estimate. However, it is not clear how to obtain statistically valid estimates of the uncertainty in the chronogram age when different errors for widely varying dates need to be combined into a single error estimate. We believe that the estimate used above gives reasonable chronogram age errors. The ratio of any two chronogram errors also provides a plausible estimate of the relative precision of the ages.

For the chronograms (Figures A4.1–A4.125) we have used the convention that time runs numerically from left to right: the older stratigraphic stage is always on the right of the chronogram. This has the effect of reversing geologic convention where, for example, it is usual to refer to the Jurassic/Cretaceous boundary, rather than the Cretaceous/ Jurassic boundary, as we have here. The horizontal and vertical ticks are in units corresponding to the autoscaling. The time units in particular may be in rather arbitrary values. However, the central point is always the minimum of the error function or the mid-point of any run of zeros in the error function. The two vertical lines on either side of the minimum are the estimated range in chronogram error for the boundary. The value of the minimum, best estimate and maximum are always given.

5.5 Glauconite and non-glauconite dates

A major problem in calibrating the time scale is deciding how to assess glauconite dates. The age calibration program has four options in it: (1) to include all relevant data; (2) to use only glauconite; (3) to regard all glauconite as minimum dates and (4) to ignore all glauconite.

The differences in Ma between the chronogram ages for Cenozoic, Cretaceous and the later Jurassic periods obtained by using only glauconite and no glauconite are shown in Table 5.2. A double asterisk means that there are no significant glauconite or non-glauconite dates in a time span of 10 Ma for the boundary concerned. For example, there are no significant non-glauconite dates in the time interval 136 to 152.5 Ma that define the 30/31 (=Valanginian/Berriasian) boundary, though glauconite dates exist in this time range.

The glauconite dates in the database systematically give younger chronogram ages than the non-glauconite dates (Table 5.2). The difference exceeds 5 Ma at the Aptian/ Barremian and older Cretaceous boundaries (Figure 5.4). It is the major cause of the consistently younger Jurassic and Early Cretaceous calibration ages in the time scale of Odin (1982b) and Snelling (1985b) compared with GTS 82. If glauconites are excluded, then the age estimates obtained for the Early Cretaceous interval are very similar to those found by regarding glauconites as minimum dates. We consider that all glauconite dates older than 115 Ma should be regarded as minimum dates only. We therefore treat glauconite dates as minimum dates in all age calibrations that are older than the initial Aptian boundary.

There appears to be no systematic reduction in difference between no glauconite and glauconite dates with decreasing age (Table 5.2, column 2 minus column 3) for the interval 0–115 Ma. The average value of the difference for those stage boundaries where there are significant high-temperature and glauconite data in a time interval smaller than 10 Ma is –1.0 Ma, with a range of –4.5 to 3.4 Ma. We therefore take the chronogram ages of nearly all boundaries younger than the Aptian/Barremian boundary to be the value given by using all the data. The difference between the ages given by using all the data and glauconite only (Table 5.2, column 1 minus column 2) data averages to 0.7 Ma, which is usually less than the uncertainty range of the chronogram age. The relative influence of glauconite dates and dates from other minerals on the chronograms is shown in Table 5.3 under the heading 'Data controlling age range'. Two columns are given: both give the number of high-temperature dates (= H) and glauconite (= G) that control the older and younger ranges of the chronogram age. Thus one can see the relative importance of glauconite and high-temperature dates, together with the number of dates available.

Table 5.2. *Comparison of glauconite with non-glauconite dates*

Code	Abbrev.	All (1)	Glauc (2)	No. Gl (3)	Difference Glauc − No Gl	Data controlling age range using all all data option: H = high-T mineral, G=glauconite	
						Younger	Older
Chronostratic error = 1 Ma							
3/4	Pia/Zan	3.5	2.8**	3.5	−0.7	3H	5H
4/5	Zan/Mes	6.0	3.0**	6.0	−3.0	3H	4H
5/6	Mes/Tor	8.4	8.8**	8.4	0.4	3H	6H
6/7	Tor/Srv	10.4	13.8	10.4	3.4	4H,1G	9H
7/8	Srv/Ln2	14.6	13.8	14.7	−0.9	2H	6H
8/9	Ln2/Ln1	Not used					
9/10	Ln1/Bur	16.0	13.8	16.0	−2.2	8H	1H
10/11	Bur/Aqt	22.0	21.0	22.0	−1.0	1H,1G	3H,3G
11/12	Aqt/Cht	23.8	22.4	24.7	−2.3	1H,1G	2H,3G*
Chronostratic error = 2.5 Ma							
12/13	Cht/Rup	30.6	27.5	32.0	−4.5	4H,1G	3H,3G
13/14	Rup/Prb	36.4	37.5	35.6	1.9	4H,2G	7H,4G
14/15	Prb/Brt	39.4	39.0	40.5	−1.5	3H,2G	1H,8G*
15/16	Brt/Lut	41.4	41.4	42.4	−1.0	2H,3G*	1H,6G*
16/17	Lut/Ypr	48.0	47.4	51.0	−3.6	2H,8G*	1H,10G*
17/18	Ypr/Tha	53.4	53.4	51.7	1.7	2G*	1H,5G*
18/19	Tha/Dan	61.0	61.5	60.7	0.8	1H,3G*	5H,3G
19/20	Dan/Maa	66.0	65.0	66.0	−1.0	4H	3H,2G
20/21	Maa/Cmp	71.5	71.5	71.5	0.0	4H,2G	6H,1G
21/22	Cmp/San	82.4	83.5	82.0	1.5	6H,2G	3H,2G
22/23	San/Con	86.6	84.7	87.5	−2.8	4H,1G	3H,12G*
23/24	Con/Tur	88.5	88.5	89.0	−0.5	1H,4G*	12G*
24/25	Tur/Cen	90.4	89.0	93.5	−4.5	5H,4G	2H,8G*
25/26	Cen/Alb	97.0	97.0	97.5	−0.5	1H,5G*	3H,8G*
26/27	Alb/Apt	110.4	110.0	112.0	−2.0	1H,6G*	2H,5G*
27/28	Apt/Brm	118.0	118.0	124.0**	−6.0	1H,4G*	2G*
28/29	Brm/Hau	122.5	121.4	131.0**	−9.6	2H,2G	2H,3G
29/30	Hau/Vlg	128.5	125.0	135.5**	−10.5	3H,2G	2H,4G*
30/31	Vlg/Ber	132.4	130.0	137.5**	−7.5	3H,2G	1H,4G*
31/32	Ber/Tth	134.0	131.0	148.0**	−17.0	4H,4G	2H,5G*

* = data dominated by glauconites

** = no data in a time interval > 10 Ma

Figure 5.4. Comparison of glauconite (Gl) and non-glauconite (non-Gl) calibration ages for Cenozoic and Cretaceous stages.

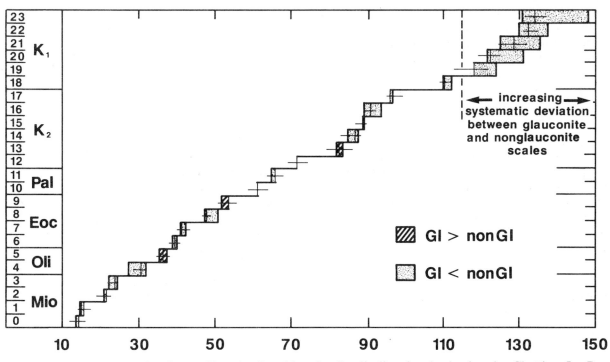

Key to stratigraphic boundaries: 0 = Serravallian; 1 = Langhian; 2 = Burdigalian; 3 = Aquitanian; 4 = Chattian; 5 = Rupelian; 6 = Priabonian; 7 = Bartonian; 8 = Lutetian; 9 = Ypresian; 10 =Thanetian; 11 = Danian; 12 = Maastrichtian; 13 = Campanian; 14 = Santonian; 15 = Coniacian; 16 = Turonian; 17 = Cenomanian; 18 = Albian; 19 = Aptian; 20 = Barremian; 21 = Hauterivian; 22 =Valanginian; 23 = Berriasian. Key to abbreviations: Mio = Miocene; Oli = Oligocene; Eoc = Eocene; Pal = Paleocene; K_2 = Late Cretaceous; K_1 = Early Cretaceous.

The dot shaded boxes with stippled ornament, which predominate, are those for which the glauconite ages are younger than the non-glauconite ages. The diagonally lined boxes are those for which glauconite ages are older. The age range of each boundary has been estimated from the chronograms using all the data for that boundary. For convenience, the mean age and age range have been plotted within the box immediately below the relevant boundary. As an example, the cross in the lower left-hand corner refers to the boundary immediately above it, i.e. the 0/1 boundary, or Serravallian/Langhian boundary. The vertical bar of the cross has no significance other than acting as a reference point for the mean age. Where the mean age lies close to the glauconite mean age, as in the 8/9, or Lutetian/Ypresian boundary, then glauconites dominate the chronogram; where it lies close to the non-glauconite mean age, as in 13/14, or Campanian/Santonian boundary, then non-glauconites dominate the chronogram. Reproduced by permission of the Geological Society of America from *Geology* (Craig, Smith & Armstrong 1989).

Table 5.3. *Initial boundaries where (glauconite minus no glauconite) age differs by more than 2 Ma and where the time interval for which there are no significant data is less than 10 Ma* * = data dominated by glauconites

Code	Abbrev.	All (1)	Glauc (2)	No Gl (3)	Difference Glauc − No Gl	Data controlling age range using all data option: H = high-T mineral, G = glauconite	
						Younger	Older
9/10	Ln1/Bur	16.0	13.8	16.0	−2.2	8H	1H
11/12	Aqt/Cht	23.8	22.4	24.7	−2.3	1H,1G	2H,3G
12/13	Cht/Rup	30.6	27.5	32.0	−4.5	4H,1G	3H,3G
16/17	Lut/Ypr	48.0	47.4	51.0	−3.6	2H,7G*	1H,10G*
22/23	San/Con	86.6	84.7	87.5	−2.8	4H,1G	3H,12G*
24/25	Tur/Cen	90.4	89.0	93.5	−4.5	5H,4G	2H,8G*
26/27	Alb/Apt	110.4	110.0	112.0	−2.0	1H,6G*	2H,5G*
27/28	Apt/Brm	118.0	118.0	124.0	−6.0	1H,4G*	2G*
28/29	Brm/Hau	122.5	121.4	131.0	−9.6	2H,2G	2H,3G
29/30	Hau/Vlg	128.5	125.0	135.5	−10.5	3H,2G	2H,4G*
30/31	Vlg/Ber	132.4	130.0	137.5	−7.5	3H,2G	1H,4G*
31/32	Ber/Tth	134.0	131.0	148.0	−17.0	4H,4G	2H,5G*

5.6 Chronogram data

A total of 127 chronograms were prepared for this work. The first two, defining the Holocene/Pleistocene and Pleistocene/Piacenzian boundaries, have not been used because the ages of both these boundaries are better estimated by other means. The remaining 125 chronograms appear in Appendix 4 as Figures A4.1 to A4.125. The essential data, as used in this work, are abstracted below in Table 5.4.

1. The columns headed 'Stages', e.g. 127(Tom) to 128(Pou), give the age and three-letter abbreviation of the stratigraphic boundary listed in Appendix 2 and Table 4.3.

2. 'Min' is the least age given by the chronogram for which the error function rises by 1 above its value at 'Best'.

3. 'Best' in Ma is the minimum age given by the chronogram.

4. 'Max' is the highest age given by the chronogram for which the error function rises by 1 above its value at 'Best'.

5. 'Range' is the difference between Min and Best, i.e. (Min−Best); and Max and Best, i.e. (Max−Best). Both are given because of the asymmetry of some chronograms.

6. 'Diff' is the difference between Max and Min, i.e. (Max−Min).

7. 'No data' is the age range for which the error function is 0.1 or less in value, i.e. the time range for which there is effectively no significant data.

8. The letters in the last column give some idea of the quality of the chronogram: For 'A' Diff is less than 5 Ma; for 'B' it is between 5 and 10 Ma; for C it is greater than 10 Ma.

Table 5.4 *Chronogram results*

CENOZOIC

Glauconites treated as normal, assumed chronostratic error = 1.0 Ma

1(Hol) to 2(Ple)   Chronograms not used
2(Ple) to 3(Pia)   Chronograms not used

Stages			Min	Best	Max	Range		Diff	No data	
3(Pia)	to	4(Zan)	2.5	3.5	4.5	−1.0	1.0	2.0	3.3	4.0 A
4(Zan)	to	5(Mes)	4.4	6.0	7.6	−1.6	1.6	3.2	5.2	6.8 A
5(Mes)	to	6(Tor)	6.0	8.4	10.8	−2.4	2.4	4.8	7.2	10.0 A
6(Tor)	to	7(Srv)	9.2	10.4	11.9	−1.2	1.5	2.7	10.1	11.0 A
7(Srv)	to	8(Lan2)	12.6	14.6	16.6	−2.0	2.0	4.0	13.8	15.6 A
8(Lan2)	to	9(Lan1)	13.0	14.6	16.6	−1.6	2.0	3.6	14.2	15.4 A
9(Lan1)	to	10(Bur)	15.0	16.0	17.5	−1.0	1.5	2.5	15.7	16.3 A
10(Bur)	to	11(Aqt)	19.6	22.0	23.2	−2.4	1.2	3.6	21.2	22.0 A
11(Aqt)	to	12(Cht)	22.7	23.8	24.7	−1.0	1.0	2.0	23.5	24.0 A

Glauconites treated as normal, assumed chronostratic error = 2.5 Ma

12(Cht)	to	13(Rup)	29.0	30.6	32.2	−1.6	1.6	3.2	30.6	31.0 A
13(Rup)	to	14(Prb)	34.8	36.4	38.0	−1.6	1.6	3.2	36.0	36.8 A
14(Prb)	to	15(Brt)	37.8	39.4	41.0	−1.6	1.6	3.2	39.0	39.8 A
15(Brt)	to	16(Lut)	39.0	41.4	43.0	−2.4	1.6	4.0	40.6	41.8 A
16(Lut)	to	17(Ypr)	46.4	48.0	49.2	−1.6	1.2	2.8	47.6	48.0 A
17(Ypr)	to	18(Tha)	51.0	53.4	55.8	−2.4	2.4	4.8	53.4	54.0 A
18(Tha)	to	19(Dan)	58.5	61.0	64.0	−2.5	3.0	5.5	60.5	62.0 B
19(Dan)	to	20(Maa)	64.0	66.0	68.5	−2.0	2.5	4.5	65.5	66.5 A

CRETACEOUS

Stages			Min	Best	Max	Range		Diff	No data	
20(Maa)	to	21(Cmp)	69.0	71.5	74.5	−2.5	3.0	5.5	70.5	72.5 B
21(Cmp)	to	22(San)	78.8	82.4	86.0	−3.6	3.6	7.2	81.8	83.6 B
22(San)	to	23(Con)	83.0	86.6	89.0	−3.6	2.4	6.0	86.0	87.2 B
23(Con)	to	24(Tur)	86.5	88.5	90.0	−2.0	1.5	3.5	88.0	88.5 A
24(Tur)	to	25(Cen)	88.8	90.4	92.0	−1.6	1.6	3.2	90.0	90.8 A
25(Cen)	to	26(Alb)	95.0	97.0	99.0	−2.0	2.0	4.0	96.5	97.5 A
26(Alb)	to	27(Apt)	108.4	110.4	112.0	−2.0	1.6	3.6	110.0	110.8 A

Glauconites treated as minimum ages, assumed chronostratic error = 2.5 Ma

Stages			Min	Best	Max	Range		Diff	No data	
27(Apt)	to	28(Brm)	113.0	125.5	138.0	−12.5	12.5	25.0	116.7	134.2 C
28(Brm)	to	29(Hau)	123.0	131.0	138.0	−8.0	7.0	15.0	128.0	135.0 C
29(Hau)	to	30(Vlg)	128.0	135.5	141.7	−7.5	6.3	13.8	133.0	138.0 C
30(Vlg)	to	31(Ber)	131.3	137.5	145.0	−6.3	7.5	13.8	135.0	140.0 C
31(Ber)	to	32(Tth)	139.0	148.0	157.0	−9.0	9.0	18.0	144.0	153.0 C

JURASSIC

Stages			Min	Best	Max	Range		Diff	No data	
32(Tth)	to	33(Kim)	139.0	151.0	161.5	−12.0	10.5	22.5	145.0	157.0 C
33(Kim)	to	34(Oxf)	151.0	156.0	162.0	−5.0	6.0	11.0	154.0	158.0 C
34(Oxf)	to	35(Clv)	151.2	156.3	165.0	−5.0	8.8	13.8	153.7	160.0 C
35(Clv)	to	36(Bth)	152.0	159.0	165.0	−7.0	6.0	13.0	157.0	162.0 C
36(Bth)	to	37(Baj)	151.7	159.2	165.5	−7.5	6.3	13.8	156.7	161.7 C
37(Baj)	to	38(Aal)	166.5	177.0	187.5	−10.5	10.5	21.0	171.0	183.0 C
38(Aal)	to	39(Toa)	167.5	178.0	188.5	−10.5	10.5	21.0	172.0	184.0 C
39(Toa)	to	40(Plb)	168.0	182.0	198.0	−14.0	16.0	30.0	172.0	192.0 C
40(Plb)	to	41(Sin)	182.0	189.5	197.0	−7.5	7.5	15.0	187.0	192.0 C
41(Sin)	to	42(Het)	198.5	203.5	211.0	−5.0	7.5	12.5	202.2	206.0 C
42(Het)	to	43(Rht)	201.7	210.5	216.7	−8.8	6.3	15.0	208.0	213.0 C

TRIASSIC

Stages			Min	Best	Max	Range		Diff	No data	
43(Rht)	to	44(Nor)	202.0	209.5	218.5	−7.5	9.0	16.5	208.0	212.5 C
44(Nor)	to	45(Crn)	214.0	226.0	233.5	−12.0	7.5	19.5	223.0	230.5 C
45(Crn)	to	46(Lad)	232.0	235.0	239.0	−3.0	4.0	7.0	235.0	236.0 B
46(Lad)	to	47(Ans)	233.2	239.5	245.7	−6.3	6.3	12.5	237.0	242.0 C
47(Ans)	to	48(Spa)	235.0	241.2	250.0	−6.3	8.8	15.0	238.7	245.0 C
48(Spa)	to	49(Smi)	235.0	242.5	251.2	−7.5	8.8	16.3	238.7	246.2 C
49(Smi)	to	50(Die)	235.0	242.5	251.2	−7.5	8.8	16.3	238.7	246.2 C
50(Die)	to	51(Gri)	235.0	242.5	251.2	−7.5	8.8	16.3	238.7	246.2 C
51(Gri)	to	52(Tat)	234.5	245.0	254.0	−10.5	9.0	19.5	239.0	251.0 C

PERMIAN

Stages			Min	Best	Max	Range		Diff	No data	
52(Tat)	to	53(Kaz)	237.5	248.0	254.0	−10.5	6.0	16.5	245.0	251.0 C
53(Kaz)	to	54(Ufi)	247.0	259.0	273.0	−12.0	14.0	26.0	253.0	267.0 C
54(Ufi)	to	55(Kun)	247.0	259.0	273.0	−12.0	14.0	26.0	253.0	267.0 C
55(Kun)	to	56(Art)	252.0	262.5	274.5	−10.5	12.0	22.5	256.5	270.0 C
56(Art)	to	57(Sak)	252.0	262.0	274.0	−10.0	12.0	22.0	256.0	270.0 C
57(Sak)	to	58(Ass)	268.0	280.0	294.0	−12.0	14.0	26.0	274.0	286.0 C
58(Ass)	to	59(Nog)	281.5	292.0	301.0	−10.5	9.0	19.5	287.5	296.5 C

continued

CARBONIFEROUS

Stages			Min	Best	Max	Range		Diff	No data	
59(Nog)	to	60(Kla)	281.5	292.0	302.5	−10.5	10.5	21.0	287.5	296.5 C
60(Kla)	to	61(Dor)	288.0	295.5	301.7	−7.5	6.3	13.8	294.2	296.7 C
61(Dor)	to	62(Chv)	288.0	295.5	301.7	−7.5	6.3	13.8	294.2	296.7 C
62(Chv)	to	63(Kre)	295.0	301.2	307.5	−6.3	6.3	12.5	300.0	303.7 C
63(Kre)	to	64(Mya)	298.0	303.0	308.0	−5.0	5.0	10.0	302.0	304.0 B
64(Mya)	to	65(Pod)	301.0	305.0	309.0	−4.0	4.0	8.0	304.0	306.0 B
65(Pod)	to	66(Ksk)	305.2	311.5	317.7	−6.3	6.3	12.5	310.2	312.7 C
66(Ksk)	to	67(Vrk)	305.5	316.0	326.5	−10.5	10.5	21.0	311.5	320.5 C
67(Vrk)	to	68(Mel)	307.0	316.0	326.5	−9.0	10.5	19.5	311.5	320.5 C
68(Mel)	to	69(Che)	307.0	316.0	326.5	−9.0	10.5	19.5	311.5	320.5 C
69(Che)	to	70(Yea)	307.0	316.0	326.5	−9.0	10.5	19.5	311.5	320.5 C
70(Yea)	to	71(Mrd)	307.0	316.0	326.5	−9.0	10.5	19.5	311.5	320.5 C
71(Mrd)	to	72(Kin)	307.0	316.0	326.5	−9.0	10.5	19.5	311.5	320.5 C
72(Kin)	to	73(Alp)	308.5	320.5	326.5	−12.0	6.0	18.0	317.5	320.5 C
73(Alp)	to	74(Cho)	309.0	323.0	333.0	−14.0	10.0	24.0	319.0	327.0 C
74(Cho)	to	75(Arn)	313.5	324.0	333.0	−10.5	9.0	19.5	319.5	328.5 C
75(Arn)	to	76(Pnd)	313.5	324.0	333.0	−10.5	9.0	19.5	319.5	328.5 C
76(Pnd)	to	77(Bri)	319.2	326.7	333.0	−7.5	6.3	13.8	324.2	328.0 C
77(Bri)	to	78(Asb)	333.5	341.0	350.0	−7.5	9.0	16.5	338.0	344.0 C
78(Asb)	to	79(Hlk)	337.2	343.5	351.0	−6.3	7.5	13.8	342.2	344.7 C
79(Hlk)	to	80(Aru)	337.2	343.5	351.0	−6.3	7.5	13.8	342.2	344.7 C
80(Aru)	to	81(Chd)	337.2	343.5	351.0	−6.3	7.5	13.8	342.2	344.7 C
81(Chd)	to	82(Ivo)	345.7	349.5	354.5	−3.8	5.0	8.8	349.5	350.7 B
82(Ivo)	to	83(Has)	350.0	357.5	367.5	−7.5	10.0	17.5	353.7	362.5 C
83(Has)	to	84(Fam)	356.3	362.5	367.5	−6.3	5.0	11.3	362.5	363.7 C

DEVONIAN

Stages			Min	Best	Max	Range		Diff	No data	
84(Fam)	to	85(Frs)	361.0	367.0	371.0	−6.0	4.0	10.0	365.0	368.0 B
85(Frs)	to	86(Giv)	365.5	376.0	386.5	−10.5	10.5	21.0	370.0	382.0 C
86(Giv)	to	87(Eif)	367.0	382.0	391.0	−15.0	9.0	24.0	377.5	386.5 C
87(Eif)	to	88(Ems)	381.0	386.0	391.0	−5.0	5.0	10.0	385.0	387.0 B
88(Ems)	to	89(Pra)	381.0	391.0	405.0	−10.0	14.0	24.0	385.0	397.0 C
89(Pra)	to	90(Lok)	382.0	392.5	404.5	−10.5	12.0	22.5	386.5	400.0 C
90(Lok)	to	91(Prd)	403.5	408.5	412.2	−5.0	3.8	8.8	407.2	408.5 B

SILURIAN

Glauconites treated as minimum ages, assumed chronostratic error = 1.7 Ma

Stages			Min	Best	Max	Range		Diff	No data	
91(Prd)	to	92(Ldf)	405.7	409.5	414.5	−3.8	5.0	8.8	408.2	409.5 B
92(Ldf)	to	93(Gor)	407.5	412.5	417.5	−5.0	5.0	10.0	411.2	412.5 B
93(Gor)	to	94(Gle)	421.0	424.0	428.0	−3.0	4.0	7.0	423.0	425.0 B
94(Gle)	to	95(Whi)	422.7	426.5	432.7	−3.8	6.3	10.0	426.5	427.7 B
95(Whi)	to	96(She)	422.5	426.2	432.5	−3.8	6.3	10.0	426.2	427.5 B
96(She)	to	97(Tel)	424.0	431.5	440.5	−7.5	9.0	16.5	427.0	436.0 C
97(Tel)	to	98(Aer)	425.0	432.5	440.0	−7.5	7.5	15.0	429.5	437.0 C
98(Aer)	to	99(Rhu)	427.0	437.0	447.0	−10.0	10.0	20.0	435.0	439.0 C
99(Rhu)	to	100(Hir)	431.5	439.0	446.5	−7.5	7.5	15.0	436.0	440.5 C

ORDOVICIAN

Stages			Min	Best	Max	Range		Diff	No data	
100(Hir)	to	101(Raw)	432.2	438.5	447.2	−6.3	8.8	15.0	436.0	441.0 C
101(Raw)	to	102(Cau)	433.0	443.5	448.0	−10.5	4.5	15.0	440.5	443.5 C
102(Cau)	to	103(Pus)	433.0	443.5	448.0	−10.5	4.5	15.0	440.5	443.5 C
103(Pus)	to	104(Onn)	434.0	443.0	449.0	−9.0	6.0	15.0	443.0	444.5 C
104(Onn)	to	105(Act)	439.0	445.0	448.0	−6.0	3.0	9.0	444.0	445.0 B
105(Act)	to	106(Mrb)	439.5	444.5	448.2	−5.0	3.8	8.8	444.5	444.5 B
106(Mrb)	to	107(Lon)	439.5	444.5	448.2	−5.0	3.8	8.8	444.5	444.5 B
107(Lon)	to	108(Sou)	449.5	457.0	461.5	−7.5	4.5	12.0	457.0	457.0 C
108(Sou)	to	109(Har)	451.2	457.5	462.5	−6.3	5.0	11.3	457.5	458.7 C
109(Har)	to	110(Cos)	453.7	458.7	462.5	−5.0	3.8	8.8	457.5	458.7 B
110(Cos)	to	111(Llo3)	455.0	462.5	466.2	−7.5	3.8	11.3	461.2	462.5 C
111(Llo3)	to	112(Llo2)	455.0	462.5	466.2	−7.5	3.8	11.3	461.2	462.5 C
112(Llo2)	to	113(Llo1)	455.0	462.5	466.2	−7.5	3.8	11.3	461.2	462.5 C
113(Llo1)	to	114(Lln2)	465.2	471.5	475.2	−6.3	3.8	10.0	470.2	472.7 B
114(Lln2)	to	115(Lln1)	465.2	471.5	475.2	−6.3	3.8	10.0	470.2	472.7 B
115(Lln1)	to	116(Arg)	471.5	482.0	486.5	−10.5	4.5	15.0	480.5	483.5 C
116(Arg)	to	117(Tre)	482.0	492.0	500.0	−10.0	8.0	18.0	490.0	494.0 C
117(Tre)	to	118(Dol)	500.5	513.0	528.0	−12.5	15.0	27.5	508.0	520.5 C

CAMBRIAN

Glauconites treated as minimum ages, assumed chronostratic error = 5.0 Ma

Stages			Min	Best	Max	Range		Diff	No data	
118(Dol)	to	119(Mnt)	498.0	513.0	530.5	−15.0	17.5	32.5	505.5	520.5 C
119(Mnt)	to	120(Men3)	504.0	522.0	540.0	−18.0	18.0	36.0	513.0	531.0 C
120(Men3)	to	121(Men2)	502.0	520.0	538.0	−18.0	18.0	36.0	514.0	529.0 C
121(Men2)	to	122(Men1)	503.0	520.5	538.0	−17.5	17.5	35.0	513.0	530.5 C
122(Men1)	to	123(Sol2)	516.0	530.0	538.0	−14.0	8.0	22.0	528.0	532.0 C
123(Sol2)	to	124(Sol1)	527.0	535.0	543.0	−8.0	8.0	16.0	535.0	537.0 C
124(Sol1)	to	125(Len)	528.5	536.0	542.0	−7.5	6.0	13.5	534.5	537.5 C
125(Len)	to	126(Atb)	528.0	536.0	542.0	−8.0	6.0	14.0	536.0	536.0 C
126(Atb)	to	127(Tom)	557.0	565.0	575.0	−8.0	10.0	18.0	563.0	567.0 C
127(Tom)	to	128(Pou)	567.0	575.0	583.0	−8.0	8.0	16.0	573.0	577.0 C

Illites included – see end of Table 4.2

Stages			Min	Best	Max	Range		Diff	No data	
126(Atb)	to	127(Tom)	564.2	569.2	575.5	−5.0	6.3	11.3	568.0	570.5 C
127(Tom)	to	128(Pou)	579.2	584.2	590.5	−5.0	6.3	11.3	584.2	585.5 C

Note: Because of rounding errors, the ranges quoted may in some cases not be the precise differences between Min and Best or Max and Best.

5.7 Interpolation methods

We now have a set of chronogram ages and estimated errors (Table 5.4). How do we use these in the construction of a time scale? Ideally, each chronogram age would have such a small error range that the age could be used directly as a calibration point in the scale. However, this ideal is far from achievement. In fact, it may never be achieved, because as current standards are surpassed new, more rigorous, standards become the norm until they in turn are replaced. Three interpolation methods have been considered: (1) relative thicknesses of sediments; (2) the 'equal stage' and equal chron hypothesis; (3) relative spacing of ocean-floor magnetic anomalies.

5.7.1 Relative stratigraphic thicknesses

This method has been used for some parts of the time scale (e.g. Boucot 1975, Churkin, Carter & Johnson 1977, Ziegler 1978, Gale *et al.* 1980, McKerrow *et al.* 1980) and has the greatest potential value for its future overall refinement, although some of these authors now reject this method as being too subjective (Gale 1985, McKerrow, Lambert & Cocks 1985). However, since there is no systematic stage-by-stage global compilation of relative thicknesses from a variety of depositional environments, we have not been able to use this method for interpolation.

5.7.2 Relative biostratigraphic discrimination

The degree of biostratigraphic discrimination would give an indication of the relative durations of such divisions adopted for any time scale if it could be assumed that the erection of taxa was consistent and that evolutionary rates were approximately constant. This is suspected not to be the case; yet methods so based may give some coarse indication of relative time spans. In GTS 82, from lack of opportunity to do better, stages were taken as of equal duration for interpolation of values for stage boundaries between tie-points. In GTS 89, with more time, relative stage durations have been based on approximate numbers of chrons. To attempt to justify how many chrons constitute each stage would be a major research undertaking. We have done no more than make a notional apportionment where necessary or appropriate of 'equal chrons'.

The following discussion indicates that, however good or bad our estimate of number of chrons per stage may be, it is probably a significant improvement on the 'equal stage' method used in GTS 82. Others have addressed this problem at length. We do no more than state our assumptions which can easily be varied according to alternative opinions on relative duration.

Stages and their constituent chrons are, or will be, defined by GSSP (Global Stratotype Sections and Points, Sections 1.3 and 3.2) and are therefore not defined biostratigraphically. Nevertheless, Phanerozoic stages have become established as successively refined biostratigraphically characterized divisions to which the GSSP should approximate where possible. Stages have been generally regarded as the shortest chronostratic divisions suitable for global correlation, and this has been borne out by experience (see discussion in Arkell 1956).

Chrons are chronostratic divisions ranking below stages so that two or more chrons would comprise a stage. They should in principle be defined in the same way by GSSP but this procedure may be some way off. It is somewhat accidental which rank various divisions have assumed as they are all of the same nature.

Pending the establishment of chrons, the time equivalent of various biozones (i.e. biochrons) has been used. For some intervals more than one biozonal scheme is used, with one supplementing the others (e.g. for Paleogene divisions). For the most intensively worked period (Jurassic), the ammonite biozones have provided a pattern for convenient scale divisions (potential chrons and subchrons). Whether or not one or more biochrons apply to the same time interval, there should be only one standardized chron sequence. In favourable circumstances global correlation by biologic means may be effected at chron or subchron level.

For each stratigraphic interval it is a matter of historical accident, depending on the opinions of the scientists involved, as to what degree of biostratigraphic discrimination may be adopted for epochs, stages and chrons. For example, Tremadoc (once considered a stage), with as yet no formal divisions, spans approximately 17 Ma. The Scythian Epoch (also once a stage) now comprises three stages with a total span of about 4 Ma while the Norian Stage spans about 13 Ma and the Rhaetian no more than 2 Ma – the former divisible into three sub-stages and seven chrons while Rhaetian can hardly be divided at all. Likewise the Albian Stage spans about 15 Ma and the Coniacian Stage perhaps less than 2 Ma. There is at present no case for reclassifying these time-honoured divisions according to their duration.

It may be remarked, however, that the average duration of the stages when viewed through a 60 Ma sliding window is, except for Carboniferous stages, remarkably constant back to Early Paleozoic. The seven youngest Cretaceous stages average about 6.5 Ma and that of the next twenty older stages averages 6.4 Ma. This gives some confidence in applying the above method.

5.7.3 Relative ocean-spreading rates

During periods of constant spreading, the relative spacing of the ocean-floor anomalies represents their relative duration. The biostratigraphic ages of the anomalies can be found from deep-ocean (e.g. IPOD) cores or land sections. The chronometric age of stage boundaries can then be found from the chronograms. In this way the magnetostratigraphic time scale can be linked to isotopic dates. We have used magnetostratigraphy to refine several Cenozoic, Cretaceous and Jurassic stage boundaries, by amounts ranging up to 3 Ma, all of which are within the uncertainties of the isotopic dates and chronograms. Details are given in Chapter 6. However, magnetostratigraphy has not proved useful for most periods, either because field reversals are absent in the ocean-floor record (mid-Cretaceous), or the ocean-floor has been subducted (pre-mid-Jurassic and older floor).

5.7.4 'Tie-points'

In GTS 82 we coined the notion of 'tie-points', i.e. relatively well-controlled chronograms. Tie-points were sometimes consecutive, as in the Late Cretaceous, but in older rocks they were irregularly distributed. In GTS 89 the chronograms have been divided into 'A, B and C' categories. In A the error range is less than 5 Ma; in B from 5 to 10 Ma; and in C it is greater than 10 Ma. Sometimes it has been necessary to use class C chronograms: comments are made in individual cases below.

In certain periods, for example the Triassic Period, two stages have been combined, e.g. the Smithian and Dienerian stages used in GTS 82 have been replaced approximately by Nammalian (see Section 3.15.3). Throughout the Early Triassic or Scythian Epoch (= Spathian + Nammalian + Griesbachian) it is not easy to correlate 'Scythian' strata with the marine Early Triassic stages. Most dates are simply assigned to the Scythian Epoch in the literature. The result is that each stage boundary within the Scythian Epoch will have approximately the same chronogram value. The effect is even more noticeable in the Silurian Period. Here the discriminations occur at epoch rather than at stage level and the resulting chronograms are either poor or meaningless. Clearly, we need some method for interpolating between calibration points.

5.8   Interpolation between tie-points using Phanerozoic chrons

The numbers of chrons adopted here in each 'stage' are listed in the tables which follow. The number of Late Permian chrons was modified as explained in Chapter 3 with newly proposed divisions to which isotopic determinations have not yet been applied. Interpolation by chrons is unnecessary for the Cenozoic because most boundaries are tie-points or can be related to the magnetostratigraphic time scale.

The method assumes that geologic time is proportional to the chrons between two tie-points or pseudo-tie-points. To apply it, the chrons listed in the tables below are assumed to represent equal time durations between two tie-points. For example, the initial ages of the Devonian stages between the initial Eifelian and initial Famennian stages have been obtained by linear interpolation between the Eifelian/Emsian tie-point at 386 Ma (see Devonian section below) and the Famennian/Frasnian tie-point at 367 Ma.

Because we have decided to interpolate using chronostratic chrons, it is essential to have tie-points at chronostratic boundaries. Where the chronostratic chrons between two tie-points are based on the same class of fossil, e.g. ammonites, the age of the chronostratic boundary can be obtained by interpolation between two tie-points that lie above and below that boundary. However, although the equal duration of chron assumption appears to hold remarkably well **on average**, it does not hold in detail. In some cases we have chosen to interpolate from one period to the next by using the boundary age chronogram, even if it is poor, to estimate an age. In other cases the data give the same age ranges for several stratigraphic boundaries. In some of these we have inspected the data and assigned a visually selected tie-point. Such tie-points, based as they are on poor chronograms and subjective judgement, are referred to as 'pseudo-tie-points'. All the points are listed in the tables that follow.

General note for tables of interpolation (pages 118 to 138)

The results of interpolation are tabulated and discussed period by period in Subsections 5.8.1 to 5.8.10 according to the following conventions. In Chapter 6 some of the interpolations and some of the tie-points are further modified in the light of ocean-floor spreading data.

1. 'Stage' is the number code and three-letter abbreviation of the stage (Table 4.3) for which the **initial** boundary is being estimated.

2. 'Chrons' contains two columns. The first number is the number of chrons in the 'stage'; the second is the total number of chrons in the time scale to the base of that stage. For example, there are 4 chrons in the Tommotian (stage 127) and the total number of chrons to the base of the Tommotian is 365.3.

3. A non-zero value in the 'tie-point' column gives the value in Ma of the chronogram that has been used as a tie-point. Its quality can be assessed by reference to Table 5.4. A zero value in the tie-point column indicates that the chronogram has not been used as a tie-point. If the tie-point value differs from the chronogram value, the point is a 'pseudo-tie-point', i.e. a value which has been adopted to override the chronogram value or a value interpolated between two chronograms.

4. (C) is the chronogram value (= 'C-gram'), i.e. is the best value of the boundary from the chronograms in Table 5.4. If it has been used as a tie-point it is identical to the 'tie-point' value to the left.

5. (I) is the interpolated value (='Interpol.') between two tie-points. If the point is a tie-point then the interpolated value is identical to the tie-point value and the chronogram value (C).

6. 'Per cent' is the difference between the 'Min' and 'Max' chronogram estimates of Table 5.4, divided by the chronogram value, expressed as a percentage. It gives an estimate of the uncertainty in the chronogram value as a percentage of the chronogram age.

7. (C−I) is the difference between the chronogram and interpolated values. It is zero for a tie-point.

General caption for all figures in Section 5.8 of chronogram ages and chrons (Figures 5.5 to 5.23 – odd numbers)

Each stratigraphic boundary shows the minimum, best and maximum chronogram values plotted as three crosses on the horizontal boundary lines, which are labelled. A line connecting the best values is shown. In addition, tie-points and pseudo-tie-points are shown as squares connected by a line. Tie-points are selected from chronograms having a small range. Pseudo-tie-points do not correspond to a chronogram value for reasons set out in the appropriate section below. Tie-points and pseudo-tie-points have been chosen so as to visually minimize the differences between the chronogram values and linear interpolations between tie-points and pseudo-tie-points.

General caption for all figures in Section 5.8 of isotopic dates and stratigraphic boundaries (Figures 5.6 to 5.24 – even numbers)

The stepped line connects the best chronogram values together. The squares represent a date from non-glauconitic material and crosses represent glauconite dates. When a date is bracketed by more than one stratigraphic boundary it is plotted twice. Where one of the plotted points lies far above or below the stepped line it is the point nearer the stepped line that provides a useful constraint.

5.8.1   CENOZOIC
Boundaries not defined by chronograms:

1 (Hol)   Base of Holocene defined as 0.01 Ma (Section 3.21.5).
2 (Ple)   Base of Pleistocene defined as 1.64 Ma (Section 3.21.1).

Boundaries defined by chronograms:

Stage	Chrons In stage	Sum	Tie-point	C-gram (C)	Interpol.(I)	Per cent	C−I (Ma)
3 (Pia)	2.0	5.0	3.5	3.5	3.5	57.1	0.0
4 (Zan)	2.0	7.0	6.0	6.0	6.0	53.3	0.0
5 (Mes)	0.8	7.8	8.4	8.4	8.4	57.1	0.0
6 (Tor)	1.8	9.5	10.4	10.4	10.4	26.0	0.0
7 (Srv)	5.5	15.0	14.6	14.6	14.6	27.4	0.0
8 (Lan2)	1.0	16.0	14.6	14.6	14.6	24.7	0.0
9 (Lan1)	1.0	17.0	16.0	16.0	16.0	15.6	0.0
10 (Bur)	2.0	19.0	22.0	22.0	22.0	16.4	0.0
11 (Aqt)	2.0	21.0	23.8	23.8	23.8	8.4	0.0
12 (Cht)	3.0	24.0	30.6	30.6	30.6	10.5	0.0
13 (Rup)	2.5	26.5	36.4	36.4	36.4	8.8	0.0
14 (Prb)	2.5	29.0	39.4	39.4	39.4	8.1	0.0
15 (Brt)	2.0	31.0	41.4	41.4	41.4	9.7	0.0
16 (Lut)	3.0	34.0	48.0	48.0	48.0	5.8	0.0
17 (Ypr)	3.5	37.5	53.4	53.4	53.4	9.0	0.0
18 (Tha)	3.5	41.0	61.0	61.0	61.0	9.0	0.0
19 (Dan)	2.0	43.0	66.0	66.0	66.0	6.8	0.0

General comment:

Only a few dates for the youngest Cenozoic interval (0–3.4 Ma) have been put into the isotopic database in Table 4.2. There are several hundred K–Ar determinations for this interval, referenced in Chapter 6, where they are linked directly to the magnetic reversal time scale. All Cenozoic chronograms except Thanetian/Danian (just class B) are class A and have been used as chronogram tie-points. Parts of the Cenozoic have been further calibrated using biostratigraphic data tied directly to magnetostratigraphy (Chapter 6).

Discussion:

Holocene/Pleistocene: Taken at 10 000 years BP or at a climatic event about then to be standardized in a varved sequence in Sweden (Section 3.21.5).

Pleistocene/Pliocene (= Pleistocene/Piacenzian): Aguirre & Pasini (1985) and Berggren *et al.* (1985b) place the initial Pleistocene boundary slightly above the Olduvai Event. They use a magnetic reversal date of 1.66 and give 1.64 rounded to 1.6 Ma (see Sections 3.21.1, 6.4.2 and 6.4.4, Tables 6.4 and 6.7). Our revised magnetic reversal age is 1.65 Ma, in agreement with the precision of the interpretation. We retain the value of 1.64 for the definitive scale.

Piacenzian/Zanclian; Zanclian/Messinian; Messinian/ Tortonian; Tortonian/Serravallian; Serravallian/Langhian: All class A chronograms, reasonable distribution of high-temperature data; no conflicts with previously published scales.

Late Langhian/Early Langhian: Not used in GTS 89 (see Section 3.20.2 (ii)).

Langhian/Burdigalian; Burdigalian/Aquitanian; Aquitanian/Chattian: All class A chronograms, reasonable distribution of high-temperature data; no conflicts with previously published scales.

Chattian/Rupelian: 30.6 Ma agrees with other timescales: DNAG 30 Ma, GSL 85 30 Ma, but not with Odin (NDS 27 +2,−1 Ma). Odin used mostly glauconites. Glauconites-only option gives 27 Ma, in agreement with Odin.

Rupelian/Priabonian; Priabonian/Bartonian; Bartonian/ Lutetian: All class A chronograms. **Rupelian/Priabonian** and **Priabonian/Bartonian** have reasonable distribution of high-temperature data; no conflicts with previously published scales.

Bartonian/Lutetian at 41.4 Ma largely controlled by glauconite dates. NDS gives 39 Ma; DNAG and GSL 85 43.6 Ma. See also Chapter 6.

Lutetian/Ypresian: 48.0 Ma age controlled by ten glauconites. NDS gives 45 +1,−0.5 Ma; DNAG and GSL 85 give 52 Ma; GTS 82 is 50.5 Ma. GTS 89 lies between NDS and GSL 85 as it is controlled mainly by glauconites. The age of this boundary is the most uncertain and the most controversial in the Cenozoic. The uncertainties are well above the analytical uncertainties for material of this age. See also Chapter 6.

Ypresian/Thanetian: 53.4 Ma age from three glauconites in the middle of a range 46.5–60.9 Ma with no high-temperature data in it. DNAG chose 57.8 and GSL 85 chose 58 Ma; NDS gives 53 ± 1 Ma. See also Chapter 6.

Thanetian/Danian: 61.0 Ma is controlled mainly by glauconites, although seven high-temperature dates on older side have some influence. High-temperature dates alone give 60.7 Ma, lower than the glauconite age of 61.5 Ma. NDS gives 59 +1,−2 Ma; DNAG gives 63.6 Ma; GSL 85 gives 64 Ma. See also Chapter 6.

Cenozoic/Cretaceous (= Danian/Maastrichtian): Chronogram gives 66.0 Ma. There is discussion within USGS about dates obtained which disagree with those found by other laboratories on same tuff horizon. USGS preferred age is 66 Ma; Alberta preferred age is 64 Ma. No data between 65.5 and 66.5 Ma. 65 Ma was adopted for GTS 82. NDS gave 65 +1, −1.5 Ma. Berggren *et al.* in GSL 85 adopt 66.0 Ma, while Snelling (1985a) in GSL 85 adopts 65.0 Ma. DNAG adopts 66.4 Ma. See also Chapter 6.

Figure 5.5. Chronogram age in Ma plotted against Cenozoic chrons. Squares are tie-points. The three crosses associated with each boundary show the minimum, best and maximum age of the boundary.

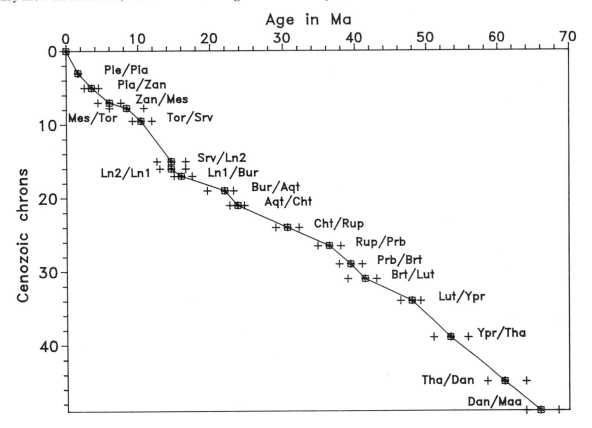

Figure 5.6. Dates in the isotopic database plotted against Cenozoic stage boundaries. Crosses are glauconite dates; squares are non-glauconite dates.

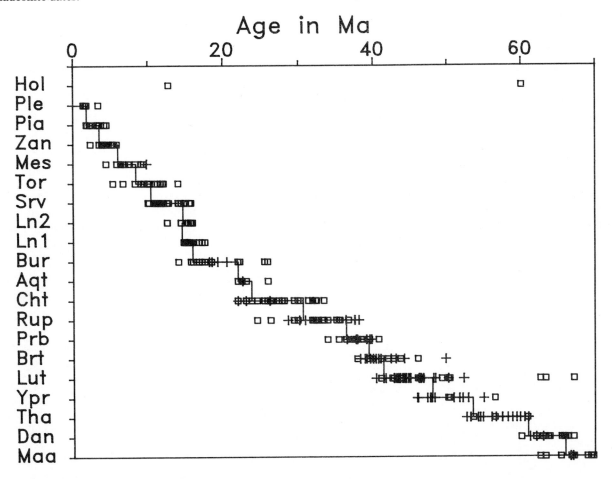

5.8.2 CRETACEOUS

Stage	Chrons		Tie-point	C-gram (C)	Interpol. (I)	Per cent	C−I (Ma)
	In stage	Sum					
[19 (Dan)	2.0	43.0	66.0	66.0		Tertiary	]
20 (Maa)	2.0	45.0	71.5	71.5	71.5	8.4	0.0
21 (Cmp)	4.0	49.0	82.4	82.4	82.4	8.7	0.0
22 (San)	2.0	51.0	86.6	86.6	86.6	6.9	0.0
23 (Con)	2.0	53.0	88.5	88.5	88.5	4.0	0.0
24 (Tur)	3.0	56.0	90.4	90.4	90.4	3.5	0.0
25 (Cen)	4.0	60.0	97.0	97.0	97.0	4.1	0.0
26 (Alb)	7.0	67.0	112.0	110.4	112.0	3.2	−1.6
27 (Apt)	5.0	72.0	0.0	125.5	120.8	19.9	4.7
28 (Brm)	4.0	76.0	0.0	131.0	127.9	11.5	3.1
29 (Hau)	4.0	80.0	135.5	135.5	135.5	10.4	0.0
30 (Vlg)	6.0	86.0	0.0	137.5	142.5	10.0	−5.0
31 (Ber)	2.0	88.0	145.0	148.0	145.0	12.2	3.0

Discussion:

Tie-points: Maastrichtian/Campanian; Campanian/ Santonian; Santonian/Coniacian; Coniacian/Turonian; Turonian/Cenomanian; Cenomanian/Albian: The initial ages of the six youngest Cretaceous stages give three class B and three class A chronograms using all the data and have been taken as chronogram tie-points.

Albian/Aptian: GTS 82 used the Albian/Aptian boundary as a tie-point at 113 Ma; the 'all data' option gives 110.4 Ma, the 'no glauconite' option at 112 Ma is preferred. The large difference between glauconite dates and dates excluding glauconites for older initial boundaries has led to the treatment of glauconites as young dates for all boundaries older than 115 Ma, i.e. for all stages older than Aptian (Section 5.5). NDS age is 107 ± 1 Ma; DNAG is 113 Ma; GSL 85 is 107 Ma.

Aptian/Barremian; Barremian/Hauterivian: Chron interpolation between Albian/Aptian and Hauterivian/ Valanginian gives 120.8 and 127.9 Ma.

Hauterivian/Valanginian: Taken as a reasonable tie-point at 135.5 Ma. The next reasonable tie-point is the Kimmeridgian/Oxfordian boundary at 156 Ma. Because of the large number of intervening chrons, a pseudo-tie-point has been placed by visual interpolation of the remaining Cretaceous chrons at the Berriasian/Tithonian boundary at 145 Ma.

Valanginian/Berriasian: Chron interpolation between Hauterivian/Valanginian and Berriasian/Tithonian gives 142.5 Ma.

Cretaceous/Jurassic (= Berriasian/Tithonian): The chronogram is controlled by a poor whole-rock date and a poor high-temperature date. It gives a value of 148 Ma. A pseudo-tie-point of 145 Ma is used. From USGS data (Lanphere & Jones 1978), the value must be older than 137 Ma. 144 Ma was used in GTS 82. GSL 85 give 135 Ma and NDS 130 ± 3 Ma. Further modifications (to 145.6 Ma) are made in Chapter 6 using the M-anomalies.

Further modifications using M-anomalies: All stage boundaries in Aptian to Berriasian interval were subsequently modified by the best linear fit of the M-anomaly sequence to chronograms (Section 6.5.3). The chronogram values are generally higher than all previous time scales, e.g. Barremian/ Hauterivian is at 131 Ma; GTS 82 125 Ma; NDS 114 +2,−1 Ma; DNAG 124 Ma and GSL 85 116 Ma.

Chrons as equal time intervals: Figure 5.7 shows a plot of the age in Ma against the number of chrons in each Cretaceous stage. If chrons represent equal time intervals then the graph should be a straight line. Clearly, the graph is non-linear. If the plot is made of age in Ma against Cretaceous stages, then the plot is even more non-linear. This suggests that, while chrons themselves are of unequal duration, their use as one of the axes for time scale interpolation produces better values than does interpolation based on the equal duration of stages hypothesis used in GTS 82 interpolation.

Figure 5.7. Chronogram age in Ma plotted against Cretaceous chrons. Squares that do not coincide with the best chronogram estimate are pseudo-tie-points.

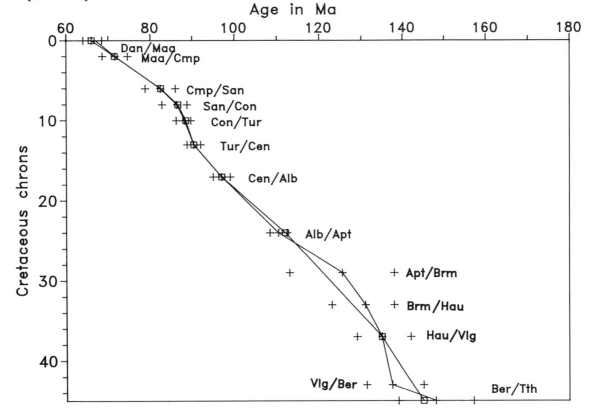

Figure 5.8. Dates in the isotopic database plotted against Cretaceous stage boundaries.

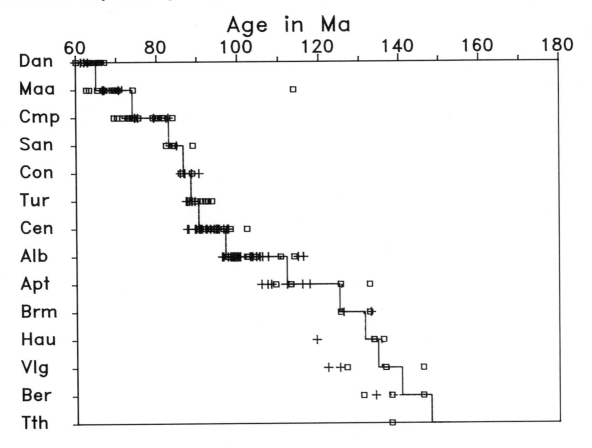

5.8.3 JURASSIC

| Stage | Chrons | | Tie-point | C-gram (C) | Interpol.(I) | Per cent | C−I (Ma) |
	In stage	Sum					
[31 (Ber)	2.0	88.0	145.0	148.0	145.0	Cretaceous	]
32 (Tth)	17.0	105.0	0.0	151.0	153.5	14.9	−2.5
33 (Kim)	5.0	110.0	156.0	156.0	156.0	7.1	0.0
34 (Oxf)	8.0	118.0	0.0	156.3	159.8	8.8	−3.5
35 (Clv)	6.0	124.0	0.0	159.0	162.7	8.2	−3.7
36 (Bth)	8.0	132.0	0.0	159.2	166.6	8.7	−7.4
37 (Baj2)	3.5	135.5	168.2	168.2	168.2	0.0	0.0
37 (Baj1)	3.5	139.0	0.0	177.0	173.5	11.9	3.5
38 (Aal)	3.0	142.0	0.0	178.0	178.0	11.8	0.0
39 (Toa)	6.0	148.0	0.0	182.0	187.0	16.5	−5.0
40 (Plb)	5.0	153.0	0.0	189.5	194.5	7.9	−5.0
41 (Sin)	6.0	159.0	203.5	203.5	203.5	6.1	0.0
42 (Het)	3.0	162.0	0.0	210.5	208.0	7.1	2.5
[43 (Rht)	1.0	163.0	209.5	209.5	209.5	Triassic	]

Discussion:

General comment: Five tie-points are used for Jurassic chron interpolation: **Berriasian/Tithonian** boundary provides a pseudo-tie-point (see Cretaceous, above); **Kimmeridgian/ Oxfordian**; the **mid-Bajocian** provides a pseudo-tie-point; there are tie-points at the **Sinemurian/Hettangian** boundary and at the top of the Triassic at the **Rhaetian/Norian** boundary.

Chrons are based on ammonite zones of Cope *et al.* (1980a, 1980b) in GTS 89, Figure 3.10. These chrons give approximately the same age as those derived by interpolation using the chrons given by Hallam in GSL 85.

Tithonian/Kimmeridgian: From chron interpolation the age is 153.5 Ma. The Tithonian/Kimmeridgian boundary is difficult to correlate with some other time scales as different definitions of Tithonian are in use (Section 3.16.4).

Kimmeridgian/Oxfordian: The Kimmeridgian/ Oxfordian age is the same as DNAG and GTS 82 at 156 Ma, but high compared with other time scales: NDS 140 ± 4 Ma; GSL 85 144 Ma. Data from California are definitive and leave little doubt that this high value is correct (Schweickert *et al.* 1984).

Tithonian/Kimmeridgian; Kimmeridgian/Oxfordian: These two boundaries have been modified from best linear fit of chronograms to M-anomaly sequence (Section 6.5.3).

Oxfordian/Callovian; Callovian/Bathonian; Bathonian/ Bajocian: Chronograms similar due to absence of data. Dominated by NDS102, a biotite which, although a high-temperature date, has the effect of keeping the chronogram ages low. Interpolated chron ages are Oxfordian/Callovian at 159.8 Ma; Callovian/Bathonian at 162.7 Ma; and Bathonian/ Bajocian at 166.6 Ma. They are lower than those adopted for GTS 82.

Mid-Bajocian pseudo-tie-point: A mid-Bajocian pseudo-tie-point at 168.2 Ma is used.

Bajocian/Aalenian; Aalenian/Toarcian: Chronograms dominated by NDS182, a Bajocian hornblende, and WEST2, a feldspar. Both have the effect of keeping the chronogram ages low. The chronograms are poor with no significant data in the range 172 to 183 Ma. Interpolated chron ages are Bajocian/Aalenian at 173.5 Ma and Aalenian/Toarcian at 178 Ma. For the Bajocian/Aalenian boundary other ages are: NDS 181 Ma; GSL 85 180 Ma; DNAG 187 Ma; GTS 82 181 Ma.

Toarcian/Pliensbachian: The younger side is controlled by WEST2 and NDS182 in particular, i.e. the same data as controlled the two boundaries above. The chronogram is poor with no significant data in the range 172 to 192 Ma. The interpolated chron age is 187 Ma. GSL 85 chose 188 Ma; NDS 189 Ma; DNAG 193 Ma; GTS 82 194 Ma.

Pliensbachian/Sinemurian: Chronogram interpolation gives 194.5 Ma. Other time scales chose: GSL 85 195 Ma; NDS 195 Ma; DNAG 198 Ma; GTS 82 200 Ma.

Sinemurian/Hettangian: The younger boundary is controlled by NDS181, a hornblende, the older by NDS177, a biotite. Both are poorly controlled stratigraphically and a C class chronogram results. Interpolation by chrons gives 203.5 Ma; GSL 85 chose 201 Ma; NDS 201 Ma; DNAG 204 Ma; GTS 82 206 Ma.

Jurassic/Triassic boundary (= Hettangian/Rhaetian): Poor chronogram at 210.5 Ma. Chron interpolation from a better chronogram at the **Rhaetian/Norian** boundary at the top of the Triassic gives an age of 208 Ma for this boundary. Other timescales give younger ages: NDS 204 ± 4 Ma; GSL 85 205 Ma and DNAG 206 Ma. Westerman (1984) chrons the Jurassic period and gets 208 Ma.

Figure 5.9. Chronogram age in Ma plotted against Jurassic chrons.

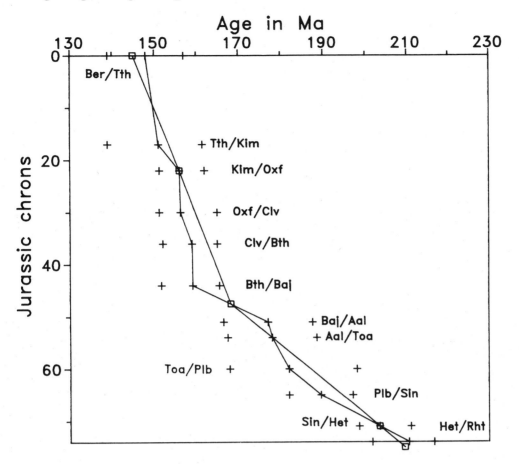

Figure 5.10. Dates in the isotopic database plotted against Jurassic stage boundaries.

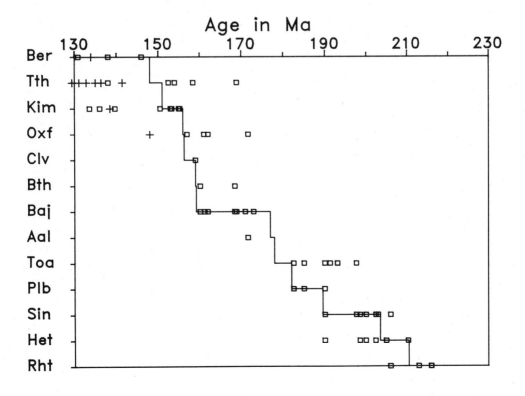

5.8.4 TRIASSIC

Stage	Chrons		Tie-point	C-gram (C)	Interpol. (I)	Per cent	C−I (Ma)
	In stage	Sum					
[42 (Het)	3.0	162.0	0.0	210.5	208.0	Jurassic	]
43 (Rht)	1.0	163.0	209.5	209.5	209.5	7.9	0.0
44 (Nor)	6.0	169.0	0.0	226.0	223.4	8.6	2.6
45 (Crn)	5.0	174.0	235.0	235.0	235.0	3.0	0.0
46 (Lad)	5.0	179.0	239.5	239.5	239.5	5.2	0.0
47 (Ans)	4.0	183.0	0.0	241.2	241.1	6.2	0.1
48 (Spa)	2.0	185.0	0.0	242.5	241.9	6.7	0.6
(Note: Smithian (=Smi) and Dienerian (=Die) now amalgamated into Nammalian)							
49 (Smi)	2.0	187.0	0.0	242.5	242.6	6.7	−0.1
50 (Die)	2.0	189.0	0.0	242.5	243.4	6.7	−0.9
51 (Gri)	4.0	193.0	245.0	245.0	245.0	8.0	0.0

Discussion:

General remarks: 31 chrons recognized (biozones in Figure 3.8, GTS 89). Four tie-points are used: Rhaetian/Norian; Carnian/Ladinian; Ladinian/Anisian and Griesbachian/Tatarian. All except the Carnian/Ladinian, which is a class B tie-point, are class C tie-points. Apart from initial Triassic age, all are older than other time scales, but are consistently supported by chronograms.

Rhaetian/Norian: This boundary is absent from some time scales, e.g. DNAG. Both NDS and GSL 85 recognize that the Carnian and Norian stages are longer and hence the Rhaetian shorter than in GTS 82 where all stages in the Rhaetian to Ladinian interval were assigned a duration of 6 or 7 Ma. The chronograms also suggest that the Carnian and Norian stages are longer and the Rhaetian stage shorter. The chronogram gives 209.5 Ma; NDS gives 208 Ma; GSL 85 210 Ma. NDS assigns about 4 Ma to the Rhaetian; GSL 85 assigns 5 Ma; and GTS 89 assigns 1.5 Ma. Lack of data for the Rhaetian stage leads to very similar chronograms for the Hettangian/Rhaetian and the Rhaetian/Norian boundaries.

Norian/Carnian: There are no significant data in the range 223 to 230.5 Ma and the chronogram age is taken as the mid-point of this interval, 226 Ma; chron interpolation gives 223.4 Ma. Some time scales give younger ages: NDS 220 ± 8 Ma; GSL 85 220 Ma; while DNAG 225 Ma and GTS 82 225 Ma are close to GTS 89.

Carnian/Ladinian: Class B chronogram at 235 Ma. GSL 85 chose 230 Ma; NDS 229 ± 5 Ma; DNAG 230 Ma; and GTS 82 231 Ma.

Ladinian/Anisian: Reasonable chronogram at 239.5 Ma. The range of 12 Ma makes it a C class chronogram. It is used as a tie-point, as in GTS 82. GTS 82 was 238 Ma; other values are NDS 233 ± 4 Ma; DNAG and GSL 85 235 Ma. This

boundary is used to interpolate ages to the base of the Triassic Period.

Anisian/Scythian (= Spathian): A poor chronogram is given with no data in the range 238.7 to 245 Ma. The boundary age of 241.1 Ma is calculated by chron interpolation from the Ladinian/Anisian and Griesbachian/Tatarian tie-points. Some time scales chose younger ages for this boundary: NDS 239 ±5 Ma; DNAG and GSL 85 use 240 Ma and GTS 82 243 Ma.

Spathian/Nammalian: Chron interpolation gives 241.9 Ma.

Nammalian/Griesbachian: Chron interpolation gives 243.4 Ma.

Scythian (= Spathian + Nammalian + Griesbachian): In general it is not easy to correlate 'Scythian' strata with the marine Early Triassic stages. Most dates are simply assigned to the Scythian in the literature, hence the Early Triassic is subdivided by chron interpolation. The chronograms for the boundaries within the Scythian all use the same data and hence give the same results. They are all high-temperature dates. There are no significant data in the range 238.7 to 246.2 Ma, the range within which the boundary ages would be expected to lie. Other time scales do not assign ages to subdivisions of the Scythian Epoch.

Permian/Triassic (= Griesbachian/Tatarian) boundary: Poor chronogram, with no data in range 239 to 251 Ma. 245 Ma is mid-point of this range. GSL 85 suggested age of 250 Ma, which is permitted by the chronogram data, but in authors' opinion this may not give adequate recognition for the distribution of ages in the Late Permian recently recognized in China. For this reason, although from a class C chronogram, the age of 245 Ma is taken as the initial Triassic boundary. NDS gave 245 ± 5 Ma and DNAG chose 245 Ma.

Figure 5.11. Chronogram age in Ma plotted against Triassic chrons.

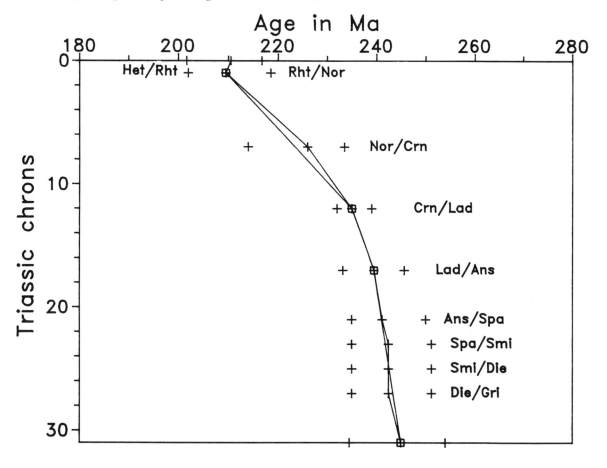

Figure 5.12. Dates in the isotopic database plotted against Triassic stage boundaries.

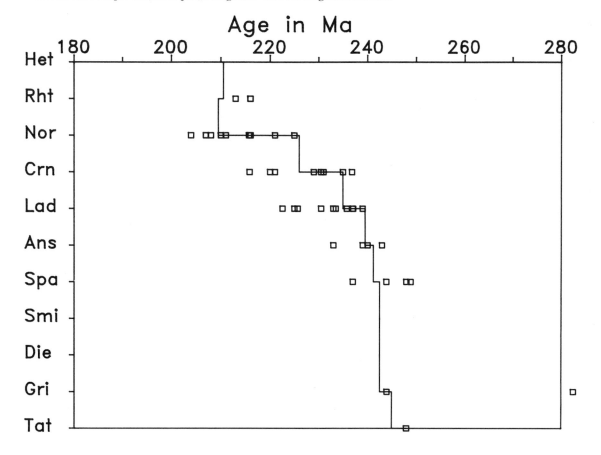

5.8.5 PERMIAN

Stage	Chrons		Tie-point	C-gram (C)	Interpol. (I)	Per cent	C−I (Ma)
	In stage	Sum					
[51 (Gri)	4.0	193.0	245.0	245.0	245.0	Triassic	]
52 (Tat)	3.0	196.0	0.0	248.0	250.5	6.7	−2.5
53 (Kaz)	2.0	198.0	0.0	259.0	254.2	10.0	4.8
54 (Ufi)	1.0	199.0	0.0	259.0	256.1	10.0	2.9
55 (Kun)	2.0	201.0	0.0	262.5	259.7	8.6	2.8
56 (Art2)	1.5	202.5	262.5	262.5	262.5	0.0	0.0
56 (Art1)	1.5	204.0	0.0	262.0	268.8	8.4	−6.8
57 (Sak)	3.0	207.0	0.0	280.0	281.5	9.3	−1.5
58 (Ass)	2.0	209.0	290.0	292.0	290.0	6.7	2.0

Discussion:

Late Permian classification and nomenclature is undergoing extensive revision (Section 3.14.3). The divisions used in the table above do not take account of this revision but they are converted in the definitive scale (Figure 1.7). Like the Early Triassic dates, most Late Permian dates in the literature are correlated to stratigraphic sections which have not been correlated with, or are difficult to correlate with, the new standards. In GTS 82 tie-points were accepted at the Artinskian/Sakmarian and Asselian/Noginskian boundaries. The Permian ages are very poorly controlled (Figure 5.14). A weak tie-point lies at the **Griesbachian/Tatarian** boundary. Pseudo tie-points are recognized in **mid-Artinskian** time and at the **Asselian/Noginskian** boundary.

The GTS 82 time scale for the Permian period (in Ma) and the major changes in GTS 89 are shown below. These changes have not been incorporated into the isotopic database of Table 4.2.

GTS 82			Changes in GTS 89
P2			
initial Tatarian		253	(= mid-point of Capitanian stage)
initial Kazanian		255.5	(= initial Wordian stage)
Ufimian		258	(see Section 3.14.3)
P1			
Kungurian		263	(see Section 3.14.3)
Artinskian		268	
Sakmarian		277	
Asselian		286	

No attempt has been made to fit the existing isotopic data into the three new stages at the top of the Permian, i.e. Changxingian (latest Permian), Longtanian and Capitanian.

Changxingian/Longtanian; Longtanian/Capitanian; Capitanian/Wordian; Wordian/Kungurian; Kungurian/ Artinskian; Artinskian/Sakmarian; Sakmarian/Asselian: Obtained by interpolation according to the chron scheme of Table 5.1. All these boundaries lie within the chronograms, and only just within the range for the Artinskian/Sakmarian boundary. The dominant data for this chronogram is a hornblende/biotite date FW18; this date is poor (Forster & Warrington in Snelling 1985 p.105) and it should perhaps be excluded from the database.

Carboniferous/Permian (=Asselian/Noginskian): There are no data in the range 287.5 to 296.5 Ma, with a mid-point at 292 Ma. GSL 85 accepts 290 Ma and NDS gives 290 +10,−5 Ma. DNAG and GTS 82 choose 286 Ma. Back extrapolation in the Carboniferous (see below) from a pseudo-tie-point in the mid-Noginskian and at the Krevyakinskian/ Myachkovskian boundary gives an age close to 290 Ma, which is adopted here as a pseudo-tie-point. This boundary is a pseudo-tie-point because its value differs slightly from that of the chronogram.

Figure 5.13. Chronogram age in Ma plotted against Permian chrons.

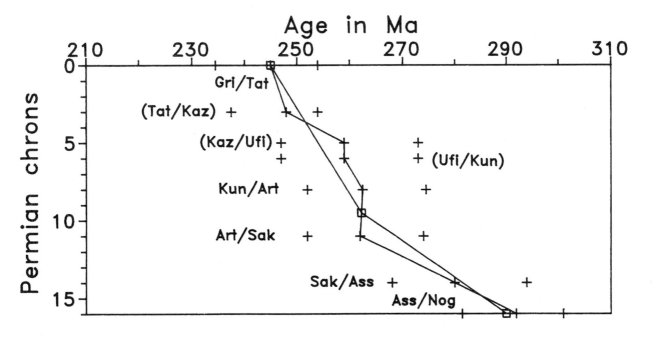

Figure 5.14. Dates in the isotopic database plotted against Permian stage boundaries.

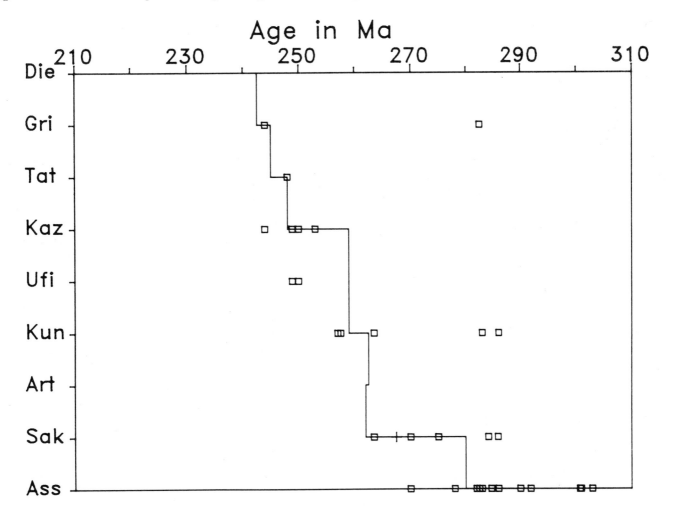

5.8.6 CARBONIFEROUS

Stage	Chrons In stage	Chrons Sum	Tie-point	C-gram (C)	Interpol. (I)	Per cent	C−I (Ma)
[58 (Ass)	2.0	209.0	290.0	292.0	290.0	Permian	]
59 (Nog2)	1.0	210.0	292.0	292.0	292.0	0.0	0.0
59 (Nog1)	1.0	211.0	0.0	292.0	293.6	7.2	−1.6
60 (Kla)	1.0	212.0	0.0	295.5	295.1	4.6	0.4
61 (Dor)	2.0	214.0	0.0	295.5	298.3	4.6	−2.8
62 (Chv)	1.0	215.0	0.0	301.2	299.9	4.2	1.3
63 (Kre)	2.0	217.0	303.0	303.0	303.0	3.3	0.0
64 (Mya)	2.0	219.0	305.0	305.0	305.0	2.6	0.0
65 (Pod)	2.0	221.0	0.0	311.5	307.1	4.0	4.4
66 (Ksk)	2.0	223.0	0.0	316.0	309.2	6.2	6.8
67 (Vrk)	2.0	225.0	0.0	316.0	311.3	6.2	4.7
68 (Mel)	2.0	227.0	0.0	316.0	313.4	6.2	2.6
69 (Che2)	2.5	229.5	316.0	316.0	316.0	0.0	0.0
69 (Che1)	2.5	232.0	0.0	316.0	318.3	6.2	−2.3
70 (Yea)	2.5	234.5	0.0	316.0	320.6	6.2	−4.6
71 (Mrd)	1.0	235.5	0.0	316.0	321.5	6.2	−5.5
72 (Kin)	1.5	237.0	0.0	320.5	322.8	5.6	−2.3
73 (Alp)	3.0	240.0	0.0	323.0	325.6	7.4	−2.6
74 (Cho)	3.0	243.0	0.0	324.0	328.3	6.0	−4.3
75 (Arn)	3.0	246.0	0.0	324.0	331.1	6.0	−7.1
76 (Pnd)	2.0	248.0	0.0	326.7	332.9	3.8	−6.2
77 (Bri2)	1.0	249.0	333.8	341.0	333.8	0.0	7.2
77 (Bri1)	1.0	250.0	0.0	341.0	336.0	4.8	5.0
78 (Asb)	1.5	251.5	0.0	343.5	339.4	4.0	4.1
79 (Hlk)	1.5	253.0	0.0	343.5	342.8	4.0	0.7
80 (Aru)	1.0	254.0	0.0	343.5	345.0	4.0	−1.5
81 (Chd)	2.0	256.0	349.5	349.5	349.5	2.5	0.0
82 (Ivo)	1.0	257.0	0.0	357.5	353.8	4.9	3.7
83 (Has)	2.0	259.0	362.5	362.5	362.5	3.1	0.0

Discussion:

Fifty chrons from the foraminiferal zones in the Donetz Basin were used for interpolation (GTS 89, Figure 3.6); 28 in the Pennsylvanian Period and 22 in the Mississippian Period. As in the Early Triassic and Late Permian, nearly all the available dates are from sequences that cannot readily be correlated with the international standard succession. Thus the foraminiferal and ammonoid zones of the marine Russian standard are not easy to correlate with the non-marine plant successions in western Europe and North America.

For the database of isotopic dates, the **Stephanian/Westphalian** boundary is correlated with the **mid-Myachkovskian** stage; the **Westphalian/Namurian** boundary with the **Cheremshanskian/Yeadonian** boundary; the **Namurian/Visean** with the **Pendleian/Brigantian** boundary and the **Visean/Tournaisian** with the **Chadian/Ivorian** boundary. More chrons may well be established in the future in the Visean and Tournaisian epochs.

For many boundaries within this period the results of the chronograms are very similar because of the lack of data apart from the dates from the non-marine plant successions. For example, the Vereiskian/Melekesskian to Marsdenian/

Kinderscoutian boundaries all lie in the range 315.2 to 316 Ma.

Four tie-points and four pseudo-tie-points are used. The tie-points are **Krevyakinskian/Myachkovskian; Myachkovskian/Podolskian; Chadian/Ivorian;** and **Hastarian/Famennian.** Pseudo-tie-points are at the **Asselian/Noginskian, mid-Noginskian, mid-Cheremshanskian** and **mid-Brigantian**.

Noginskian/Klazminskian; Klazminskian/Dorogomilovskian; Dorogomilovskian/Chamovnicheskian; Chamovnicheskian/Krevyakinskian: Ages are found by chron interpolation.

Krevyakinskian/Myachkovskian; Myachkovskian/Podolskian: Class B chronogram at 303 Ma gives a tie-point for the Krevyakinskian/Myachkovskian boundary. Similarly a class B chronogram at 305 Ma gives a tie-point for the Myachkovskian/Podolskian boundary. Both boundaries are controlled by the same dates with the exception of PTS8a and LI1 which influence the younger side of the Krevyakinskian/Myachkovskian boundary and B2 which influences all of the Myachkovskian/Podolskian boundary. The Stephanian/Westphalian boundary is taken as the mid-point between these

Figure 5.15. Chronogram age in Ma plotted against Carboniferous chrons.

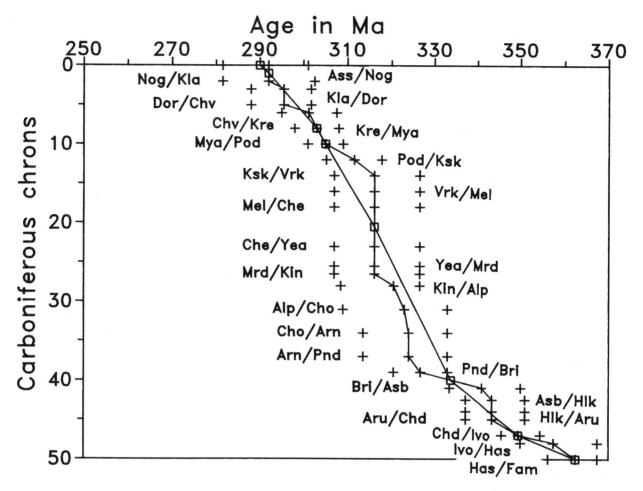

Figure 5.16. Dates in the isotopic database plotted against Carboniferous stage boundaries.

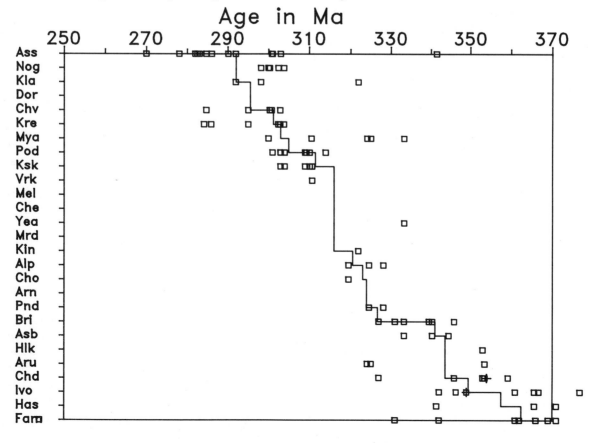

two boundaries at 304 Ma. This is older than NDS 300 ± 5, GSL 85 300 Ma, and GTS 82 and DNAG at 296 Ma.

Podolskian/Kashirskian, Kashirskian/Vereiskian; Vereiskian/Melekesskian; Melekesskian/Cheremshanskian: Ages found by chron interpolation between initial Myachovskian and mid-Cheremshanskian boundaries (Figure 5.15).

Mid-Cheremshanskian: Pseudo-tie-point at 316 Ma.

Cheremshanskian/Yeadonian; Yeadonian/Marsdenian; Marsdenian/Kinderscoutian; Kinderscoutian/Alportian; Alportian/Chokerian; Chokerian/Arnsbergian; Arnsbergian/ Pendleian: Ages found by chron interpolation between mid-Cheremshanskian and mid-Brigantian pseudo-tie-points (Figure 5.15).

Pendleian/Brigantian: Class C chronogram at 326.7 Ma, though not used as a tie-point. Chron interpolation gives 332.9 Ma. The PTS and LI dates keep the younger side of this boundary young. Most time scales put it in the 320 to 330 Ma range, but 320 Ma violates all data. NDS167 is causing it to be young – a young date from an older stage. See discussion in NDS (pp.848–851) about ages in this part of the time scale. NDS age is 320 +10, –5 Ma; GSL 85 325 Ma; DNAG and GTS 82 333 Ma. This is the Namurian/Visean boundary.

Brigantian/Asbian; Asbian/Holkerian; Holkerian/ Arundian; Arundian/Chadian: Ages found by chron interpolation between initial Pendleian boundary and initial Chadian boundary. Other time scales do not differentiate between the individual boundaries.

Chadian/Ivorian: Class B chronogram, although lifted slightly off the base, at 349.5 Ma. GTS 82 and DNAG use 352 Ma. All published ages lie within 0.5 Ma of the chronogram age range for this boundary: NDS gives 355 ± 5 Ma. Equivalent to the Visean/Tournaisian boundary.

Ivorian/Hastarian: Age found by chron interpolation.

Carboniferous/Devonian (= Hastarian/Famennian) boundary: Class C chronogram at 362.5 Ma. Dominated by an old date PTS360f (see discussion on Item NDS167 in NDS). GTS 82 value was 360 Ma.

See discussion by Sandberg *et al.* (1982), who give much more time to late Devonian. NDS gave 360 +5,–10 Ma.

5.8.7 DEVONIAN

Stage	Chrons		Tie-point	C-gram (C)	Interpol. (I)	Per cent	C−I (Ma)
	In stage	Sum					
[83 (Has)	2.0	259.0	362.5	362.5	362.5	Carboniferous	]
84 (Fam)	7.0	266.0	367.0	367.0	367.0	2.7	0.0
85 (Frs)	6.0	272.0	0.0	376.0	377.4	5.6	−1.4
86 (Giv)	2.0	274.0	0.0	382.0	380.8	6.3	1.2
87 (Eif)	3.0	277.0	386.0	386.0	386.0	2.6	0.0
88 (Ems)	5.0	282.0	0.0	391.0	390.4	6.1	0.6
89 (Pra2)	1.5	283.5	391.7	391.7	391.7	0.0	0.0
89 (Pra1)	1.5	285.0	0.0	392.5	396.3	5.7	−3.8
90 (Lok)	4.0	289.0	408.5	408.0	408.5	2.0	−0.5

Discussion:

Chrons are the 30 conodont biozones (Figure 3.5). There are no tie-points in the Devonian Period in GTS 82; four are used in GTS 89. There is good agreement among published Devonian time scales.

There are three tie-points: **Famennian/Frasnian**; **Eifelian/Emsian; Lochkovian/Pridoli**; and one pseudo-tie-point in the **mid-Pragian**.

Famennian/Frasnian: Class B chronogram at 367 Ma, as in GTS 82 and DNAG. NDS235 has the effect of keeping the younger boundary at an older value. GSL 85 chose 365 Ma.

Frasnian/Givetian; Givetian/Eifelian: Ages found by chron interpolation from the initial Famennian to the Eifelian boundary. For the Frasnian/Givetian chron interpolation gives

377.4 Ma; GTS 82 and DNAG chose 374 Ma; NDS and GSL 85 chose 375 ± 5 Ma. For the Givetian/Eifelian, GTS 89 chron interpolation gives 380.8 Ma, GTS 82 gave 380 Ma; DNAG chose 380 Ma; and GSL 85 chose 383 Ma.

Eifelian/Emsian: Class B chronogram at 386 Ma. NDS chose 385 ± 8 Ma; GTS 82 and DNAG chose 387 Ma; and GSL 85 chose 390 Ma.

Emsian/Pragian: Chronogram age at 391 Ma similar to chron interpolation at 390.4 Ma.

Pragian/Lochovian: Chronogram age at 392.5 is much younger than interpolated chron age at 396.3 Ma.

Devonian/Silurian (= Lochkovian/Pridoli) boundary: Good chronogram at 408.5 Ma, as in GTS 82. NDS chose 400 +10,−5 Ma; DNAG 408 Ma; and GSL 85 405 Ma.

Figure 5.17. Chronogram age in Ma plotted against Devonian chrons.

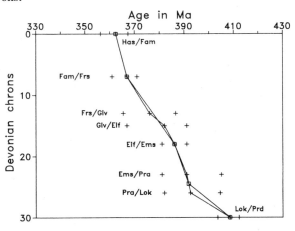

Figure 5.18. Dates in the isotopic database plotted against Devonian stage boundaries.

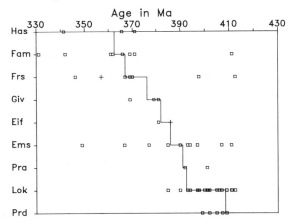

5.8.8 SILURIAN

| Stage | Chrons | | Tie-point | C-gram (C) | Interpol. (I) | Per cent | C−I (Ma) |
	In stage	Sum					
[90 (Lok)	4.0	289.0	408.5	408.0	408.5	Devonian	]
91 (Prd)	1.0	290.0	0.0	409.5	410.7	2.1	−1.2
92 (Ldf)	2.0	292.0	0.0	412.5	415.1	2.4	−2.6
93 (Gor)	4.0	296.0	424.0	424.0	424.0	1.7	0.0
94 (Gle)	2.0	298.0	0.0	426.5	425.4	2.3	1.1
95 (Whi)	1.0	299.0	0.0	426.2	426.1	2.3	0.1
96 (She)	6.0	305.0	0.0	431.5	430.4	3.8	1.1
97 (Tel)	3.0	308.0	0.0	432.5	432.6	3.5	−0.1
98 (Aer)	6.0	314.0	0.0	437.0	436.9	4.6	0.1
99 (Rhu)	3.0	317.0	439.0	439.0	439.0	3.4	0.0

Discussion:

General comment: Most correlations discriminate at epoch rather than stage level, i.e. they use Pridoli, Ludlow, Wenlock and Llandovery, rather than the stages. The chronograms for stage boundaries within these epochs are either poor or meaningless and must be estimated using chron interpolation. Chrons are the 30 graptolite biozones (Figure 3.4).

Three tie-points are used: Lochkovian/Pridoli (Silurian/ Ordovician boundary); Ludlow/Wenlock (= Gorstian/ Gleedonian); and Rhuddanian/Hirnantian.

Pridoli/Ludlow (= Pridoli/Ludfordian): Interpolation between Lochkovian/Pridoli and Ludlow/Wenlock (= Gorstian/Gleedonian) gives 410.7 Ma. GTS 82, DNAG and GSL 85 give 414 Ma.

Ludfordian/Gorstian: Class B chronogram at 412.5 Ma. Interpolated to 415.1 Ma. The data used for this chronogram are very similar to those for the Pridoli/Ludlow boundary.

Ludlow/Wenlock (= Gorstian/Gleedonian): Class B chronogram used as tie-point at 424 Ma. GTS 82 gave 421 Ma, based on equal duration of zone hypothesis. NDS and DNAG gave 421 Ma; GSL 85 420 Ma.

Gleedonian/Whitwellian; Whitwellian/Sheinwoodian; Sheinwoodian/Telychian; Telychian/Aeronian; Aeronian/ Rhuddanian: Ages found by chron interpolation between the initial Gorstian and the initial Rhuddanian boundary age. The Sheinwoodian/Telychian boundary, a tie-point in GTS 82, is a class C chronogram with too large an age range to be considered as a tie-point here. For this boundary, with a chronogram at 431.5 Ma, GTS 82 and DNAG chose 428 Ma and GSL 85 chose 425 Ma. Chron interpolation gives 430.4 Ma.

Silurian/Ordovician (= Rhuddanian/Hirnantian): Class C chronogram at 439 Ma. 439 Ma is given by Kunk et al. (1985) in GSL, but they also quote 436 Ma for top of cyphus zone, which is the fourth chron above the initial boundary. GTS 82 and DNAG chose 438 Ma; GLS 85 chose 435 Ma; and NDS was much younger at 418 +5,−10 Ma.

Figure 5.19. Chronogram age in Ma plotted against Silurian chrons.

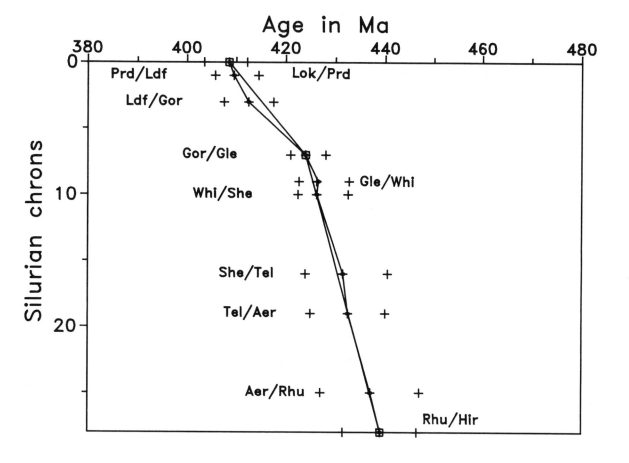

Figure 5.20. Dates in the isotopic database plotted against Silurian stage boundaries.

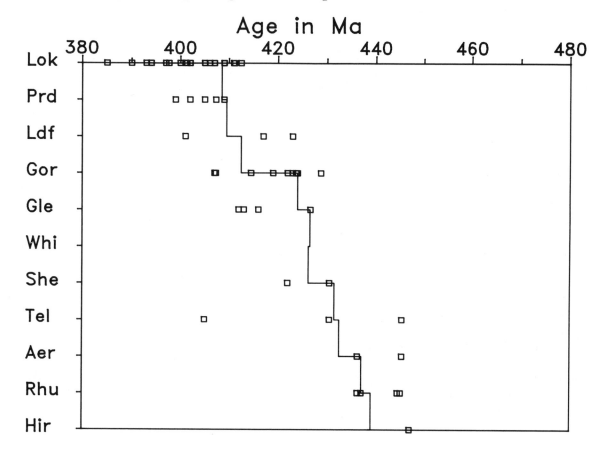

5.8.9 ORDOVICIAN

Stage	Chrons		Tie-point	C-gram (C)	Interpol. (I)	Per cent	C−I (Ma)
	In stage	Sum					
[99 (Rhu)	3.0	317.0	439.0	439.0	439.0	Silurian	]
100 (Hir)	0.3	317.3	0.0	438.5	439.5	3.4	−1.0
101 (Raw)	0.3	317.7	0.0	443.5	440.1	3.4	3.4
102 (Cau)	0.3	318.0	0.0	443.5	440.6	3.4	2.9
103 (Pus)	1.5	319.5	0.0	443.0	443.1	3.4	−0.1
104 (Onn)	0.5	320.0	0.0	445.0	444.0	2.0	1.0
105 (Act)	0.3	320.3	444.5	444.5	444.5	2.0	0.0
106 (Mrb)	0.3	320.6	0.0	444.5	447.1	2.0	−2.6
107 (Lon)	0.3	321.0	0.0	457.0	449.7	2.6	7.3
108 (Sou)	1.0	322.0	457.5	457.5	457.5	2.5	0.0
109 (Har)	1.0	323.0	0.0	458.7	462.3	1.9	−3.6
110 (Cos)	0.3	323.3	0.0	462.5	463.9	2.4	−1.4
111 (Llo3)	0.3	323.6	0.0	462.5	465.4	2.4	−2.9
112 (Llo2)	0.3	324.0	0.0	462.5	467.0	2.4	−4.5
113 (Llo1)	0.3	324.3	0.0	471.5	468.6	2.1	2.9
114 (Lln2b)	0.5	324.8	471.0	471.0	471.0	0.0	0.0
114 (Lln2a)	0.5	325.3	0.0	471.5	472.7	2.1	−1.2
115 (Lln1)	1.0	326.3	0.0	482.0	476.1	3.1	5.9
116 (Arg)	5.0	331.3	0.0	492.0	493.0	3.7	−1.0
117 (Tre)	5.0	336.3	510.0	513.0	510.0	5.4	3.0

Discussion:

Chrons are the 20 graptolite biozones in Figure 3.3. Most correlations discriminate at epoch rather than stage level, i.e. they use Ashgill, Caradoc, Llandeilo and Llanvirn rather than the stages. The chronograms for stage boundaries within these epochs are either poor or meaningless and must be estimated using chron interpolation.

Three tie-points and two pseudo-tie-points are used: **Silurian/Ordovician (= Rhuddanian/Hirnantian); Actonian/ Marshbrookian; Soudleyan/Harnagian;** with pseudo-tie-points in **mid-Llanvirn–2** and at the **Ordovician/Cambrian (= Tremadoc/Dolgellian)**.

Hirnantian/Rawtheyan; Rawtheyan/Cautleyan; Cautleyan/Pusgillian; Pusgillian/Onnian; Onnian/Actonian: Ages found by chron interpolation between **Rhuddanian/ Hirnantian** and **Actonian/Marshbrookian** boundary. For the **Pusgillian/Onnian (Ashgill/Caradoc)** boundary the interpolated chron age is 443.1 Ma; NDS chose 425 ± 8 Ma; GLS 85 chose 440 Ma; and DNAG and GTS 82 448 Ma.

Actonian/Marshbrookian: Class B chronogram at 444.5 Ma. This chronogram is essentially the same as several other chronograms at this stratigraphic level. For example, three of them lie within the *Dicranograptus clingani* zone. Unless shelly faunas or conodonts are present, all dates will be assigned to this zone.

Marshbrookian/Longvillian; Longvillian/Soudleyan: Found by chron interpolation.

Soudleyan/Harnagian: Class C chronogram at 457.5 Ma. The boundary appears to be a good chronogram with the data used being very similar to the Harnagian/Costonian boundary.

Harnagian/Costonian: Class B chronogram at 458.7 Ma, mostly from fission-track dates, but visual inspection suggests that chron interpolation at 462.3 Ma from the overlying **Soudleyan/Harnagian** boundary to the pseudo-tie-point in **Llanvirn–2** gives an equally plausible age.

Llandeilo/Llanvirn boundary: The internal Llandeilo and Llandeilo/Llanvirn boundaries are estimated using chron interpolation as the data are not differentiated within the stages. The Llandeilo/Llanvirn boundary is a class C chronogram at 471.5 Ma; chron interpolation gives 468.6 Ma. The NDS value for this boundary is much younger at 455 Ma; GSL 85 at 460 Ma; GTS 82 and DNAG at 468 Ma.

Mid-Llanvirn pseudo-tie-point: Taken as 471 Ma.

Llanvirn/Arenig: The chronogram estimate of 482 Ma is considered high for this boundary. The chron interpolation of 476.1 Ma is preferred. NDS chose 470 ± 10 Ma; GSL chose 473 Ma; GTS 82 and DNAG 478 Ma.

Arenig/Tremadoc: The chronogram estimate of 492 Ma is a reasonable estimate for this boundary, but chron interpolation at 493.0 Ma is preferred. NDS chose 475 +10,−5 Ma; GTS 82 and DNAG chose 488 Ma; and GSL 85 chose 491 Ma. The GTS 82 age of 488 Ma is considered too young.

Ordovician/Cambrian (= Tremadoc/Dolgellian): Poor class C chronogram at 513 Ma, but with no significant data in range of 500.5 to 528.0 Ma. NDS chose 495 +10,−5 Ma; GTS 82 and DNAG 505 Ma; and GSL 85 510 Ma. We arbitrarily choose 510 Ma as a pseudo-tie-point. The next stage boundary within the Cambrian – **Dolgellian/Maentwrogian** – also has a 513 Ma chronogram.

Figure 5.21. Chronogram age in Ma plotted against Ordovician chrons.

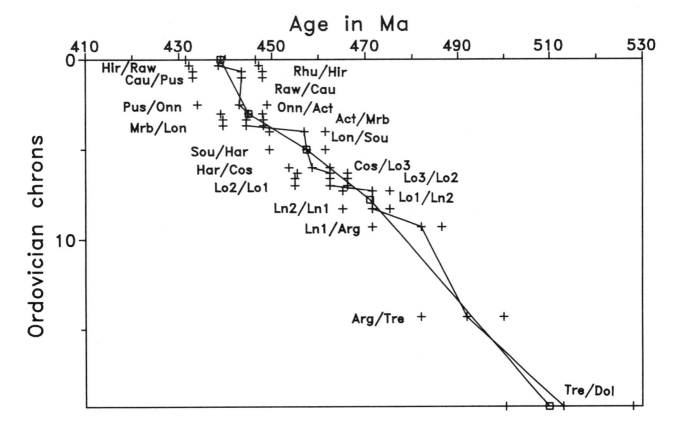

Figure 5.22. Dates in the isotopic database plotted against Ordovician stage boundaries.

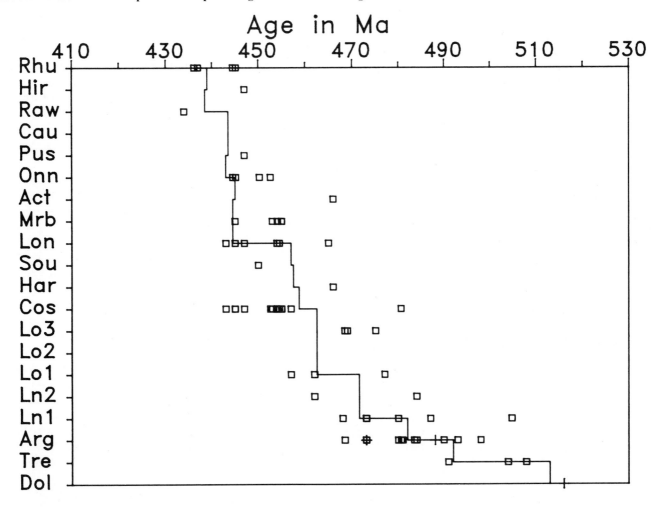

5.8.10 CAMBRIAN

Stage	Chrons In stage	Sum	Tie-point	C-gram (C)	Interpol. (I)	Per cent	C−I (Ma)
[117 (Tre)	5.0	336.3	510.0	513.0	510.0	Ordovician	]
118 (Dol)	4.0	340.3	0.0	513.0	514.1	6.3	−1.1
119 (Mnt)	3.0	343.3	0.0	522.0	517.2	6.9	4.8
120 (Men3)	2.7	346.0	520.0	520.0	520.0	6.9	0.0
121 (Men2)	2.7	348.6	0.0	520.5	525.1	6.7	−4.6
122 (Men1)	2.7	351.3	0.0	530.0	530.2	4.2	−0.2
123 (Sol2)	1.5	352.8	0.0	535.0	533.1	3.0	1.9
124 (Sol1)	1.5	354.3	536.0	536.0	536.0	2.5	0.0
125 (Len)	5.0	359.3	0.0	536.0	553.7	2.6	−17.7
126 (Atb)	2.0	361.3	0.0	565.0	560.8*	3.2	4.2
127 (Tom)	4.0	365.3	575.0	575.0	575.0*	2.8	0.0

* These ages have been modified to 560 and 570 Ma respectively. See Section 5.9.

Discussion:

A mixed trilobite and archeocyathid 33-chron scheme is used. The mixed zone is a convenience only. It can be justified by the fact that the Tommotian contains only four archeocyathid chrons which in total time span are probably not grossly dissimilar to four trilobite zones (S. Conway Morris, personal communication).

A pseudo-tie-point is used at the **Ordovician/Cambrian (= Tremadoc/Dolgellian)** boundary; tie-points at the **Menevian–3/Menevian–2** and **Solvan–1/Lenian** boundaries and a pseudo-tie-point at the **Tommotian/Poundian** boundaries.

Most correlations discriminate at epoch rather than stage level, i.e. they use Merioneth, Saint David's and Caerfai or their equivalents. Chronograms for stages within these epochs are either poor or meaningless.

Dolgellian/Maentwrogian: Found by chron interpolation at 514.1 Ma.

Maentwrogian/Menevian–3: Found by chron interpolation at 517.2 Ma.

Menevian–3/Menevian–2: Class C tie-point at 520 Ma.

Menevian–2/Menevian–1; Menevian–1/Solvan–2; Solvan–2/Solvan–1: Found by chron interpolation.

Solvan–1/Lenian: Class C tie-point at 536 Ma.

Lenian/Atdabanian: Found by chron interpolation at 553.7 Ma.

Atdabanian/Tommotian: Found by chron interpolation at 560.8 Ma.

Phanerozoic/Proterozoic (= Cambrian/Precambrian and Tommotian/Poundian): There is a range of possibilities which cannot be resolved from present data.

(a) The chronogram age, excluding all illites and treating glauconites as minimum ages, is 575 Ma, with a range of 567 to 583 Ma.

(b) Including Chinese whole-rock Rb–Sr data on illites in shales (see last entries in Table 4.2) gives a chronogram age of 584.2 Ma, with a range of 579.2 to 590.5 Ma. See also end Table 5.4, p.115.

(c) A regression line through the first three Cambrian tie-points

and pseudo-tie-points gives 550.3 Ma for the **Tommotian/Poundian** boundary (Figure 5.23) and provides an alternative estimate, more in line with proposals by Odin and others.

If the Caerfai chrons are equivalent in an evolutionary sense, the high age estimates of the Chinese data imply a rate of evolution three times as slow in the Caerfai Epoch compared with the Saint David's and younger Cambrian epochs.

The NDS estimate of 530 ± 10 Ma seems too young.

5.9 Initial Cambrian and Vendian chronometry

The systematic determination of Phanerozoic numerical ages was explained in Section 1.9 and this chapter. The values are also listed in the 'definitive' scale in Figure 1.7. However, because the database does not contain a good spread of chronostratically distributed Late Precambrian dates, we may expect that the initial Phanerozoic boundary will be less well constrained than other boundaries. This weakness is exacerbated by the notorious failure to achieve agreed results from rocks at about the position of the boundary.

While GTS 82 gave a value of 590 Ma for the initial Cambrian boundary, Odin and his coworkers (e.g. Odin 1982b) found the best value to be 530 Ma. The higher value took account of whole-rock Rb–Sr dates on early Cambrian shales and the lower value derived from the greater weight attached to apparently Precambrian or pre-part-Cambrian plutons dated by K–Ar methods. The current position is still unsatisfactory. Applying the procedure outlined in Section 1.9 gave an age of 575 Ma to the initial Tommotian boundary. Applying the chronogram program to one version of the database gave 552 Ma. One of us estimated a value of about 584 Ma based on slightly different data.

While admitting that insufficient data exist for the initial Cambrian boundary, Krasnobayev & Semikhatov (1986) used improved analytical methods on glauconites of Vendian and Late Riphean rocks. They agreed that the initial Cambrian

Figure 5.23. Chronogram age in Ma plotted against Cambrian chrons. The long-dashed line is a regression line through the first three Cambrian tie-points. It gives 550.3 Ma for the Tommotian/Poundian boundary. The smoother solid line is a visual fit to the data through selected chronograms. It gives 575 Ma for the same boundary. The eventual age chosen is 570 Ma (see Discussion 5.9).

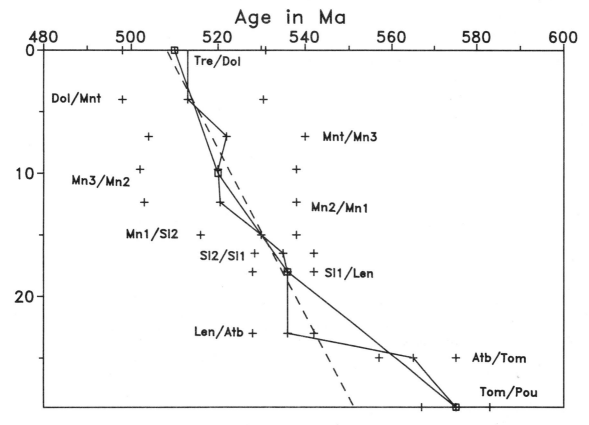

Figure 5.24. Dates in the isotopic database plotted against Cambrian stage boundaries.

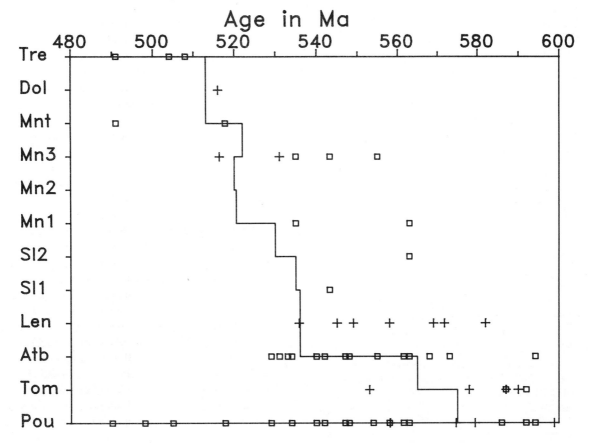

boundary could lie between 615 ± 10 Ma and 530 Ma and placed the initial Vendian boundary at 650 ± 20 Ma. This value is similar to earlier Soviet estimates of 650 ± 10 Ma for the same boundary (e.g. Sokolov & Fedonkin 1984). Such a value would accommodate older initial Cambrian ages. Sokolov and Fedonkin suggested 560 ± 10 Ma for the initial Tommotian boundary and 570 Ma for the initial Nemakit-Daldyn boundary – a division (possibly equivalent to Etcheminian) not yet internationally recognized and not yet assignable to the earliest Cambrian or latest Precambrian.

Support for a 650 Ma initial Varanger boundary came from a U–Pb zircon age of 660 ± 15 Ma for the Barangulovo granites which cut 'upper upper Riphean' deposits and whose pebbles are found in basal conglomerates of the Archa Formation which correlates with allotillites of the Kungashley Formation of Varanger age (Krasnobayev 1986).

In addressing this problem Cowie & Harland (1989) listed relevant data and showed that plausible cases could be made for higher values (say 570 to 600 Ma) or lower values (say 560 to 530 Ma). They selected 570 Ma (once again) as an interim convention only on the basis that it would avoid a greater error that might result if the age were mistakenly taken at a higher or lower value.

In the meantime, good quality determinations have become available from the critical section in the Avalon Terrane of Newfoundland (Krogh *et al.* 1988). These were a whole range of U–Pb determinations on zircons in the sedimentary volcanics of the Harbour Main Group, the Holywood granite and of other associated igneous rocks. The results suggested to Conway Morris (1989) that the initial Cambrian boundary could be placed at, say, 550 or 540 Ma. The argument may be restated for Newfoundland for the purpose of the value to be adopted in GTS 89 as follows.

The varied successions of the Vendian–Cambrian Avalon sequence were conveniently tabulated by Conway Morris (1989). Critical to this argument is the following succession.

Early Cambrian

Smith Point Fm with trilobites (Atdabanian)
Bona Vista Fm with small shelly fossils (Tommotian)
Random Fm

Chapel Island Fm with shelly fossils
Paleozoic type trace fossils
Early trace fossils

Vendian

Signal Hill Gp
St Johns Gp with *Aspidella*

Conception Gp ⎰ Mistaken Point Fm with Ediacara fossils
⎹ Drook Fm with Ediacara fossils
⎹ Gaskiers Fm with (?early or late)
⎱ Varanger tillite

Pre-Vendian or pre-part-Varanger

Harbour Main Gp with intrusives (mainly volcanics) including the Holyrood granite
Burin Group   [763 + 2.2, –1.8]

Krogh *et al.* (1988) gave values as follows. For the Holyrood granite, which is probably, but not certainly, coeval with the Harbour Main rocks, 620.5 ± 2 Ma and for a rhyolite dyke, also intrusive into the Harbour Main Group, 585.9 ± 3 Ma.

The Harbour Main Group gave ages for different volcanic strata of 606 ± 3, 622 ± 2, 590 ± 3 and 608 ± 20 Ma.

Older than all these is the Burin Group dated at 763 ± 2 Ma (errors quoted are approximated here).

In addition, Benus (1988) referred to the Mistaken Point Formation with Ediacara fauna dated at 565 ± 3 Ma from U–Pb on zircons from a tuff within the formation.

A further determination of a post-Atdabanian rock from West Africa (Culver *et al.* 1988) gave 555 Ma.

The above information became available at too late a stage to incorporate fully into the database. It demands some rethinking of the problem. By inspection only, we propose the following scale for this part of Earth history on the basis that each number is just consistent with the available data allowing the stated range of uncertainty. The numbers in parentheses give the time intervals, the other numbers the dates (all in Ma).

			536 ± 5
	Lenian	Tojonian	
			(18)
		Botomian	
			554 ± 5
Caerfai (25)	Atdabanian		(6)
			560 ± 10
	Tommotian (including Etcheminian/Rovnian or Kotlinian)		(10)
			570 ± 15
	Poundian		(10)
Ediacara (25)			580 ± 15
	Wonokan		(10)
			590 ± 15
	Mortensenes		(10)
Varanger (25)			600 ± 15
	Smalfjord		(10)
			610 ± 20

Accordingly, we insert these figures into our definitive scale.

5.10 Concluding remarks

Except for the Pliocene, Pleistocene and younger boundaries, the resulting Phanerozoic time scale is based entirely on isotopic dating and a paleontological apportionment of stratigraphic time. Chapter 6 modifies parts of the Cenozoic and Mesozoic time scale in the light of ocean-floor spreading data. The resulting GTS 89 scale appears in Chapter 1 as Figure 1.7 (and in summary as Table 1.2).

Because of the uncertainties inherent in isotopic dating it is unwise to become preoccupied with fitting the time scale to any one or a few tie-points. Only a large body of internally consistent calibration points or a reasonable compromise between conflicting dates will give the most useful and stable calibration. Consequently our approach has been to take all reasonable data into account, keeping in mind all the sources of random and systematic error discussed in Chapter 4.

The method of age calibration used here has undoubtedly degraded the precision of the paleontological dating of some samples. For example, by assigning all dates to stages, we have ignored whether the date can be assigned a more precise chronostratigraphic position within that stage. One reason for this is that the paleontological position of most samples has not been determined to better than a stage. A second reason is that it is not clear that continued refinement of the biostratigraphic scale by the recognition of more and more zones and the tying in of the dated samples to these zones is necessarily the only way to proceed. In a future time scale it may perhaps be as useful to describe the stratigraphic position of a dated sample in relation to chron or stage boundaries, e.g. that it lies 5 m below the position of initial boundary A and 10 m above initial boundary B, with the implication that its age is two-thirds the duration of B.

If methods can be devised to integrate such data into a general scheme, perhaps by using the relative stratigraphic thicknesses of paleontological chrons together with numerical models of sedimentary basin formation, we can envisage a piecewise accumulation of the relative durations of paleontological ages which will eventually yield a much more precise time scale than we now have.

It is arguable that the precision of the scale may be understated because of the individual chronogram errors which appear as uncertainties of the scale. It is actually better than these errors suggest, because its general trend is controlled by the better chronograms, not the average nor the worst ones. The details are controlled by the quality of the assumptions used for interpolation but these should have little effect on the general values for stage boundaries.

6

The magnetostratigraphic time scale

6.1 Geomagnetic polarity reversals

6.1.1 Global synchroneity

The basis of magnetostratigraphy is the retention by rocks of a magnetic imprint acquired in the geomagnetic field that existed at the time the rocks formed. The processes that produce the geomagnetic field occur in the Earth's core, where fluid motions driven by convection comprise a dynamo that generates a global magnetic field which is roughly symmetrical about the Earth's rotation axis. For reasons that are still not well understood, at irregular times the currents flowing in the core reverse their direction, producing a reversal in the polarity of the dipole magnetic field.

By convention the present-day polarity is **normal**: lines of magnetic force at the Earth's surface are by convention directed towards the north magnetic pole. The N-seeking pole of a compass needle points north. The dip or **inclination** of the field is directed downward in the northern hemisphere and upward in the southern hemisphere. When the polarity is **reversed**, the lines of force are directed in the opposite direction and a compass needle would point to the south. The sign of the inclination is then reversed in both hemispheres.

These 180 degree changes in the direction of the magnetic field imprinted on rocks forming at the Earth's surface provide the physical basis for magnetic polarity stratigraphy. Because polarity reversals are potentially recorded simultaneously in rocks all over the world, magnetostratigraphic divisions, unlike lithostratigraphic and biostratigraphic divisions, are not time-transgressive. However, the age of magnetization might not be the same as the age of other geologic events in the history of the rock. In plutonic rocks, the magnetization is acquired after the crystallization of the rock and before the setting of the K–Ar clock in biotite. In chemically altered rocks, the magnetization

is generally acquired at the time of chemical alteration. In either case the acquisition may be complex and span several polarity intervals.

When a reversal occurs, the time required for the change is known from detailed paleomagnetic studies of transition zones to be about 5000 years (Figure 6.1). Because strata within transition zones cannot generally be correlated globally, the ultimate resolving power of magnetostratigraphy is roughly equal to the duration of transitions. Strata adjacent to a transition zone could thus be correlated globally with a precision of 0.5% for one-million-year-old rocks and 0.005% for 100-million-year-old rocks were it not for problems introduced by other factors.

6.1.2 Excursions

Even when the dipole field has a steady polarity, it undergoes swings in direction with typical amplitudes of 15 degrees and periods of 10^2 to 10^4 years. This **geomagnetic secular variation** is generally too small to be mistaken for the 180 degree changes in field direction which characterize polarity reversals. Occasionally, however, the field appears to undergo an **excursion** characterized by a large change in direction that may approach 180 degrees. Since excursions are thought to have durations of about 1000 years, they offer the

Figure 6.1. Polarity chrons, polarity subchrons, transition zones and excursions.

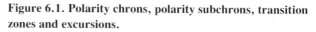
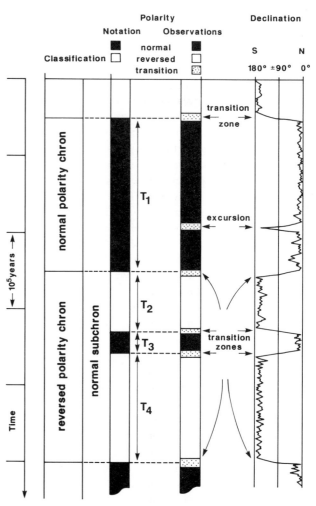

potential of providing very precise, but rarely observable, stratigraphic markers. However, excursions have not proved easy to identify with certainty, partly because of the necessity for closely spaced sampling and partly on account of the problems of distinguishing an excursion from paleomagnetic overprinting. Excursions have proven disappointingly difficult to trace globally for several reasons: excursions are so short that they are missing in many stratigraphic sections; some excursions are probably local rather than global geomagnetic phenomena; and anomalous paleomagnetic directions are sometimes due not to excursions but rather to deformation of the rocks studied (Verosub & Banerjee 1977, Banerjee, Lund & Levi 1979), or to the effects of lightning strikes, bioturbation and the like.

With the objective of including in the present work well-documented reversals, but not excursions, the following criteria were used to distinguish between the two phenomena. The first is whether the field changed direction to within several tens of degrees of a complete 180 degree reversal. The second is whether the field remained locked into the reversed direction for a measurable length of time (reversal) or simply moved through the reversed direction (excursion). The third is whether the reversal was recorded at different sites around the world. The duration of the shortest possible polarity interval consistent with these criteria would be somewhat greater than 0.01 Ma, which is twice the length of a polarity transition interval. A recent review of short-lived reversals for the past 1.1 Ma is given by Champion, Lanphere & Kuntz (1988) and is discussed further below.

6.1.3 Polarity intervals, chrons and subchrons

The time interval that elapses between two successive reversals in the polarity of the geomagnetic dynamo is generally referred to as a **polarity interval** (Cox 1968), the latter term being used as a description of a physical phenomenon and not as a chronostratigraphic unit. This usage accords with the statement in the *International Stratigraphic Guide* that 'interval' may refer to either time or space intervals and therefore should be used as a general term and not as a formal stratigraphic unit or division (Hedberg 1976, p.15). The lengths of polarity intervals vary from about 0.01 Ma to several tens of millions of years. In the course of paleomagnetic research, each new discovery of a short polarity interval changes the local polarity structure. This may be seen in Figure 6.1, where prior to the discovery of the short polarity interval labelled τ_3 only one reversed interval would have been recognized spanning the intervals τ_2, τ_3 and τ_4. Therefore, in naming or numbering polarity intervals for stratigraphic purposes, a hierarchical set of names is needed that does not change drastically with the discovery of additional short polarity intervals. Figure 6.1 demonstrates that a scheme of simply numbering polarity intervals in the sequence of their occurrence does not provide such a system.

According to the chapter on magnetostratigraphy of the *International Stratigraphic Guide* prepared by the IUGS International Subcommission on Stratigraphic Classification (ISSC)/IAGA Subcommission on the Magnetic Polarity Time Scale (Anon. 1979), the terms recommended for describing subdivisions of time based on geomagnetic polarity are

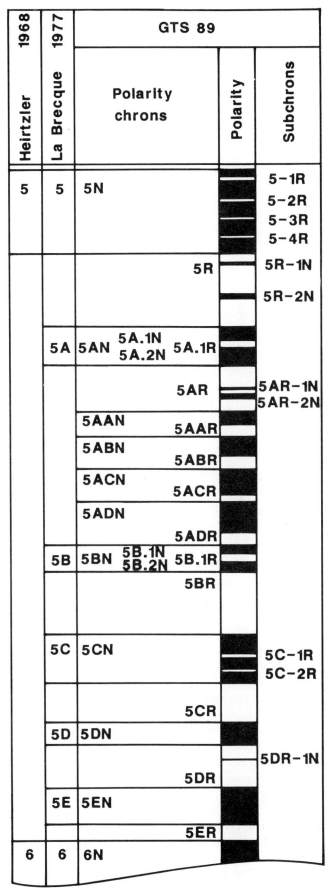

Figure 6.2. Numerical schemes for numbering chrons and subchrons derived from numbered marine magnetic anomalies. Redrawn from Figure 1 of Berggren, Kent & Flynn (1985).

polarity subchrons, **polarity chrons** and **polarity superchrons** (Figure 6.2). The corresponding chronostratigraphic terms for describing all rocks, magnetic or not, that formed during these time intervals are **polarity subchronozone**, **polarity chronozone** and **polarity superchronozone**. Magnetic lithostratigraphic intervals based on the measured magnetic properties of rocks are **polarity subzone, zone** and **superzone**. In earlier works, the Subcommission recommended the durations tabulated below for different levels in the hierarchy (McElhinny 1978).

Time term	Approximate duration
Polarity subchron	10^4–10^5 years
Polarity chron	10^5–10^6 years
Polarity superchron	10^6–10^7 years
Polarity hyperchron	10^7–10^8 years

In accord with these recommendations, we use **polarity chron** as the geochronologic term to describe the main subdivisions of time recognized on the basis of polarity. For example, the term 'Matuyama reversed polarity chron' replaces the earlier term '. . . epoch' (Anon. 1979). The term 'subchron' will be used in place of the word 'event' to describe very short (<0.1 Ma) polarity intervals occurring within a chron. For example, the Jaramillo normal subchron occurs within the Matuyama reversed chron (Figure 6.4).

6.2   Isotopically dated time scale 0 to 3.4 Ma

For the interval from the present back to 3.4 Ma, K–Ar dates and magnetic polarities have been measured for more than 350 extrusive rocks from many parts of the world. The construction of a self-consistent reversal time scale using these data (Section 6.4.2) has firmly established the global character of reversals. In addition, this research has established the existence of very short polarity intervals (subchrons) and the great variation in the lengths of polarity intervals (Section 6.4.2).

6.3   Marine magnetic anomalies: background to earlier work

6.3.1   Introduction

Marine magnetic anomalies have provided the richest single source of information about magnetic reversals. The age range of ocean floor that preserves a record of geomagnetic reversals is from Middle Jurassic to the present. The main reason for the high fidelity of the marine magnetic record is the remarkable continuity of the geologic processes by which new crust is formed along mid-ocean ridges.

The normal and reversed polarity intervals recorded on the ocean floor produce **marine magnetic anomalies** in the form of peaks and troughs that appear on magnetic profiles. Some noise is generally present on these profiles in the form of spurious small anomalies produced by seamounts and other geologic irregularities. Moreover, gaps and duplications are commonly present on the profiles because of the jumping of ridges to new positions. In addition, oceanic plates change velocity from time to time, so that marine anomalies must be calibrated using isotopic dating and biostratigraphy. However, despite problems of noise, gaps and non-linearity, the overall history of geomagnetic polarity comes through more clearly in marine magnetic profiles than in any other type of geologic record. Deep-sea cores and other sedimentary sections occasionally give details of polarity changes for small intervals of time.

6.3.2   Resolving power

Although no single magnetic profile provides a perfect record of reversals it has been possible, by comparing profiles from different parts of the world, to identify those anomalies that appear consistently on most high-quality profiles, and in this way to determine which anomalies are due to geologic noise and which anomalies correspond to the history of geomagnetic reversals. All the polarity chrons ($\tau > 0.1$ Ma) in the marine anomaly record from Middle Jurassic onwards have probably now been identified, as have many (but not all) of the subchrons ($\tau < 0.1$ Ma).

The minimum polarity interval length that can be resolved on individual profiles depends upon several factors. Interpretation of a profile begins with the use of a geophysical model, the most common model being a crustal layer made up of blocks that are alternately reversely and normally magnetized. The various mathematical procedures for finding the optimum set of blocks to fit a given magnetic profile all yield a sequence of widths, w_i, for the blocks. These widths, which in stratigraphic terms comprise a set of magnetic polarity zones, are related to the durations, τ_i, of the corresponding polarity chrons and subchrons by

$$\tau_i = w_i/v$$

where v is the half velocity of ocean-floor spreading at the time the oceanic crust formed. The length of the shortest detectable polarity interval depends upon the rate of spreading and magnetic latitude, the length of the shortest block that can be detected on a magnetic profile and the length scale of irregularities in the geologic processes by which new ocean floor forms. At water depths of 3 km the last two lengths are typically about 1 km. For very rapid spreading at a half velocity of 50 mm/yr (= 50 km/Ma), this corresponds to a minimum detectable polarity length of 0.02 Ma. This resolution appears to have been achieved for those parts of the reversal time scale for which high quality profiles are available over high-latitude ocean floor where spreading was rapid. In a few ideal situations, polarity intervals as short as 0.01 Ma appear to have been captured. Elsewhere the cut-off of observable polarity intervals appears to range from 0.02 to about 0.10 Ma, depending on the rate of ocean-floor spreading and the regularity of the spreading process.

6.3.3   Names and numbers of polarity chrons

Two systems for labelling polarity chrons are widely used. The first is the set of names (Brunhes, Matuyama, Gauss, Gilbert) used for the isotopically dated part of the reversal time scale (Figure 6.5). These have been the standard chron names used for global correlation in Pliocene and Pleistocene stratigraphy for more than a decade and their continued use as informal units is recommended by the IUGS/IAGA Subcommission (Anon. 1979). The chrons were numbered so that Brunhes is chron 1, Matuyama is chron 2, and so on. Older chrons are defined by being predominantly normally or predominantly reversely magnetized intervals in sediment cores and are numbered accordingly to at least chron 11.

Quite independently, a second numbering system was established when marine geophysicists informally gave numbers to 32 of the most prominent positive anomaly peaks appearing on magnetic profiles over ocean basins. The numbers begin with 1 at presently spreading mid-oceanic ridges, where new oceanic crust is being generated (Pitman, Herron & Heirtzler 1968). These numbers were then associated with the normally magnetized blocks used to model the positive anomalies (Le Pichon & Heirtzler 1968). In effect, the numbers that had been assigned to anomaly peaks had now become the informal names of magnetic polarity zones. The time spans of these positive anomalies were also referred to as chrons. The initial set of 32 numbered anomalies provided labels for only a fraction of the known chrons, two-thirds of the time covered by the numbering system lying outside any numbered chron. With the growing use of the marine anomaly sequence as a standard for global stratigraphic correlation, several workers have found it useful to label additional chrons representing the smaller positive anomalies in ocean-floor magnetic profiles by adding letters, decimals and prime marks to the original set of numbers (LaBrecque, Kent & Cande 1977, Ness *et al.* 1980).

In GTS 82 we used a composite of the numbering systems employed for ocean-floor anomalies, extending them slightly in order to provide numbers for all chrons in a manner that is consistent with prior usage and amenable to change in the course of future discoveries of subchrons. Our approach in extending earlier numbering systems is similar to that used in a library when new books are added. Subdivisions of presently numbered chrons such as 5A were described by additional decimal numbers, e.g. 5A.1, 5A.2. Unlabelled chrons were described by adding letters to the next youngest numbered chron, e.g. 5A and 5B follow 5 and 5AA and 5AB follow 5A. Finally, subchrons ($\tau < 0.1$ Ma) were labelled with the number of the chron in which they occur followed by '–1', '–2' etc., the numbers increasing in order of increasing age for presently known subchrons and in order of discovery for subchrons yet to be discovered. These logical rules may be confusing to the non-specialist.

Hailwood (1989) has proposed a more systematic magnetic chron numbering scheme in which the chrons are prefixed by S, for systematized. The chrons are renumbered from S1N to S50 to eliminate C2A, C2AN, C3A, etc., in the present numbering scheme. This simpler and less confusing nomenclature has yet to be adopted.

To avoid another source of confusion – between the chron numbering scheme based on magnetic properties of Neogene sediment cores and the ocean-floor spreading chrons – LaBrecque *et al.* (1983) suggested prefixing the ocean-floor spreading chrons by 'C'. Thus the time interval represented by ocean-floor magnetic anomaly 3 becomes known as C3: it

corresponds to only the middle of the Gilbert chron, or the middle of chron 4 in the scheme devised for sediment cores. In GTS 82 previously unlabelled reversed chrons were given the number of the next youngest normal chron with the letter 'r' appended, but no suffix was given to the normal chron. LaBrecque *et al.* (1983) also recommended referring to normally magnetized ocean-floor chrons by the suffix N; reversely magnetized by the suffix R. We adopt these conventions in this work: they are the only significant changes to be made to the labelling schemes in GTS 82. This convention would allow the whole of a chron, including its normal and reversely magnetized components, to be referred to simply as C12, for example, instead of C12N–C12R.

An alternative scheme for numbering events within a chron has been suggested by LaBrecque *et al.* (1983). Here the location of a particular event, such as an excursion, is defined by the name of the chron to which the event is correlated, followed by a decimal number representing the position of the event within the chron and terminated with N or R, representing the polarity at or before the event. For example, if the Cretaceous/Tertiary boundary were placed three-quarters of the way down from the level where the younger end of C29N had been recognized in sediments to the level where the older limit of C29R could be identified, and the polarity at the boundary were reversed, then its position could be described as C29.75R. An alternative, which might be more useful, would be to give its fractional position within the normal or reversed chron, rather than within the entire chron. Neither method has been formally adopted here, though the second method has been used to estimate the precise age of biostratigraphic boundaries that have been located within a chron (e.g. Table 6.6).

As in GTS 82 we use in this work the convention that the older end of a chron, for example C29N, is referred to as C29N(o), the (o) standing for 'older boundary'. The younger limit of C29N is referred to as C29N(y), the (y) standing for 'younger boundary'. Coming forward in time chrons form pairs of predominantly positive polarity that succeed predominantly negative polarity. Thus C29N(y) is the younger end of chron 29 and could be unambiguously written as C29(y). However, until the convention suggested above becomes general, the older end of chron 29 could be taken to mean either C29N(o) or C29R(o). For this reason we use the longer form when referring to the older end of a chron. The modified decimal notation suggested above would make C29N(o) equal to C29.0R, and C29N(y) equal to C29.0N. C29R(o) would become C30.0N.

Ocean floor that formed from Aptian to Santonian time is known as the 'Cretaceous quiet zone' because of the absence of globally traceable magnetic anomalies over oceanic crust of this age. The generally accepted explanation is that during this time the polarity of the Earth's field was normal except possibly for a few short reversed polarity intervals, the durations of which were probably less than 0.03 Ma and the ages of which are somewhat uncertain (for review see Lowrie, Channell & Alvarez 1980).

The polarity chrons of the pre-Aptian sequence are generally described by the designations 'M0' through 'M29' assigned to marine anomalies in order of increasing age by Larson & Pitman (1972), Larson & Hilde (1975) and Cande,

Larson & LaBrecque (1978). The prefix 'M' stands for Mesozoic. LaBrecque *et al.* (1983) proposed adding 'C' as a prefix to the M-anomalies. While recognizing the uniformity of such a scheme, we suggest it is unnecessary: the prefix 'C' was introduced to resolve an ambiguity between two Cenozoic chron numbering schemes: no such ambiguity exists for the M-anomalies.

The present numbering system of the M-sequence is highly confusing to those who are not specialists. In contrast with the practice of numbering **normal** chrons in the Cretaceous–Tertiary–Quaternary reversal sequence, **reversed** chrons were usually numbered in the Jurassic–Cretaceous reversal sequence! An additional problem is caused by the fact that, alone among the M-sequence, chrons M2 and M4 have **normal** polarity.

In GTS 82 Allan Cox tried to dispel some of the confusion by extending the numbering system adopted for the younger anomalies to all the chrons in the M-sequence (Table 6.10 and Figure 6.9). Previously unnumbered normal chrons were given the number of the older reversed chron with 'N' appended, so that M16N is younger and stratigraphically above M16. As in the Cretaceous–Tertiary–Quaternary reversal sequence, normal follows reversed – coming forwards in time – in a doublet with the same number. For consistency and clarity, we extend here the convention adopted above for the reversed chrons, i.e. we refer to M0 as M0R since it has predominantly reversed polarity. The only exceptions are M2N–M3R and M4N–M5R, both of which are doublets. Logically the number of one of the members of each of these doublets should be dropped, leaving, say M2N–M2R and M4N–M4R, but we do not adopt such a radical suggestion here. There are precedents for such a course inasmuch as C14 is no longer a recognized chron, though for different reasons.

6.3.4 Calibration by direct isotopic dating

The original marine anomaly reversal time scale of Heirtzler *et al.* (1968) was based on the assumption that ocean-floor spreading has occurred at a constant rate in part of the south Atlantic (see Section 6.4.1). Two methods have been used to calibrate this scale. The first is to drill the ocean floor beneath a well-defined magnetic anomaly and determine the K–Ar date of the basalt layer or the biostratigraphic age of the sediment immediately overlying the basalt layer. Since the hiatus between the time of basalt extrusion and the time of first sedimentation on the ocean floor is generally no more than a few million years, the sediment age provides a reasonably accurate estimate of the age of the polarity zone unless, of course, the basalt encountered is a sill.

Young polarity zones near spreading centres have been dated by matching them directly with the polarity chrons determined by isotopic dating (Vine 1966, Mankinen & Dalrymple 1979).

6.3.5 Calibration by indirect biostratigraphic correlation

The second method, which is less direct but potentially more precise, is based on the only property that distinguishes polarity zones from each other, namely their lengths. These lengths vary widely and randomly from one zone to the next.

Because of this variation, a sequence of 4 to 6 polarity intervals comprises a magnetic fingerprint or signature which can be correlated with corresponding signatures elsewhere. The widths of the polarity zones in the oceanic crust are vastly different from the thicknesses of the zones in sedimentary sections. However, if rates of ocean-floor spreading and sediment deposition do not vary greatly over intervals of the order of 10 Ma, the ratios of the lengths of polarity zones will be about the same and the magnetic fingerprints will be recognizably similar.

The field of biostratigraphic–magnetostratigraphic correlation is so large that space permits mention of only a few recent studies. These include Ogg *et al.* (1984) [Late Jurassic]; Berggren, Kent & Flynn (1985), Berggren, Kent & Van Couvering (1985) [Cenozoic]; Kent & Gradstein (1985) [Cretaceous and Jurassic]; Kent & Gradstein (1986) [Jurassic to Recent]; Bralower (1987) [Early Cretaceous]; Channell & Grandesso (1987) [Cretaceous and Jurassic]; Channell, Bralower & Grandesso (1987) [Cretaceous and Jurassic]; Bralower, Monechi & Thiestein (1989) [Cretaceous and Jurassic]; Ogg & Steiner (in press) [Early Cretaceous and Late Jurassic] and Steiner & Ogg (in press) [Middle and Early Jurassic]. These publications include references to older literature.

The Cretaceous and Early Cenozoic marine anomaly scale has been correlated with marine biostratigraphy in the Umbrian Apennines in Italy (Alvarez & Lowrie 1978, Lowrie *et al.* 1980, Lowrie & Alvarez 1981). In the early stages of this research, polarity zones were correlated regionally using the local terms such as 'Gubbio A' and 'Gubbio B'. As these polarity zones have become more firmly correlated with the marine anomaly sequence, the polarity zones in the Apennines have been correlated both locally and globally using chron numbers (Lowrie *et al.* 1980). Recent work suggests that some of the earlier correlations may be erroneous (Bralower 1987, Channell *et al.* 1987). The Late Cretaceous and Early Cenozoic marine anomaly sequence has also been correlated with land mammal stages in the southwestern United States by Butler *et al.* (1977). Correlations have been proposed between the magnetic anomaly sequence, isotopic ages and the Eocene land mammal stages of western Wyoming in the western United States (Flynn 1986).

The growing use of chron numbers derived from marine magnetic anomalies rather than names derived from magnetic stratotypes, reflects the limited usefulness of the stratotype concept in magnetostratigraphy. There are two reasons why this has turned out to be the case.

The first is that although the ratio of the lengths of polarity zones is the single property upon which identification and correlation of polarity chrons are based, because of variations in rates of deposition, these ratios vary considerably between different sedimentary sections, even in pelagic sections that are very good magnetic recorders (e.g. figure 1 of Lowrie & Alvarez 1981). Adopting any one sedimentary section as a magnetic stratotype would generally yield as an international standard a set of ratios less accurate than the ratios obtained from a composite based mainly on marine anomaly profiles.

The second reason is that because of hiatuses in deposition, some subchrons are commonly absent from the sedimentary record. Analogy with chronostratigraphic practice might suggest that a section should be adopted as a stratotype. But a single section is unlikely to record the polarity fine structure that is potentially very important for correlation and would be unsuitable as an international standard. The analogy is not quite appropriate inasmuch as standard sections in chronostratigraphy define boundaries rather than intervals. For these reasons the most generally accepted time scale has evolved as a composite based on the global consistency of many different magnetozones, most of which are found in oceanic crust. The importance of the marine anomaly record may be noted from the observation that it is only recently that it has proved possible to use stratotypes to deduce a detailed reversal time scale for Middle and Early Jurassic time (Steiner & Ogg in press), just prior to the beginning of the marine anomaly record.

Older polarity zones are currently being dated by correlating them with sedimentary or volcanic sections that also have well-known biostratigraphic ages (Butler et al. 1977, Lowrie & Alvarez 1981) and isotopic ages (e.g. McDougall et al. 1976, Montanari et al. 1985). Isotopic ages for the entire marine anomaly sequence are summarized in Haq, Hardenbohl & Vail (1989).

6.3.6 Calibration problems

Two extreme approaches can be made to the problem of calibrating the ocean-floor magnetic anomalies. In the first we would ideally take a perfect magnetic anomaly profile from one ocean basin that had spread uniformly throughout its history, stretching it between a few selected points. The Cenozoic to Late Cretaceous time scale of Heirtzler et al. (1968) [=HDHPL68] assumes that the E–W Vema–20 profile at about 28 degrees S in the south Atlantic approximates to this ideal. The time scale of LaBrecque et al. (1977) [=LKC77] is the HDHPL68 scale modified to take into account anomalies not recognized in the HDHPL68 scale, as well as making some other changes. It uses two isotopic control points: one at the older end of the Gauss chron C2A(o), taken as 3.32 Ma; the other at C29N(o), whose age was taken as 64.9 Ma. The value of 64.9 Ma was obtained from the position of C29R relative to the Cretaceous/Tertiary boundary, which was assigned an age of 65 Ma. When standardized to the new constants these ages become 3.4 Ma and 66.7 Ma. This approach assumes that spreading rates are constant and that the errors in individual calibration points are relatively small.

The second approach uses as many isotopic control points as possible. The problem with this second approach is that every time a new control point is used, a new degree of freedom is introduced. The fit of the anomalies becomes better but the internal consistency of different data sets can no longer be checked. For example, the Cenozoic to Late Cretaceous time scale of Ness et al. (1980) [=NLC80] probably has too many control points, though it is readily traceable to its original sources. GTS 82 was a modification of

the time scale of Lowrie & Alvarez (1981), which used 11 biostratigraphic control points to stretch linearly the reversal time scale of LKC77. When numerous control points are available it is necessary to adopt some form of smoothing procedure. In GTS 82 the ages of the stage boundaries were adjusted within the limits of chronogram or estimated error so as to reduce changes in spreading rate.

When the ages of the control points of Lowrie & Alvarez (1981) were plotted against the corresponding ages of LKC77 several deviations were found. Since the latter corresponds to a model of ocean-floor spreading close to the original HDHPL68 time scale, in turn believed to reflect an approximately constant spreading rate, changes in slope of the line connecting calibration points imply changes in the velocity of south Atlantic spreading.

In GTS 82 we attributed these large apparent changes in the velocity of spreading to artifacts resulting from uncertainties in the isotopic ages of intra-epoch boundaries within the Eocene and Paleocene, which were not well constrained by the available isotopic data [figure 5 of Hardenbol & Berggren (1978)]. In GTS 82 we noted that an age of 57–58 Ma for the Eocene–Paleocene boundary would greatly reduce the apparent speed-up in plate motions during the Eocene and their apparent slow-down during the Paleocene that were implied by the choice of 54.9 Ma.

That there may not be such large changes in spreading velocity as was implied by the GTS 82 scale is also suggested by recent substantial revisions of the previous stratigraphic correlations and nomenclature in the Paleocene/Eocene interval. Berggren et al. (1985a, 1985b) sought to restrict the Thanetian stage so as to insert a new post-Danian pre-'Thanetian' division for the time interval of a few Ma hitherto regarded as early Thanetian (see Section 3.19.3). They also note that the Ypresian/Lutetian boundary has been substantially revised and propose an age of 57.8 Ma for the Eocene/Paleocene boundary. Thus it could be argued that while the numerical ages used for the Eocene/Paleocene and other chronograms may be analytically correct, they may well have been assigned to different stratigraphic positions. After reviewing the evidence we do in fact conclude that there have been changes in the south Atlantic spreading rates. This important problem is discussed in detail below.

The recent Cenozoic and Late Cretaceous time scales of Berggren, Kent & Flynn (1985) [=BKF85] and Berggren, Kent & Van Couvering (1985) [=BKV85] use a slightly different technique to calibrate the time scale (Figure 6.3). The LKC77 profile is taken as the reference profile and divided into three segments, each of which is a linear fit to the data. Three calibration points on the first segment are used to determine the slope of the first line segment: the present day, the older end of anomaly C2AN dated at 3.4 Ma, and the younger end of anomaly C5 (= C5N(y)) dated at 8.87 Ma. The best linear fit through these calibration points gives anomaly C5N(o) an age of 10.42 Ma. With the origin as the first control point and 10.42 as the second control point, the ages of all anomalies between anomaly 1 (the present day) and C5N(o) are found by linear interpolation of the LKC77 profile between 0 and 10.42 Ma. Stretching does not alter the value

Figure 6.3. The Cenozoic time scale of Berggren, Kent, Flynn & Van Couvering (1985) redrawn with modifications from
Geological Society of America Bulletin **96, Figure 1, 1408. The horizontal axis shows the ages assigned to the marine magnetic**
reversal sequence of LaBrecque *et al.* **(1977) using the standardized decay constants (=LKC77). The vertical axis shows the**
isotopic age in Ma. The standardized LKC77 profile is divided into three segments, whose terminations are the origin and the
three points labelled with large type. Three calibration points are used for the first line segment: the present-day, the older end of
anomaly C2AN (=C2AN(o)) [3.4 Ma], and the younger end of anomaly C5 (=C5N(y)) [8.87 Ma]. The best linear fit through
these calibration points gives anomaly C5N(o) an age of 10.42 Ma. With the origin as the first control point and 10.42 as the
second control point, the ages of all anomalies between anomaly 1 (the present-day) and C5N(o) are found by linear interpolation.
Stretching does not alter the value of C2AN(o) at 3.4, but does modify the 8.87 age of C5N(y) to 8.92 Ma.

      The second linear segment is defined by the second control point of the first line segment C5N(o) [10.42 Ma]; C12N(y) at
32.4 Ma; C13N(y) at 34.6 Ma and C21N(y) at 49.5 Ma. The best-fit line through C5N(o)gives an estimated age of 56.14 Ma for
C24N(o). All anomalies between C5N(o) and C24N(o) are linearly stretched according to their relative spacing on the LKC77
profile. The stretching modifies the values of the three points used to determine the slope of this segment to 32.46, 35.29 and 48.75
Ma respectively (Table 2, Berggren *et al.* **1985a).**

      The third segment is defined by C24N(o) and C34N(y), dated at 84.0 Ma. All anomalies between C24N(o) and C34N(y) are
linearly stretched according to their relative spacing on the LKC77 profile.

      The dashed line is the LKC77 calibration line.

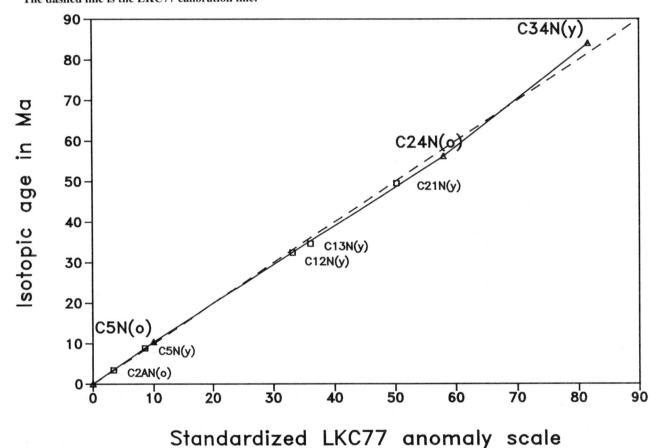

of C2A(o) at 3.4, but does modify the 8.87 age of C5N(y) to
8.92 Ma.

    The beginning of the second linear segment is defined by
the second control point of the first line segment C5N(o).
Three other points are chosen to determine the slope of the
second line segment: C12N(y) at 32.4 Ma, C13N(y) at 34.6 Ma
and C21N(y) at 49.5 Ma. The best-fit line through C5N(o) and
the three other points is extrapolated to give an estimated age
of 56.14 Ma for C24N(o), which is taken as the second control
point for the second line segment. All anomalies between
C5N(o) and C24N(o) are linearly stretched according to their
relative spacing on the LKC77 profile. The stretching modifies
the values of the three points used to determine the slope of

this segment to 32.46, 35.29 and 48.75 Ma respectively (Table
2, Berggren *et al.*1985b).

    Finally, the third segment is defined as beginning at
C24N(o), which becomes the first control point for the
segment. The line ends at the second control point, taken as
C34N(y), dated as 84.0 Ma. All anomalies between C24N(o)
and C34N(y) are linearly stretched according to their relative
positions on the LKC77 profile.

    In essence, the Berggren *et al.* time scale is a piecewise
linear fit to the data. There are two problems with this
method: (1) it is not clear what the basis is for choosing the
control points; (2) only a small fraction of the available
isotopic data are used to calibrate the time scale. The second

problem gives rise to an unstable time scale because a recalibration of the age of any one of the points used to control the slopes of the segments or the control points may significantly alter the scale. An alternative way of calibrating the time scale that uses all the available data is described below (Sections 6.4 and 6.5).

6.4   This work: the GTS 89 time scale 0 to 83 Ma
6.4.1   Relative spacings of anomalies

There are two steps in setting up a magnetic anomaly time scale. The first is to construct a magnetic anomaly sequence from ocean-floor profiles giving the relative spacing of the anomalies; the second is to calibrate the sequence in time at particular points, or by particular line segments, and interpolate the ages of the anomalies between the calibration points or on each line segment.

All existing time scales in the time range 0 to 83 Ma start with the HDHPL68 profile and modify it in some way. This profile was determined on the south Atlantic ridge, which spread relatively slowly, at 20 mm yr^{-1}. Surveys over fast-spreading ridges commonly reveal a fine structure that cannot be resolved by sea surface measurements over slow spreading ridges. Thus most of the changes that have been made to HDHPL68 since it was published are the substitution of well-defined segments showing a more detailed reversal pattern for the equivalent part of the original reversal sequence. The relative positions of the anomalies within a substituted segment can readily be found by linearly stretching (or compressing) the anomaly sequence between two control points marking the ends of the substituted segment and the corresponding points on HDHPL68. The values associated with the control points are the relative spacings of the anomalies on HDHPL68 and on the new magnetic profile.

Let the control points at the ends of a substituted segment on the new scale be **new1** and **new2**, and the corresponding points on HDHPL68 be **old1** and **old2**. Let **p** be the point on the new scale to be inserted into a modified HDHPL68 with a value **mod68**, then

$$\mathbf{mod68} = \mathbf{old1} + (\mathbf{p} - \mathbf{new1}) \times (\mathbf{old2} - \mathbf{old1})/(\mathbf{new2} - \mathbf{new1})$$

For example, if the new control points are **new1** = 10 and **new2** = 20, and the corresponding values of the control points on the old scale are **old1** = 11 and **old2** = 21, and **p** is the point on the new scale with a value of 12, then

$$\mathbf{mod68} = 11 + (12 - 10) \times (21 - 11)/(20 - 10) = 11 + 2 = 13$$

Instead of using a previously modified HDHPL68 profile as we did in GTS 82, we have returned to the original HDHPL68 magnetic anomaly sequence (Table 6.2), modifying it as set out in Table 6.1. The modified sequence is referred to as H68MOD.

Note that the last interval from 69.95 to 85.29 represents the incorporation of north Pacific data as recommended by LaBrecque *et al.* (1977). They assumed that spreading in the north Pacific was at a uniform rate from anomalies 23 to 34. They then used the control points on the HDHPL68 scale for C23N(y) at 58.04 and for C29N(o) at 69.44 to modify the HDHPL68 scale in the interval 69.44 to 108.19, using control points on their own scale at 54.29 and 64.9. We have reversed

Table 6.1. *Modifications to the relative spacing of anomalies on the Heirtzler et al. (1968) time scale*

Interval* in H68MOD	new1	new2	old1	old2	Reference†
0–5.01	0	3.32	0	3.35	KHMP75
5.01–7.91		Unchanged			HDHPL68
7.91–8.79	7.39	8.71	7.91	8.79	B74
8.79–20.19	8.71	20.19	8.79	20.19	B74
20.19–38.26		Unchanged			HDHPL68
38.68–39.47		Remove anomaly 14			LKC77
39.77–58.04		Unchanged			HDHPL68
58.29–58.33		Scaled into HDHPL68 by BC72			BC72
58.94		Unchanged			HDHPL68
59.22–59.27		Scaled into HDHPL68 by BC72			BC72
59.43–69.44		Unchanged			HDHPL68
69.95–85.29	58.04	69.44	54.29	64.9	LKC77

*The last interval is taken from LKC77 and scaled back into HDHPL68.
†B74 = Blakely (1974); BC72 = Blakely & Cox (1972); HDHPL68 = Heirtzler *et al.* (1968); KHMP75 = Klitgord *et al.* (1975); LKC77 = LaBrecque *et al.* (1977).

their procedure and calculated the modified HDHPL68 scale from LKC77. The differences between the original HDHPL68 scale and the modified scale are small (Table 6.2).

The revised values of the anomaly boundaries for HDHPL68 are given in Table 6.2, which will be referred to as H68MOD. It is important to note that Table 6.2 gives the revised **relative spacing** of the anomalies, rather than their ages. Although Heirtzler *et al.* (1968) gave these numbers a time connotation by assigning a numerical age of 3.35 Ma to C2AN(o), we could, for example, have given the older end of C29N(o), i.e. the anomaly boundary closest to the Cretaceous/Tertiary boundary, an arbitrary value of 1000 and rescaled all the anomaly boundaries accordingly.

We have compared the relative spacings of the anomalies in H68MOD with the relative spacings between calibration points in LKC77 and NLC80. The only significant differences between H68MOD and LKC77 are for anomalies C3.3N(y) to C5N(y) inclusive, where they approach 5% at the younger end of this interval. Elsewhere the differences are at most 0.6% and generally less than 0.1%. The differences reflect the fact that we have reverted to the original HDHPL68 relative spacings for C3.3N(y) to C4AN(y), and have tied the initial part of the Blakely (1974) time scale from C4AN(y) to C5N(y) into HDHPL68 using different calibration points (Table 6.1). Similar differences exist between H68MOD and NLC80, but extend to C5N(o) on the Blakely (1974) time scale. In short, all three time scales use essentially the same relative anomaly spacings between calibration points except for anomalies C3.3N(y) to C5N. There still appear to be unresolved differences in the relative magnetic anomaly spacing in this interval.

Table 6.2. *Relative spacings of magnetic anomalies in Heirtzler et al.'s (1968) ocean-floor magnetic anomaly time scale (original) and its modification used here*

Original			Modification		
Anomaly[1]	Normal polarity interval		Anomaly1	Normal polarity interval[2]	
1	0.00	0.69	1	0.00	**0.71**
1R–1	0.89	0.93	1R–1	**0.90**	**0.96**
2	1.78	1.93	2	**1.63**	**1.85**
2R–1	Not recognized		2R–1	See Table 6.4	
2A	2.48	2.93	2A	**2.42**	**2.87**
2A	Not recognized		2A	**2.94**	**3.03**
2A	3.06	3.35	2A	**3.13**	3.35
3.1	4.04	4.22	3.1	**3.82**	**3.93**
3.2	4.35	4.53	3.2	**4.06**	**4.20**
3.2R–1	4.66	4.77	3.2R–1	**4.35**	**4.42**
3.3	4.81	5.01	3.3	**4.72***	5.01
3A	5.61	5.88	3A	5.61	5.88
3A	5.96	6.24	3A	5.96	6.24
3B	6.57	6.70	3B	6.57	6.70
4	6.91	7.00	4	6.91	7.00
4	7.07	7.46	4	7.07	7.46
4	7.51	7.55	4	7.51	7.55
4A	7.91	8.28	4A	7.91	**8.18**
4AR–1	8.37	8.51	4AR–1	**8.36**	**8.42**
4AR–2	Fine structure of 4 and 5 very different		4AR–2	**8.66**	**8.71**
			5	**8.79**	**8.96**
			5	**9.01**	**9.35**
			5	**9.36**	**9.66**
			5	**9.69**	**9.99**
			5	**10.01**	**10.29**
			5R–1	**10.42**	**10.48**
			5R–2	**10.96**	**11.03**
			5A.1	**11.55**	**11.74**
	8.79	9.94	5A.2	**11.88**	**12.17**
	10.77	11.14	5AR–1	**12.54**	**12.58**
	11.72	11.85	5AR–2	**12.67**	**12.73**
	11.93	12.43	5AA	**12.96**	**13.16**
	12.72	13.09	5AB	**13.38**	**13.66**
	13.28	13.71	5AC	**13.92**	**14.36**
	13.96	14.28	5AD	**14.48**	**14.99**
	14.51	14.82	5B.1	**15.21**	**15.32**
	14.98	15.45	5B.2	**15.50**	**15.65**
	15.71	16.00	5C	**16.71**	**17.04**
	16.03	16.41	5C	**17.08**	**17.27**
	17.33	17.80	5C	**17.35**	**17.55**
	17.83	18.02	5D	**18.20**	**18.57**
	18.91	19.26	5DR–1	**18.80**	**18.83**
	19.62	19.96	5E	**19.30**	**19.88**
6	20.19	21.31	6	20.19	21.31
6A.1	21.65	21.91	6A.1	21.65	21.91
6A.2	22.17	22.64	6A.2	22.17	22.64
6AA	22.90	23.08	6AA	22.90	23.08
6AAR–1	23.29	23.40	6AAR–1	23.29	23.40
6B	23.63	24.07	6B	23.63	24.07
6C.1	24.41	24.59	6C.1	24.41	24.59
6C.2	24.82	24.97	6C.2	24.82	24.97
6C.3	25.25	25.43	6C.3	25.25	25.43
7	26.86	26.98	7	26.86	26.98
7	27.05	27.37	7	27.05	27.37
7A	27.83	28.03	7A	27.83	28.03
8	28.35	28.44	8	28.35	28.44
8	28.52	29.33	8	28.52	29.33
9	29.78	30.42	9	29.78	30.42
9	30.48	30.93	9	30.48	30.93
10	31.50	31.84	10	31.50	31.84
10	31.90	32.17	10	31.90	32.17
11	33.15	33.55	11	33.15	33.55

Table 6.2. *continued*

Original			Modification		
Anomaly[1]	Normal polarity interval		Anomaly1	Normal polarity interval[2]	
11	33.61	34.07	11	33.61	34.07
12	34.52	35.00	12	34.52	35.00
13	37.61	37.82	13	37.61	37.82
13	37.89	38.26	13	37.89	38.26
14	38.68	38.77		Anomaly 14 no	
14	38.83	38.92		longer recognized	
14	39.03	39.11		Anomaly 14 no	
14	39.42	39.47		longer recognized	
15	39.77	40.00	15	39.77	40.00
15	40.03	40.25	15	40.03	40.25
15A	40.71	40.97	15A	40.71	40.97
16	41.15	41.46	16	41.15	41.46
16	41.52	41.96	16	41.52	41.96
17	42.28	43.26	17	42.28	43.26
17	43.34	43.56	17	43.34	43.56
17	43.64	44.01	17	43.64	44.01
18	44.21	44.69	18	44.21	44.69
18	44.77	45.24	18	44.77	45.24
18	45.32	45.79	18	45.32	45.79
19	46.76	47.26	19	46.76	47.26
20	47.91	49.58	20	47.91	49.58
21	52.41	54.16	21	52.41	54.16
22	55.92	56.66	22	55.92	56.66
23	58.04	Not recognized	23	58.04	**58.29**
23	Not recognized	58.94	23	**58.33**	58.94
23R–1	Not recognized		23R–1	**59.22**	**59.27**
24	59.43	59.69	24	59.43	59.69
24	60.01	60.53	24	60.01	60.53
25	62.75	63.28	25	62.75	63.28
26	64.14	64.62	26	64.14	64.62
27	66.65	67.10	27	66.65	67.10
28	67.77	68.51	28	67.77	68.51
29	68.84	69.44	29	68.84	69.44
30	69.93	71.12	30	**69.95**	**71.44**
31	71.22	72.11	31	**71.52**	**72.31**
32.1	74.17	74.30	32.1	**74.06**	**74.31**
32.2	74.64	76.33	32.2	**74.54**	**75.99**
32R–1		Not given	32R–1	**76.36**	**76.40**
33		Not given	33	**76.66**	**81.88**
34		Not given	34	**85.29**	**115.95**

Notes:

1. Since all the anomaly numbers refer to normal polarity intervals in the ocean-floor magnetic anomaly scale, the prefix C and the suffix N have been omitted. Thus 24 is C24N.
2. Bold type shows where numerical values of the modified scale differ from the HDHPL68 scale.
* This is a compromise value necessitated by splicing HDHPL68 to KHMP75.

Figure 6.4. The distribution in time of ten short-lived reversals in the past 1.2 Ma. Based on Champion *et al.* (1988). Note that estimates for the Brunhes/Matuyama chron boundary vary from 730 to 790 ka (see text). 730 ka has been used here.

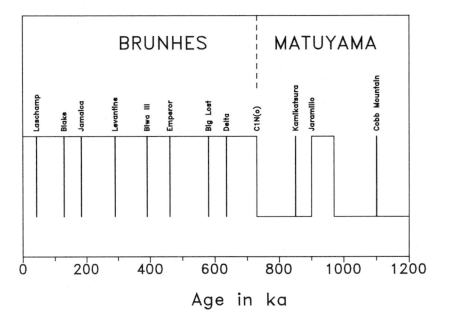

6.4.2 Direct isotopic dating of the magnetic anomalies

As noted above, there are two methods by which the magnetic anomaly sequence can be calibrated in time. In this work we use results from both direct dating of the magnetic reversals and from chronogram boundaries that can be tied to the reversal time scale. Dated anomalies predominate in the time interval 0–15 Ma, but chronogram boundaries become increasingly important thereafter and are the only data available for numerical dating of the Mesozoic anomalies.

The polarity reversal scale 0 to 1.2 Ma. The main chron boundary in the interval 0 to 1.2 Ma is the C1N(o) boundary marking the polarity change from C1R to C1N. C1N is known as the Brunhes polarity chron; C1R is the younger end of the Matuyama chron (Figures 6.4 and 6.5). The CIN (o) boundary was dated isotopically at 0.73 Ma (Mankinen & Dalrymple 1979). However, based on correlations with the Milankovitch cycles, Johnson (1982) gave an age of 788 ka, i.e. 0.79 Ma. More recently, the normally magnetized Bishop Tuff of California and adjacent regions has been dated at 738 ± 8 ka (Izett, Obradovitch & Mehnert 1988). The tuff lies stratigraphically a metre or so above the chron boundary, suggesting the boundary itself is about 0.75 Ma old (Izett *et al.* 1988).

Champion *et al.* (1988) show that there is good evidence for ten short-lived subchrons within the Brunhes chron and the upper part of the Matuyama chron (Table 6.3), in addition to the well-established Jaramillo subchron (C1R–1), dated at 0.90 to 0.97 Ma (Mankinen & Dalrymple 1979). These data are plotted in Figure 6.4.

The polarity reversal scale 1.2 to 3.4 Ma. The current estimates of the ages of chron and subchron boundaries determined from isotopic dating (Mankinen & Dalrymple 1979) for the period 1.2 to 3.4 Ma, i.e. from the Cobb Mountain subchron to the younger end of the Gilbert chron, are listed in Table 6.4 and shown in Figure 6.5. All the ages determined by Mankinen and Dalyrymple (1979) were

estimated using chronograms similar to those described in Chapter 4 (Cox & Dalrymple 1967, Mankinen & Dalrymple 1979). Isotopic dating is the major source of the age estimates for short-lived events in the 0 to 3.4 Ma interval because these are difficult to resolve on marine magnetic profiles.

The polarity reversal scale 3.4 to 5 Ma. Mankinen & Dalrymple (1979) note that estimates of the ages and durations of the shorter chrons, Cochiti (C3.1N), Nunivak (C3.2N) and Thevra (C3.3N), and of the longer subchron Sidufjall (C3.2R–1N), in the time range 3.4 to 5 Ma, are not well constrained. Though values have been assigned by them to the chrons in this interval, partly from ocean-floor magnetic anomalies and deep-sea sediment cores, they differ significantly in places from those obtained by calibrating H68MOD against other well-defined calibration points. These values have not been used in this report.

Dated anomalies older than 5.0 Ma. There are major problems in the direct calibration of the magnetic anomaly time scale for rocks older than 5 Ma. While the anomaly age of ocean-floor rocks may be well known at the sea surface, obtaining a sample by dredging or drilling is difficult. Even if one is obtained, it is commonly altered. By contrast, samples abound on land but their position in the magnetic reversal sequence is commonly uncertain or ambiguous: there are no geologic environments where igneous activity or sedimentation takes place at a steady rate for periods of millions of years, uninterrupted by hiatuses.

A long-standing estimate of the age of C5N(y) is the standardized value of 8.87 Ma by Evans (1970). Two recent studies have attempted to calibrate the reversal time scale for rocks in the age range 9 to 14 Ma. The first, in Iceland (McDougall, Kristjansson & Saemundsson 1984), sampled about 1000 lavas for paleomagnetic measurements in an eastern and a western section and obtained about 70 K–Ar whole-rock dates. The eastern section can clearly be correlated with the ocean-floor anomaly sequence, but, as the authors note,

Table 6.3. *Dated magnetic reversals and short-lived magnetic events in age range 0 to 1.2 Ma*

Chron	Subchron	Isotopic age of mid-point (ka)	Comment
	C1N(y)	0	Present day
	Laschamp	42	Dated isotopically
	Blake	128	Dated isotopically
B	Jamaica (=Biwa I)	~182	Dated isotopically
R			
U	Levantine (=Biwa II)	~290	Found in sediments
N			
H	Biwa III	~390	Found in sediments
E			
S	[Emperor	~460	Found in sediments]
	Big Lost	580	Dated isotopically
	[Delta	~635	Found in sediments]
	Transition from C1N(o) to C1R(y)	730	Mankinen & Dalrymple (1979) dated isotopically
M	[C1N(o)	788	Johnson (1982) astronomical dating]
A			
T	Kamikatsura	850	Dated isotopically
U	Jaramillo CR1R–1(y)	900	Duration is 70 ka
Y	(Age of top)		Mankinen & Dalrymple (1979)
A			
M			
A	Cobb Mountain	1100	Dated isotopically

Notes:
  1. Except where noted, all data from Champion *et al.* (1988). Events that are not well established are enclosed in [ ].
  2. Some of the alternative names available are enclosed in ( ).
  3. Except for the Jaramillo subchron, the duration of each event is probably about 10 ka.
  4. The Emperor and Big Lost events may be the same event (D. Wilson, personal communication).

for reasons that are unclear, how to correlate the western section is not at all obvious. Based on a review of these data and earlier data from Iceland, McDougall *et al.* conclude that the most reliable date is of 9.64 Ma for C5N(y). They reject an earlier estimate of 10.3 Ma from Icelandic lavas (McDougall *et al.* 1976) which was used in GTS 82 for C5N(o), noting that two other independent estimates exist from Icelandic lavas. The first is 10.47 Ma (Saemundsson *et al.* 1980) and the second is their preferred value of 11.07 Ma.

The second study in the East African rift valley (Tauxe *et al.* 1985) used about 50 paleomagnetic pole positions and 9 sanidines in volcanogenic sediments to suggest an age of about 10 Ma for C5N(o). The reversal pattern of the African sequence is similar to that of the H68MOD except that C5R–1 is missing.

There is also a date of 10.3 Ma for C5N(o) from biotites in continental sequences in the western United States (McDowell, Wilson & Clark 1973).

We have listed all values for C5N(y) and C5N(o) in Table 6.5, except the East African estimate, for which a specific value is not given by the authors. Table 6.5 lists other isotopic ages that are available for calibrating the magnetic anomaly scale directly. These calibration ages have not been determined by chronograms.

Correlation of the dated sequences to the magnetic anomaly time scale is reasonably straightforward in Iceland, less so in marine sections, though the section at Gubbio (Italy) has been intensively studied and is in a marine pelagic facies. Correlation is least certain in continental sequences because of the abrupt discontinuities in continental sedimentation.

Figure 6.5. K–Ar dates of the major reversals in the past 4 Ma (Table 6.4), plotted against H68MOD values from Table 6.2. The regression line has an equation close to Age = 1.014*H68MOD Ma. The reversal pattern is represented as the rectilinear pattern at the top of the diagram. N and R are normally and reversely magnetized polarities. Note that the Emperor, Cobb Mountain, Reunion and 'X'-event have not been plotted on the regression line because they have not yet been unambiguously related to the H68MOD anomaly profile. Their positions in the reversal pattern have been obtained by using the equation of the regression line.

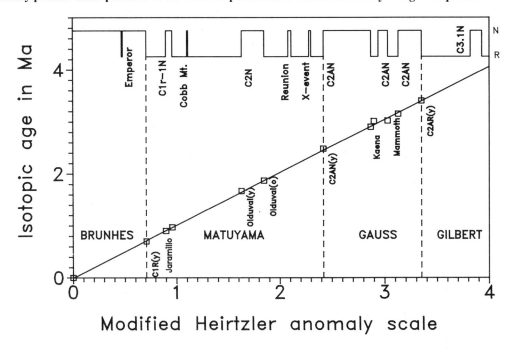

Table 6.4. *Dated magnetic reversals and short-lived magnetic events in age range 1.2 to 5 Ma*

Magnetic anomaly	Chron/subchron	Isotopic age (Ma)	References
C2N(y)	Olduvai young	1.67	Mankinen & Dalrymple 1979
C2N(o)	old	1.87	
C2R–1(y)	Reunion young	2.06*	Emilia & Heinrichs 1969
C2R–1(o)	old	2.09†	McDougall & Watkins 1973, Rea & Blakely 1975, Mankinen & Dalrymple 1979, Vine 1968
C2R–2(y)	'X' event young	2.31‡	Rea and Blakely 1975
C2R–2(o)	old	2.33	Heirtzler *et al.* 1968, Emilia & Heinrichs 1969, Mankinen & Dalrymple 1979
C2AN(y)	Matuyama/Gauss	2.48	Mankinen & Dalrymple 1979
C2A–1R(y)	Kaena young	2.92	Mankinen & Dalrymple 1979
C2A–1R(o)	old	3.01	
C2A–2R(y)	Mammoth young	3.05	Mankinen & Dalrymple 1979
C2A–2R(o)	old	3.15	
C2AR(y)	Gauss/Gilbert	3.40	Mankinen & Dalrymple 1979

Notes:

* Based on the bimodal distribution of K–Ar ages, Mankinen & Dalrymple (1979) proposed two Reunion events, but the 0.06 Ma duration of reversed polarity between the two subchrons proposed by Mankinen & Dalrymple (1979) is not consistent with marine magnetic anomalies (D. Wilson, personal communication). As a result the 'X' event has been named C2R–2 in Table 6.4.

† The age adopted is taken from Rea & Blakely (1975).

‡ Age has been standardized.

Table 6.5. *Dated magnetic reversals older than 5 Ma*

Magnetic anomaly	Material and method	Isotopic age (Ma)	References
*C5N(y)	K–Ar on lavas New Zealand	8.87	Evans 1970
(Used as calibration point for DNAG scale)			
*C5N(y)	K–Ar on lavas Iceland	9.64	McDougall *et al.* 1984
*C5N(o)	K–Ar on lavas Iceland	10.30	McDougall *et al.* 1976
*C5N(o)	K–Ar biotite (tuff) Utah, S Dakota	10.30	McDowell *et al.* 1973
*C5N(o)	K–Ar on lavas Iceland	10.47	Saemundsson *et al.* 1980
*C5N(o)	K–Ar on lavas Iceland	11.07	McDougall *et al.* 1984
C9N(o)	K–Ar/Rb–Sr biotite Gubbio, Italy	28.0	Montanari *et al.* 1985
*C12N(y)	K–Ar biotite (tuff) Wyoming	32.4	Prothero *et al.* 1982
(Used as calibration point for DNAG scale)			
C12R(upper)	K–Ar biotite (tuff) Gubbio, Italy	32.0	Montanari *et al.* 1985
*C13N(y)	K–Ar biotite (tuff) Wyoming	34.6	Prothero *et al.* 1982
(Used as calibration point for DNAG scale)			
C13N(y)	K–Ar biotite (tuff) Gubbio, Italy	35.4	Montanari *et al.* 1985
C16N(y)	K–Ar/Rb–Sr biotite Gubbio, Italy	36.0	Montanari *et al.* 1985
C17N(upper)	K–Ar biotite (tuff) Gubbio, Italy	36.4	Montanari *et al.* 1985
*C21N(y)	More than 20 K–Ar lavas and tuffs Wyoming	49.3	Flynn 1986
(Used as calibration point for DNAG scale with age of 49.5)			

* dated material is from a non-marine sequence

6.4.3 Indirect dating via biostratigraphic correlation

The second calibration method involves dating the anomaly sequence biostratigraphically and then relating the biostratigraphy to isotopic ages obtained elsewhere. Table 6.6 shows the magnetic anomalies that have been biostratigraphically dated and that also span a chronogram boundary. These anomalies and the chronogram ages form a set of control points relating the anomaly sequence to isotopic ages.

6.4.4 Calibration of the modified Heirtzler et al. (1968) relative spacing of anomalies with isotopic ages

The H68MOD values of the dated magnetic anomalies (Table 6.5) can be read off the H68MOD scale and plotted against time in Ma. A similar plot can be made of the chronogram ages and their H68MOD values (Table 6.6). Inspection shows that, except for some points noted below, there is reasonable agreement between the two data sets. The two sets are therefore merged for the purposes of calibrating the time scale. Inspection also shows that if linear segments are to be fitted to the data, then at least three segments are required: the first in the interval from 0 to 3.4 Ma, the second in the interval from about 9 Ma to about 50 Ma and the third from about 50 Ma to about 85 Ma. We have used no data in the interval 3.4 to about 9 Ma.

If several magnetic profiles are available for the same anomaly sequence in different ocean basins, it may be possible to determine when there are significant breaks in the spreading rates in a given ocean basin without referring to a time scale (e.g. Aubry et al. 1988). Let there be four profiles A, B, C and D. Then if the spreading rates in A and B are constant, a graph of the corresponding profiles on A and B plotted against each other will be a straight line. If the spreading rate of one profile is not constant, the graph will be non-linear. Suppose only A changes its spreading rate in the interval. By plotting all possible pairs of profiles, i.e. AB, AC, AD, BC, BD and CD, the graphs AB, AC and AD will all show changes at the same part of the graph, whereas BC, BD and CD will be straight lines.

Comparisons of the South Atlantic ocean-floor anomaly profiles with those in other ocean basins (e.g. Aubry et al. 1988) suggest that the South Atlantic spreading rate changed between C24 and C25 – Late Paleocene, between C26 and C27 – mid-Paleocene, and between C31 and C32A – mid-Maastrichtian (Aubry et al. 1988). Between C11 and C24 – between mid-Oligocene and Early Eocene – spreading was at a constant rate. Thus the spreading data suggest that a linear fit should exist between H68MOD and the chronogram/dated magnetic anomaly data for at least the Eocene to mid-Oligocene interval. For pre-Eocene to Maastrichtian time the data suggest that spreading consisted of at least four episodes of short duration.

0 to 3.4 Ma. The data set for the first linear segment are the present day, the Brunhes/Matuyama, Jaramillo, Olduvai, Kaena, Mammoth and the Gauss/Gilbert boundaries as listed in Tables 6.3 and 6.4. The data are an excellent fit to a regression line with the equation Age = 1.014 × H68MOD value (Figure 6.5).

9 to about 50 Ma. Inspection shows that while there is reasonable agreement between the magnetic and chronogram data from 9 Ma to the Bartonian/Lutetian boundary at 41.4 Ma (Table 6.6), thereafter there is a divergence between the chronogram dates for the Lutetian/Ypresian and Ypresian/Thanetian boundaries at 48 and 53.4 Ma and the magnetically correlated boundary of C21N(y) isotopically dated at 49.3 Ma. The Lutetian/Ypresian boundary is placed at C21R(o). The duration of Chron 21 is about 3 Ma. Thus we have a chronogram estimate for C21N(y) of about 45 Ma, compared with 49.3 Ma for the dated boundary. This is more than 4 Ma younger than the alternative estimate and considered to be well outside the error limits.

The chronogram age estimates for the Lutetian/ Ypresian and Ypresian/Thanetian chronograms are dominated by glauconites. Berggren et al. (1985a) and Aubry et al. (1988) have criticized the use of glauconites for boundary determination, as well as making specific criticisms of the correlations made for many of the glauconites used in estimating the age of these two stratigraphic boundaries. We have therefore eliminated all three data points from the regression data for the second linear segment of the time scale. The discrepancies are discussed further below. The regression line fitting the remaining points has an equation Age = 0.861 × H68MOD + 1.88 Ma.

About 50 to about 85 Ma. There are no dated magnetic anomalies in the 50 to 85 Ma interval. Were there tight control provided by the four chronograms in this time interval, each could be used to calibrate a short linear segment of the magnetic anomaly scale, as was done in GTS 82, figure 4.4. However, the control is poor and the Maastrichtian/ Campanian point clearly lies well off a best-fit line to the data. We prefer to fit a single line to the data in the Paleocene to Campanian interval. Many such lines are possible. Our choice is to some extent arbitrary, guided in part by a wish not to change previous values in GTS 82 when there are no data to support such a change. We therefore fix the Tertiary/ Cretaceous boundary at 65 Ma and the Campanian/Santonian boundary at 83 Ma. Both boundaries are well within the ranges suggested by the chronograms – 64 to 68.5 and 78.8 to 86.2 Ma respectively. The adopted values are also very close to what is obtained by a linear regression through the three chronograms that lie closest to a straight line, i.e. the Thanetian/Danian, Danian/Maastrichtian and Campanian/ Santonian. Its equation is Age = 1.145 × H68MOD – 14.66 Ma.

Line intersections. The first two lines intersect at H68MOD = 12.29 at 12.46 Ma within 5A.2R; the second and third lines intersect at H68MOD = 58.24 at 52.02 Ma in the younger half of C23N. The modified Heirtzler et al. scale can be converted into a numerical time scale using the equations of the lines given above (Figure 6.6). In order to represent the relative lengths of polarity intervals accurately, the ages of the beginning and end of all polarity intervals are listed in Table 6.7 with a precision of 0.01 Ma. The actual uncertainty in the chronometric ages of the boundaries is about two orders of magnitude greater, or ~1 Ma over much of the time scale beyond 20 Ma, but the relative errors, i.e. the errors for the

Table 6.6. *Biostratigraphically dated magnetic anomalies that have been correlated with chronogram stage boundaries*

Stratigraphic age	Magnetic anomaly	H68MOD value	Chronogram age in Ma
M/Lt Miocene (Tortonian to Serravallian)	earliest C5N	C5N(o) is 10.29	10.4
mid-M Miocene (Serravallian to Langhian)	C5AA–C5AB or C5BN(o)	C5AA(y) is 12.96; C5B.2N(o) is 15.65; mean is 14.31	14.6
Miocene/Oligocene (Aquitanian to Chattian)	mid-C6CN, taken as mid-C6C.2N	C6C.2N ranges from 24.82 to 24.97, i.e. mid-point is 24.90	23.8
(Umbria)	C6CR(y) + estimated 0.2 × C6C3R	C6CR ranges from 25.43 to 26.86, i.e. 25.43+ 0.2 × 1.43 or 25.72	23.8
Lt/E Oligocene (Chattian to Rupelian)	C10N	C10N(y) is 31.50; C10N(o) is 32.17; mid-point is 31.84	30.6
Oligocene/Eocene (Rupelian to Priabonian) (Umbria)‡	C13R(y) + estimated 0.48 × C13R	C13R ranges from 38.26 to 39.77, i.e. 38.26 + 0.48 × 1.51 or 38.98	36.4
(Forams, mammals)	mid-C13R	i.e. 39.01	36.4
Lt/M Eocene (Priabonian to Bartonian)	within late C17	take younger half C17N, i.e. from 39.4 42.28 to 43.26, i.e. 42.77	
M Eocene (Bartonian to Lutetian)	uncertain, but near C19N(y)	C19N(y) is 46.76	41.4
M/E Eocene (Lutetian to Ypresian) (forams)	latest C22N	C22N(y) is 55.92	48.0
Eocene/Paleocene (Ypresian to Thanetian)	C24R(y) + estimated 0.72 × C24R	C24R ranges from 60.53 to 62.75, i.e. 60.53 +0.72 × 2.22 or 62.13	53.4
Lt/E Paleocene (Thanetian to Danian)	C26R	C26R ranges from 64.62 to 66.65, mid-point is 65.64	61.0
CENOZOIC/CRETACEOUS (Danian to Maastrichtian) (marine)	C29R(y) + estimated 0.26 × 29R	C29R ranges from 69.44 to 69.95, i.e. 69.44 + 0.26 × 0.51 or 69.57	66.0
Maastrichtian to Campanian (Umbria)	C33N(y) + 0.15 × C33N	C33N ranges from 76.66 to 81.88, i.e. 76.66+0.15 × 5.22 or 77.44	71.5
Campanian to Santonian	C34N(y)	85.29	82.4

Notes:

1. Biostratigraphic correlation from Berggren *et al.* (1985a, 1985b, 1985c).
2. Chronogram ages from Section 5.4.
3. Interpolations estimated from Berggren *et al.* (1985a) or Lowrie & Alvarez (1981).

* Preferred chron assignment.

‡ Premoli-Silva *et al.* (1988) placed boundary near top of 13R at ~0.80 × C13R.

Figure 6.6. Calibration of Cenozoic time scale with ocean-floor magnetic anomalies. Dated magnetic anomalies on land-based sections taken from Tables 6.3–6.5 shown by smaller crosses; chronogram ages from Table 6.6 shown by squares. Large open stars are dates from Koko seamount and DSDP 516. The large joined crosses are times of spreading rate changes in the South Atlantic (from Aubry _et al._ 1988).

Visually, the data can be subdivided into three linear segments. The first and second segments intersect at the position of the first asterisk on the left of the graph; the second and third segments intersect at the second asterisk in the centre of the graph. The third segment fits the remaining three asterisks. See text for details.

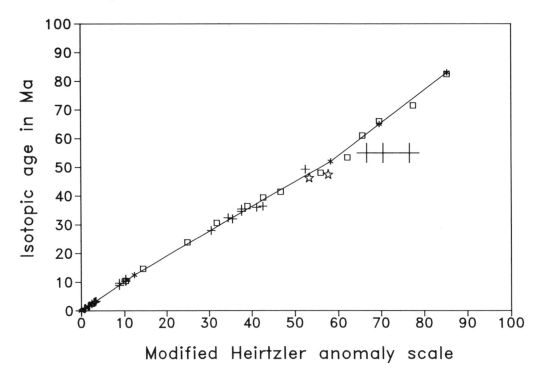

durations of the chrons, are probably of the order of a fraction of 1 Ma.

Chronogram ages for each stage boundary embody the influence of all the available isotopic dates, weighted according to the individual analytical errors (Chapter 5). When combined with the isotopically dated magnetic anomalies, they provide a data set that draws on all the available isotopic and magnetic information rather than that from a few key points. Thus a new age determination of a particular stage boundary or a magnetic anomaly will not greatly alter the time scale unless many of the pre-existing data are rejected or their error estimates substantially increased. Such a time scale calibration will change, but it will do so slowly as new data accumulate.

6.4.5 Comparison with other recent time scales

0 to 12 Ma. There are significant differences between the DNAG scale of Kent & Gradstein (1986) and GTS 89 only where major changes have been made to HDHPL68.

12 to 52 Ma. We compare GTS 89 in the 12–52 Ma interval with four recent Cenozoic time scales: Odin & Curry (1982); Berggren _et al._ (1985a, 1985b); Haq, Hardenbol & Vail (1987, 1989) (Figure 6.7). The largest differences among all these scales are in estimates of the Paleocene to Eocene scales (Table 6.8). Berggren _et al._ (1985a, 1985b) give two estimates of the boundaries based on isotopic ages and on paleontologically correlated magnetostratigraphic ages. The

two methods are not independent since the magnetostratigraphic ages have themselves been calibrated by reference to isotopically dated control points.

In particular, the M/E Eocene, Eocene/Paleocene and Lt/E Paleocene boundaries are consistently younger on HHV89 and OC82 compared with their estimated ages on the BKV85 and BKF85. The primary reason that the BKV85 and BKF85 scales give older ages for this interval is that, as discussed above, they use more than 20 K–Ar dates to estimate the age of C21N(y) as 49.5 Ma. This point controls the slope of the calibration line between isotopic ages and the magnetic anomaly scale. The magnetic chron age is derived by correlation of Wyoming mammal remains interbedded with the volcanics with a marine sequence in the San Diego basin, California (Flynn 1986).

There are at least two independent high-temperature dates which have not been used in the construction of any of the time scales but which suggest that the lower estimates of the M/E Eocene boundary age based on the ocean-floor calibrations given in Table 6.8 here are more likely to be correct. The Koko Guyot, part of the Hawaiian Islands/ Emperor Seamounts chain, has been dated by K–Ar whole-rock methods at 46.4 ± 1.1 Ma (Clague & Dalrymple 1973), or about 47.5 Ma using new constants. Additional data gave an estimate of 46.9 ± 0.8 Ma (Dalrymple & Clague 1976), or 48 Ma using standardized constants. It is probably one of the best dated seamounts. Shallow-water sediments associated

Table 6.7. *Revised ages of normal polarity chrons for ocean-floor magnetic anomalies in the time range 0 to 83 Ma*

Anomaly	Normal polarity interval		Anomaly	Normal polarity interval	
1	0.00	0.72	6C.1	22.90	23.05
1R–1	0.91	0.97	6C.2	23.25	23.38
2	1.65	1.88	6C.3	23.62	23.78
*2R–1	2.06	2.09	7	25.01	25.11
(2R–2 not recognized)			7	25.17	25.45
2A	2.45	2.91	7A	25.84	26.01
2A	2.98	3.07	8	26.29	26.37
2A	3.17	3.40	8	26.44	27.13
3.1	3.87	3.99	9	27.52	28.07
3.2	4.12	4.26	9	28.12	28.51
3.2R–1	4.41	4.48	10	29.00	29.29
3.3	4.79	5.08	10	29.35	29.58
3A	5.69	5.96	11	30.42	30.77
3A	6.04	6.33	11	30.82	31.21
3B	6.66	6.79	12	31.60	32.01
4	7.01	7.10	13	34.26	34.44
4	7.17	7.56	13	34.50	34.82
4	7.62	7.66	15	36.12	36.32
4A	8.02	8.29	15	36.35	36.54
4AR–1	8.48	8.54	15A	36.93	37.16
4AR–2	8.78	8.83	16	37.31	37.58
5	8.91	9.09	16	37.63	38.01
5	9.14	9.48	17	38.28	39.13
5	9.49	9.80	17	39.20	39.39
5	9.83	10.13	17	39.45	39.77
5	10.15	10.43	18	39.94	40.36
5R–1	10.57	10.63	18	40.43	40.83
5R–2	11.11	11.18	18	40.90	41.31
5A.1	11.71	11.90	19	42.14	42.57
5A.2	12.05	12.34	20	43.13	44.57
5AR–1	12.68	12.71	21	47.01	48.51
5AR–2	12.79	12.84	22	50.03	50.66
5AA	13.04	13.21	23	51.85	52.08
5AB	13.40	13.64	23	52.13	52.83
5AC	13.87	14.24	23R–1	53.15	53.20
5AD	14.35	14.79	24	53.39	53.69
5B.1	14.98	15.07	24	54.05	54.65
5B.2	15.23	15.35	25	57.19	57.80
5C	16.27	16.55	26	58.78	59.33
5C	16.59	16.75	27	61.65	62.17
5C	16.82	16.99	28	62.94	63.78
5D	17.55	17.87	29	64.16	64.85
5DR–1	18.07	18.09	30	65.43	67.14
5E	18.50	19.00	31	67.23	68.13
6	19.26	20.23	32.1	70.14	70.42
6A.1	20.52	20.74	32.2	70.69	72.35
6A.2	20.97	21.37	32R–1	72.77	72.82
6AA	21.60	21.75	33	73.12	79.09
6AAR–1	21.93	22.03	34	83.00	
6B	22.23	22.60			

with the Guyot are in the *Discoaster lodoensis* zone (Bukry 1975), i.e. NP13 (Chapter 3, Figures 3.13 and 3.14). From inspection of Berggren *et al.* l985a, (p.162), NP13 ranges from about C22R(y) to 0.75 × (duration of C22R). From Table 6.7 C22R(y) is about 50.7 Ma and C22R(o) is 51.8 Ma. The estimated age range of NP13 is from 50.7 to 51.6 Ma, with a mean age of about 51.2 Ma.

Analogy with modern Hawaiian volcanoes suggests that any shallow-water sediments should be 1 to 5 Ma younger than the alkaline volcanics. Thus the K–Ar age should be 1 to 5 Ma older than the age of the oldest sediments. If the maximum errors on the sediments and the magnetic correlations are taken into account, then on GTS 89 the K–Ar age is 2.2 Ma younger than the youngest age that could be assigned to the basal sediments, which is unlikely.

Similarly, Bryan & Duncan (1983) date some volcanic and plutonic detritus from DSDP hole 516 F at 47.0 ± 0.3 Ma. It is unclear whether the material includes contemporaneous ash. Its paleontological age is probably about P10 or NP15, which would give it a magnetic chron age of about the range of C21N. The isotopic date places a maximum age on the horizon. From Table 6.7 the age of C21N(y) is about 46.9 Ma, that of C21N(o) is about 48.5 Ma with a mean of 47.7 Ma. The K–Ar age is 0.6 Ma younger than the youngest age that could be assigned to the basal sediments.

Of course, in both cases the K–Ar dates could be low because of alteration. The Koko samples in particular would be more prone to alteration because they are dredge rather than drill samples.

The chronogram ages that have not been used in the calibration of this linear segment are the Lutetian/Ypresian boundary at 48 Ma and the Ypresian/Thanetian boundary at 53.4 Ma. The estimated upper chronogram ages are 49 Ma and 55.8 Ma respectively. Both are controlled entirely by glauconites. While we have argued in Section 5.6 that, in the present state of knowledge, glauconites should not be rejected for time scale calibration in the interval 0 to 100 Ma, we did show that on average glauconite chronograms were 1 Ma lower than non-glauconite chronograms. If allowance is made for this effect, then the values adopted in Table 6.8 of 50.2 and 56.5 Ma are within the limits, admittedly extreme, permitted by the chronogram data.

Finally, inclusion of the estimated DNAG age of C21N(y) at 49.5 Ma in the current data set scarcely changes the slope of the regression line. The regression line that includes this point gives an age of 49.5 Ma, a chron age somewhere in C21R. Table 6.7 gives C22N(y) an age of 50.0 Ma, which is very close to 49.5 Ma. Thus if the K–Ar date of 49.5 Ma is correct, a linear regression of all other data suggests that the horizon dated is C22N(y), rather than C21N(y), which has an age of 47.0 Ma in Table 6.7.

Notes to Table 6.7:
Since all the anomaly numbers refer to normal polarity intervals in the ocean-floor magnetic anomaly scale, the prefix C and the suffix N have been omitted. Thus 24 is C24N.

* Because of the age uncertainty of C2R–1 (the Reunion event, or events), the age assignment is from Table 6.4 rather than from the regression line.

Figure 6.7. Comparison of Cenozoic calibrations. Three time scales are shown: DNAG as diagonal crosses; Curry & Odin (1982) and Odin (1982b) as triangles; and GTS 89 as squares. GTS 89 lies between the other scales, but is closer to DNAG.

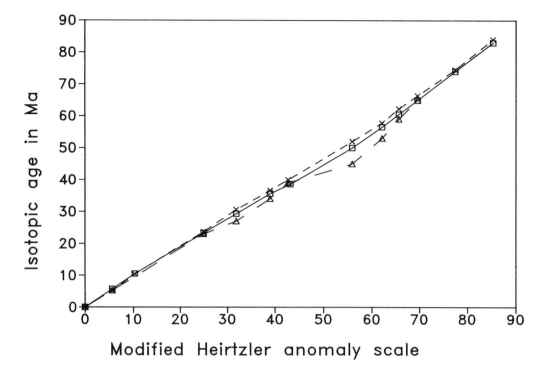

52 to 83 Ma. The calibration ages of GTS 89 are about 2 Ma less than DNAG at 52 Ma, about 1.5 Ma less at 56 Ma and 1 Ma less at 83 Ma.

Relative spreading rates. The inverse slopes of the three line segments in Figure 6.6 – i.e. the ratio of HDHPL68 values to age span – are measures of the relative spreading rates in the south Atlantic. They are 0.98 (0–12 Ma); 1.17 (12–52 Ma); and 0.87 (52–83 Ma). If the value of 1 is assigned to the inverse slope for the 0–12 Ma interval, then the relative rates implied by the other line segments are 1.19 and 0.89 respectively. According to these data there was a speeding up of south Atlantic spreading in Early Eocene time and a slowing down in mid-Miocene time. Similar changes are implied by Haq *et al.* (1989). Smaller changes in south Atlantic

spreading rates are also implied by the DNAG scale. If unit rate is assigned to the DNAG interval 0–10.42 Ma, the rate for the second linear segment is about 1.1 and for the third segment it is 0.94. The revised ages of stratigraphic boundaries in the time interval 0 to 83 Ma are given in Table 6.9.

6.5 This work: the GTS 89 time scale 83 to 200 Ma
6.5.1 Relative spacing of anomalies 83 to 158 Ma

The Oxfordian to Barremian interval was a time of rapid reversals. The basic data used for constructing our time scale for this interval are the lengths of polarity intervals given by Kent & Gradstein (1985), based on Larson & Hilde's (1975) Hawaiian lineations (Table 6.10). There are still unresolved problems about reliability of the Hawaiian sequence. As

Table 6.8. *Comparison of recent Paleocene to Eocene time scales. All ages in Ma*

	This work Chronograms		HHV89	OC82	BKV85, Anoms +fossils	BKF85 Isotopic ages
	only	+anoms +adjustments				
Olig./Eoc.	36.4	35.4	36	34	36.6	37
Lt Eoc./M Eoc.	39.4	38.6	39.4	39	40.0	—
M Eoc./E Eoc.	48.0	50.0	49	45	52.0	—
Eoc./Paleoc.	53.4	56.5	54	53	57.8	56.5
Lt/E Paleoc.	61.0	60.5	60.2	59	62.3	—
Paleoc./Maastr.	66.0	65.0	66.5	65	66.4	66

References are: OC82 = Odin & Curry (1982); HHV89 = Haq, Hardenbol & Vail (1989); BKF85 = Berggren, Kent & Flynn (1985); BKV85 = Berggren, Kent & Van Couvering (1985).

Table 6.9. *Revised ages of stratigraphic boundaries based on piecewise linear interpolation of magnetic anomalies 0 to 83 Ma*

Stratigraphic boundary	C-gram	Revised age	Diff.
Pleistocene/Pliocene (Taken as C2N(y) at the younger end of the Olduvai chron. Dated at 1.67 Ma by Mankinen & Dalrymple (1979) and 1.66 by Berggren *et al.* (1985b), and calibrated here at 1.65 Ma. For details see Sections 3.21.1 and 5.8.2 where a date of 1.64 Ma is specified.)	—	1.65	—
Piacenzian/Zanclian (Taken as C2A(o))	3.5	3.4	0.1
Zanclian/Messinian (Taken as C3A(y), Berggren *et al.* 1985b) (Taken as near C3R(y): Hilgren & Langereis 1988; Channell *et al.* 1988)	6.0	5.7*	0.3
		5.2*	
Messinian/Tortonian (Taken as mid-C3BN)	8.4	6.7	1.7
Tortonian/Serravallian (Taken as C5N(o))	10.4	10.4	0.0
Serravallian/Langhian†	14.6	14.2	0.4
Langhian/Burdigalian (Taken as C5CN(y))	16.0	16.3	−0.3
Burdigalian/Aquitanian (Taken as mid-C6A.2R)	22.0	21.5	0.5
Aquitanian/Chattian	23.8	23.3	0.5
Chattian/Rupelian	30.6	29.3	1.3
Rupelian/Priabonian	36.4	35.4	1.0
Priabonian/Bartonian	39.4	38.6	0.8
Bartonian/Lutetian (Taken as C19N(y))	41.4	42.1	−0.7
Lutetian/Ypresian	48.0	50.0	−2.0
Ypresian/Thanetian	53.4	56.5	−3.1
Thanetian/Danian	61.0	60.5	0.5
Danian/Maastrichtian	66.0	65.0	1.0
Maastrichtian/Campanian	71.5	74.0	−2.5
Campanian/Santonian	82.4	83.0	−0.6

Magnetic quiet zone

reviewed by Channell *et al.* (1987), M5R to M14R in this sequence cannot be correlated with their presumed equivalents in the Keathley sequence of the central Atlantic. The difficulties of correlation may spring either from their absence in the central Atlantic, or from ridge jumps in the Pacific ocean floor which have duplicated anomalies in the Hawaiian sequence (Channell *et al.* 1987).

Chrons M4R–M9R, which had not been clearly identified in any sedimentary sections at the time Channell *et al.* wrote their review, now appear to have been recognized by Ogg & Steiner (1989). However, chrons M25A to M28 have not yet been recognized in stratigraphic sections (Ogg & Steiner in press), though M29 has been provisionally identified as straddling the Oxfordian/Callovian boundary (Steiner & Ogg (in press), Steiner, personal communication).

6.5.2 Direct isotopic dating of the magnetic anomalies

The M-sequence oceanic crust has not yet been dated: there is therefore no direct evidence of the age of the M-anomalies.

6.5.3 Indirect dating via biostratigraphic correlation

Bralower (1987) and Channell *et al.* (1987) have reappraised the positions of the M-sequence anomalies and most of the biostratigraphically dated Early Cretaceous stage boundaries in sections in the Umbrian Apennines and a site in the southern Alps. Channell *et al.* (1987) have extended the reappraisal down to the Tithonian/Kimmeridgian boundary. Their estimates differ substantially from previous work. Their work has been reviewed and extended by Ogg & Steiner (1989). The Berriasian/Tithonian (also equal to the Cretaceous/Jurassic boundary), the Tithonian/Kimmeridgian and the Kimmeridgian/Oxfordian boundaries have also been recognized in an ammonite-bearing sequence in the Betics of

Notes: to Table 6.9

Some stratigraphic boundaries have been explicitly related to the magnetic reversal sequence (Table 6.6). Where an explicit correlation is absent, we have read the magnetic correlations from the charts in Berggren *et al.* (1985a, 1985b) and converted the magnetic chron age into Ma via the calibration table (Table 6.7). The correlations are given above.

*The initial Pliocene boundary is placed slightly below the Thevra N Event in recent papers. Hilgen & Langereis (1988) estimate the boundary to lie 0.09 to 0.10 Ma below the reversal. Channell *et al.* (1988) estimate the boundary to lie approximately 0.15 Ma below the reversal. They used a calibration of the reversal time scale that put the reversal at about 4.77 Ma and thus came up with a boundary date of slightly less than 5.0 Ma. Berggren *et al.* (1985b) placed the base of the Pliocene near the older end of the Gilbert chron (C3R(o)), a reversal assigned an age of 5.35 Ma. Thus they came up with a round number of 5.3 for the boundary age. Mankinen & Dalrymple (1979) placed the older end of the Thevra event (C3.3N(o)) at about 5.00 Ma. Our new calibration places C3.3N(o) at 5.08 Ma. Based on our new calibration the base of the Pliocene should be about 5.2 Ma. This is conveniently close to the Berggren *et al.* scale but only because they used a different definition and different magnetic reversal date and the effects cancel one another.

†This boundary lies either within C5AA–C5AB or at C5BN(o). The age of the first boundary is 13.0 Ma, that of the second is 15.3 Ma. The mean age of 14.2 Ma is given in the Table.

Table 6.10. *Relative spacing of M-sequence anomalies based on Hawaiian lineations (Kent & Gradstein 1985)*

Anomaly (this work)	Normal polarity interval		Anomaly (KG85) (rev)	Anomaly (this work)	Normal polarity interval		Anomaly (KG85) (rev)
C34N	84.00	118.00	M0	M20N	146.44	146.75	M19
M1N	118.70	121.81	M1	M20N	146.81	147.47	M20
M2N	122.25	123.03	M3	M21N	148.33	149.42	M21
M4N	125.36	126.46	M5	M22N	149.89	151.46	M21
M6N	127.05	127.21	M6	M22N	151.51	151.56	M21
M7N	127.34	127.52	M7	M22N	151.61	151.69	M22
M8N	127.97	128.33	M8	M22AN	152.53	152.66	M22
M9N	128.60	128.91	M9	M23N	152.84	153.21	M22
M10N*	129.43	129.82	M10	M23–1	153.49	153.52	M23
M10NN	130.19	130.57	M10	M24N	154.15	154.48	M23
M10NN	130.63	131.00	M10	M24–1	154.85	154.88	M24
M10NN	131.02	131.36	M10N	M24AN	155.08	155.21	M24
M11N	131.65	132.53	M11	M24BN	155.48	155.84	M24
M11–1	133.03	133.08	M11	M25N	156.00	156.29	M25
M11AN	133.50	134.31	M11	M25AN	156.55	156.70	M25
M12N	134.42	134.75	M12	M25AN	156.78	156.88	M25
M12.2N	135.56	135.66	M12	M25AN	156.96	157.10	M25
M12AN	135.88	136.24	M12	M26N	157.20	157.30	M25
M13N	136.37	136.64	M13	M26N	157.38	157.46	M25
M14N	137.10	137.39	M14	M26N	157.53	157.61	M25
M15N	138.30	139.01	M15	M26N	157.66	157.85	PM26
M16N	139.58	141.20	M16	M27N	158.01	158.21	PM27
M17N	141.85	142.27	M17	M28N	158.37	158.66	PM28
M18N	143.76	144.33	M18	M29N	158.87	159.80	PM29
M19N	144.75	144.88	M18	J-QZ†	160.33	169.00	
M19N	144.96	145.98	MJ9				

* M10N is the name given to the dominantly normal interval following M10 (here named M10R). Logically, in our scheme it could have been named M10A, and subdivided into M10AN and M10AR. However, we have preserved the present nomenclature even though it gives rise to the awkward numbering M10NN.

† PM means Pacific Mesozoic anomaly.

‡ J-QZ = Jurassic quiet zone, based on ocean-floor data. Land-based work shows the interval older than M29N to be one of frequent reversals, rather than one of no reversals as implied by the term quiet zone (Steiner & Ogg in press).

southern Spain (Ogg *et al.* 1984). Since ammonites form the basis for identifying the Jurassic stages, these calibration points, as modified by Ogg & Steiner (in press) are adopted here in preference to those based on other fossil groups such as calpionellids (Table 6.11). Revised correlations of the Mesozoic polarity chrons and calpionellid zones are presented in Channell & Grandesso (1987).

6.5.4 Calibration of the Kent & Gradstein (1985) ages with chronogram ages

The ages assigned by Kent & Gradstein (1985) to the M-sequence anomalies (Table 6.10) are based on fixing the Barremian/Aptian boundary at 118 Ma and the Oxfordian/Kimmeridgian boundary at 156 Ma. The relative separations of the magnetic anomalies within this interval are those observed by Larson & Hilde (1975) in the Hawaiian M-lineations. Figure 6.8 shows a plot of the Kent & Gradstein (1985) ages against chronogram ages. The horizontal extent of each error box is the estimated range of uncertainty in the magnetic anomaly (Table 6.11) expressed in the units of the Kent & Gradstein (1985) scale of Table 6.10. The vertical

extent of each error box expresses the age uncertainty in Ma using the chronogram values of Chapter 5. The central cross in the box represents the chronogram minimum value; the crosses above and below it in each box represent the age range for which there are no significant data in the chronogram.

The chronograms in the Aptian to Callovian interval are poorly constrained. It is not clear how to obtain the best estimate of the corresponding ages of the magnetic lineations from these data. In many cases the central value for the age of a stratigraphic boundary lies within a time span for which there are no isotopic dates. In addition the chronograms may be asymmetric. As a compromise a linear regression line has been drawn through the data set consisting of the central, maximum and minimum values of the chronograms and the corresponding ages of the magnetic lineations on the Kent & Gradstein scale. The line has the equation: Age = 0.795 × KG85 + 30.52 Ma. The Kent & Gradstein (1985) scale was then converted into a numerical time scale (Table 6.12). Other regressions were made but in all cases the maximum difference between the values in Table 6.12 and those from other regressions was about 0.5 Ma.

Table 6.11. *Biostratigraphically dated M-anomalies that can be correlated with chronogram stage boundaries*

Stratigraphic age	Magnetic anomaly	Kent & Gradstein (1985) age in Ma	Chronogram age in Ma
Aptian/ Barremian	M0R†	M0 ranges from 118.0 to 118.7 Mean is 118.35	125.5
	M0R(y)+0.35 × M0R§	M0(y) is 118.00, 118.0 + 0.35 × 0.7, i.e. 118.25	
Barremian/ Hauterivian	Within chrons M6N, M6R, M7N or M7R*†	M6N(y) is 127.05, M7R(o) is 127.97, Mean is 127.51	131.0
	Estimated mid-M5 to older end of M7R‡	M5(y) is 126.46, M5(o) is 127.05, i.e. 126.76 to 127.97 Mean is 127.36	
Hauterivian/ Valanginian	Within chron M10N or M10NN*†	M10N(y) is 129.43, M10NN(o) is 131.36 Mean is 130.40	135.5
	Oldest M10NN(y) to estimated M11N +0.25 × M11N‡	Oldest M10NN(y) is 131.02, M11N(y) is 131.65, M11N(o) is 132.53, i.e. 131.87 Mean is 131.45	
Valanginian/ Berriasian	Early M14R to late M15N†	M14R(o) is 138.3	137.5
	mid-M15N‡	M15N ranges from 138.30 to 139.01 Mean is 138.66	
JURASSIC/ CRETACEOUS Berriasian/Tithonian	Base of M18R†	M18R(o) is 144.75	148
Tithonian/ Kimmeridgian	M22N†	M22N(y) is 149.89, M22N(o) is 151.69 Mean is 150.79	151
	M22AR to M23N‡	M22AR(y) is 152.66, M23N(o) is 153.21 Mean is 152.94	
Kimmeridgian/ Oxfordian	mid-M24BR to mid–25R‡	M24BR(y) is 155.84, M24BR(o) is 156.0, i.e. mid-M24BR=155.92 M25R(y) is 156.29, M25R(o) is 156.55, i.e. mid-M25R=156.42 Mean is 156.17	156
Oxfordian/ Callovian	M29N(y)‡ + estimated 0.4 × duration of M29N‡	M29N(y) is 158.87, M29N(o) is 159.80, duration is 0.93 Mean is 159.24	156.3

Notes:

References: * = Bralower (1987), † = Channell *et al.* (1987), ‡ = Ogg & Steiner (in press), § = Steiner & Ogg (in press).

   1. Magnetic anomalies read from original figures; nomenclature varies from author to author and is sometimes ambiguous. Names used in this work given in Kent & Gradstein age column and taken from Table 6.10. They originate from A. V. Cox in GTS 82. In all cases Ogg & Steiner (in press) or Steiner & Ogg (in press) values have been adopted for calibration. Where no range is given an arbitrary error of ± 0.1 has been assigned to the value.

   2. Chronogram ages from Chapter 5.

Figure 6.8. Calibration of later Mesozoic time scale with M-anomaly sequence. The horizontal axis is the Kent & Gradstein (1985) Hawaiian anomaly age. The vertical axis is the GTS 89 calibration age. The rectangular boxes are error boxes showing the estimated uncertainty in the position of the stratigraphic boundary in the M-anomaly sequence (horizontal range) and the chronometric uncertainty in the chronogram age (vertical range). The central cross in each error box represents the estimated chronogram age; the two crosses above and below the central cross represent the age range for which there are no significant isotopic data. The linear regression line is taken as the calibration line for converting Kent & Gradstein (1985) ages into isotopic ages. See text for details.

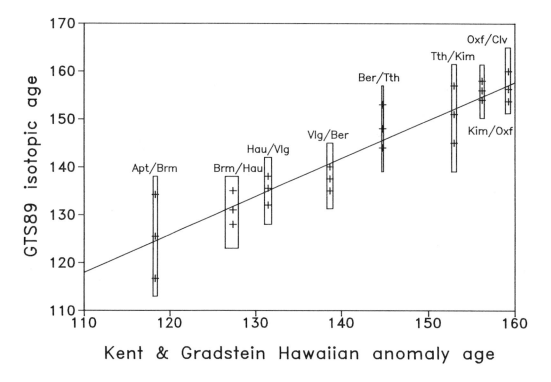

6.5.5 Comparison with other recent time scales

The chronograms for all stage boundaries older than 115 Ma regard glauconite ages as minimum ages (Chapter 5.5). Thus the main difference between GTS 89 and scales based on glauconites (e.g. Odin & Curry 1982) is that the stage boundaries are systematically older in GTS 89. The main differences between the DNAG scale and the GTS 89 ages are that, while the GTS 89 ages are close to the DNAG ages at the older end of the M-sequence scale, they are systematically older than DNAG ages as one moves forward in time. GTS 89 also assigns different values to the relative durations of the Kimmeridgian to Barremian stages.

The DNAG scale uses the GTS 82 estimates of 119 Ma for the Barremian/Aptian boundary and 156 Ma for the Oxfordian/Kimmeridgian boundary. The GTS 82 chronograms for both of these boundaries had considerable uncertainties and were not used in estimating these ages in GTS 82. In GTS 82 ages were assigned to all stages in the Aptian to Ladinian (Triassic) interval by fixing the Albian/Aptian boundary at 113 Ma and the Ladinian/Anisian boundary at 238 Ma and distributing the time equally among the 19 intervening stratigraphic stages. The method was an application of the 'equal-age' hypothesis in which the rate of morphological change of ammonoids was assumed to be constant. The hypothesis is discussed by House (1985b). Thus the assignment of 119 and 156 Ma was to some extent arbitrary.

Although the chronograms have not greatly improved since GTS 82, the relationships of the biostratigraphic ages to the magnetic chrons are now very well known. Thus the relative durations of most of the Late Jurassic and Early Cretaceous stages can be determined from the relative spacing of the M-anomalies: it is no longer necessary to use the 'equal-age' or even the equal duration of chron hypothesis – which are both clearly of limited value. This calibration assumes that spreading was at a uniform rate throughout the time span of the M-anomaly sequence. The only remaining major problem in assessing relative durations in the Middle Jurassic to present-day interval lies in the Aptian to Santonian stages in the mid-Cretaceous. All of these are in chron C34N (which could just as readily be named M0N, or M0 could be renamed C34R). The chronograms are reasonable, but there is as yet no independent check on them, such as might be provided by relative sediment accumulation in a constant tectonic and sedimentary setting. The deviations of chronogram mid-points from the regression line are mostly less than 2 Ma, suggesting considerable consistency among the chronogram calibrations.

As in the C1–C33 anomaly sequence, in order to represent the relative lengths of polarity intervals accurately, the ages of the beginning and end of all polarity intervals are listed in Table 6.12 with a precision of 0.01 Ma. The uncertainty in the chronometric ages of the boundaries is probably still in the range 3 to 5 Ma. A small additional

Table 6.12. *Revised ages of normal polarity chrons for ocean-floor magnetic anomalies in the time range 124 to 158 Ma*

Anomaly	Normal polarity interval		Anomaly	Normal polarity interval	
C34N	(83.00)	124.32	M20N	146.93	147.17
M1N	124.88	127.35	M20N	147.22	147.75
M2	127.70	128.32	M21N	148.43	149.30
M4	130.17	131.05	M22N	149.67	150.92
M6N	131.51	131.64	M22N	150.96	151.00
M7N	131.74	131.89	M22N	151.04	151.10
M8N	132.25	132.53	M22AN	151.77	151.87
M9N	132.75	132.99	M23N	152.01	152.31
M10N[1]	133.41	133.72	M23–1	152.53	152.56
M10NN[1]	134.01	134.31	M24N	153.06	153.32
M10NN	134.36	134.65	M24–1	153.61	153.64
M10NN	134.67	134.94	M24AN	153.80	153.90
M11N	135.17	135.87	M24BN	154.11	154.40
M11–1N	136.27	136.31	M25N	154.53	154.76
M11AN	136.64	137.30	M25AN	154.96	155.08
M12N	137.37	137.63	M25AN	155.15	155.23
M12–2N	138.28	138.36	M25AN	155.29	155.40
M12AN	138.53	138.82	M26N	155.48	155.56
M13N	138.92	139.14	M26N	155.62	155.69
M14N	139.50	139.73	M26N	155.74	155.81
M15N	140.46	141.02	M26N	155.85	156.00
M16N	141.47	142.76	M27N	156.12	156.28
M17N	143.28	143.61	M28N	156.41	156.64
M18N	144.80	145.25	M29N	156.81	157.55
M19N	145.58	145.69	M29R(o)	157.98	
M19N	145.75	146.56			

[1] For explanation see text.

Table 6.13. *Revised ages of stratigraphic boundaries based on piecewise linear interpolation of magnetic anomalies 125 to 158 Ma*

Stratigraphic boundary	C-gram	Revised age	Diff.
CRETACEOUS			
Quiet magnetic zone			
Aptian/Barremian	125.5	124.5	1.0
Barremian/Hauterivian	131.0	131.8	−0.8
Hauterivian/Valanginian	135.5	135.0	0.5
Valanginian/Berriasian	137.5	140.7	−3.2
Berriasian/Tithonian	148.0	145.6	2.4
JURASSIC			
Tithonian/Kimmeridgian	151.0	152.1	−1.1
Kimmeridgian/Oxfordian	156.0	154.7	1.3
Oxfordian/Callovian	156.3	157.1	−0.8
End ocean-floor anomalies			
Revised chron interpolations			
Callovian/Bathonian	159.0	161.3	
Bathonian/Bajocian	159.2	166.1	

Note: The chron interpolations of Chapter 5 for the Oxfordian to Bathonian interval have been revised by taking 157.1 Ma as the base of the Oxfordian.

number of biostratigraphically controlled isotopic ages in the M-sequence time interval would greatly reduce the errors in the ages of the M-anomalies but would probably not show major shifts from the values adopted here. Revised chronogram ages are given in Table 6.13.

6.5.6 The polarity reversal scale 158 to 200 Ma

Steiner & Ogg (in press) present a very detailed polarity reversal scale for Middle to Early Jurassic time. The relationship to biostratigraphic zones is well established but it is not yet possible to relate these zones to isotope ages with the same precision as for Later Jurassic time, when ocean-floor data are available.

6.6 Summary of polarity reversal time scale 0 to 200 Ma

The conclusions from the discussion above are summarized in Figures 6.9a, b and c, which show the magnetic anomaly scale back to about 158 Ma (Callovian time). Appendix 5, with Figures A5.1 to A5.16, shows the same time scale. The convention here is that normally magnetized intervals plot to the right of the figures and reversely magnetized intervals to the left. Intervals of frequent reversals too detailed to show on Figures 6.9a,b and c are indicated by a line halfway between normal and reversed positions.

6.7 Polarity bias superchrons
6.7.1 The phenomenon of polarity bias

If the reversal time scale is viewed through a sliding window 25 Ma wide, the character of the polarity pattern seen in the window undergoes marked changes as the window is moved from the present back to the beginning of geologic time. A typical change occurred at 83 Ma at the younger end of the Santonian Stage (Figure 6.9b). From the present back to 83 Ma the field reversed rapidly and symmetrically, spending approximately equal amounts of time in the normal and reversed polarity states. For several tens of millions of years prior to that time the field remained in the normal state with at most a few brief, scattered intervals of reversed polarity.

Paleomagnetic research has shown that throughout geologic time the field has been characterized by long intervals of time during which the **polarity bias** has remained constant. During times of **normal polarity bias** the field remains normal all or almost all of the time. **Reversed polarity bias** describes the opposite state. During times of **mixed polarity** the field alternates symmetrically between the normal and reversed polarity states.

Figure 6.9. Summary of Cenozoic and Mesozoic reversal time scale: (a) 0–60 Ma; (b) 50–110 Ma; (c) 100–158 Ma. In this summary, normally and reversely magnetized intervals that are shorter than 0.15 Ma have been treated in three different ways. If the short-lived interval is bracketed by intervals that are both longer than 0.15 Ma, the three intervals are amalgamated into a single interval of the dominant polarity. Amalgamation may also continue to the next short-lived interval. Two or more short-lived intervals whose total duration is less than 0.15 Ma are split equally between intervals that are longer than 0.15 Ma, but if the total duration of the short-lived intervals is greater than 0.15 Ma then the interval is treated as a mixed interval.

Normally magnetized intervals lie to the right of the polarity graph; reversely magnetized intervals to the left of the graph and mixed intervals lie in the central portions of the graph.

The beginnings of the normal polarity intervals for Chrons C1–C34 are labelled, except where they are no longer recognized, as in C14, or where space precludes an assignment, or where the initial position has been modified in some way. The beginning of the normal polarity interval for most of the M-anomalies is shown except where they are closely spaced or have been modified. The nomenclature for the first few M-anomalies is confusing: M0, logically equivalent to C34R (which is not formally recognized) is the reversely magnetized chron immediately preceding M1N; M2R is known formally as M3; M4R is known formally as M5. Thus M3N and M5N do not exist. See text for details. M11AN should be of longer duration and M11AR correspondingly shorter than plotted here.

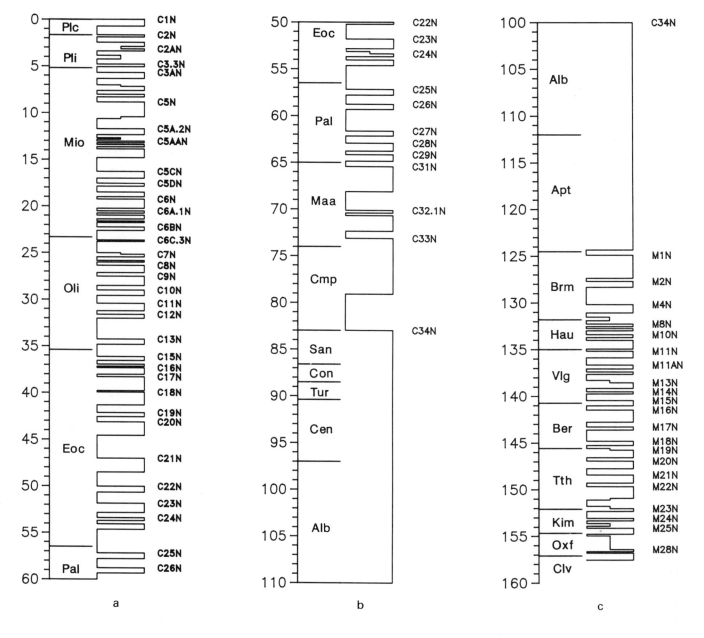

The duration of intervals of constant polarity bias ranges from 30 Ma to about 100 Ma, which is more than an order of magnitude larger than the duration of chrons and subchrons during the Cenozoic. This difference suggests that the physical origin of polarity bias may be different from that of individual reversals. Quite possibly, individual reversals are the result of perturbations in the motion of fluid in the Earth's core, whereas changes in polarity bias reflect longer-term changes in the boundary conditions at the core–mantle interface (Irving & Pullaiah 1976, Cox 1981, McFadden & Merrill 1984). Whatever their origin, polarity bias intervals comprise a distinct geomagnetic phenomenon that is useful for global stratigraphic correlation.

6.7.2 Nomenclature

The name 'superchron' is the next level above 'chron' in the hierarchy of magnetostratigraphic names recommended for international usage (Anon. 1979) and will be used in the present work to describe intervals of polarity bias. It should be noted that superchrons may reflect a large increase in the time between reversals rather than an inherent bias favouring one polarity or the other. Polarity bias superchrons have been named in different ways by different workers. Some have left them unnamed (Sasajima & Shimada 1966, Helsley & Steiner 1969). Some have named them for type localities (Irving & Parry 1963, Khramov 1967, Pechersky & Khramov 1973). Some have a mixture of type locality names and the names of distinguished scientists (McElhinny & Burek 1971). Some have simply referred polarity bias superchrons to the geologic period or periods in which they occur (Irving & Couillard 1973, Irving & Pullaiah 1976). The latter use the expression 'Cretaceous normal quiet interval' and the symbol 'KN' to describe the interval of normal polarity bias that occurred during the Cretaceous Period. Similarly, they refer to the 'Permo-Carboniferous quiet reversed interval' with the symbol 'PCR'. In the present work we use the nomenclature of Irving & Pullaiah (1976) with several minor modifications and extensions (Table 6.14). Three types of polarity bias superchrons are recognized: normal, reversed and mixed polarity superchrons (versus Irving & Pullaiah's normal quiet intervals, reversed quiet intervals, and disturbed intervals). Particular superchrons are identified by the names of the period or periods in which they occur, as in Table 6.14.

The relationship of these names to others in the literature is reviewed by Irving & Pullaiah (1976). The K–N superchron is equivalent to the Mercanton interval of McElhinny & Burek (1971), to the Jalal interval of Pechersky & Khramov (1973) and to the KN normal interval of Irving & Pullaiah (1976). The JK–M polarity superchron is equivalent to the Hissar interval of Pechersky & Khramov (1973). Between the JK–M and the PTr–M polarity superchrons was a time predominantly of mixed polarity. Steiner & Ogg (in press) have shown that mixed polarity, with frequent reversals, characterizes the Middle and Early Jurassic. There may have been short normal superchrons in the Triassic (Irving & Pullaiah 1976). However, the ages of these possible normal superchrons are not known well enough for global correlation and, in fact, even the age of the upper boundary of the Permo-Triassic mixed superchron is uncertain. The PTr–M

Table 6.14. *Superchron polarities*

KTQ-M	Cretaceous–Tertiary–Quaternary Mixed Polarity Superchron
K–N	Cretaceous Normal Polarity Superchron
JK–M	Jurassic–Cretaceous Mixed Polarity Superchron
	Later Triassic polarity bias uncertain
PTr–M	Permo-Triassic Mixed Polarity Superchron
PC–R	Permo-Carboniferous Reversed Polarity Superchron
C–M	Carboniferous Mixed Polarity Superchron

superchron is the equivalent of the Illawarra interval of Pechersky & Khramov (1973), the PC–R superchron is the equivalent of the Kiaman reversed interval of Irving & Parry (1963), and the C–M superchron is the equivalent of the Debal Tseve interval of Khramov (1967).

6.7.3 Ages of polarity superchrons

Our fragmentary knowledge of the reversal history of the magnetic field prior to Jurassic time limits analysis of the pre-Jurassic field mostly to the large-scale features such as superchrons. However, there are some detailed magnetostratigraphic studies of individual sedimentary sequences with continuous or nearly continuous records of the magnetic field. It was such a study in Australia that resulted in the discovery of the Permo-Carboniferous reversed polarity superchron (Irving & Parry 1963). Global syntheses of all available paleomagnetic polarity data including information from isotopically and paleontologically dated rocks allow the broad features of the field to be estimated for other periods (McElhinny 1971, Irving & Pullaiah 1976). These global sets of paleomagnetic data are analysed statistically by calculating for rocks within a specified age window the fraction of samples (and presumably the fraction of time) with normal polarity. Values near 1 indicate normal polarity, values near $\frac{1}{2}$ indicate mixed polarity, and values near 0 indicate reversed polarity. The ages of Triassic and older superchrons were determined using both approaches, as described below. The ages of the three post-Early Jurassic superchrons were determined by the methods described previously.

The initial and terminal boundaries of the Carboniferous–Permian reversed superchron are well dated by stratigraphic studies in Australia, North America and the USSR. The initial boundary is either within the Namurian Epoch or between the Namurian and Westphalian epochs and the terminal boundary is either late or mid-Tatarian (Irving & Pullaiah 1976). In the present analysis we have placed the initial boundary of the Kinderscoutian stage at 323 Ma and the terminal boundary at the mid-point of the Tatarian Stage at

Figure 6.10. Polarity bias superchrons.

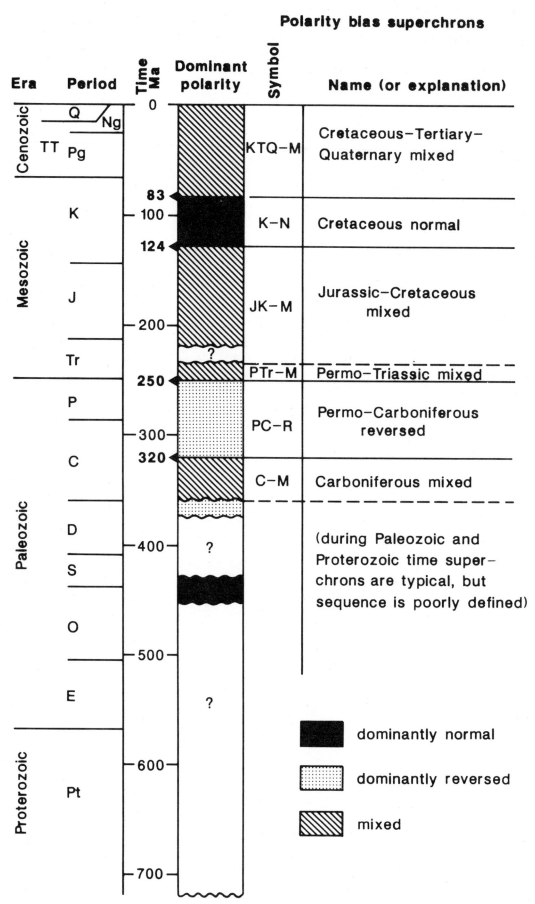

248 Ma (= early Lopingian) (Figure 6.10). The difference
between these values and the dates of 313 and 227 Ma used by
Irving & Pullaiah (1976) reflects revisions in the geologic time
scale and not different interpretations of the paleomagnetic
data.

The Carboniferous–Permian reversed superchron PC–R
is preceded by the Carboniferous mixed polarity superchron
C–M, the beginning of which is not well dated. The early
Paleozoic paleomagnetic record, although fragmentary, points
to the presence of superchrons similar to those in the younger
part of the geologic record. The Devonian field appears to
have been predominantly reversed and the Silurian and
Ordovician fields to have been predominantly normal (Irving
& Pullaiah 1976).

The phenomenon of polarity bias also appears to have
occurred during Proterozoic time but as yet the polarity
structure is poorly defined. Paleomagnetic data from
Laurentia (Irving & McGlynn 1976) suggest that during
Proterozoic time the field was 'normal' 77% of the time.
However, knowing whether a given Proterozoic
paleomagnetic direction corresponds to the polarity we now
call normal depends upon tracing continuously back from the
present to Proterozoic time the path along which the spin axis
has moved as viewed from North America. There are several
gaps in the record across which different pole paths can be
drawn, corresponding to different polarity histories.
Therefore, although the available data demonstrates that
polarity bias occurred during the Proterozoic Eon, it is still
uncertain whether the bias was normal or reversed.

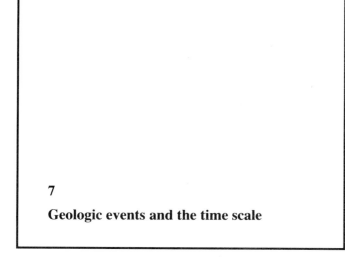

7

Geologic events and the time scale

7.1 Introduction

The function of a geologic time scale is to serve the purposes of correlation and thus build our understanding of Earth history. Earlier chapters of this book have documented the definition and numerical calibration of the time scale, and at the end of this chapter we set out our time scale as it stood in late 1988. Before doing so, we wish to emphasize the role of the GTS as a tool in correlation by providing charts of geologic events plotted against a linear time scale – the scale that we have developed for this work (Figures 7.2 to 7.9).

7.2 Natural time scales

Geologic events are what have created the stratigraphic record, and it is geologic events that we aim to discover and interpret in that record. The geologic time scale had its origin as a composite scale of such events at the regional level, but it has necessarily become increasingly conventional because research on a global scale has revealed the greater importance of inter-regional differences in the succession of events through geologic time. We have taken the view in this work that the chronostratic scale should be regarded as entirely conventional, and as touching the actual rock record only at GSSPs, namely Global Stratotype Sections and Points (see Section 1.3). This leaves the geoscientist free to erect whatever scales of local or regional or, indeed, global events that he chooses. The chronostratic scale stands as an entirely independent reference scale through which to compare the relative ages of geologic events of all kinds. These event scales are thus 'natural' or (better) phenomenal, in being based on successions of whatever events are interpreted from the record with the resources and techniques available at a particular time. Continuing changes both in scientific perspective and technology necessarily bring continuing changes in interpretation. Event scales are thus very much a product of the purpose and circumstances of their creation. They are of

necessity impermanent, by contrast with our intended greater durability of the conventional Geologic Time Scale.

It could be said that there is only one 'natural' geologic time scale, and that is the 'true' succession of geohistorical events that have shaped the stratigraphic record. Such a scale is the undoubted goal of stratigraphic research, but our concern is to keep such a conceptual ideal separate from the conventional scale, which has value only insofar as it remains completely independent from geohistorical interpretation. We are aware of the existence of alternative points of view. Some natural phenomenal sequences are discussed in Section 1.6 of the Introduction.

7.3 Geologic event charts

Figures 7.2 to 7.9 show our geologic time scale and selected geologic events plotted on a linear scale of geologic time. This presentation is similar to that of GTS 82, in which eight similar figures were published (Harland *et al.* 1982). All eight figures were combined into a single wall chart (Harland *et al.* 1982), to which was added additional information. Many of the events herein are taken from these works, along with others referenced below. There is neither the space nor the intention to justify the selection of events. Indeed, not even all the authors of this work could be agreed on the selection or dating of some events; the figures are indicative rather than definitive. Our primary purpose is to demonstrate the application of GTS 89 by portraying its relationship with key events in the stratigraphic record. The figures also summarize our calibration of the entire chronostratic scale, rather than the non-linear summaries in Chapters 2 and 3, and the partial linear summary in Chapter 6.

In addition to GTS 82 (Harland *et al.* 1982), we acknowledge the following sources of information: a Wallchart of Evolution (Fewtrell Smith 1981); the Mesozoic–Cenozoic Cycle Correlation Chart of Haq *et al.* (1987); and the Geological Time Table of Haq & Van Eysinga (1987). Valuable comments and additional contributions were received from M. House, P. Skelton and R. Spicer on biologic events, from D. Rowley on plate-tectonic events and from M. J. Hambrey on glacial events.

7.4 The geochronologic time scale GTS 89

The scale adopted for this publication is finalized in Figure 1.7 and the book cover and summarized in Table 1.2. The data are plotted in Figure 7.1.

In spite of its entirely independent construction and refinement, the new time scale is similar to GTS 82, made over six years ago, in turn largely based on a 1976 database. Few boundary ages have changed from 1982 by as much as 1%. Change has been in the form of a dramatic reduction in the tie-point chronogram errors, which are now about 2% of their age. We think this is the current uncertainty in the calibration of the Phanerozoic time scale.

Although the geoscientist can erect scales of local, regional or global events that have an age uncertainty of a few thousand years – as in the case of a change in magnetic polarity, or some geochemical anomaly such as an iridium-rich layer – only in the past hundred thousand years or so of geologic time is it possible to assign a numerical value to such

events with a similar precision. We foresee a great increase in
the precision of event correlation by a variety of methods –
magnetic, geochemical, climatic, stratigraphic, tectonic
modelling and the like – but no obvious method, or methods,
whereby the uncertainties in the numerical ages assigned to
events related to the chronostratic scale or the uncertainties in
the chronostratic scale names applied to precisely measured
ages can be dramatically reduced.

7.5 A new wall chart

GTS 1982 was published simultaneously as a book and as
a wall chart. The figures in that book (analogous to Figures 7.2
to 7.9 here) were adapted from the wall chart. For GTS 89 the
figures were drawn for the book first and the wall chart was
constructed later. See page i for contents and size of chart. See
also Appendix 6.

171

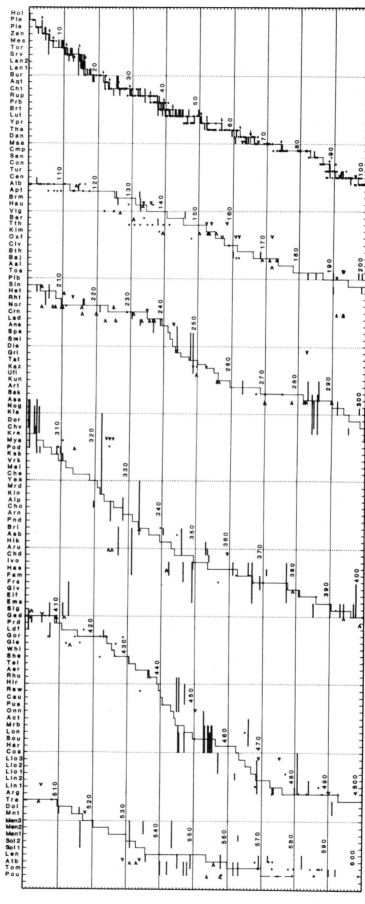

Figure 7.1. Comparison of the items used for calibration of the geologic time scale and the final scale summarized in Figure 1.7. The sloping zig-zag line is the final adjusted time scale, the vertical parts being the boundaries between stages and the horizontal parts the duration of the stages. Where the data are closely crowded the positions of the stage boundaries are emphasized by small arrows. The time scale can be compared with the distribution of data that defined (solid dots or vertical bars) or bracketed (arrowheads) the time scale. With few exceptions, and those are discussed at length in the text, the scale is entirely compatible with the data presented. Alternative interpretations are limited to a 1 to 2 per cent shift of the general position of the zig-zag line. The relative distribution of time between stages can be modified only to the extent that the hypothesis of equal duration of chrons is untrue or if sea-floor spreading rate changes are global, abrupt and numerous.

STRATIGRAPHIC POSITION OR RANGE OF ITEM

ISOTOPIC DATE (Ma)

Figure 7.2. Cenozoic linear time scale.

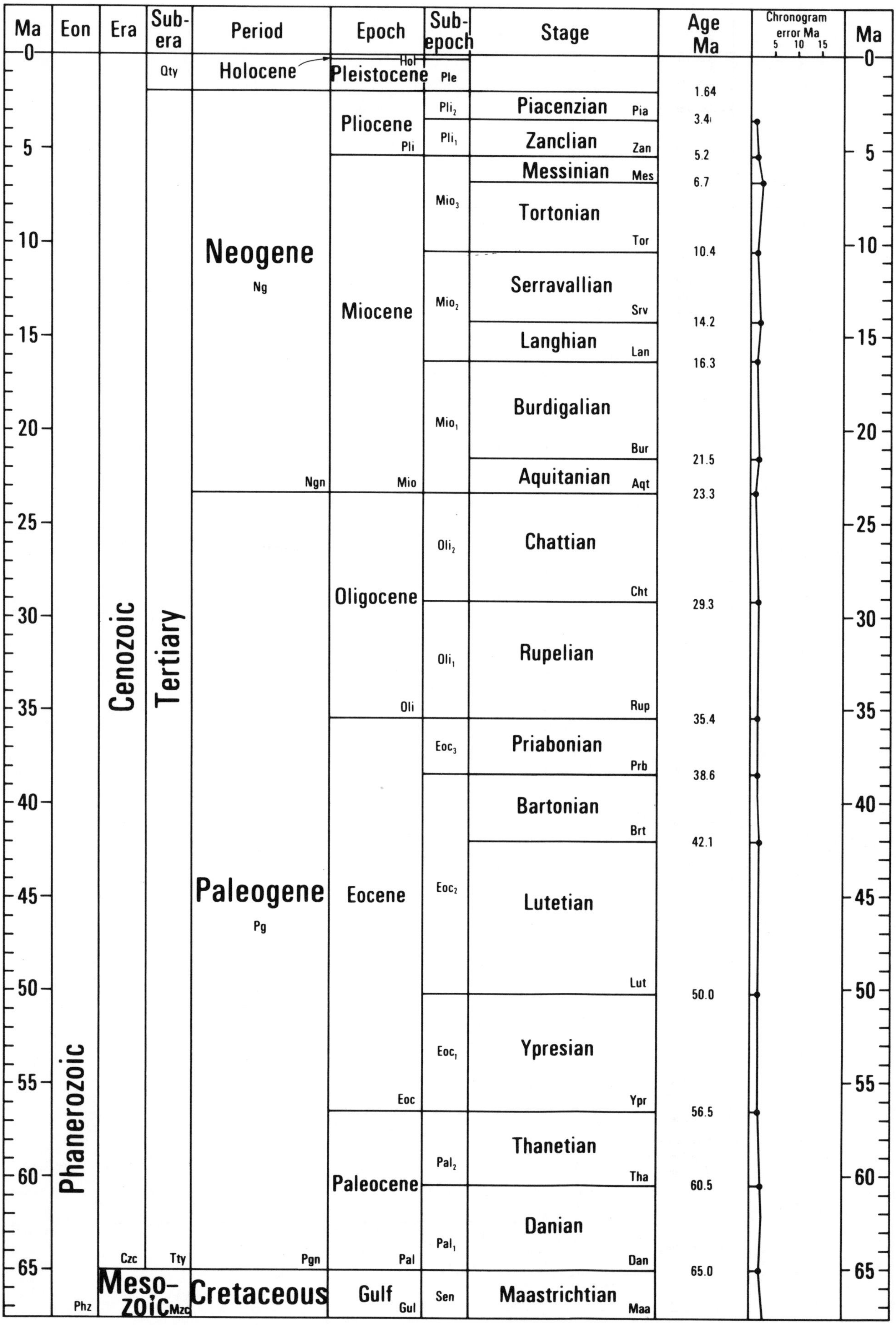

Ma	Eon	Era	Sub-era	Period	Epoch	Sub-epoch	Stage	Age Ma	Chronogram error Ma 5 10 15	Ma
0			Qty	Holocene	Pleistocene Hol	Ple		1.64		0
5				Neogene Ng	Pliocene Pli	Pli₂	Piacenzian Pia	3.4		5
						Pli₁	Zanclian Zan	5.2		
						Mio₃	Messinian Mes	6.7		
							Tortonian Tor	10.4		
10					Miocene Mio₂		Serravallian Srv	14.2		10
15						Mio₂	Langhian Lan	16.3		15
20						Mio₁	Burdigalian Bur			20
					Mio Ngn		Aquitanian Aqt	21.5 / 23.3		
25					Oligocene	Oli₂	Chattian Cht			25
30	Phanerozoic	Cenozoic	Tertiary					29.3		30
35				Paleogene Pg	Oli	Oli₁	Rupelian Rup	35.4		35
					Eocene	Eoc₃	Priabonian Prb	38.6		
40							Bartonian Brt	42.1		40
45						Eoc₂	Lutetian Lut			45
50							Ypresian Ypr	50.0		50
55					Eoc	Eoc₁		56.5		55
60					Paleocene	Pal₂	Thanetian Tha	60.5		60
65			Czc Tty	Pgn	Pal	Pal₁	Danian Dan	65.0		65
	Phz	Meso-zoic Mzc		Cretaceous	Gulf Gul	Sen	Maastrichtian Maa			

Figure 7.3. Some Cenozoic events plotted on linear time scale. Triangles indicate glacial events.

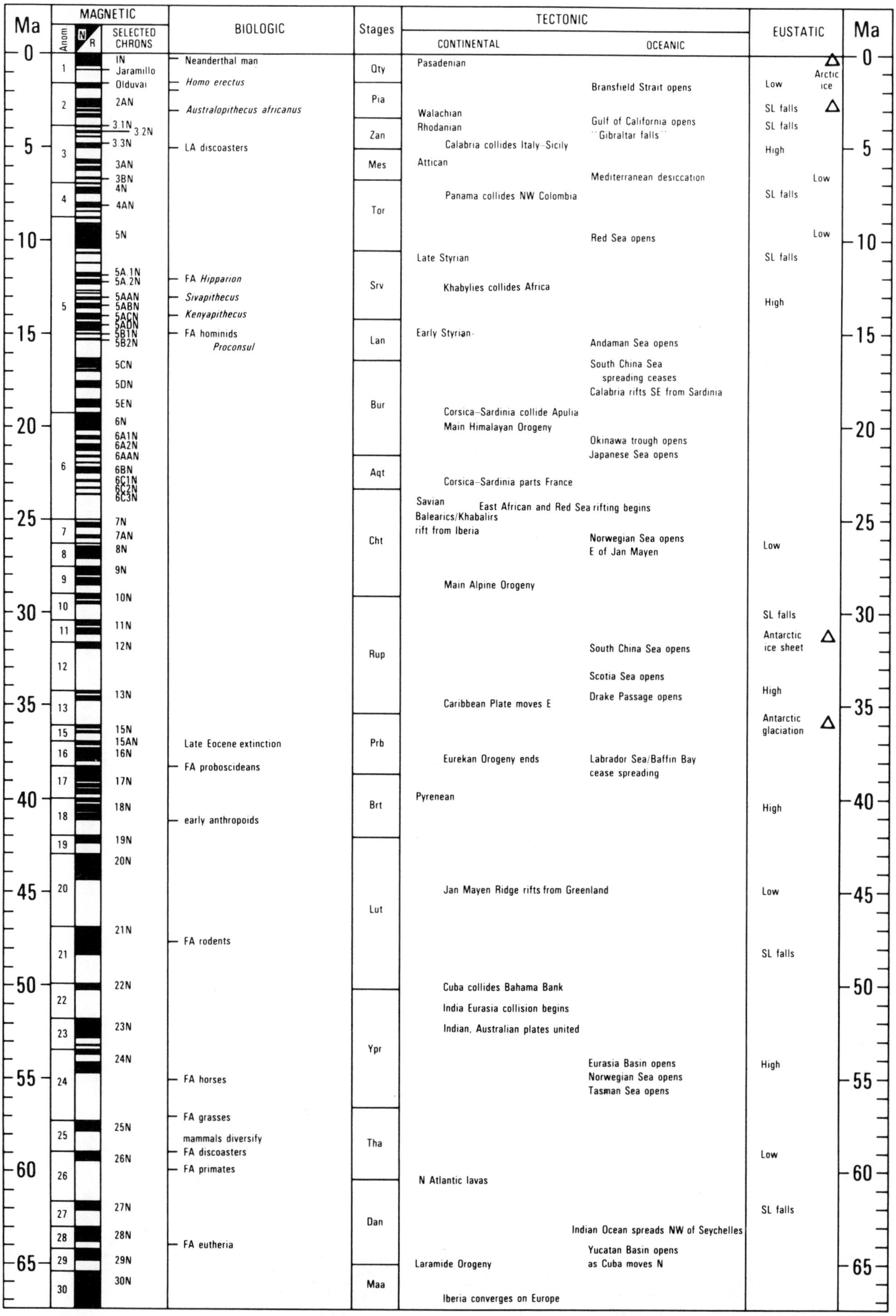

Ma	MAGNETIC Anom	MAGNETIC N/R	SELECTED CHRONS	BIOLOGIC	Stages	TECTONIC CONTINENTAL	TECTONIC OCEANIC	EUSTATIC	Ma
0	1		1N / Jaramillo / Olduvai	Neanderthal man / *Homo erectus*	Qty	Pasadenian		Low / Arctic ice △	0
	2		2AN	*Australopithecus africanus*	Pia	Walachian	Bransfield Strait opens	SL falls / SL falls △	
	3		3.1N 3.2N / 3.3N / 3AN / 3BN / 4N	LA discoasters	Zan / Mes	Rhodanian / Attican / Calabria collides Italy–Sicily	Gulf of California opens / "Gibraltar falls"	High	
5			4AN		Tor	Panama collides NW Colombia	Mediterranean desiccation	Low / SL falls	5
10			5N			Late Styrian	Red Sea opens	Low / SL falls	10
	5		5A.1N / 5A.2N / 5AAN / 5ABN / 5ACN / 5ADN / 5B1N / 5B2N	FA *Hipparion* / *Sivapithecus* / *Kenyapithecus* / FA hominids / *Proconsul*	Srv / Lan	Khabylies collides Africa / Early Styrian	Andaman Sea opens	High	
15			5CN / 5DN / 5EN		Bur	Corsica–Sardinia collide Apulia / Main Himalayan Orogeny	South China Sea spreading ceases / Calabria rifts SE from Sardinia / Okinawa trough opens / Japanese Sea opens		15
20	6		6N / 6A1N / 6A2N / 6AAN / 6BN / 6C1N / 6C2N / 6C3N		Aqt	Corsica–Sardinia parts France			20
25	7 / 8 / 9		7N / 7AN / 8N / 9N		Cht	Savian East African and Red Sea rifting begins / Balearics/Khabalirs rift from Iberia / Main Alpine Orogeny	Norwegian Sea opens E of Jan Mayen	Low	25
30	10 / 11 / 12		10N / 11N / 12N		Rup		South China Sea opens / Scotia Sea opens / Drake Passage opens	SL falls / Antarctic ice sheet △ / High	30
35	13 / 15 / 16		13N / 15N / 15AN / 16N	Late Eocene extinction / FA proboscideans	Prb	Caribbean Plate moves E / Eurekan Orogeny ends	Labrador Sea/Baffin Bay cease spreading	Antarctic glaciation △	35
40	17 / 18 / 19		17N / 18N / 19N / 20N	early anthropoids	Brt	Pyrenean		High	40
45	20 / 21		21N	FA rodents	Lut	Jan Mayen Ridge rifts from Greenland		Low / SL falls	45
50	22 / 23 / 24		22N / 23N / 24N	FA horses	Ypr	Cuba collides Bahama Bank / India Eurasia collision begins / Indian, Australian plates united	Eurasia Basin opens / Norwegian Sea opens / Tasman Sea opens	High	50
55	25 / 26		25N / 26N	FA grasses / mammals diversify / FA discoasters / FA primates	Tha	N Atlantic lavas		Low	55
60	27 / 28 / 29		27N / 28N / 29N	FA eutheria	Dan	Laramide Orogeny	Indian Ocean spreads NW of Seychelles / Yucatan Basin opens as Cuba moves N	SL falls	60
65	30		30N		Maa	Iberia converges on Europe			65

Figure 7.4. Mesozoic linear time scale.

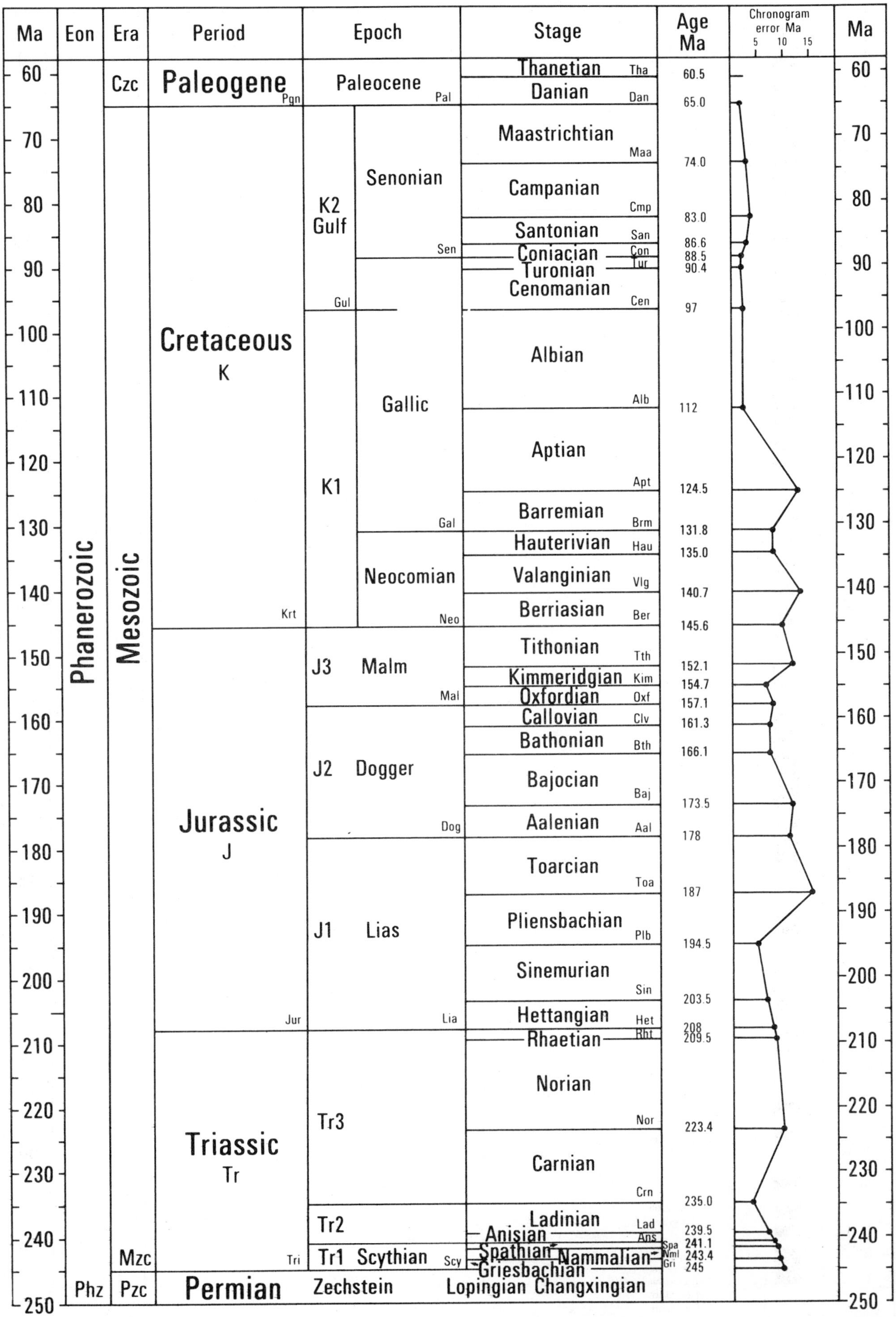

Ma	Eon	Era	Period	Epoch	Stage	Age Ma	Chronogram error Ma 5 10 15	Ma
60		Czc	**Paleogene** Pgn	Paleocene Pal	Thanetian Tha	60.5		60
					Danian Dan	65.0		
70					Maastrichtian Maa	74.0		70
80				K2 Gulf Senonian Sen	Campanian Cmp	83.0		80
					Santonian San	86.6		
90				Gul	Coniacian Con	88.5		90
					Turonian Tur	90.4		
					Cenomanian Cen	97		
100			**Cretaceous** K	Gallic	Albian			100
110					Alb	112		110
120				K1	Aptian Apt	124.5		120
130				Gal	Barremian Brm	131.8		130
					Hauterivian Hau	135.0		
				Neocomian	Valanginian Vlg	140.7		
140				Krt Neo	Berriasian Ber	145.6		140
150	Phanerozoic	Mesozoic		J3 Malm Mal	Tithonian Tth	152.1		150
					Kimmeridgian Kim	154.7		
					Oxfordian Oxf	157.1		
					Callovian Clv	161.3		
160				J2 Dogger	Bathonian Bth	166.1		160
170				Dog	Bajocian Baj	173.5		170
			Jurassic J		Aalenian Aal	178		
180				J1 Lias	Toarcian Toa	187		180
190					Pliensbachian Plb	194.5		190
200					Sinemurian Sin	203.5		200
				Jur Lia	Hettangian Het	208		
210					Rhaetian Rht	209.5		210
220			**Triassic** Tr	Tr3	Norian Nor	223.4		220
230					Carnian Crn	235.0		230
240		Mzc		Tr2	Ladinian Lad	239.5		240
					Anisian Ans	241.1		
				Tri Tr1 Scythian Scy	Spathian Spa / Nammalian Nml	243.4		
					Griesbachian Gri	245		
250	Phz	Pzc	**Permian**	Zechstein	Lopingian Changxingian			250

Figure 7.5. Some Mesozoic events plotted on linear time scale. Encircled numbers 1, 2 and 3 refer to three Cretaceous anoxic events.

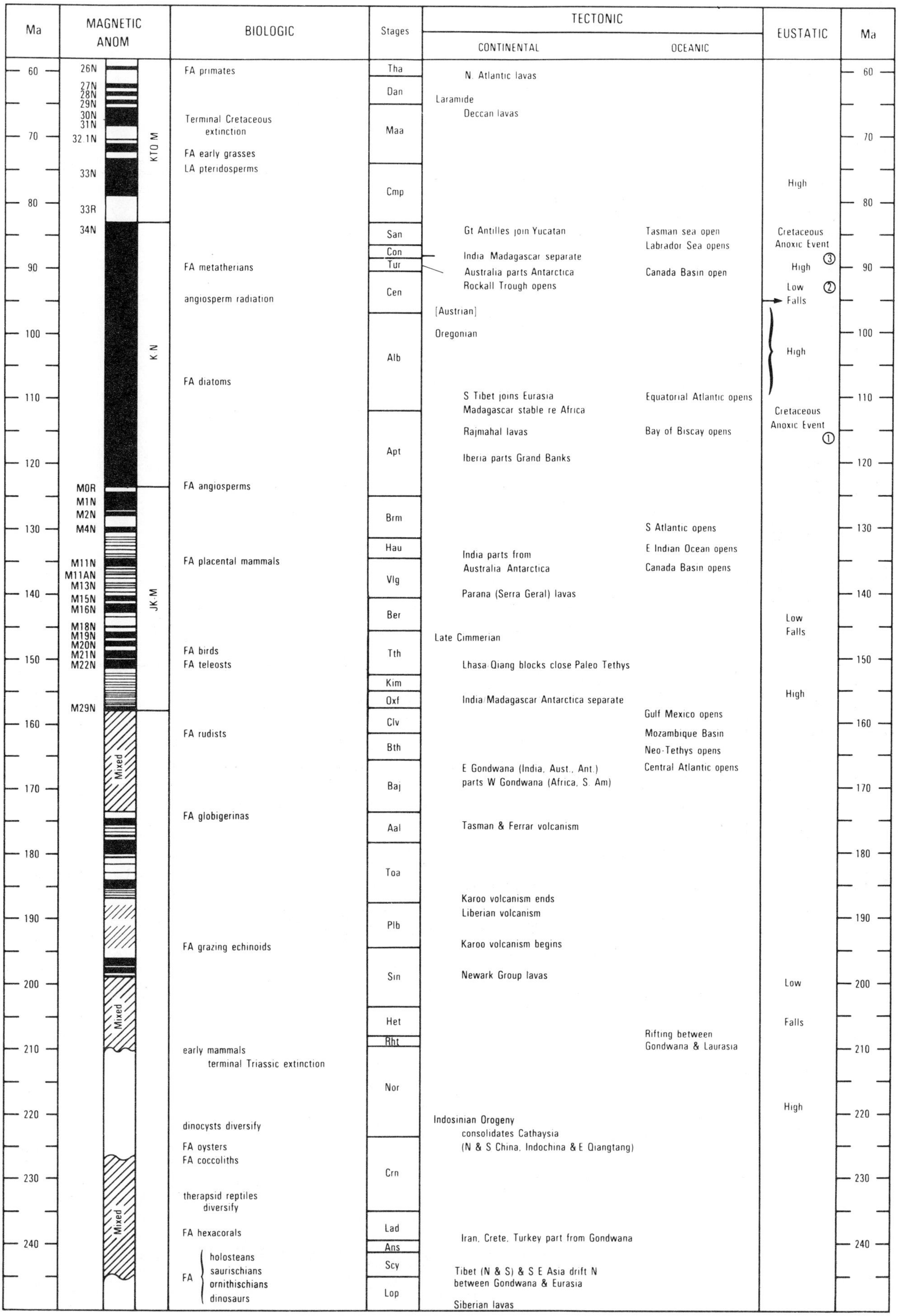

Figure 7.6. Paleozoic linear time scale.

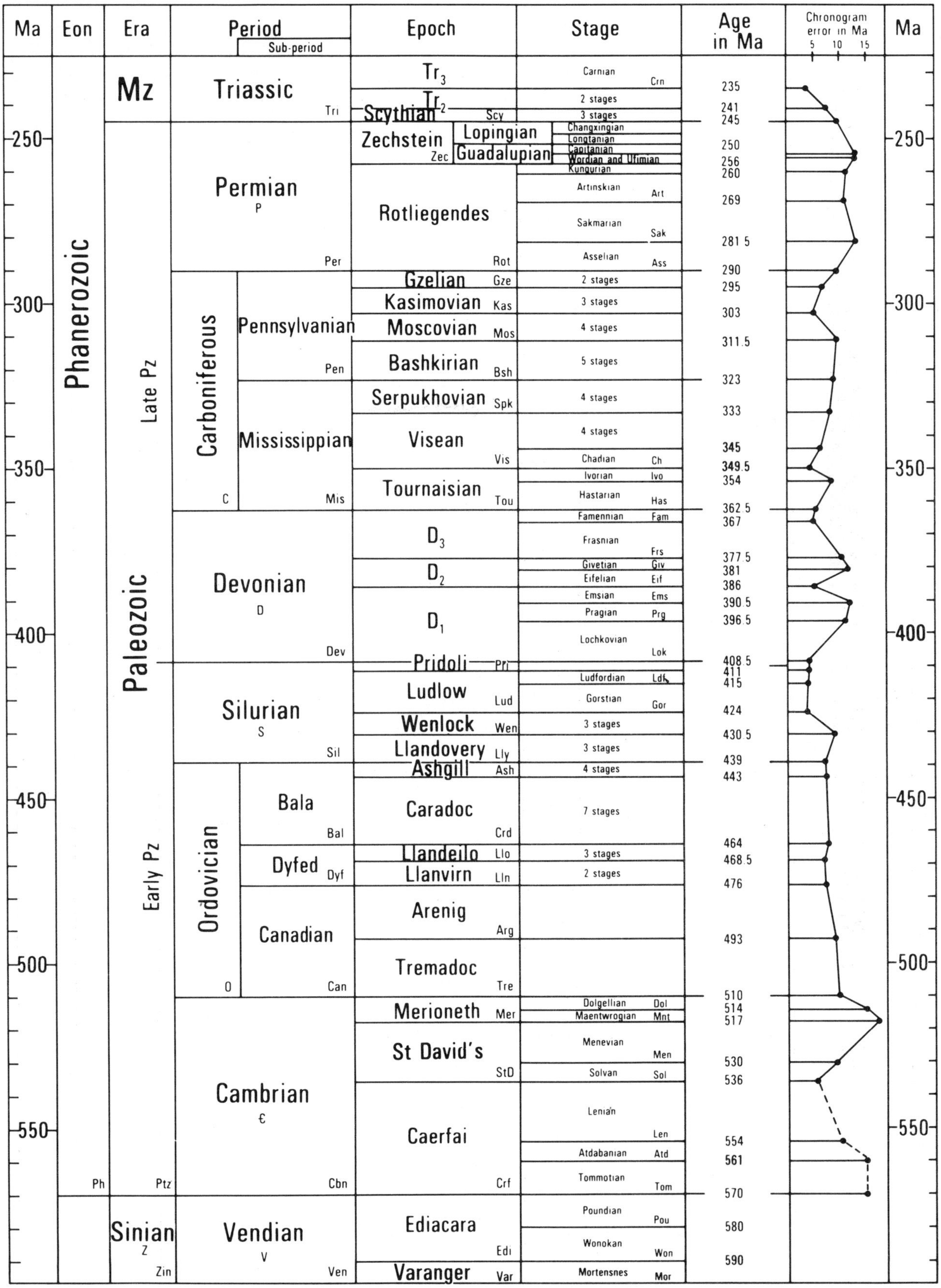

Ma	Eon	Era	Period / Sub-period	Epoch	Stage	Age in Ma	Chronogram error in Ma 5 10 15	Ma
		Mz	Triassic / Tri	Tr₃	Carnian / Crn	235		
				Tr₂	2 stages	241		
			Scythian / Scy		3 stages	245		
-250			Permian / P / Per	Zechstein / Zec	Lopingian — Changxingian / Longtanian	250		-250
				Guadalupian — Capitanian / Wordian and Ufimian / Kungurian	256 / 260			
				Rotliegendes / Rot	Artinskian / Art	269		
					Sakmarian / Sak	281.5		
					Asselian / Ass	290		
-300			Carboniferous / C	Pennsylvanian / Pen	Gzelian / Gze	2 stages	295		-300
					Kasimovian / Kas	3 stages	303		
					Moscovian / Mos	4 stages	311.5		
					Bashkirian / Bsh	5 stages	323		
		Late Pz		Mississippian / Mis	Serpukhovian / Spk	4 stages	333		
					Visean / Vis	4 stages	345		
						Chadian / Ch	349.5		
-350				Tournaisian / Tou	Ivorian / Ivo	354		-350
					Hastarian / Has	362.5		
					Famennian / Fam	367		
			Devonian / D / Dev	D₃	Frasnian / Frs	377.5		
				D₂	Givetian / Giv	381		
					Eifelian / Eif	386		
	Phanerozoic			D₁	Emsian / Ems	390.5		
-400		Paleozoic			Pragian / Prg	396.5		-400
					Lochkovian / Lok	408.5		
			Silurian / S / Sil	Pridoli / Pri		411		
				Ludlow / Lud	Ludfordian / Ldf	415		
					Gorstian / Gor	424		
				Wenlock / Wen	3 stages	430.5		
				Llandovery / Lly	3 stages	439		
			Ordovician / O	Bala / Bal	Ashgill / Ash	4 stages	443		
-450		Early Pz			Caradoc / Crd	7 stages	464		-450
				Dyfed / Dyf	Llandeilo / Llo	3 stages	468.5		
					Llanvirn / Lln	2 stages	476		
				Canadian / Can	Arenig / Arg		493		
					Tremadoc / Tre		510		
-500				Merioneth / Mer	Dolgellian / Dol	514		-500
					Maentwrogian / Mnt	517		
				St David's / StD	Menevian / Men	530		
			Cambrian / Є / Cbn		Solvan / Sol	536		
-550				Caerfai / Crf	Lenian / Len	554		-550
					Atdabanian / Atd	561		
					Tommotian / Tom	570		
	Ph / Ptz	Sinian / Z / Zin	Vendian / V / Ven	Ediacara / Edi	Poundian / Pou	580		
					Wonokan / Won	590		
				Varanger / Var	Mortensnes / Mor			

Figure 7.7. Some Paleozoic events plotted on linear time scale. Triangles indicate glacial events, those with dots are well defined. Names in square brackets are Stille's (1924) orogenic phases.

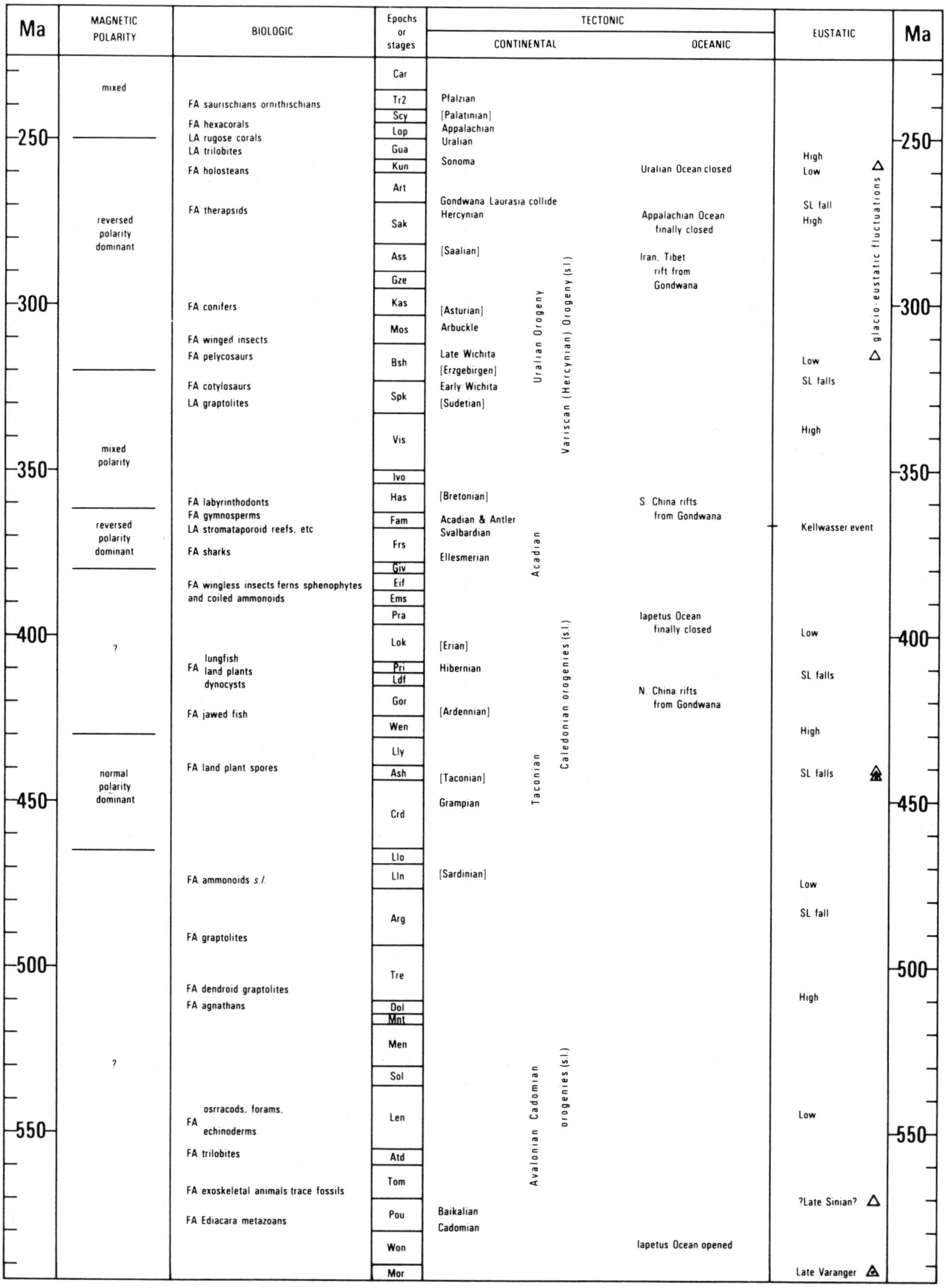

Ma	Magnetic Polarity	Biologic	Epochs or stages	Tectonic – Continental	Tectonic – Oceanic	Eustatic	Ma
	mixed		Car				
		FA saurischians ornithischians	Tr2	Pfalzian			
250		FA hexacorals	Scy	[Palatinian]			
		LA rugose corals	Lop	Appalachian			
		LA trilobites	Gua	Uralian			
		FA holosteans	Kun	Sonoma	Uralian Ocean closed	High / Low	250
			Art				
	reversed polarity dominant	FA therapsids	Sak	Gondwana Laurasia collide / Hercynian	Appalachian Ocean finally closed	SL fall / High	
			Ass	[Saalian]	Iran, Tibet rift from Gondwana		
			Gze				
300		FA conifers	Kas	[Asturian]			300
			Mos	Arbuckle			
		FA winged insects	Bsh	Late Wichita		Low	
		FA pelycosaurs		[Erzgebirgen]		SL falls	
		FA cotylosaurs	Spk	Early Wichita			
		LA graptolites		[Sudetian]			
			Vis			High	
	mixed polarity						
350			Ivo				350
		FA labyrinthodonts	Has	[Bretonian]	S. China rifts from Gondwana		
	reversed polarity dominant	FA gymnosperms	Fam	Acadian & Antler		Kellwasser event	
		LA stromatoporoid reefs, etc		Svalbardian			
		FA sharks	Frs	Ellesmerian			
			Giv				
		FA wingless insects ferns sphenophytes and coiled ammonoids	Eif				
			Ems				
			Pra		Iapetus Ocean finally closed		
400	?	FA lungfish	Lok	[Erian]		Low	400
		FA land plants	Pri	Hibernian		SL falls	
		FA dynocysts	Ldf		N. China rifts from Gondwana		
		FA jawed fish	Gor	[Ardennian]			
			Wen			High	
			Lly				
	normal polarity dominant	FA land plant spores	Ash	[Taconian]		SL falls	
450			Crd	Grampian			450
			Llo				
		FA ammonoids s.l.	Lln	[Sardinian]		Low	
			Arg			SL fall	
		FA graptolites					
500		FA dendroid graptolites	Tre			High	500
		FA agnathans	Dol				
			Mnt				
			Men				
	?		Sol				
		FA osrracods, forams, echinoderms	Len			Low	
550		FA trilobites	Atd				550
		FA exoskeletal animals trace fossils	Tom				
			Pou	Baikalian		?Late Sinian? △	
		FA Ediacara metazoans		Cadomian			
			Won		Iapetus Ocean opened		
			Mor			Late Varanger △	

Spanning tectonic labels: *Uralian Orogeny*, *Variscan (Hercynian) Orogeny (s.l.)*, *Acadian*, *Taconian*, *Caledonian orogenies (s.l.)*, *Avalonian Cadomian orogenies (s.l.)*. Eustatic right margin: *glacio-eustatic fluctuations*.

Figure 7.8. Precambrian linear time scale.

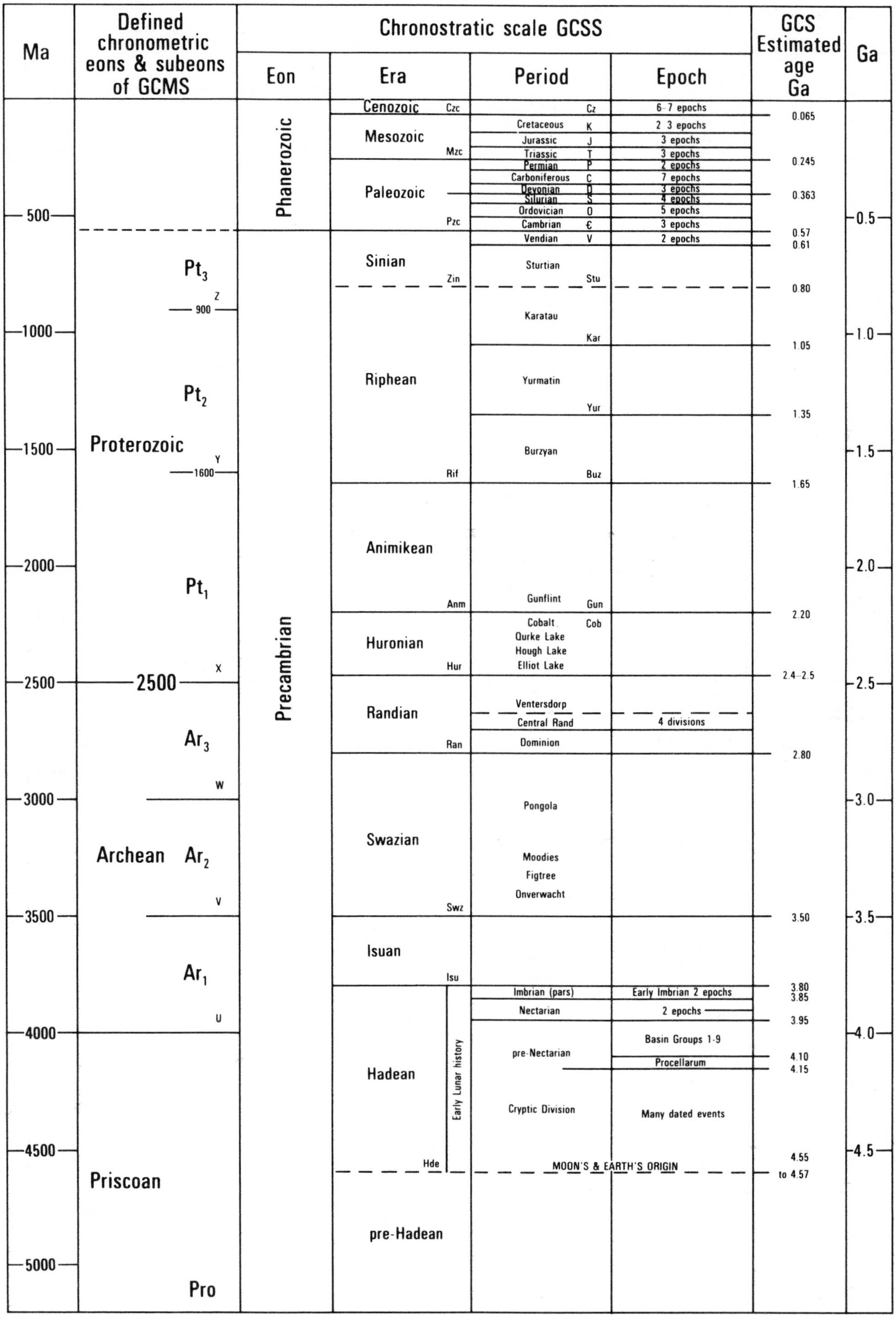

Ma	Defined chronometric eons & subeons of GCMS	Chronostratic scale GCSS				GCS Estimated age Ga	Ga
		Eon	Era	Period	Epoch		
		Phanerozoic	Cenozoic Czc		Cz 6–7 epochs	0.065	
			Mesozoic Mzc	Cretaceous K	2 3 epochs		
				Jurassic J	3 epochs		
				Triassic T	3 epochs	0.245	
			Paleozoic Pzc	Permian P	2 epochs		
				Carboniferous C	7 epochs		
				Devonian D	3 epochs		
				Silurian S	4 epochs	0.363	
				Ordovician O	5 epochs		
500				Cambrian Є	3 epochs	0.57	0.5
		Precambrian	Sinian Zin	Vendian V	2 epochs	0.61	
	Pt₃			Sturtian Stu		0.80	
	Z 900			Karatau			
1000				Kar		1.05	1.0
	Pt₂		Riphean	Yurmatin			
				Yur		1.35	
1500	Proterozoic			Burzyan			1.5
	Y 1600		Rif	Buz		1.65	
			Animikean				
2000	Pt₁						2.0
			Anm	Gunflint Gun		2.20	
			Huronian	Cobalt Cob			
				Qurke Lake			
				Hough Lake			
	X		Hur	Elliot Lake		2.4–2.5	
2500	2500		Randian	Ventersdorp			2.5
				Central Rand	4 divisions		
	Ar₃		Ran	Dominion		2.80	
3000	W		Swazian	Pongola			3.0
	Archean Ar₂			Moodies			
				Figtree			
	V		Swz	Onverwacht			
3500			Isuan			3.50	3.5
	Ar₁		Isu				
				Imbrian (pars)	Early Imbrian 2 epochs	3.80 / 3.85	
	U			Nectarian	2 epochs	3.95	
4000			Hadean (Early Lunar history)	pre-Nectarian	Basin Groups 1-9	4.0	
					Procellarum	4.10 / 4.15	
				Cryptic Division	Many dated events		
4500	Priscoan		Hde	MOON'S & EARTH'S ORIGIN		4.55 to 4.57	4.5
			pre-Hadean				
5000	Pro						

Figure 7.9. Some global events plotted on linear time scale. The following symbols qualify the adjacent names of events: small circles with dots = distinctive biotas; circles with crosses = extinction events; the letter Omega = orogeny; v = volcanic; inverted v = intrusive; triangle = glaciation; asterisk = bolide impact.

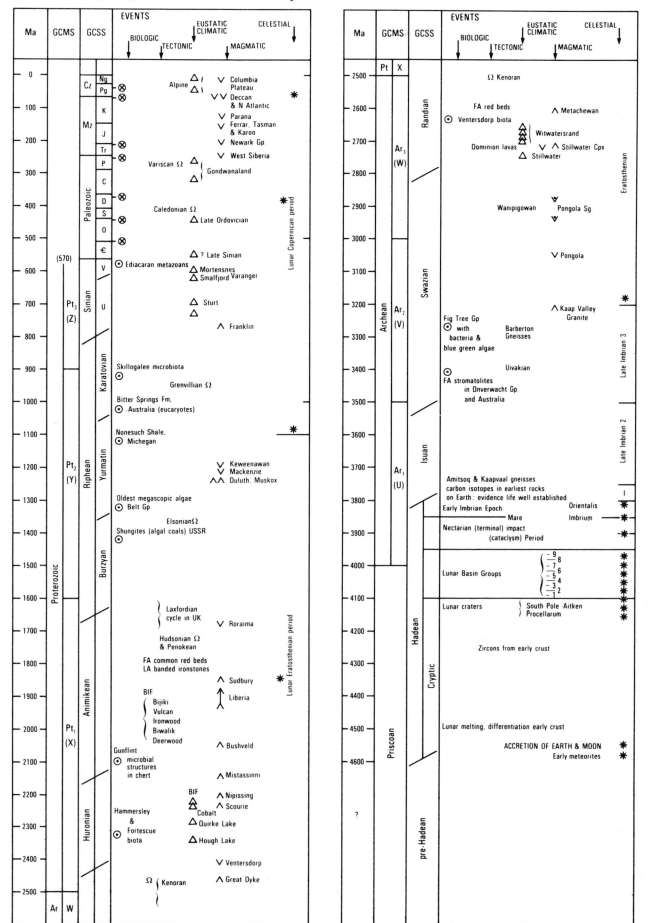

APPENDIX 1

Origins of some stage names
Reprinted with permission from *General Stratigraphy* by J. W. Gregory & B. H. Barrett, Methuen's Geological Series (General Editor J. W. Gregory), London 1931, pages 240–257 as **'List of formations'**.

Name	System	Author	Year	Locality or Derivation	Reference	Use
Aalenian	M. Jur.	Mayer-Eymar	1864	Aalen, Wurtemburg	Tabl. synchr.	
Acadian	M. Camb.	Dawson	1855	Acadia, Canada	Acad. Geol.	
Acheulian	L. Pleist.	Mortillet	1878	St. Acheul, France	Congr. géol. Paris, p.179	
Aftonian	Pleist.	Chamberlin and Salisbury	1906	Afton, Iowa	Geol., iii, p.383	1st American interglacial
Albian	M. Cret.	d'Orbigny	1842	Aube, France	Pal. fr., Crét., ii	
Alexandrian	L. Silur.	Savage	1908	Alexander Co., U.S.A.	Am. Jour. Sci (4), 25, pp.433–4	= Llandovery
Algonkian	U. Pampal.	Van Hise	1892	Algonquin, Canada	Bull. U.S.G.S., 86, p.475	
Alleghenian	M. Carb.	Prosser	1901	Allegheny, U.S.A.	Am. Jour. Sci. (4), xi, p.199	pt. Moscovian
Animikean	U. Pampal	J. S. Hunt	1873	Animikie, L. Superior	Tr. Amer. Inst. Min. Eng., i, pp.331–45, ii, pp.58–9	
Anisian	M. Trias.	Waagen and Diener	1895		Akad. Wiss. Wien, civ.	
Anversian	U. Mio.	Cogels	1879	Antwerp, Belgium	Explic. planchettes d'Hoboken, etc., Carte géol. Belg.	
Aptian	L. Cret.	d'Orbigny	1843	Aptia, France	Pal. fr., Crét., ii	
Aquilonian	U. Jur.	Pavlov	1892	Aquilo, France	Argile de Speeton, p.192	
Aquitanian	U. Olig.	Mayer-Eymar	1858	Aquitaine, France	Acta Schw. Nat. Ges. Trogen, p.188	
Arenig	L. Ord.	Sedgwick	1847	Arenig, Wales		
Argovian	M. Jur.	Marcou	1848	Argovie, Switz.	Jura Salinois, p.116	= Corallian
Arnusian	L. Pleist.	Mayer-Eymar	1884	Arno, Italy	Classif. de Terre	= Sicilian
Artinskian	L. Perm.	Karpinsky	1874	Artinsk, Russia	Gorn. Journ., ii	
Ashgillian	U. Ord.	Marr	1905	Ashgill, Lake District	Q. J. G. S., lxi, p.lxxxiv	
Astian	M. Plio.	Rouville	1853	Asti, Italy	Descr. géol. Montpellier, p.155	
Aturian	U. Cret.	de Lapparent and Munier	1893	Aturia, Spain	Geol., 3rd ed., p.1150.	Campanian
Autunian	L. Perm.	de Lapparent	1893	Autun, France	*ibid.*, p.886	= Artinskian
Auversian	U. Eoc.	Dollfus	1880	Auvers, France	Expos. géol.	= Ledian
Bajocian	M. Jur.	d'Orbigny	1847	Bayeux, France	Pal. fr., Jura, i, p.606	Inf. Ool.
Balcombian	Olig.	Hall and Pritchard	1902	Balcombe, Vict.	Proc. R. S. Vict., n.s. xiv, p.78	
Barremian	L. Cret.	Coquand	1861	Barrême, France	Mem. Soc. Emul. Provence, i, p.127	
Bartonian	U. Eoc.	Mayer-Eymar	1857	Barton, England	Verh. Schweiz. Nat. Ges., Trogen, p.178	
Bathonian	M. Jur.	Omalius	1843	Bath, England	Précis Géol., p.470	
Bedoulian	Cret.	Toucas	1888	La Bedoule, France	Bull. Soc. géol. Fr. (3), xvi, p.921	
Bernician	L. Carb.	Woodward	1856	Bernicia, S. Scotland	Man. Moll.	
Berriasian	L. Cret.	Coquand	1876	Berrias, France	Bull. Soc. géol. Fr. (3), iii, p.685	
Bolderian	U. Mio.	Dumont	1849	Bolderberg, Belgium	Bull. Acad. Sci., Belg.	
Bononian	U. Jur.	Blake	1888	Bononia, Boulogne	Cited in de Lapparent, Bull. Soc. Géol. Fr. (3), xxi, p.462	Portlandian

Name	System	Author	Year	Locality or derivation	Reference	Use
Bradfordian	M. Jur.	Desor	1859	Bradford, Wilts.	Etud. Jura Neuchat, p.85	L. Bath.
Burdigalian	L. Mio.	Depéret	1892	Bordeaux, France	Bull. Soc. géol. Fr. (3), xx, p.155; xxi, p.170	
Butleyan	M. Plio.	Harmer	1900	Butley, England	Q. J. G. S., xvi, p.721	= U. Astian
Callovian	M. Jur.	d'Orbigny	1849	Kelloway, England	Pal. fr., Jura, i, p.608	
Campanian	Cret.	Coquand	1857	Campania, N. France	Bull. géol. Fr. (2), xiv, p.887	
Canadian	L. Ord.	Dana	1874	Canada	Am. Jour. Sci. (3), viii, p.214	Arenig
Caradocian	U. Ordov.	Murchison	1839	Caradoc, Wales	Silurian System	
Cartennian	L. Mio.	Pomel	1858	Tenes, Algeria	C. R. Acad. Sci., 1858, p.480	Burdigalian
Casselian	U. Olig.	Dollfus	1910	Cassel, Hesse	Bull. Soc. géol. Fr. (4), x, p.582	Chattian
Casterlian	L. Plio.	Dumont; van den Broeck	1874	Casterlé, Belgium	Ann. Soc. R. Mal. Belg., xvii, iii–viii	= Plaisancian
Cayugan	U. Silur.	Clark and Schuchert	1898	L. Cayuga, N. Y.	Science, n.s., x, p.876	
Cenomanian	L. Cret.	d'Orbigny	1852	Le Mans, France	Cours él., Pal., ii	
Champlainian	M. Ord.	Emmons	1842	L. Champlain, U.S.A.	Geol. N. Y., pp.100–1	
Charmouthian	L. Jur.	Mayer-Eymar	1864	Charmouth, England	Tabl. synchr.	
Chatauquan	U. Dev.	Clark and Schuchert	1898		Science, n.s., x, pp.874–78	
Chattian	U. Olig.	Fuchs	1894	Chatti, tribe in Hesse	K.-ungar. geol. Anstalt. Mitt. x, p.173	
Chillesfordian	L. Pleist.	Prestwich	1849	Chillesford, Suffolk	Q. J. G. S. v, p.345	
Cincinnatian	U. Ord.	Meek and Worthen	1865	Cincinnati, U.S.A.	Pr. Acad. Nat. Sci. Philad., xvii, p.155	
Coblenzian	L. Dev.	Dumont	1848	Coblenz, Germany	Mem. Ter. Ard., 2nd pt., p.183	
Comanchean	L. Cret.	R. T. Hill	1887	Comanche, U.S.A.	Am. Jour. Sci. (3), xxxiv, pp.287–309	
Conemaughan	U. Carb.	Prosser	1901	Conemaugh, U.S.A.	Ibid., (4), xi, p.199	pt. Uralian
Coniacian	U. Cret.	Coquand	1857	Cognac, France	Bull. Soc. géol. Fr. (2), xiv, p.882	
Corallian	M. Jur.	Thurmann	1832	England		
Couvinian	M. Dev.	Dupont	1885	Couvin, Belgium	Carte géol. Belge	
Croixian	U. Camb.	Walcott	1912	St. Croix, Min.	Smithson, Misc. Coll., 57, x, pp.306–7	Pacific coast
Cromerian	L. Pleist.	Harmer	1900	Cromer, England	Q. J. G. S., lvi, 725	= Cromer Forest, Bed.
Cuisian	L. Eoc.	Dollfus	1880	Cuise-la-Motte, Fr.	Soc. géol. Norm., p.589	= London Clay
Danian	U. Cret.	Desor	1850	Denmark		
Delémontian	U. Olig.	Greppin	1867	Bernese Jura	Essai sur Jura, p.128	Aquitanian
Demetian	M. Carb.	S. P. Woodward	1856		Man. Moll., p.409	Coal Meas. and M. Grit.
Deutozoic	U. Paleoz.	Lapworth	1888	Deuteros = second	Intro. Textbk. Geol., p.512	
Diestian	L. Plio.	Dumont	1839	Diest, Belgium	Acad. Sci. Belg.	Plaisancian
Dimetian	Pampal.	Hicks	1878	Dimetia, Wales	Rep. Brit. Assoc.	
Dinantian	L. Carb.	Lapparent	1893	Dinant, Belgium	Traité Géol., p.819	
Dittonian	Up. Dev.	W. W. King	1921		Proc. G.S., 1921, p.124	Up. O.R.S.
Divesian	M. Jur.	Renevier	1874	Dives, France	Tabl. Terr.	Oxfordian
Doinyan	Eoc.	Gregory	1896	Doinyo Lersubugo, Kenya Col.	Great Rift Valley, p.235	
Domerian	L. Jur.	Bonarelli	1894	Mt. Domero, Italy	Acad. Turin, xxx	
Donzérian	Cret.	Torcapel	1882	Donzère, France	Urg. du Languedoc, p.4	Barremian
Dordonian	U. Cret.	Coquand	1857	Dordogne, France	Bull. Soc. géol. Fr. (2), xiv, p.882	Danian
Downtonian	U. Silur.	Lapworth	1879	Downton Castle, Eng.	Ann. Mag. N. H. (5), iii, p.455	
Dubisian	U. Jur.	Desor	1859	Doubs, Switzerland	Jura Neuch., p.45	Purbeckian
Durntenian	M. Pleist.	Mayer-Eymar	1881	Durnten, Switzerland	Classif. internat.	Interglacial
Dyas	Permian	Marcou	1859		Bibliot. Univers. Geneve, 1859	
Eifelian	M. Dev.	Dumont	1848	Eifel, Germany	Mem. Ard., p.382	
Elberfeldian	U. Carb.	Mayer-Eymar	1881	Elberfeld, Germany	Classif. internat.	Uralian
Emscherian	U. Cret.	de Lapparent	1893		Géol., 3rd ed., p.1150	
Erian	Dev.	Dawson	1871	L. Erie, N. America Indian Tribe	Rep. Geol. Surv. Can., p.10	
Etcheminian	L. Carb.					
Etrurian	L. Olig.	Pareto	1865	Etruria, Italy	Bull. Soc. géol. Fr., xxii, p.215	Tongrian
Falunian	Mio.	d'Orbigny	1852	Faluns of France	Cours. élém., ii, p.775	
Famennian	U. Dev.	Gosselet	1880	Fammene, Belgium	Esq. géol. N. France, p.107	
Firmitian	U. Olig.	Dollfus	1880	Ferté-Alais, France	Expos. géol. Havre, p.600	Aquitanian
Flandrian	L. Eoc.	Mayer-Eymar	1881	Flanders	Classif. internat.	Montian
Forestian	U. Pleist.	Jas. Geikie	1895	Forest Beds	Q. J. G. S., p.250	4th and 5th Interglacials
Fossanian	U. Plio.	Sacco	1886	Fossans, Italy	Bull. Soc. géol. Fr. (3), xv, p.27	Astian

Name	System	Author	Year	Locality or derivation	Reference	Use
Franconian	M. Trias.	de Lapparent	1883	Franconia, Germany	Géol. p.793	Muschelkalk
Frasnian	U. Dev.	Gosselet	1880	Frasne, Belgium	Esq. géol. N. France, p.95	
Gargasian	L. Cret.	Kilian	1887	Gargas, France	Ann. géol. univers., iii, p.314	Aptian
Garumnian	U. Cret.	Leymerie	1862	Garonne, France	Bull. Soc. géol. Fr. (2), xix, p.1107	Part of Danian
Gedgravian	M. Plio.	Harmer	1900	Gedgrave	Q. J. G. S. lvi, p.707	Plaisancian
Gedinnian	L. Dev.	Dumont	1848	Gédinne, Belgium	Mem. Terr. Ard., p.167	
Georgian	L. Camb.	Hitchcock	1861	Georgia, N. America	Bull. U.S. Geol. Surv., No. 81	
Givetian	M. Dev.	Gosselet	1880	Givet, France	Esq. géol. N. France, p.88	
Glyptician	M. Jur.	Etallon	1861	Zone of *Glypticus hieroglyphicus*	Mem. Emul. Doubs, vi, p.53	Corallian
Gothlandian	Silur.	de Lapparent	1893	Gothland, Baltic	Géol., p.748	
Gshelian	U. Carb.	Nikitin	1890	Gshel, Russia	Mem. géol. Russ., v, p.156	Uralian
Guadalupian	L. Perm.	Girty	1902	Guadalupe Mts. U.S.A.	Am. Jour. Sci. (4), xiv, p.368	= Artinskian
Hauterivian	L. Cret.	Renevier	1874	Hauterive, Switz.	Tabl. Terr. sed.	Part of Neocomian
Heathcotian	Up. Camb.	Gregory	1902	Heathcote, Victoria	Proc. R.S. Vict., n.s., xv, p.148	
Heersian	L. Eocene	Dumont	1851	Heers, Belgium		Thanetian
Helderbergian	L. Dev.	Clark and Schuchert	1898	Helderberg Mts., N.Y.	Science, n.s., x, pp.874–78	
Helvetian	M. Mio.	Mayer-Eymar	1857	Helvetia, Switzerland	Verh. Schweiz. Nat. Ges., Trogen, Table	
Helvetian	Pleist.	J. Geikie	1895	Helvetia, Switzerland	Jour. Géol., p.248	
Hettangian	L. Jur.	Renevier	1864	Hettange, France	Not. Alp. vaud., i, p.51	
Humbletonian	U. Perm.	Mayer-Eymar	1888	Humbleton, Yorks	Tabl. Terr. séd.	Thuringian
Hunsruckian	L. Dev.	Dumont	1848	Hunsruck, Germany	Mem. Terr. Ard., p.194	Siegenian
Huronian	U. Pampal.	Logan	1850	L. Huron, Canada	Rep. Geol. Surv., Canada	
Icenian	U. Plio.	Harmer	1900	Norwich Crag	Q. J. G. S., lvi, p.721	= Sicilian
Igualadian	M. Eoc.	Vézian	1858	Igualada, Spain	Bull. Soc. géol. Fr. (2), xv, p.438	Lutetian
Illinoian	Pleisto.	Chamberlin and Salisbury	1906	Illinois	Geol., iii, p.383	= Rissian
Iowan	Pleisto.	Chamberlin and Salisbury	1906	Iowa	*Ibid.*, iii, p.383	4th Amer. Glaciation
Jacksonian	U. Eocene	Heilprin	1888	Jackson, Alabama	Congr. géol. inter. Rep. Amer. Com., p.814	
Janjucian	Mio.	Hall and Pritchard	1902	Janjuc, Victoria	Proc. R. Soc. Vict., n.s., xiv, p.78	
Jeurian	M. Olig.	Dollfus	1880	Jeurre, France	Expos. géol. Havre, p.599	Tongrian
Jovarian	M. Olig.	Dollfus	1880	Jonarre, France	Expos. géol. Havre, p.590	Tongrian
Juvavian	U. Trias.	Mojsisovics	1892	Juvavo, Salzburg	Sitzb. Ak. Wiss. Wien, p.777	
Kansan	Pleist.	Chamberlin and Salisbury	1906	Kansas	Geol., iii, p.383	= Mindelian
Kapitian	Cret. Eoc.	Gregory	1896	Kapiti, Kenya Col.	Great Rift Valley, p.235	
Karnian	U. Trias.	Mojsisovics	1869	Carnic Alps	Verh. géol. Reichs., p.65	
Keewatin	Pampal.	Lawson	1888	Indian word for W. wind	Archaean Geol., p.70	
Keweenawan	Pampal.	Brooks	1876	Keweenaw, America	Am. Jour. Sci. (3), xi, pp.206–11	
Kimeridgian	U. Jur.	Thurmann	1832	Kimeridge, England	*Cf.* d'Orbigny in Pal. fr. Jur., i, p.610	
Ladinian	L. Trias.	Bittner	1892	Ladini, people of	Jahr. Reichs. Wien, xlii, p.387	Bunter: pt
Laedonian	M. Jur.	Marcou	1846	Laedo, Switzerland	Jura Salin., p.70	Bajocian
Laekenian	M. Eoc.	Dumont	1851	Laeken, Belgium		Lutetian
Laikipian	Mio.	Gregory	1896	Laikipia, Kenya Col.	Great Rift Valley, p.235; Rift Valley and E. Afr., p.204	
Lanarkian	M. Carb.	Kidston	1905	Lanark, Scotland	Q. J. G. S., lxi, pp.308–21	Pt. Moscovian
Landenian	L. Eoc.	Dumont	1849	Landen, Belgium	Bull. Acad. Belg., xvi, p.16	= Thanetian
Langhian	M. Mio.	Pareto	1865	Langhe, Italy	Bull. Soc. geol. Fr. (2), xxii, p.229	= Burdigalian
Latdorfian	L. Olig.	Mayer-Eymar	1863	Latdorf, Germany	*Ibid.*, xxi, p.7	Tongrian
Lausannian	M. Mio.	Rollier	1892	Lausanne, Switz.	Eclog. geol. Helv., iii, p.83	Burdigalian
Ledburian	U. Silur.	Renevier	1874	Ledbury, England	Tabl. Terr.	Downtonian
Ledian	U. Eoc.	Mourlon	1880	Lede, Belgium	Soc. malac. Belg., xviii, p.10	Bartonian
Lenhamian	L. Plioc.	Harmer	1900	Lenham, Kent	Q. J. G. S., lvi, p.708	= Diestian
Lennoxian	U. Pampal.	Gregory	1928	Lennox, Scotland	Trans. Geol. Soc. Glasgow, xviii, p.305	
Liburnian	U. Cret.	Stache	1889	Dalmatia	Abh. geol. Reichs.	= Danian
Ligerian	M. Mio.	Rouville	1853	Loire, France	Géol. Montpellier, p.180	Helvetian
Ligurian	L. Olig.	Mayer-Eymar	1857	Liguria, Italy	Verh. Schweiz. Nat. Ges. Trogen, p.182	Tongrian

Name	System	Author	Year	Locality or derivation	Reference	Use
Lingulian	U. Camb.	Renevier	1874	*Lingula*	Tabl. Terr. séd., 1st ed.	
Llandeilian	M. Ord.	Murchison	1839	Llandeilo, Wales	Silurian System	
Llandoverian	L. Sil.	Murchison	1839	Llandovery, Wales	*Ibid.*	
Llanvirnian	L. Ord.	Marr	1905	Llanvirn, S. Wales	Q. J. G. S., lxi, Proc. lxxxi	
Lodevian	M. Perm.	Renevier	1874	Lodève, France	Tabl. Terr. séd. 1st ed.	Punjabian
Loganian	Pampal.	Lawson	1913	Sir William Logan	Congr. géol. inter. Canada	
Londinian	L. Eoc.	Mayer-Eymar	1857	London, England	Verh. Schweiz. Nat. Ges., Trogen, p.175	
Longmyndian	L. Camb.	Mayer-Eymar	1874	Longmynd, England	Class. Méthod.	
Lotharingian	L. Jur.	Haug	1911	Lorraine, France	Traité, p.961	Pt. Charmouthian
Ludian	L. Olig.	de Lapparent	1893	Ludes, France	Géol., 3rd edit., p.1219	Tongrian
Ludlovian	U. Silur.	Murchison	1839	Ludlow, England	Silurian System	
Lusitanian	U. Jur.	Choffat	1885	Lusitania, Portugal	Faun. Jur. Portugal	Corallian
Lutetian	M. Eoc.	de Lapparent	1883	Lutetia, Paris	Géol., p.989	
Maastrichtian	U. Cret.	Dumont	1849	Maastricht, Holland	Bull. Acad. sc. Belg.	Pt. Danian
Magdalenian	Pleisto.	Mortillet	1878	Madelaine, France	Congr. géol., Paris, p.179	
Manresian	U. Eoc.	Vézian	1858	Manresa, Spain	Bull. Soc. géol. Fr. (2) xv, p.439	Bartonian
Mareniscan	Pampal.	Van Hise	1892	Marenisco, N. Amer.	Bull. U.S. Geol. Surv., No. 86	
Marquettian	Pampal.	Winchell	1888	Marquette, America	Congr. géol. inter. Rep. Am. Com., p.14	
Marylandian	M. Mio.	Heilprin	1882	Maryland, America	Proc. Acad. sc. Philad.	
Maudunian	L. Eoc.	de Lapparent	1883	Meudon, France	Géol. p.989	Montian
Mayencian	M. Mio.	Mayer-Eymar	1857	Mayence, Germany	Verh. Schweiz. Nat. Ges., Trogen, Table	Burdigalian
Mecklenburgian	M. Pleisto.	J. Geikie	1895	Mecklenburg, Ger.	Journ. Geol., p.250	4th Glaciation
Melbournian	Up. Sil.	Gregory	1902	Melbourne, Victoria	Proc. R. Soc. Vict., n.s., xv, p.171	
Menevian	M. Camb.	Salter and Hicks	1865	Menevia, St. David's, Wales	Q. J. G. S., xxiv	
Messinian	L. Plio.	Mayer-Eymar	1867	Messina, Sicily	Cat. Foss. Mus. Zurich, p.13	
Mississippian	L. Carb.	H. S. Williams	1891	Mississippi, America	Bull. U.S. Geol. Surv., No. 80	
Modenian	L. Olig.	Pareto	1865	Modena, Italy	Bull. Soc. géol. Fr. (2), xxii, p.216	Tongrian
Mohawkian	M. Ord.	Hall	1842	R. Mohawk, N. Y.	Am. Jour. Sci., xliii. p.52	= Llandeilian
Monian	L. Camb.	Blake	1888	Mona, Anglesey	Congr. géol. inter. London, p.36	
Monongahelan	U. Carb.	Prosser	1901	R. Monongahela, U.S.A.	Am. Jour. Sci. (4), xi, p.199	Pt. Uralian
Montian	L. Eoc.	Dewalque	1868	Mons, Belgium	Prodr. Géol. Belgique, p.185	
Morfontian	U. Eoc.	Dollfus.	1880	Mortefontaine, France	Expos. géol. Havre, p.592	Bartonian
Mornasian	U. Cret.	Coquand	1862	Mornas, France	Bull. Soc. géol. Fr. (2), xx, p.50	Pt. Senonian
Moscovian	M. Carb.	Nikitin	1890	Moscow, Russia	Mem. Com. géol. Russ., v, p.147	
Mousterian	Pleisto.	Mortillet	1878	Moustier, France	Congr. géol. inter. Paris. p.179	
Naivashan	Plio.	Gregory	1896	Naivasha, Kenya Col.	Great Rift Valley, p.235	
Nemausian	L. Cret.	Saruan	1875	Nemausum, Nimes, France	Bull. sc. nat. Nimes	Valanginian
Neocomian	L. Cret.	Thurmann	1835	Neocomium, Neuchatel	Bull. Soc. géol. Fr., vii, p.209	
Neptodunian	M. Eoc.	Dollfus	1850	Neptodunum, Paris	Expos. géol. Havre, p.592	Lutetian
Nervian	U. Cret.	Dumont	1849	Nervians, people of Belgium	Bull. Acad. Belg., xvi, p.360	Pt. Senonian
Neudeckian	Pleisto.	J. Geikie	1895		Jour. Geol., p.249	3rd Interglacial
Newark System	U. Tr. and L. Jur.	Redfield	1856	Newark, N. Y.	*Cf.* J. E. Russell, 1892, Bull. U. S. G. S., 85	
Newbournian	L. Plio.	Harmer	1900	Newbourne, Suffolk	Q. J. G. S., lvi, p.270	= M. Astian
Niagaran	Silur.	Sterry-Hunt		Niagara R., N. Amer.		M. Silur.
Norfolkian	Pleisto.	J. Geikie	1895	Norfolk, England	Jour. Geol. p.247	1st Interglacial
Norian	Pampal.	Sterry-Hunt	1870		Congr. géol. inter. London, p.73	Lower Trias
Norian	U. Trias.	Mojsisovics	1869	Alps	Verh. geol. Reichs., p.65	
Nyasan	Olig.	Gregory	1896	L. Nyasa	Great Rift Valley, p.235	
Oranian	U. Mio.	Welsch	1895	Oran, Algeria	Bull. Soc. géol. Fr. (3), xxiii, P. v, p.60	Pontian
Oriskanian	L. Dev.	Clark and Schuchert	1898	Oriskany	Science, n.s., x, pp.874–78	
Oswegan	L. Sil.	Clark and Schuchert	1898	Oswego, N. Y.	Science, x, p.876	
Oxfordian	M. Jur.	Brogniart	1829	Oxford, England	Tabl. Terr.	
Paleocene	L. Eoc.	Schimper	1874	*Palaois* and Eocene	Pal. veget.. iii. p.680	
Paleolithic	Pleisto.	Lubbock	1865	Ancient stone	Prehistoric Times	
Paniselian	M. Eoc.	Dumont	1851	Mt. Panisel, Belgium	Bull. Acad. sc. Belg.	Lutetian

Name	System	Author	Year	Locality or derivation	Reference	Use
Parisian	Eoc. and L. Olig.	Brongniart	1820	Paris		Eocene and Tongrian
Parnian	M. Eoc.	Dollfus	1880	Parnes, near Paris	Expos. géol. Havre, p.591	Lutetian
Patagonian	Mio.	d'Orbigny	1842	Patagonia, S. Amer.	Voy. Amér. Mérid., iii	
Pauletian	U. Cret	Dumas	1852	St. Paulet, France	Carte géol. d'Uzès 1874	Pt. Cenomanian
Pebidian	Pampal.	Hicks	1878	Pebidia, Wales	Rep. Brit. Assoc., 1878	
Pennsylvanian	Mid. Carb.	H. S. Williams	1898	Pennsylvania, U.S.A.	Bull. U. S. G. S., No. 80	
Peorian	Pleisto.	Chamberlin and Salisbury	1906		Geol. iii, p.383	4th Amer. Interglacial
Petchorian	L. Cret.	Nikitin	18??	Petchora, Russia		Berriasian
Pilatan	M. Eoc.	Kaufmann	1872	M. Pilatus, Switz.	Mat. Carte Suisse 11th Livre, p.158	Lutetian
Plaisancian	L. Plio.	Mayer-Eymar	1857	Plaisance, Italy	Verh. Nat. Ges. Trogen, Table	
Platian	Pleisto.	Ameghino	1889	La Plata, S. Amer.	Mam. Foss. Arg., p.106	
Pliensbachian	L. Jur.	Oppel	1858	Pliensbach, Germany	Juraformation, p.815	Charmouthian
Poecilitic	Permian and Trias	Conybeare	1832		Rep. Brit. Ass., 1832, p.379	
Poederlian	U. Plio.	Vincent	1889	Poederlé, Belgium		
Polandian	Pleisto.	J. Geikie	1895	Poland	Jour. Geol. p.249	3rd Glaciation
Pontian	U. Mio.	Marny	1869	Pontus Euxinus	Géol. de Cherson	
Portlandian	U. Jur.	Brongniart	1829	Portland, England	Tabl. Terr.	
Priabonian	U. Eoc.	de Lapparent	1893	Priabona, Italy	Géol. 3rd ed., p.149	Bartonian
Proterozoic	L. Paleoz.	Lapworth	1888	*Proteros* = first	Intro.Textbk.Geol., p.152	
Punjabian	M. Perm.	de Lapparent	1893	Punjab, India	Géol., 3rd ed., p.886	
Purbeckian	U. Jur.	Brongniart	1829	Purbeck, England	Tabl. Terr.	
Radstockian	M. Carb.	Kidston	1905	Radstock, England	Q. J. G. S., lxi, pp.308–21	Pt. Moscovian
Raiblian	U. Trias.	Stoppani	1860	Raibl, Austria	Pal. Lomb., pp.226, 229	
Rauracian	M. Jur.	Gressly	1867	Rauracia, Jura	Essai sur Jura, p.72	Corallian
Revinian	M. Carb.	Dumont	1847	Revin, France	Bull. Acad. Belg.	Acadian
Rhenan	L. Dev.	Dumont	1848	R. Rhine, Germany	*Ibid.*	Siegenian
Rhaetian	L. Jur.	Guembel	1861	Rhaetian Alps	Bay. Alp., p.122	L. Jurassic or U. Trias
Rhodanian	U. Cret.	Renevier	1854	Perte-du-Rhône, France	Mem. sur Perte-du-Rhône, p.68	
Rognacian	U. Cret.	Caziot	1890	Rognac, France	Bull. Soc. géol. Fr. (3), xviii, p.227	Pt. Danian
Rotomagian	U. Cret.	Coquand	1857	Rothomagus, Rouen	*Ibid.* (2), xiv, p.882	Pt. Cenomanian
Roubian	L. Olig.	Vézian	1858	Rubio, Spain	*Ibid.* (2), xv, p.440	Tongrian
Rupelian	M. Olig.	Dumont	1849	Rupel, Belgium	Bull. Acad. Belg., xvi, p.367	Regarded by Continental authors as Upper Olig.
Sahelian	U. Mio.	Pomel	1858	Sahel, Algeria	C. R. Acad. Sci., xlvii, p.479	Pontian
Saliferian	U. Trias	d'Orbigny	1852	Salt bearing	Cours élém., p.404	Keuper
Sallomacian	M. Mio.	Fallot	1893	Salles, France	Bull. Soc. géol. Fr. (3), xxi, Pr. v, p.77	Helvetian
Salmian	U. Camb.	Dumont	1847	Salm, France	Bull. Acad. Belg.	Potsdamian
Salopian	M. Silur.	Lapworth	1879	Salop, England	Ann. Mag. Nat. Hist. (5), iii, opp. p.455	Wenlock
Sangamon	Pleisto.	Leverett	1899	Sangamon Co., Ill.	Mon. U.S. Geol. S., xxxviii	3rd Amer. Interglac.
Sannoisian	L. Olig.	de Lapparent	1893	Sannoise, France	Géol., 3rd ed., p.1263	Tongrian
Santonian	U. Cret.	Coquand	1857	Santes, France	Bull. Soc. géol. Fr. (2), xiv, p.882	Pt. Senonian
Sarmatian	U. Mio.	Barbot de Marny	1869	Sarmatia	Esq. géol. de Cherson	Phase of Pontian
Saxonian	Pleisto.	J. Geikie	1895	Saxony, Germany	Journ. Géol., p.247	2nd Glaciation
Saxonian	M. Perm.	de Lapparent	1893	Saxony, Germany	Géol., 3rd ed., p.886	
Scaldisian	U. Plio.	Dumont	1849		Bull Acad. Belg.	Astian
Scandinavian	U. Camb.	de Lapparent	1883	Scandinavia	Géol., p.732	
Scanian	Pleisto.	J. Geikie	1895	Scania, Sweden	Journ. Géol., p.246	1st Glaciation
Senecan	U. Dev.	Clarke and Schuchert	1898	Seneca, U.S.A.	Science, n.s., x, pp.874–8	
Senonian	U. Cret	d'Orbigny	1843	Senones, a tribe in France	Pal. fr., Crét., ii, Table, pl. 236 bis.	
Sequanian	M. Jur.	Thurmann and Marcou	1848	Sequania, Jura	Mem. Soc. géol. Fr., iii, p.96	Corallian
Sicilian	L. Pleisto	Döderlein	1872	Sicily	Nat. sur. Carte géol. de Moden, p.14	
Siderolitique	L. Olig.	Gressly	1841	Ferruginous	Jur. sol., p.251	Tongrian
Sinemurian	L. Jur.	d'Orbigny	1849	Sémur, France	Pal. fr., Jur., i, p.604	
Sinian	Camb.	Richthofen	1882	China	Nord. Chin	U. Pampalozoic
Skiddavian	L. Ord.	Marr	1905	Skiddaw	Q. J. G. S., lxi, Proc. p.lxxxvi	
Skytian	L. Trias	Waagen and Diener	1895		Ak. Wis. Wien, civ	
Snowdonian	U. Ord.	S. P. Woodward	1856	Snowdon, Wales	Mon. Moll., p.409	= Bala
Soissonian	L. Eoc.	Mayer-Eymar	1857	Soissons, France	Verh. Schweiz. Nat. Ges., Trogen, Table	
Solutrian	Pleisto.	Mortillet	1878	Solutré, France	Congr. géol. inter. Paris, p.179	

Name	System	Author	Year	Locality or derivation	Reference	Use
Sparnacian	L. Eoc.	Dollfus	1880	Epernay, France	Bull. Soc. géol. Norm., vi, p.588	
Staffordian	M. Carb.	Kidston	1905	Stafford, England	Q. J. G. S., lxi, pp.308–21	Pt. Moscovian
Stampian	L. Olig.	Rouville	1853	Etampes, France	Géol. de Montpel., p.180	Tongrian
Stephanian	U. Carb.	Mayer-Eymar	1878	St. Etienne, France	Class. inter., 1881	Uralian
Suessonian	L. Eoc.	d'Orbigny	1852	Soissons, France	Cours, élém. pal., p.712	
Taconian	L. Camb.	Emmons	1842	Taconic Mts., Amer.	Geol. N. Y., pp.135–64	Lower Ordovician
Taunusian	L. Dev.	Dumont	1848	Taunus, Belgium	Mem. Ter. Arden., p.183	Siegenian
Tennessean	L. Carb.	Ulrich	1911	Tennessee	Geol. Soc. Am. Bull., vol. xxii, pp.581–2	Visean
Thanetian	L. Eoc.	Renevier	1873	Thanet, England	Tabl. Terr.	
Thurgovian	U. Mio.	Rollier	1892	Thurgovia, Switz.	Eclog. geol. Helv., iii, p.83	Tortonian
Thuringian	U. Perm.	Renevier	1874	Thuringia, Germany	Tabl. Terr.	
Toarcian	L. Jur.	d'Orbigny	1849	Thouars, France	Pal. fr., Jur., i, p.106	
Tongrian	L. Olig.	Dumont	1839	Tongres, Belgium	Bull. Acad. sc. Belg., vi, p.773	
Tortonian	U. Mio.	Mayer-Eymar	1857	Tortona, Italy	Verh. Schweiz. Nat. Ges., Trogen, Table	
Tournaisian	L. Carb.	Koninck	1872	Tournai, Belgium	Mem. Ac. R. Sc. L.B.A. Belge., xxxix, pp.1–178	
Tremadocian	U. Camb.	Renevier	1874	Tremadoc, Wales	Tabl. Terr.	Potsdamian
Turbasian	Pleisto.	J. Geikie	1895		Jour. Geol.	Geikie's 5th and 6th Glaciations
Turonian	U. Cret.	d'Orbigny	1843	Touraine, France	Pal. fr., Crét, ii, Table, pl. 236 bis.	Pt. Senonian
Tyrolian	U. Trias.	de Lapparent	1885	Tyrol, Austria	Geol., 2nd ed., p.905	Keuper
Ulsterian	M. Dev.	Clarke and Schuchert	1898		Science, n.s., x, pp.874–8	
Uralian	U. Carb.	de Lapparent	1893	Ural Mountains	Géol., 3rd ed., p.819	
Urgonian	L. Cret.	d'Orbigny	1850	Orgon, France	Cours. élém., ii, p.606	Barremian and Aptian
Valanginian	L. Cret.	Desor	1854	Valangin, Switzerland	Bull. Sc. nat. Neuchat., iii, p.177	Pt. Neocomian
Valdonnian	U. Cret.	Matheson	1878	Valdonne, France	Rech. pal. Midi	Pt. Senonian
Valentian	L. Silur.	Lapworth	1879	Valentia, S. Scotland	Ann. Mag. Nat. Hist. (5), iii, opp. p.455	
Vasconian	L. Mio.	Fallot	1893	Vasconia, France	Bull. Soc. géol. Fr. (3), xxi, Pr. v, p.79	Burdigalian
Vectian	L. Cret.	Jukes-Brown	1885	Vectium, I.O.W.	Geol. Mag., iii, p.298	Urgonian
Vesulian	M. Jur.	Marcou	1848	Vesoul, France	Mem. Soc. géol. Fr., iii, p.73	L. Bathonian
Villafranchian	L. Pleisto.	Pareto	1865	Villafranca, Italy	Bull. Soc. géol. Fr. (2), xxii, p.262	Sicilian
Vindobonian	U. Mio.	Depéret	1895	Vindobona, Vienna	Ibid. (3), xxiii, Pr. v, p.34	Helvetian and Tortonian
Virginian	M. Mio.	Heilprin	1882	Virginia, America	Acad. Nat. Sc. Philad.	
Virglorian	L. Trias	Renevier	1874	Virgloria, Switzerland	Tabl. Terr.	Pt. Bunter
Virgulian	U. Jur.	Thurmann	1852	Exogyra virgula	Mitth. Bern. naturf. Ges., p.217	Kimeridgian
Virtonian	L. Jur.	Mourlon	1880	Virton, Belgium	Geol. Belg., i, p.143	Charmouthian
Visean	M. Carb.	Dupont	1883	Visé, Belgium	Bull. Acad. Belg., xv, p.212	Moscovian
Volgian	U. Jur.	Nikitin	1881	Volga, Russia	Mem. Acad. Imp. Sci. St. Petersb., 7th Ser., xxviii, p.98	Portlandian
Vosgian	L. Trias	de Lapparent (after de Beaumont)	1885	Vosges, Germany	Géol., 2nd ed., p.905	Bunter
Vraconnian	U. Cret.	Renevier	1867	Vraconne, Jura	Faun. Chevill, p.201	Pt. Cenomanian
Waltonian	U. Plioc.	Harmer	1900	Walton, Essex	Q. J. G. S., lvi, p.709	=
Waucoban	L. Camb.	Walcott	1912	Waucoba Springs, Cal.	Smithson. Misc. Coll., vol. 57, No. 10, pp.305–6	
Waverlian	L. Carb.	C. Briggs, jr. (Mather)	1838		Ohio Geol. Surv., 1st Ann. Rep., pp.74, 79–80	Tournaisian
Wealden	L. Cret.	Mantell		Weald, England	Foss. S. Downs	Pt. Neocomian
Wenlock	M. Silur.	Murchison	1839	Wenlock, England	Silurian System	
Werfenian	L. Trias	Renevier	1874	Werfen, Austria	Tabl. Terr.	Bunter
Westphalian	M. Carb.	de Lapparent	1893	Westphalia, Germany	Géol., 3rd ed. p.819	= Moscovian
Weybournian	U. Plio.	Harmer	1900	Weybourne, Suffolk	Q. J. G. S., lvi, p.724	
Wisconsin	Pleisto.	Chamberlin and Salisbury	1906	Wisconsin, U.S.A.	Geol., iii, p.383	= Wurmian
Yarmouthian	Pleisto.	Chamberlin and Salisbury	1906	Yarmouth, Iowa	Geol., iii, p.383	2nd Amer. Interglac.
Yeovilian	L. Ool.	Buckman		Yeovil		Midford Sands
Yeringian	U. Sil.	Gregory	1902	Yering, Victoria	Proc. R. Soc. Victoria, n.s., xv, p.172	
Yorkian	M. Carb.	Watts		Ypres, Belgium		Pt. Moscovian
Ypresian	L. Eoc.	Dumont	1849	Ypres, Belgium	Bull. Acad. sc. Belg., xvi, p.369	= London Clay
Zechstein	U. Perm.	German miners' term		? from Zähe, tough		Thuringian

APPENDIX 2

Recommended three-character abbreviations for chronostratic names with alternative symbols. Subdivisions are indicated by subscript number. Combinations would be six character expressions.

Chronostratic division	Recommended three-character expression	Alternative traditional symbols: (3) International (4) British Geological Survey	
		Int.	BGS
(1)	(2)	(3)	(4)
Aalenian	Aal		
Actonian	Act		
Aegean	Aeg		
Aeronian	Aer		
Alavnian	Ala		
Albertian	Abt		
Albian	Alb		
Aldanian	Ald		
Alexandrian	Alx		
Alportian	Alp		
Animikean	Anm		
Anisian	Ans		
Aphebian	Aph		
Aptian	Apt		
Aquitanian	Aqt		
Archean	Arc	Ar	
Arenig	Arg		
Arnsbergian	Arn		
Artinskian	Art		
Arundian	Aru		
Asbian	Asb		
Ashgill	Ash		
Asselian	Ass		
Atdabanian	Atb		
Bajocian	Baj		
Bala	Bal		
Barremian	Brm		
Bartonian	Brt		
Bashkirian	Bsh		
Basin Groups 1–9	Bg1 etc.		

Chronostratic division	Recommended three-character expression	Alternative traditional symbols: (3) International (4) British Geological Survey	
		Int.	BGS
(1)	(2)	(3)	(4)
Bathonian	Bth		
Berriasian	Ber		
Botomian	Bot		
Brigantian	Bri		
Burdigalian	Bur		
Burzyan	Buz		
Caerfai	Crf		
Calabrian	Clb		
Callovian	Clv		
Cambrian	Cbn	ЄE	a
Campanian	Cmp		
Canadian	Cnd		
Cantabrian	Ctb		
Caradoc	Crd		
Carboniferous	Cbs	C	d
Carnian	Crn		
Cautleyan	Cau		
Cayugan	Cay		
Cenomanian	Cen		
Cenozoic	Czc		
Central Rand	Crd		
Chadian	Chd		
Chamovnicheskian	Chv		
Champlainian	Chp		
Changxingian	Chx		
Chattian	Cht		
Cheremshanskian	Che		
Chokierian	Cho		
Cincinnatian	Cin		
Cobalt	Cob		
Comley	Com		
Coniacian	Con		
Costonian	Cos		
Courceyan	Cor		
Couvinian	Cov		
Cretaceous	Krt	K	h
Croixian	Crx		
Cryptic	Cry		
Danian	Dan		
Devonian	Dev	D	c
Dienerian	Die		
Dinantian	Din		
Dogger	Dog		
Dolgellian	Dol		
Dominion	Dom		
Dorogomilovskian	Dor		
Downtonian	Dow		
Dyfed	Dfd		
Dzhulfian	Dzh		

Chronostratic division (1)	Recommended three-character expression (2)	Alternative traditional symbols: (3) International — Int. (3)	(4) British Geological Survey — BGS (4)
Early	Er-		
Ediacara	Edi		
Eifelian	Eif		
Elliot Lake	Etl		
Emsian	Ems		
Eocene	Eoc		
Etcheminian	Etm		
Famennian	Fam		
Fig Tree	Fig		
Frasnian	Frs		
Fronian	Fro		
Gallic	Gal		
Gedinnian	Ged		
Givetian	Giv		
Gleedon	Gle		
Gorstian	Gor		
Government	Gvt		
Griesbachian	Gri		
Guadalupian	Gua		
Gulf	Gul		
Gunflint	Gun		
Gzelian	Gze		
Hadean	Hde		
Hadrynian	Hdy		
Harnagian	Har		
Hastarian	Has		
Hauterivian	Hau		
Helikian	Hel		
Hettangian	Het		
Hirnantian	Hir		
Holkerian	Hlk		
Holocene	Hol		
Homerian	Hom		
Hospital Hill	Hsp		
Hough Lake	Hou		
Huronian	Hur		
Idwian	Idw		
Illyrian	Ill		
Imbrian	Imn		
Imbrium	Imm		
Induan	Ind		
Isuan	Isu		
Ivorian	Ivo		
Johannesburg	Jbg		
Julian	Jul		

Chronostratic division (1)	Recommended three-character expression (2)	Alternative traditional symbols: (3) International — Int. (3)	(4) British Geological Survey — BGS (4)
Jurassic	Jur	J	g
Karatavian	Kar		
Kashirskian	Ksk		
Kasimovian	Kas		
Kazanian	Kaz		
Keeweenawan	Kee		
Kimmeridgian	Kim		
Kinderscoutian	Kin		
Klazminskian	Kla		
Kotlinian	Kot		
Krevyakinskian	Kre		
Kungurian	Kun		
Ladinian	Lad		
Langhian	Lan		
Latdorfian	Lat		
Late	Lt-		
Lenian	Len		
Leonardian	Leo		
Lias	Lia		
Llandeilo	Llo		
Llandovery	Lly		
Llanvirn	Lln		
Lochkovian	Lok		
Longtanian	Lgt		
Longvillian	Lon		
Lopingian	Lop		
Ludlow	Lud		
Ludorfian	Ldf		
Lutetian	Lut		
Maastrichtian	Maa		
Maentwrogian	Mnt		
Malm	Mlm		
Marsdenian	Mrd		
Marshbrookian	Mrb		
Meishucunian	Mei		
Melekesskian	Mel		
Menevian	Men		
Merioneth	Mer		
Mesozoic	Mzc	Mz	
Messinian	Mes		
Middle	Mid		
Miocene	Mio		
Mississipian	Mis	M	
Moodies	Moo		
Moscovian	Mos		
Myachkovskian	Mya		

continued

Chronostratic division (1)	Recommended three-character expression (2)	Alternative traditional symbols: (3) International — Int. (3)	(4) British Geological Survey — BGS (4)
Nammalian	Nml		
Namurian	Nam		
Nectarian	Nec		
Neocomian	Neo		
Neogene	Ngn	Ng	i
Neozoic	Nzc		
Niagaran	Nia		
Noginskian	Nog		
Norian	Nor		
Olenekian	Olk		
Oligocene	Oli		
Onnian	Onn		
Onverwacht	Ovt		
Ordovician	Ord	O	b1–3
Orenburgian	Orn		
Orientale	Ori		
Oxfordian	Oxf		
Paleocene	Pal		
Paleogene	Pgn	Pg	i
Paleozoic	Pzc	Pz	
Pelsonian	Pel		
Pendleian	Pnd		
Pennsylvanian	Pen		
Permian	Per	P	e
Phanerozoic	Phz		
Piacenzian	Pia		
Pleinsbachian	Plb		
Pleistocene	Ple		
Pleistogene	Ptg		
Pliocene	Pli		
Podolskian	Pod		
Pongola	Pgl		
Portlandian	Por		
post-	po-		
Poundian	Pou		
Pragian	Prg		
pre-	pr-		
Precambrian	Pr-Cbn/Prc	PC or PE	
Priabonian	Prb		
Pridoli	Prd		
Priscoan	Pro		
Procellarum	Pcl		
Proterozoic	Ptz	Pt	
Purbeckian	Pur		
Pusgillian	Pus		
Quaternary	Qty	Q	

Chronostratic division (1)	Recommended three-character expression (2)	Alternative traditional symbols: (3) International — Int. (3)	(4) British Geological Survey — BGS (4)
Quirke Lake	Qkl		
Randian	Ran		
Rawtheyan	Raw		
Rhaetian	Rht		
Rhuddanian	Rhu		
Riphean	Rif	R	
Rotliegendes	Rot		
Rupelian	Rup		
Ryazanian	Ryz		
Saint David's	StD		
Sakmarian	Sak		
Santonian	San		
Scythian	Scy		
Senonian	Sen		
Serpukhovian	Spk		
Serravallian	Srv		
Sheinwoodian	She		
Siegenian	Sig		
Silesian	Sls		
Silurian	Sil	S	b4–7
Sinemurian	Sin		
Sinian	Zin	Z	
Smalfjord	Sma		
Smithian	Smi		
Solva	Sol		
Soudleyan	Sou		
South Pole-Aitken	Spl		
Spathian	Spa		
Stampian	Sta		
Stephanian	Ste		
Strunian	Str		
Sturtian	Stu	U	
Swazian	Swz		
Tatarian	Tat		
Telychian	Tel		
Tertiary	Tty	T TT	i
Thanetian	Tha		
Timiskaming	Tim		
Tithonian	Tth		
Toarcian	Toa		
Toiotian	Toi		
Tojonian	Toj		
Tommotian	Tom		
Tortonian	Tor		
Tournaisian	Tou		
Tremadoc	Tre		

Chronostratic division	Recommended three-character expression	Alternative traditional symbols: (3) International (4) British Geological Survey		Chronostratic division	Recommended three-character expression	Alternative traditional symbols: (3) International (4) British Geological Survey	
		Int.	BGS			Int.	BGS
(1)	(2)	(3)	(4)	(1)	(2)	(3)	(4)
Triassic	Tri	Ŧ Tr	f	Wenlock	Wen		
Turffontein	Tnt			Westphalian	Wph		
Turonian	Tur			Whitwell	Whi		
Tuvalian	Tul			Witwatersrand	Wit		
				Wonokan	Won		
Ufimian	Ufi			Woolstonian	Woo		
				Wujiapingian	Wuj		
Valanginian	Vlg						
Valdaian	Vld			Xenian	Xen	X	
Varanger	Var						
Vendian	Ven	V		Yeadonian	Yea		
Ventersdorp	Vtp			Yovian	Yov	Y	
Vereiskian	Vrk			Ypresian	Ypr		
Vetternian	Vet			Yurmatin	Yur		
Visean	Vis						
Volgian	Vol			Zanclian	Zan		
				Zechstein	Zec		
Waucoban	Wau			Zedian	Zed	Z	
Weltian	Wlt	W					

APPENDIX 3

Calculation of isotopic dates using conventional decay constants

Introduction

In general an isotopic date is obtained from the equation:

$$t = (1/\lambda) \ln [(D/P) + 1]$$

where t = age, λ = decay constant (fraction of parent disintegrating per unit time), D = number or moles of daughter atoms produced by radioactive decay in the specimen, P = number or moles of parent atoms in the specimen, and $\ln = \log_e$. In cases where some daughter atoms are initially present, as in Rb–Sr, Sm–Nd and U–Pb dating, it is presumed that these have been subtracted from the amount of daughter first to give the amount solely due to radioactive decay. The conventional assumption is a single period of closed system evolution – no loss or gain of daughter or parent between the time of formation and time of analysis of the specimen. Discussion of these basic assumptions and their many exceptions and accommodations for those exceptions may be found in textbooks such as Faure (1986).

Refinements in mass spectrometers and in the measurement of radioactive decay rates have led to improvements in the precision and accuracy of isotopic dates. For Rb–Sr, Sm–Nd and U–Pb dates the relationship between dates calculated by different decay constants is precisely the ratio of the different decay constants. For K–Ar and Pb–Pb the relationship is more complex because ^{40}K has a branching decay, each branch with its own decay constant, and because the Pb dates involve two decay constants in an equation with multiple exponential terms. For conversion of these dates recalculation from original data is strongly recommended. The changes may be approximated by the use of graphs or tables but this may accumulate round-off errors leading to small discrepancies with properly recalculated dates.

Conventional decay constants

At the meeting of the IUGS Subcommission on Geochronology, at the International Geological Congress in Sydney, Australia, in 1976, a recommended set of isotopic abundance ratios and decay constants was proposed (Steiger & Jäger 1977). These are used in the present work and are, with a few additions, listed below.

Uranium

$\lambda(^{238}\text{U}) = 1.55125 \times 10^{-10}/\text{yr}$

$\lambda(^{235}\text{U}) = 9.8485 \times 10^{-10}/\text{yr}$

atomic ratio $^{238}\text{U}/^{235}\text{U} = 137.88$

Thorium

$\lambda(^{232}\text{Th}) = 4.9475 \times 10^{-11}/\text{yr}$

Rubidium

$\lambda(^{87}\text{Rb}) = 1.42 \times 10^{-11}/\text{yr}$

atomic ratio $^{85}\text{Rb}/^{87}\text{Rb} = 2.59265$

Strontium

atomic ratios $^{86}\text{Sr}/^{88}\text{Sr} = 0.1194$ and $^{84}\text{Sr}/^{86}\text{Sr} = 0.056584$

Potassium

$\lambda(^{40}\text{K}\beta\text{-}) = 4.962 \times 10^{-10}/\text{yr}$

$\lambda(^{40}\text{K}_e) + \lambda(^{40}\text{K}_e) = 0.581 \times 10^{-10}/\text{yr}$

isotopic abundances $^{39}\text{K} = 93.2581$ atom%

$^{40}\text{K} = 0.01167$ atom%

$^{41}\text{K} = 6.7302$ atom%

Argon

atomic ratio $^{40}\text{Ar}/^{36}\text{Ar}$ atmospheric = 295.5

Samarium

$\lambda(^{147}\text{Sm}) = 6.54 \times 10^{-12}\text{yr}$

isotopic abundance $^{147}\text{Sm} = 14.996$ atom%

Neodymium

Atomic ratios are reported on the basis of normalization to $^{146}\text{Nd}/^{144}\text{Nd} = 0.7219$, or alternatively $^{146}\text{Nd}/^{142}\text{Nd} = 0.636151$. This can be a source of confusion that is not easily dispelled in a few words, but as long as the initial and present-day isotopic compositions are given in the published source with a single convention applied there is no effect on age determinations.

Most publications since 1978 adhere to these conventions. Future improvements in measurement precision may bring about refinements on the order of 1% or less, but this will have only a small effect on time scale calibration.

Virtually all isotopic dates published before 1978 use obsolete decay constants. This appendix outlines how non-standardized dates may be converted to dates calculated with IUGS conventional constants. The dating methods and decay constants originally used for the dates in Table 4.2 are coded according to conventions listed in the key to Table 4.2.

K–Ar dates

In terms of IUGS conventional decay constants, K–Ar dates are calculated with the equation:

$$t(\text{Ma}) = (10^4/5.543) \times \ln (1 + 9.540 \times (^{40}\text{Ar}/^{40}\text{K}))$$

This can be easily done on any scientific calculator with an 'ln' key.

Useful conversion factors are:

$$9.540 \times (^{40}\text{Ar}/^{40}\text{K}) = 0.014261 \times (\text{cc} \; ^{40}\text{Ar} \times 10^6/\%\text{K})$$
$$= 0.03850 \times (\text{moles} \; ^{40}\text{Ar} \times 10^{10}/\%\text{K}_2\text{O})$$

and

$$\%\,K_2O = 1.2047 \times (\%\,K)$$

The tables in Dalrymple (1979) or Harland *et al.* (1982) and in this appendix may be used to convert old Western and Soviet K–Ar dates to values for the conventional IUGS constants.

Revised K–Ar dates may be **approximated** by multiplying 200 to 400 Ma dates calculated using old Western conventional constants by 1.021 (for old Soviet constants multiply by 0.9756). For younger dates the conversion is approximately 1.025 (0.9768) and for dates from 400 to 600 Ma the conversion is approximately 1.018 (0.9744).

Rb–Sr dates

The equation for calculation of conventional Rb–Sr dates is:

$$t(Ma) = (10^5/1.42) \times \ln (1 + {}^{87}Sr/{}^{87}Rb)$$

For a single sample:

${}^{87}Sr/{}^{87}Rb$ = (present-day ${}^{87}Sr/{}^{86}Sr$ of sample minus assumed

or measured initial ${}^{87}Sr/{}^{86}Sr$ ratio)/${}^{87}Rb/{}^{86}Sr$ of sample

or for a pair or suite of samples (where subscripts a and b indicate the respective samples, such as mica (a) and whole rock (b)):

${}^{87}Sr/{}^{87}Rb$ = [(${}^{87}Sr/{}^{86}Sr)_a$
  − (${}^{87}Sr/{}^{86}Sr)_b$]/[(${}^{87}Rb/{}^{86}Sr)_a$ − (${}^{87}Rb/{}^{86}Sr)_b$]
  = slope of isochron given by regression of data on ${}^{87}Sr/{}^{86}Sr$ vs ${}^{87}Rb/{}^{86}Sr$ diagram (York 1967, least squares cubic regression algorithm being the most commonly used).

The conversion of Rb/Sr ratios to ${}^{87}Rb/{}^{86}Sr$ ratios is achieved by:

${}^{87}Rb/{}^{86}Sr$ = (ppm Rb/ppm Sr)
  × (2.97645 + 0.28309 × (${}^{87}Sr/{}^{86}Sr − 1$))

Conversions for changed decay constants are simply the ratio of the decay constants. For the old 1.39 constant the conversion to new 1.42 constant involves multiplication of the old date by 0.9789. For conversion of old 1.47 constant dates to new-

constant dates multiply by 1.035.

U–Pb dates

The equations for conventional U–Pb and Pb/Pb dates are:

$$t(Ma) = (10^4/1.55125) \times \ln (1 + {}_2{}^{06}Pb/{}^{238}U)$$
$$t(Ma) = (10^4/9.8485) \times \ln (1 + {}_2{}^{07}Pb/{}^{235}U)$$

and

$({}^{207}Pb/{}^{206}Pb)$
  $= (1/137.88 \times (e^{9.8485 \times 10^{-4} \times t(Ma)} − 1)/(e^{1.55125 \times 10^{-4} \times t(Ma)} − 1)$

(which cannot be rearranged for **t** in terms of measured quantities).

A U–Pb date taken from the literature can have a variety of meanings. It could be the ${}^{206}Pb/{}^{238}U$ date alone, a combination of the ${}^{206}Pb/{}^{238}U$ and ${}^{207}Pb/{}^{235}U$ dates for one or several samples, or a concordia interpretation (consult Faure (1986) and references therein for elaboration on that subject). The analytical data presented with U–Pb dates are seldom complete enough to enable *ab initio* recalculation of dates and

errors. The ${}^{206}Pb/{}^{238}U$ and ${}^{207}Pb/{}^{235}U$ ratios are, however, often given and can be used in the above equations to calculate the revised dates.

An approximate conversion of U–Pb dates from old to new decay constants may be obtained by multiplying the old date by 0.99. Calculation or conversion of Pb/Pb dates, even when the ${}^{207}Pb/{}^{206}Pb$ ratio is given, is complicated and requires a graph (such as figure 18.4 in Faure), table (such as table 18.3 in Faure), or iterative computer algorithm.

A caution. Some papers, particularly those published before 1960, use decay constants different from those discussed above. Dates from such sources must be recalculated individually, from original analytical data if possible.

Sm–Nd dates

The equation for calculating conventional Sm–Nd dates is:

$$t(Ma) = (10^6/6.54) \times \ln (1 + {}^{143}Nd/{}^{147}Sm)$$

Where measured ${}^{143}Nd/{}^{144}Nd$ and ${}^{147}Sm/{}^{144}Nd$ ratios are reported, as is usually the case, the calculation of the ${}^{143}Nd/{}^{147}Sm$ ratio is analogous to the calculation of the ${}^{87}Sr/{}^{87}Rb$ ratio outlined above. The Sm–Nd literature is not confused by a multiplicity of decay constants so conversion to modern conventional constants is not usually an issue of concern.

Fission track

In fission-track dating the decay constant used is linked with the calibration of neutron flux used to determine U concentration, so that it is extremely hazardous, and usually wrong, simply to convert dates from one reported decay constant to another. The problems, pitfalls and proper procedures are discussed at length in recent reviews of fission-track technique (Hurford & Green 1982, 1983, Green 1985). Since the fission-track dates are ultimately standardized using minerals dated by K–Ar, the appropriate conversions of old-to new-convention dates are usually the same as for K–Ar. The analytical errors for individual fission-track dates are usually much larger than any adjustment for revised K–Ar decay constants so this is largely an insignificant issue.

Conversion table for old K–Ar ages

The table that follows gives standardized ages for ages that use the constants most common in Western laboratories (code number 1) or Soviet laboratories (code number 2) (see Table 4.2). To find a standardized Western age, look down the age tables until the published age is found; the second figure in the column gives the standardized age. To find a standardized Soviet age, look down the age tables until the published age is found; the third figure in the column gives the standardized age.

Example. Published Western age = 31.0 Ma; standardized Western age = 31.76 Ma (31.8 Ma). Published Soviet age = 690 Ma; standardized Soviet age = 671.6 Ma (672 Ma).

Note. Ages published prior to about 1960 may use constants other than those used to standardize the age. Such ages need to be recalculated individually and are not discussed here. Some Western laboratories used a constant of 0.585 instead of 0.584 for the electron capture-decay of ${}^{40}K$: the standardized ages are not significantly different from those provided by the table.

Old age	New Western	New Soviet
0.1	0.10	0.10
0.2	0.20	0.19
0.3	0.30	0.29
0.4	0.41	0.39
0.5	0.51	0.49
0.6	0.61	0.58
0.7	0.72	0.68
0.8	0.82	0.78
0.9	0.92	0.88
1.0	1.02	0.98
1.1	1.13	1.07
1.2	1.23	1.17
1.3	1.33	1.27
1.4	1.43	1.37
1.5	1.54	1.47
1.6	1.64	1.56
1.7	1.74	1.66
1.8	1.84	1.76
1.9	1.95	1.86
2.0	2.05	1.95
2.1	2.15	2.05
2.2	2.25	2.15
2.3	2.36	2.25
2.4	2.46	2.34
2.5	2.56	2.44
2.6	2.66	2.54
2.7	2.76	2.64
2.8	2.87	2.74
2.9	2.97	2.83
3.0	3.07	2.93
3.1	3.17	3.03
3.2	3.28	3.13
3.3	3.38	3.22
3.4	3.48	3.32
3.5	3.59	3.42
3.6	3.69	3.52
3.7	3.79	3.62
3.8	3.89	3.71
3.9	4.00	3.81
4.0	4.10	3.91
4.1	4.20	4.01
4.2	4.30	4.10
4.3	4.41	4.20
4.4	4.51	4.30
4.5	4.61	4.40
4.6	4.71	4.50
4.7	4.82	4.59
4.8	4.92	4.69
4.9	5.02	4.79
5.0	5.12	4.89

Old age	New Western	New Soviet
5.1	5.23	4.98
5.2	5.33	5.08
5.3	5.43	5.18
5.4	5.53	5.28
5.5	5.63	5.37
5.6	5.74	5.47
5.7	5.84	5.57
5.8	5.94	5.67
5.9	6.05	5.76
6.0	6.15	5.86
6.1	6.25	5.96
6.2	6.35	6.06
6.3	6.46	6.16
6.4	6.56	6.26
6.5	6.66	6.35
6.6	6.76	6.45
6.7	6.87	6.55
6.8	6.97	6.65
6.9	7.07	6.74
7.0	7.17	6.84
7.1	7.28	6.94
7.2	7.38	7.04
7.3	7.48	7.13
7.4	7.58	7.23
7.5	7.69	7.33
7.6	7.79	7.43
7.7	7.89	7.53
7.8	7.99	7.62
7.9	8.10	7.72
8.0	8.20	7.82
8.1	8.30	7.92
8.2	8.40	8.02
8.3	8.51	8.11
8.4	8.61	8.21
8.5	8.71	8.31
8.6	8.81	8.41
8.7	8.92	8.50
8.8	9.02	8.60
8.9	9.12	8.70
9.0	9.22	8.80
9.1	9.33	8.89
9.2	9.43	8.99
9.3	9.53	9.09
9.4	9.63	9.19
9.5	9.73	9.28
9.6	9.84	9.38
9.7	9.94	9.48
9.8	10.04	9.58
9.9	10.14	9.68
10.0	10.25	9.77

Old age	New Western	New Soviet
11	11.27	10.75
12	12.30	11.73
13	13.32	12.71
14	14.35	13.68
15	15.37	14.66
16	16.40	15.64
17	17.42	16.61
18	18.44	17.59
19	19.47	18.57
20	20.49	19.55
21	21.52	20.52
22	22.54	21.50
23	23.57	22.48
24	24.59	23.46
25	25.62	24.43
26	26.64	25.41
27	27.66	26.39
28	28.69	27.37
29	29.71	28.34
30	30.73	29.32
31	31.76	30.30
32	32.78	31.27
33	33.81	32.25
34	34.83	33.23
35	35.86	34.21
36	36.88	35.18
37	37.90	36.16
38	38.93	37.14
39	39.95	38.11
40	40.97	39.09
41	42.00	40.07
42	43.02	41.05
43	44.04	42.02
44	45.07	43.00
45	46.09	43.98
46	47.12	44.95
47	48.14	45.93
48	49.16	46.91
49	50.18	47.88
50	51.21	48.86

Old age	New Western	New Soviet
51	52.23	49.84
52	53.25	50.81
53	54.28	51.79
54	55.30	52.77
55	56.33	53.75
56	57.35	54.72
57	58.37	55.70
58	59.39	56.68
59	60.42	57.65
60	61.44	58.63
61	62.46	59.61
62	63.48	60.58
63	64.51	61.56
64	65.53	62.54
65	66.55	63.51
66	67.58	64.49
67	68.60	65.47
68	69.62	66.44
69	70.64	67.42
70	71.67	68.39
71	72.69	69.37
72	73.71	70.35
73	74.73	71.33
74	75.76	72.30
75	76.78	73.28
76	77.80	74.26
77	78.82	75.23
78	79.85	76.21
79	80.87	77.19
80	81.89	78.16
81	82.91	79.14
82	83.93	80.11
83	84.95	81.09
84	85.98	82.07
85	87.00	83.04
86	88.02	84.02
87	89.04	85.00
88	90.07	85.97
89	91.09	86.95
90	92.11	87.93
91	93.13	88.90
92	94.15	89.88
93	95.17	90.85
94	96.20	91.83
95	97.22	92.81
96	98.24	93.78
97	99.26	94.76
98	100.28	95.74
99	101.30	96.71
100	102.33	97.69

Old age	New Western	New Soviet
101	103.3	98.7
102	104.4	99.6
103	105.4	100.6
104	106.4	101.6
105	107.4	102.6
106	108.5	103.5
107	109.5	104.5
108	110.5	105.5
109	111.5	106.5
110	112.5	107.4
111	113.6	108.4
112	114.6	109.4
113	115.6	110.4
114	116.6	111.4
115	117.6	112.3
116	118.7	113.3
117	119.7	114.3
118	120.7	115.3
119	121.7	116.2
120	122.7	117.2
121	123.8	118.2
122	124.8	119.2
123	125.8	120.1
124	126.8	121.1
125	127.8	122.1
126	128.9	123.1
127	129.9	124.0
128	130.9	125.0
129	131.9	126.0
130	133.0	127.0
131	134.0	127.9
132	135.0	128.9
133	136.0	129.9
134	137.0	130.9
135	138.1	131.8
136	139.1	132.8
137	140.1	133.8
138	141.1	134.8
139	142.1	135.8
140	143.2	136.7

Old age	New Western	New Soviet
141	144.2	137.7
142	145.2	138.7
143	146.2	139.7
144	147.2	140.6
145	148.3	141.6
146	149.3	142.6
147	150.3	143.6
148	151.3	144.5
149	152.3	145.5
150	153.4	146.5
151	154.4	147.5
152	155.4	148.4
153	156.4	149.4
154	157.4	150.4
155	158.4	151.4
156	159.5	152.3
157	160.5	153.3
158	161.5	154.3
159	162.5	155.3
160	163.5	156.2
161	164.6	157.2
162	165.6	158.2
163	166.6	159.2
164	167.6	160.1
165	168.6	161.1
166	169.7	162.1
167	170.7	163.1
168	171.7	164.0
169	172.7	165.0
170	173.7	166.0
171	174.8	167.0
172	175.8	167.9
173	176.8	168.9
174	177.8	169.9
175	178.8	170.9
176	179.8	171.8
177	180.9	172.8
178	181.9	173.8
179	182.9	174.8
180	183.9	175.7
181	184.9	176.7
182	186.0	177.7
183	187.0	178.7
184	188.0	179.6
185	189.0	180.6
186	190.0	181.6
187	191.1	182.6
188	192.1	183.5
189	193.1	184.5
190	194.1	185.5

Old age	New Western	New Soviet
191	195.1	186.5
192	196.1	187.4
193	197.2	188.4
194	198.2	189.4
195	199.2	190.4
196	200.2	191.3
197	201.2	192.3
198	202.3	193.3
199	203.3	194.3
200	204.3	195.2
201	205.3	196.2
202	206.3	197.2
203	207.3	198.2
204	208.4	199.1
205	209.4	200.1
206	210.4	201.1
207	211.4	202.1
208	212.4	203.0
209	213.4	204.0
210	214.5	205.0
211	215.5	206.0
212	216.5	206.9
213	217.5	207.9
214	218.5	208.9
215	219.6	209.9
216	220.6	210.8
217	221.6	211.8
218	222.6	212.8
219	223.6	213.8
220	224.6	214.7
221	225.7	215.7
222	226.7	216.7
223	227.7	217.7
224	228.7	218.6
225	229.7	219.6
226	230.7	220.6
227	231.8	221.6
228	232.8	222.5
229	233.8	223.5
230	234.8	224.5
231	235.8	225.5
232	236.8	226.4
233	237.9	227.4
234	238.9	228.4
235	239.9	229.4
236	240.9	230.3
237	241.9	231.3
238	242.9	232.3
239	244.0	233.3
240	245.0	234.2

Old age	New Western	New Soviet
241	246.0	235.2
242	247.0	236.2
243	248.0	237.2
244	249.0	238.1
245	250.1	239.1
246	251.1	240.1
247	252.1	241.1
248	253.1	242.0
249	254.1	243.0
250	255.1	244.0
251	256.2	245.0
252	257.2	245.9
253	258.2	246.9
254	259.2	247.9
255	260.2	248.9
256	261.2	249.8
257	262.3	250.8
258	263.3	251.8
259	264.3	252.7
260	265.3	253.7
261	266.3	254.7
262	267.3	255.7
263	268.3	256.6
264	269.4	257.6
265	270.4	258.6
266	271.4	259.6
267	272.4	260.5
268	273.4	261.5
269	274.4	262.5
270	275.5	263.5
271	276.5	264.4
272	277.5	265.4
273	278.5	266.4
274	279.5	267.4
275	280.5	268.3
276	281.5	269.3
277	282.6	270.3
278	283.6	271.3
279	284.6	272.2
280	285.6	273.2
281	286.6	274.2
282	287.6	275.2
283	288.7	276.1
284	289.7	277.1
285	290.7	278.1
286	291.7	279.0
287	292.7	280.0
288	293.7	281.0
289	294.7	282.0
290	295.8	282.9

Old age	New Western	New Soviet	Old age	New Western	New Soviet	Old age	New Western	New Soviet	Old age	New Western	New Soviet	Old age	New Western	New Soviet	Old age	New Western	New Soviet	Old age	New Western	New Soviet	Old age	New Western	New Soviet
291	296.8	283.9	341	347.5	332.6	391	398.1	381.2	441	448.6	429.9	491	499.1	478.5	541	549.5	527.0	591	599.8	575.6	641	650.1	624.1
292	297.8	284.9	342	348.5	333.6	392	399.1	382.2	442	449.7	430.8	492	500.1	479.4	542	550.5	528.0	592	600.9	576.5	642	651.1	625.1
293	298.8	285.9	343	349.5	334.5	393	400.1	383.2	443	450.7	431.8	493	501.1	480.4	543	551.5	529.0	593	601.9	577.5	643	652.1	626.0
294	299.8	286.8	344	350.5	335.5	394	401.1	384.2	444	451.7	432.8	494	502.1	481.4	544	552.5	529.9	594	602.9	578.5	644	653.1	627.0
295	300.8	287.8	345	351.5	336.5	395	402.1	385.1	445	452.7	433.8	495	503.2	482.4	545	553.5	530.9	595	603.9	579.5	645	654.1	628.0
296	301.8	288.8	346	352.5	337.5	396	403.2	386.1	446	453.7	434.7	496	504.2	483.3	546	554.6	531.9	596	604.9	580.4	646	655.1	628.9
297	302.9	289.8	347	353.6	338.4	397	404.2	387.1	447	454.7	435.7	497	505.2	484.3	547	555.6	532.9	597	605.9	581.4	647	656.1	629.9
298	303.9	290.7	348	354.6	339.4	398	405.2	388.1	448	455.7	436.7	498	506.2	485.3	548	556.6	533.8	598	606.9	582.4	648	657.1	630.9
299	304.9	291.7	349	355.6	340.4	399	406.2	389.0	449	456.7	437.5	499	507.2	486.2	549	557.6	534.8	599	607.9	583.3	649	658.1	631.9
300	305.9	292.7	350	356.6	341.4	400	407.2	390.0	450	457.7	438.6	500	508.2	487.2	550	558.6	535.8	600	608.9	584.3	650	659.1	632.8
301	306.9	293.7	351	357.6	342.3	401	408.2	391.0	451	458.7	439.6	501	509.2	488.2	551	559.6	536.7	601	609.9	585.3	651	660.2	633.8
302	307.9	294.6	352	358.6	343.3	402	409.2	391.9	452	459.8	440.6	502	510.2	489.2	552	560.6	537.7	602	610.9	586.3	652	661.2	634.8
303	309.0	295.6	353	359.6	344.3	403	410.2	392.9	453	460.8	441.5	503	511.2	490.1	553	561.6	538.7	603	611.9	587.2	653	662.2	635.7
304	310.0	296.6	354	360.6	345.2	404	411.2	393.9	454	461.8	442.5	504	512.2	491.1	554	562.6	539.7	604	612.9	588.2	654	663.2	636.7
305	311.0	297.6	355	361.7	346.2	405	412.3	394.9	455	462.8	443.5	505	513.2	492.1	555	563.6	540.6	605	613.9	589.2	655	664.2	637.7
306	312.0	298.5	356	362.7	347.2	406	413.3	395.8	456	463.8	444.5	506	514.2	493.0	556	564.6	541.6	606	614.9	590.1	656	665.2	638.6
307	313.0	299.5	357	363.7	348.2	407	414.3	396.8	457	464.8	445.4	507	515.3	494.0	557	565.6	542.6	607	615.9	591.1	657	666.2	639.6
308	314.0	300.5	358	364.7	349.1	408	415.3	397.8	458	465.8	446.4	508	516.3	495.0	558	566.6	543.5	608	616.9	592.1	658	667.2	640.6
309	315.0	301.4	359	365.7	350.1	409	416.3	398.8	459	466.8	447.4	509	517.3	496.0	559	567.6	544.5	609	617.9	593.0	659	668.2	641.6
310	316.1	302.4	360	366.7	351.1	410	417.3	399.7	460	467.8	448.3	510	518.3	496.9	560	568.7	545.5	610	619.0	594.0	660	669.2	642.5
311	317.1	303.4	361	367.7	352.1	411	418.3	400.7	461	468.8	449.3	511	519.3	497.9	561	569.7	546.5	611	620.0	595.0	661	670.2	643.5
312	318.1	304.4	362	368.7	353.0	412	419.3	401.7	462	469.9	450.3	512	520.3	498.9	562	570.7	547.4	612	621.0	596.0	662	671.2	644.5
313	319.1	305.3	363	369.8	354.0	413	420.4	402.6	463	470.9	451.3	513	521.3	499.8	563	571.7	548.4	613	622.0	596.9	663	672.2	645.4
314	320.1	306.3	364	370.8	355.0	414	421.4	403.6	464	471.9	452.2	514	522.3	500.8	564	572.7	549.4	614	623.0	597.9	664	673.2	646.4
315	321.1	307.3	365	371.8	356.0	415	422.4	404.6	465	472.9	453.2	515	523.3	501.8	565	573.7	550.3	615	624.0	598.9	665	674.2	647.4
316	322.1	308.3	366	372.8	356.9	416	423.4	405.6	466	473.9	454.2	516	524.3	502.8	566	574.7	551.3	616	625.0	599.8	666	675.2	648.3
317	323.1	309.2	367	373.8	357.9	417	424.4	406.5	467	474.9	455.1	517	525.3	503.7	567	575.7	552.3	617	626.0	600.8	667	676.2	649.3
318	324.2	310.2	368	374.8	358.9	418	425.4	407.5	468	475.9	456.1	518	526.3	504.7	568	576.7	553.2	618	627.0	601.8	668	677.2	650.3
319	325.2	311.2	369	375.8	359.8	419	426.4	408.5	469	476.9	457.1	519	527.4	505.7	569	577.7	554.2	619	628.0	602.7	669	678.2	651.3
320	326.2	312.2	370	376.8	360.8	420	427.4	409.5	470	477.9	458.1	520	528.4	506.6	570	578.7	555.2	620	629.0	603.7	670	679.2	652.2
321	327.2	313.1	371	377.9	361.8	421	428.4	410.4	471	478.9	459.0	521	529.4	507.6	571	579.7	556.2	621	630.0	604.7	671	680.2	653.2
322	328.2	314.1	372	378.9	362.8	422	429.4	411.4	472	479.9	460.0	522	530.4	508.6	572	580.7	557.1	622	631.0	605.7	672	681.2	654.2
323	329.2	315.1	373	379.9	363.7	423	430.5	412.4	473	481.0	461.0	523	531.4	509.6	573	581.7	558.1	623	632.0	606.6	673	682.2	655.1
324	330.2	316.1	374	380.9	364.7	424	431.5	413.3	474	482.0	461.9	524	532.4	510.5	574	582.7	559.1	624	633.0	607.6	674	683.2	656.1
325	331.3	317.0	375	381.9	365.7	425	432.5	414.3	475	483.0	462.9	525	533.4	511.5	575	583.7	560.0	625	634.0	608.6	675	684.2	657.1
326	332.3	318.0	376	382.9	366.7	426	433.5	415.3	476	484.0	463.9	526	534.4	512.5	576	584.8	561.0	626	635.0	609.5	676	685.3	658.0
327	333.3	319.0	377	383.9	367.6	427	434.5	416.3	477	485.0	464.9	527	535.4	513.4	577	585.8	562.0	627	636.0	610.5	677	686.3	659.0
328	334.3	319.9	378	384.9	368.6	428	435.5	417.2	478	486.0	465.8	528	536.4	514.4	578	586.8	563.0	628	637.0	611.5	678	687.3	660.0
329	335.3	320.9	379	386.0	369.6	429	436.5	418.2	479	487.0	466.8	529	537.4	515.4	579	587.8	563.9	629	638.1	612.5	679	688.3	660.9
330	336.3	321.9	380	387.0	370.5	430	437.5	419.2	480	488.0	467.8	530	538.4	516.4	580	588.8	564.9	630	639.1	613.4	680	689.3	661.9
331	337.3	322.9	381	388.0	371.5	431	438.5	420.1	481	489.0	468.7	531	539.4	517.3	581	589.8	565.9	631	640.1	614.4	681	690.3	662.9
332	338.4	323.8	382	389.0	372.5	432	439.6	421.1	482	490.0	469.7	532	540.4	518.3	582	590.8	566.8	632	641.1	615.4	682	691.3	663.9
333	339.4	324.8	383	390.0	373.5	433	440.6	422.1	483	491.0	470.7	533	541.5	519.3	583	591.8	567.8	633	642.1	616.3	683	692.3	664.8
334	340.4	325.8	384	391.0	374.4	434	441.6	423.1	484	492.1	471.7	534	542.5	520.2	584	592.8	568.8	634	643.1	617.3	684	693.3	665.8
335	341.4	326.8	385	392.0	375.4	435	442.6	424.0	485	493.1	472.6	535	543.5	521.2	585	593.8	569.8	635	644.1	618.3	685	694.3	666.8
336	342.4	327.7	386	393.0	376.4	436	443.6	425.0	486	494.1	473.6	536	544.5	522.2	586	594.8	570.7	636	645.1	619.2	686	695.3	667.7
337	343.4	328.7	387	394.1	377.4	437	444.6	426.0	487	495.1	474.6	537	545.5	523.1	587	595.8	571.7	637	646.1	620.2	687	696.3	668.7
338	344.4	329.7	388	395.1	378.3	438	445.6	427.0	488	496.1	475.6	538	546.5	524.1	588	596.8	572.7	638	647.1	621.2	688	697.3	669.7
339	345.4	330.7	389	396.1	379.3	439	446.6	427.9	489	497.1	476.5	539	547.5	525.1	589	597.8	573.6	639	648.1	622.2	689	698.3	670.6
340	346.5	331.6	390	397.1	380.3	440	447.6	428.9	490	498.1	477.5	540	548.5	526.1	590	598.8	574.6	640	649.1	623.1	690	699.3	671.6

Old age	New Western	New Soviet
691	700.3	672.6
692	701.3	673.6
693	702.3	674.5
694	703.3	675.5
695	704.3	676.5
696	705.3	677.4
697	706.3	678.4
698	707.3	679.4
699	708.3	680.3
700	709.3	681.3
701	710.3	682.3
702	711.3	683.3
703	712.3	684.2
704	713.3	685.2
705	714.3	686.2
706	715.3	687.1
707	716.4	688.1
708	717.4	689.1
709	718.4	690.0
710	719.4	691.0
711	720.4	692.0
712	721.4	692.9
713	722.4	693.9
714	723.4	694.9
715	724.4	695.9
716	725.4	696.8
717	726.4	697.8
718	727.4	698.8
719	728.4	699.7
720	729.4	700.7
721	730.4	701.7
722	731.4	702.6
723	732.4	703.6
724	733.4	704.6
725	734.4	705.5
726	735.4	706.5
727	736.4	707.5
728	737.4	708.5
729	738.4	709.4
730	739.4	710.4
731	740.4	711.4
732	741.4	712.3
733	742.4	713.3
734	743.4	714.3
735	744.4	715.2
736	745.4	716.2
737	746.4	717.2
738	747.4	718.1
739	748.4	719.1
740	749.4	720.1
741	750.4	721.1
742	751.4	722.0
743	752.4	723.0
744	753.4	724.0
745	754.4	724.9
746	755.4	725.9
747	756.4	726.9
748	757.4	727.8
749	758.4	728.8
750	759.5	729.8
751	760.5	730.7
752	761.5	731.7
753	762.5	732.7
754	763.5	733.7
755	764.5	734.6
756	765.5	735.6
757	766.5	736.6
758	767.5	737.5
759	768.5	738.5
760	769.5	739.5
761	770.5	740.4
762	771.5	741.4
763	772.5	742.4
764	773.5	743.3
765	774.5	744.3
766	775.5	745.3
767	776.5	746.3
768	777.5	747.2
769	778.5	748.2
770	779.5	749.2
771	780.5	750.1
772	781.5	751.1
773	782.5	752.1
774	783.5	753.0
775	784.5	754.0
776	785.5	755.0
777	786.5	755.9
778	787.5	756.9
779	788.5	757.9
780	789.5	758.8
781	790.5	759.8
782	791.5	760.8
783	792.5	761.8
784	793.5	762.7
785	794.5	763.7
786	795.5	764.7
787	796.5	765.6
788	797.5	766.6
789	798.5	767.6
790	799.5	768.5
791	800.5	769.5
792	801.5	770.5
793	802.5	771.4
794	803.5	772.4
795	804.5	773.4
796	805.5	774.3
797	806.5	775.3
798	807.5	776.3
799	808.5	777.3
800	809.5	778.2
801	810.5	779.2
802	811.5	780.2
803	812.5	781.1
804	813.5	782.1
805	814.5	783.1
806	815.5	784.0
807	816.5	785.0
808	817.5	786.0
809	818.5	786.9
810	819.5	787.9
811	820.5	788.9
812	821.5	789.8
813	822.5	790.8
814	823.5	791.8
815	824.5	792.7
816	825.5	793.7
817	826.5	794.7
818	827.5	795.7
819	828.5	796.6
820	829.5	797.6
821	830.5	798.6
822	831.5	799.5
823	832.5	800.5
824	833.5	801.5
825	834.5	802.4
826	835.5	803.4
827	836.5	804.4
828	837.5	805.3
829	838.5	806.3
830	839.5	807.3
831	840.5	808.2
832	841.5	809.2
833	842.5	810.2
834	843.5	811.1
835	844.5	812.1
836	845.5	813.1
837	846.5	814.1
838	847.5	815.0
839	848.5	816.0
840	849.5	817.0
841	850.5	817.9
842	851.5	818.9
843	852.5	819.9
844	853.5	820.8
845	854.5	821.8
846	855.5	822.8
847	856.5	823.7
848	857.5	824.7
849	858.5	825.7
850	859.5	826.6
851	860.5	827.6
852	861.5	828.6
853	862.5	829.5
854	863.5	830.5
855	864.5	831.5
856	865.5	832.4
857	866.5	833.4
858	867.5	834.4
859	868.5	835.4
860	869.5	836.3
861	870.5	837.3
862	871.5	838.3
863	872.5	839.2
864	873.5	840.2
865	874.5	841.2
866	875.5	842.1
867	876.5	843.1
868	877.5	844.1
869	878.5	845.0
870	879.5	846.0
871	880.5	847.0
872	881.5	847.9
873	882.5	848.9
874	883.5	849.9
875	884.5	850.8
876	885.5	851.8
877	886.5	852.8
878	887.5	853.7
879	888.5	854.7
880	889.5	855.7
881	890.5	856.6
882	891.5	857.6
883	892.5	858.6
884	893.5	859.6
885	894.5	860.5
886	895.5	861.5
887	896.5	862.5
888	897.5	863.4
889	898.5	864.4
890	899.5	865.4
891	900.5	866.3
892	901.5	867.3
893	902.5	868.3
894	903.5	869.2
895	904.5	870.2
896	905.5	871.2
897	906.4	872.1
898	907.4	873.1
899	908.4	874.1
900	909.4	875.0
901	910.4	876.0
902	911.4	877.0
903	912.4	877.9
904	913.4	878.9
905	914.4	879.9
906	915.4	880.8
907	916.4	881.8
908	917.4	882.8
909	918.4	883.7
910	919.4	884.7
911	920.4	885.7
912	921.4	886.7
913	922.4	887.6
914	923.4	888.6
915	924.4	889.6
916	925.4	890.5
917	926.4	891.5
918	927.4	892.5
919	928.4	893.4
920	929.4	894.4
921	930.4	895.4
922	931.4	896.3
923	932.4	897.3
924	933.4	898.3
925	934.4	899.2
926	935.4	900.2
927	936.4	901.2
928	937.4	902.1
929	938.4	903.1
930	939.4	904.1
931	940.4	905.0
932	941.4	906.0
933	942.4	907.0
934	943.4	907.9
935	944.4	908.9
936	945.4	909.9
937	946.4	910.8
938	947.4	911.8
939	948.4	912.8
940	949.4	913.7
941	950.4	914.7
942	951.3	915.7
943	952.3	916.6
944	953.3	917.6
945	954.3	918.6
946	955.3	919.5
947	956.3	920.5
948	957.3	921.5
949	958.3	922.4
950	959.3	923.4
951	960.3	924.4
952	961.3	925.4
953	962.3	926.3
954	963.3	927.3
955	964.3	928.3
956	965.3	929.2
957	966.3	930.2
958	967.3	931.2
959	968.3	932.1
960	969.3	933.1
961	970.3	934.1
962	971.3	935.0
963	972.3	936.0
964	973.3	937.0
965	974.3	937.9
966	975.3	938.9
967	976.3	939.9
968	977.3	940.8
969	978.3	941.8
970	979.3	942.8
971	980.3	943.7
972	981.3	944.7
973	982.3	945.7
974	983.2	946.6
975	984.2	947.6
976	985.2	948.6
977	986.2	949.5
978	987.2	950.5
979	988.2	951.5
980	989.2	952.4
981	990.2	953.4
982	991.2	954.4
983	992.2	955.3
984	993.2	956.3
985	994.2	957.3
986	995.2	958.2
987	996.2	959.2
988	997.2	960.2
989	998.2	961.1
990	999.2	962.1
991	1000.2	963.1
992	1001.2	964.0
993	1002.2	965.0
994	1003.2	966.0
995	1004.2	966.9
996	1005.2	967.9
997	1006.2	968.9
998	1007.2	969.8
999	1008.2	970.8
1000	1009.2	971.8
1005	1014	976
1010	1019	981
1015	1024	986
1020	1029	991
1025	1034	995
1030	1039	1000
1035	1043	1005
1040	1048	1010
1045	1053	1015
1050	1058	1020
1055	1063	1024
1060	1068	1029
1065	1073	1034
1070	1078	1039
1075	1083	1044
1080	1088	1049
1085	1093	1053
1090	1098	1058
1095	1103	1063
1100	1108	1068
1105	1113	1073
1110	1118	1078
1115	1123	1082
1120	1128	1087
1125	1133	1092
1130	1138	1097
1135	1143	1102
1140	1148	1107
1145	1153	1111
1150	1158	1116
1155	1163	1121
1160	1168	1126
1165	1173	1131
1170	1178	1136
1175	1183	1140
1180	1188	1145
1185	1193	1150
1190	1198	1155
1195	1202	1160
1200	1207	1165
1205	1212	1169
1210	1217	1174
1215	1222	1179
1220	1227	1184
1225	1232	1189
1230	1237	1193
1235	1242	1198
1240	1247	1203
1245	1252	1208
1250	1257	1213
1255	1262	1218
1260	1267	1222
1265	1272	1227
1270	1277	1232
1275	1282	1237
1280	1287	1242
1285	1292	1247
1290	1297	1251
1295	1302	1256
1300	1307	1261
1305	1311	1266
1310	1316	1271
1315	1321	1275
1320	1326	1280
1325	1331	1285
1330	1336	1290
1335	1341	1295
1340	1346	1300
1345	1351	1304
1350	1356	1309
1355	1361	1314
1360	1366	1319
1365	1371	1324
1370	1376	1329
1375	1381	1333
1380	1386	1338
1385	1391	1343
1390	1396	1348
1395	1401	1353
1400	1405	1357

Old age	New Western	New Soviet
1405	1410	1362
1410	1415	1367
1415	1420	1372
1420	1425	1377
1425	1430	1382
1430	1435	1386
1435	1440	1391
1440	1445	1396
1445	1450	1401
1450	1455	1406
1455	1460	1410
1460	1465	1415
1465	1470	1420
1470	1475	1425
1475	1480	1430
1480	1485	1435
1485	1489	1439
1490	1494	1444
1495	1499	1449
1500	1504	1454
1505	1509	1459
1510	1514	1463
1515	1519	1468
1520	1524	1473
1525	1529	1478
1530	1534	1483
1535	1539	1488
1540	1544	1492
1545	1549	1497
1550	1554	1502
1555	1558	1507
1560	1563	1512
1565	1568	1516
1570	1573	1521
1575	1578	1526
1580	1583	1531
1585	1588	1536
1590	1593	1541
1595	1598	1545
1600	1603	1550
1605	1608	1555
1610	1613	1560
1615	1618	1565
1620	1622	1569
1625	1627	1574
1630	1632	1579
1635	1637	1584
1640	1642	1589
1645	1647	1593
1650	1652	1598
1655	1657	1603
1660	1662	1608
1665	1667	1613
1670	1672	1618
1675	1677	1622
1680	1682	1627
1685	1686	1632
1690	1691	1637
1695	1696	1642
1700	1701	1646
1705	1706	1651
1710	1711	1656
1715	1716	1661
1720	1721	1666
1725	1726	1670
1730	1731	1675
1735	1736	1680
1740	1741	1685
1745	1745	1690
1750	1750	1695
1755	1755	1699
1760	1760	1704
1765	1765	1709
1770	1770	1714
1775	1775	1719
1780	1780	1723
1785	1785	1728
1790	1790	1733
1795	1795	1738
1800	1800	1743
1805	1804	1747
1810	1809	1752
1815	1814	1757
1820	1819	1762
1825	1824	1767
1830	1829	1771
1835	1834	1776
1840	1839	1781
1845	1844	1786
1850	1849	1791
1855	1854	1795
1860	1858	1800
1865	1863	1805
1870	1868	1810
1875	1873	1815
1880	1878	1820
1885	1883	1824
1890	1888	1829
1895	1893	1834
1900	1898	1839
1905	1903	1844
1910	1907	1848
1915	1912	1853
1920	1917	1858
1925	1922	1863
1930	1927	1868
1935	1932	1872
1940	1937	1877
1945	1942	1882
1950	1947	1887
1955	1952	1892
1960	1956	1896
1965	1961	1901
1970	1966	1906
1975	1971	1911
1980	1976	1916
1985	1981	1920
1990	1986	1925
1995	1991	1930
2000	1996	1935
2005	2001	1940
2010	2005	1944
2015	2010	1949
2020	2015	1954
2025	2020	1959
2030	2025	1964
2035	2030	1968
2040	2035	1973
2045	2040	1978
2050	2045	1983
2055	2050	1988
2060	2054	1992
2065	2059	1997
2070	2064	2002
2075	2069	2007
2080	2074	2012
2085	2079	2016
2090	2084	2021
2095	2089	2026
2100	2094	2031
2105	2098	2036
2110	2103	2040
2115	2108	2045
2120	2113	2050
2125	2118	2055
2130	2123	2060
2135	2128	2065
2140	2133	2069
2145	2138	2074
2150	2142	2079
2155	2147	2084
2160	2152	2089
2165	2157	2093
2170	2162	2098
2175	2167	2103
2180	2172	2108
2185	2177	2113
2190	2182	2117
2195	2186	2122
2200	2191	2127
2205	2196	2132
2210	2201	2137
2215	2206	2141
2220	2211	2146
2225	2216	2151
2230	2221	2156
2235	2226	2160
2240	2230	2165
2245	2235	2170
2250	2240	2175
2255	2245	2180
2260	2250	2184
2265	2255	2189
2270	2260	2194
2275	2265	2199
2280	2270	2204
2285	2274	2208
2290	2279	2213
2295	2284	2218
2300	2289	2223
2305	2294	2228
2310	2299	2232
2315	2304	2237
2320	2309	2242
2325	2313	2247
2330	2318	2252
2335	2323	2256
2340	2328	2261
2345	2333	2266
2350	2338	2271
2355	2343	2276
2360	2348	2280
2365	2352	2285
2370	2357	2290
2375	2362	2295
2380	2367	2300
2385	2372	2304
2390	2377	2309
2395	2382	2314
2400	2387	2319
2405	2391	2324
2410	2396	2328
2415	2401	2333
2420	2406	2338
2425	2411	2343
2430	2416	2348
2435	2421	2352
2440	2426	2357
2445	2430	2362
2450	2435	2367
2455	2440	2372
2460	2445	2376
2465	2450	2381
2470	2455	2386
2475	2460	2391
2480	2465	2396
2485	2469	2400
2490	2474	2405
2495	2479	2410
2500	2484	2415
2505	2489	2419
2510	2494	2424
2515	2499	2429
2520	2504	2434
2525	2508	2439
2530	2513	2443
2535	2518	2448
2540	2523	2453
2545	2528	2458
2550	2533	2463
2555	2538	2467
2560	2543	2472
2565	2547	2477
2570	2552	2482
2575	2557	2487
2580	2562	2491
2585	2567	2496
2590	2572	2501
2595	2577	2506
2600	2581	2511
2605	2586	2515
2610	2591	2520
2615	2596	2525
2620	2601	2530
2625	2606	2535
2630	2611	2539
2635	2616	2544
2640	2620	2549
2645	2625	2554
2650	2630	2558
2655	2635	2563
2660	2640	2568
2665	2645	2573
2670	2650	2578
2675	2654	2582
2680	2659	2587
2685	2664	2592
2690	2669	2597
2695	2674	2602
2700	2679	2606
2705	2684	2611
2710	2688	2616
2715	2693	2621
2720	2698	2626
2725	2703	2630
2730	2708	2635
2735	2713	2640
2740	2718	2645
2745	2722	2649
2750	2727	2654
2755	2732	2659
2760	2737	2664
2765	2742	2669
2770	2747	2673
2775	2752	2678
2780	2757	2683
2785	2761	2688
2790	2766	2693
2795	2771	2697
2800	2776	2702
2805	2781	2707
2810	2786	2712
2815	2791	2717
2820	2795	2721
2825	2800	2726
2830	2805	2731
2835	2810	2736
2840	2815	2740
2845	2820	2745
2850	2825	2750
2855	2829	2755
2860	2834	2760
2865	2839	2764
2870	2844	2769
2875	2849	2774
2880	2854	2779
2885	2859	2784
2890	2863	2788
2895	2868	2793
2900	2873	2798
2905	2878	2803
2910	2883	2807
2915	2888	2812
2920	2892	2817
2925	2897	2822
2930	2902	2827
2935	2907	2831
2940	2912	2836
2945	2917	2841
2950	2922	2846
2955	2926	2851
2960	2931	2855
2965	2936	2860
2970	2941	2865
2975	2946	2870
2980	2951	2874
2985	2956	2879
2990	2960	2884
2995	2965	2889
3000	2970	2894
3005	2975	2898
3010	2980	2903
3015	2985	2908
3020	2990	2913
3025	2994	2918
3030	2999	2922
3035	3004	2927
3040	3009	2932
3045	3014	2937
3050	3019	2941
3055	3023	2946
3060	3028	2951
3065	3033	2956
3070	3038	2961
3075	3043	2965
3080	3048	2970
3085	3053	2975
3090	3057	2980
3095	3062	2985
3100	3067	2989
3105	3072	2994
3110	3077	2999
3115	3082	3004
3120	3087	3008
3125	3091	3013
3130	3096	3018
3135	3101	3023
3140	3106	3028
3145	3111	3032
3150	3116	3037
3155	3120	3042
3160	3125	3047
3165	3130	3052
3170	3135	3056
3175	3140	3061
3180	3145	3066
3185	3150	3071
3190	3154	3075
3195	3159	3080
3200	3164	3085
3205	3169	3090
3210	3174	3095
3215	3179	3099
3220	3183	3104
3225	3188	3109
3230	3193	3114
3235	3198	3118
3240	3203	3123
3245	3208	3128
3250	3212	3133
3255	3217	3138
3260	3222	3142
3265	3227	3147
3270	3232	3152
3275	3237	3157
3280	3242	3162
3285	3246	3166
3290	3251	3171
3295	3256	3176
3300	3261	3181
3305	3266	3185
3310	3271	3190
3315	3275	3195
3320	3280	3200
3325	3285	3205
3330	3290	3209
3335	3295	3214
3340	3300	3219
3345	3304	3224
3350	3309	3228
3355	3314	3233
3360	3319	3238
3365	3324	3243
3370	3329	3248
3375	3333	3252
3380	3338	3257
3385	3343	3262
3390	3348	3267
3395	3353	3272
3400	3358	3276

New Soviet	New Western	Old age	New Soviet	New Western	Old age	New Soviet	New Western	Old age	New Soviet	New Western	Old age
4236	4328	4405	4379	4472	4555	4523	4617	4705	4666	4761	4855
4241	4333	4410	4384	4477	4560	4527	4621	4710	4670	4766	4860
4246	4337	4415	4389	4482	4565	4532	4626	4715	4675	4771	4865
4251	4342	4420	4394	4487	4570	4537	4631	4720	4680	4775	4870
4255	4347	4425	4399	4491	4575	4542	4636	4725	4685	4780	4875
4260	4352	4430	4403	4496	4580	4546	4641	4730	4690	4785	4880
4265	4357	4435	4408	4501	4585	4551	4645	4735	4694	4790	4885
4270	4361	4440	4413	4506	4590	4556	4650	4740	4699	4795	4890
4274	4366	4445	4418	4511	4595	4561	4655	4745	4704	4799	4895
4279	4371	4450	4422	4516	4600	4566	4660	4750	4709	4804	4900
4284	4376	4455	4427	4520	4605	4570	4665	4755	4713	4809	4905
4289	4381	4460	4432	4525	4610	4575	4670	4760	4718	4814	4910
4294	4386	4465	4437	4530	4615	4580	4674	4765	4723	4819	4915
4298	4390	4470	4441	4535	4620	4585	4679	4770	4728	4823	4920
4303	4395	4475	4446	4540	4625	4589	4684	4775	4732	4828	4925
4308	4400	4480	4451	4544	4630	4594	4689	4780	4737	4833	4930
4313	4405	4485	4456	4549	4635	4599	4694	4785	4742	4838	4935
4317	4410	4490	4461	4554	4640	4604	4698	4790	4747	4843	4940
4322	4414	4495	4465	4559	4645	4608	4703	4795	4752	4847	4945
4327	4419	4500	4470	4564	4650	4613	4708	4800	4756	4852	4950
4332	4424	4505	4475	4568	4655	4618	4713	4805	4761	4857	4955
4337	4429	4510	4480	4573	4660	4623	4718	4810	4766	4862	4960
4341	4434	4515	4484	4578	4665	4628	4722	4815	4771	4867	4965
4346	4439	4520	4489	4583	4670	4632	4727	4820	4775	4872	4970
4351	4443	4525	4494	4588	4675	4637	4732	4825	4780	4876	4975
4356	4448	4530	4499	4593	4680	4642	4737	4830	4785	4881	4980
4360	4453	4535	4504	4597	4685	4647	4742	4835	4790	4886	4985
4365	4458	4540	4508	4602	4690	4651	4746	4840	4794	4891	4990
4370	4463	4545	4513	4607	4695	4656	4751	4845	4799	4896	4995
4375	4467	4550	4518	4612	4700	4661	4756	4850	4804	4900	5000

Old age	New Western	New Soviet	New Soviet	New Western	Old age	New Soviet	New Western	Old age	New Soviet	New Western	Old age
3405	3363	3281	3520	3604	3655	3759	3846	3905	3998	4087	4155
3410	3367	3286	3525	3609	3660	3764	3851	3910	4002	4092	4160
3415	3372	3291	3530	3614	3665	3768	3855	3915	4007	4096	4165
3420	3377	3295	3534	3619	3670	3773	3860	3920	4012	4101	4170
3425	3382	3300	3539	3624	3675	3778	3865	3925	4017	4106	4175
3430	3387	3305	3544	3628	3680	3783	3870	3930	4022	4111	4180
3435	3392	3310	3549	3633	3685	3788	3875	3935	4026	4116	4185
3440	3396	3315	3553	3638	3690	3792	3879	3940	4031	4121	4190
3445	3401	3319	3558	3643	3695	3797	3884	3945	4036	4125	4195
3450	3406	3324	3563	3648	3700	3802	3889	3950	4041	4130	4200
3455	3411	3329	3568	3653	3705	3807	3894	3955	4045	4135	4205
3460	3416	3334	3573	3657	3710	3811	3899	3960	4050	4140	4210
3465	3421	3338	3577	3662	3715	3816	3904	3965	4055	4145	4215
3470	3425	3343	3582	3667	3720	3821	3908	3970	4060	4149	4220
3475	3430	3348	3587	3672	3725	3826	3913	3975	4064	4154	4225
3480	3435	3353	3592	3677	3730	3831	3918	3980	4069	4159	4230
3485	3440	3358	3596	3682	3735	3835	3923	3985	4074	4164	4235
3490	3445	3362	3601	3686	3740	3840	3928	3990	4079	4169	4240
3495	3450	3367	3606	3691	3745	3845	3933	3995	4084	4174	4245
3500	3454	3372	3611	3696	3750	3850	3937	4000	4088	4178	4250
3505	3459	3377	3616	3701	3755	3854	3942	4005	4093	4183	4255
3510	3464	3381	3620	3706	3760	3859	3947	4010	4098	4188	4260
3515	3469	3386	3625	3711	3765	3864	3952	4015	4103	4193	4265
3520	3474	3391	3630	3715	3770	3869	3957	4020	4107	4198	4270
3525	3479	3396	3635	3720	3775	3874	3961	4025	4112	4203	4275
3530	3483	3401	3639	3725	3780	3878	3966	4030	4117	4207	4280
3535	3488	3405	3644	3730	3785	3883	3971	4035	4122	4212	4285
3540	3493	3410	3649	3735	3790	3888	3976	4040	4127	4217	4290
3545	3498	3415	3654	3739	3795	3893	3981	4045	4131	4222	4295
3550	3503	3420	3659	3744	3800	3897	3986	4050	4136	4227	4300
3555	3508	3424	3663	3749	3805	3902	3990	4055	4141	4231	4305
3560	3512	3429	3668	3754	3810	3907	3995	4060	4146	4236	4310
3565	3517	3434	3673	3759	3815	3912	4000	4065	4150	4241	4315
3570	3522	3439	3678	3764	3820	3916	4005	4070	4155	4246	4320
3575	3527	3444	3682	3768	3825	3921	4010	4075	4160	4251	4325
3580	3532	3448	3687	3773	3830	3926	4015	4080	4165	4256	4330
3585	3537	3453	3692	3778	3835	3931	4019	4085	4169	4260	4335
3590	3541	3458	3697	3783	3840	3936	4024	4090	4174	4265	4340
3595	3546	3463	3702	3788	3845	3940	4029	4095	4179	4270	4345
3600	3551	3467	3706	3793	3850	3945	4034	4100	4184	4275	4350
3605	3556	3472	3711	3797	3855	3950	4039	4105	4189	4280	4355
3610	3561	3477	3716	3802	3860	3955	4043	4110	4193	4284	4360
3615	3566	3482	3721	3807	3865	3959	4048	4115	4198	4289	4365
3620	3570	3487	3725	3812	3870	3964	4053	4120	4203	4294	4370
3625	3575	3491	3730	3817	3875	3969	4058	4125	4208	4299	4375
3630	3580	3496	3735	3822	3880	3974	4063	4130	4212	4304	4380
3635	3585	3501	3740	3826	3885	3979	4068	4135	4217	4308	4385
3640	3590	3506	3745	3831	3890	3983	4072	4140	4222	4313	4390
3645	3595	3510	3749	3836	3895	3988	4077	4145	4227	4318	4395
3650	3599	3515	3754	3841	3900	3993	4082	4150	4232	4323	4400

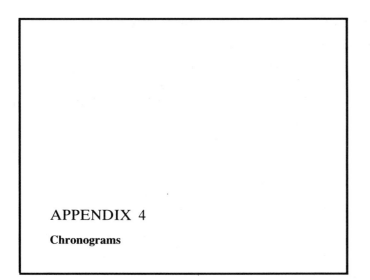

APPENDIX 4

Chronograms

The chronogram is explained in Chapter 5 (Section 5.4) and in Figures 5.1, 5.2 and 5.3. The systematic application of the chronogram method to each chronostratic boundary has resulted in 127 chronograms of which 125 are reproduced here as Figures A4.1 to A4.125. The numerical data from these 125 chronograms are listed in Table 5.4 (nos. 3 to 127).

The horizontal axis is the scanning age in Ma. The vertical axis is the numerical value of the error function (Section 5.4). Ideally, error functions should be steep-sided, symmetric and touch the time-axis. If the error function has shallow sides, then the controlling dates are sparse or have large errors, or both. If the error function is raised above the time-axis, there are overlapping age data, i.e. some dates from the stratigraphically younger unit are greater than dates from the stratigraphically older unit. If the error function is flat for an interval, then there are no age data in the flat interval.

Note that the horizontal scale varies from chronogram to chronogram. The change in scale produces deceptively similar chronograms for very different data. The vertical scale varies for some Cambrian chronograms.

The 'best age' is taken as the minimum of the error function. In a flat interval the minimum is taken as the mid-point of the age range for which there are no data. The 'minimum age' is taken as the youngest age for which the error function first exceeds the minimum value of the error function by one; likewise, the 'maximum age' is taken as the oldest age for which the error function first exceeds the maximum value by one.

The first two chronograms, omitted here, concerned the Holocene–Pleistocene and the Pleistocene–Pliocene boundaries whose ages were then estimated differently (Sections 3.21.1, 3.21.5 and 5.6).

The database of Table 4.2 was modified with new items after the chronograms had been completed. The following chronograms show values that differ slightly from those in Table 5.4: 3, 18, 31, 32, 35, 64, 74, 88, 101, 115, 119 and 120. Only in the case of A4.115 are the differences considered significant.

A4.1 Piacenzian – Zanclian

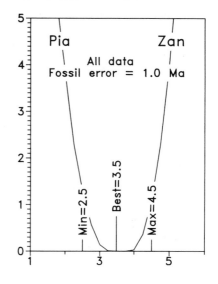

A4.2 Zanclian – Messinian

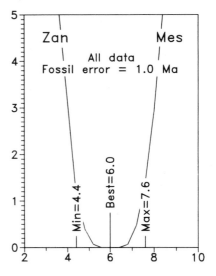

A4.3 Messinian – Tortonian

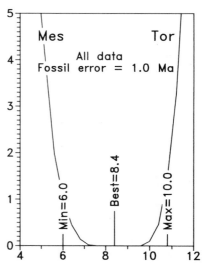

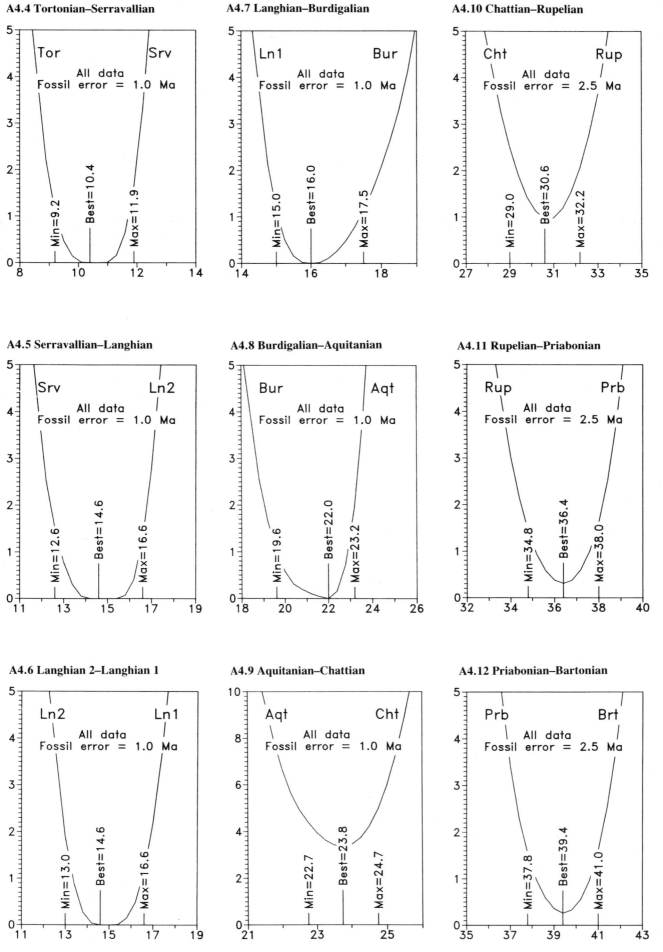

A4.4 Tortonian–Serravallian

Tor Srv
All data
Fossil error = 1.0 Ma
Min=9.2 Best=10.4 Max=11.9

A4.5 Serravallian–Langhian

Srv Ln2
All data
Fossil error = 1.0 Ma
Min=12.6 Best=14.6 Max=16.6

A4.6 Langhian 2–Langhian 1

Ln2 Ln1
All data
Fossil error = 1.0 Ma
Min=13.0 Best=14.6 Max=16.6

A4.7 Langhian–Burdigalian

Ln1 Bur
All data
Fossil error = 1.0 Ma
Min=15.0 Best=16.0 Max=17.5

A4.8 Burdigalian–Aquitanian

Bur Aqt
All data
Fossil error = 1.0 Ma
Min=19.6 Best=22.0 Max=23.2

A4.9 Aquitanian–Chattian

Aqt Cht
All data
Fossil error = 1.0 Ma
Min=22.7 Best=23.8 Max=24.7

A4.10 Chattian–Rupelian

Cht Rup
All data
Fossil error = 2.5 Ma
Min=29.0 Best=30.6 Max=32.2

A4.11 Rupelian–Priabonian

Rup Prb
All data
Fossil error = 2.5 Ma
Min=34.8 Best=36.4 Max=38.0

A4.12 Priabonian–Bartonian

Prb Brt
All data
Fossil error = 2.5 Ma
Min=37.8 Best=39.4 Max=41.0

A4.13 Bartonian–Lutetian

A4.16 Thanetian–Danian

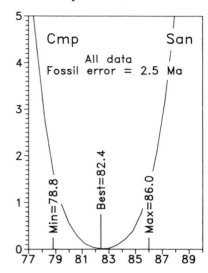
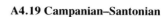

A4.19 Campanian–Santonian

A4.14 Lutetian–Ypresian

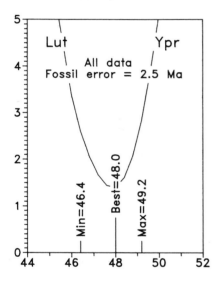

A4.17 Danian–Maastrichtian

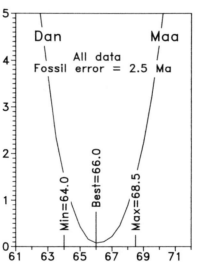

A4.20 Santonian–Coniacian

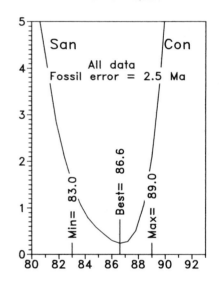

A4.15 Ypresian–Thanetian

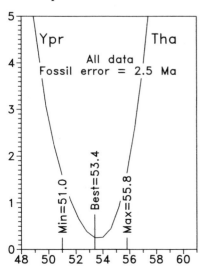

A4.18 Maastrichtian–Campanian

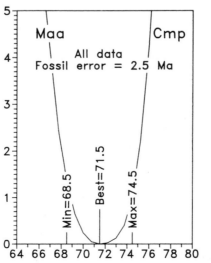

A4.21 Coniacian–Turonian

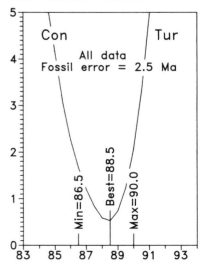

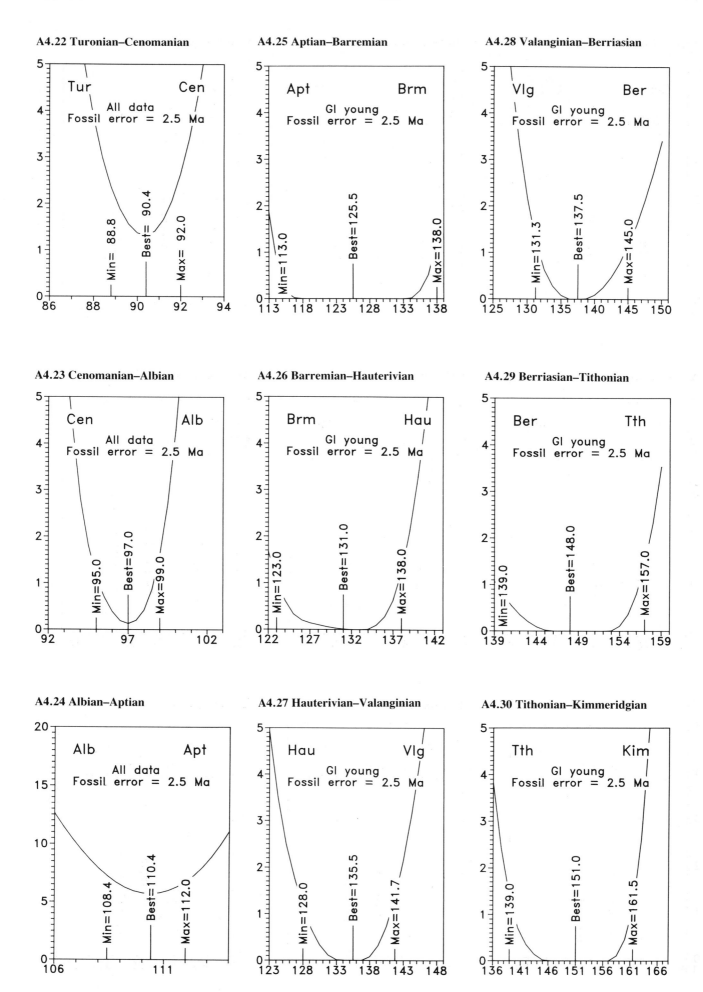

A4.22 Turonian–Cenomanian

A4.25 Aptian–Barremian

A4.28 Valanginian–Berriasian

A4.23 Cenomanian–Albian

A4.26 Barremian–Hauterivian

A4.29 Berriasian–Tithonian

A4.24 Albian–Aptian

A4.27 Hauterivian–Valanginian

A4.30 Tithonian–Kimmeridgian

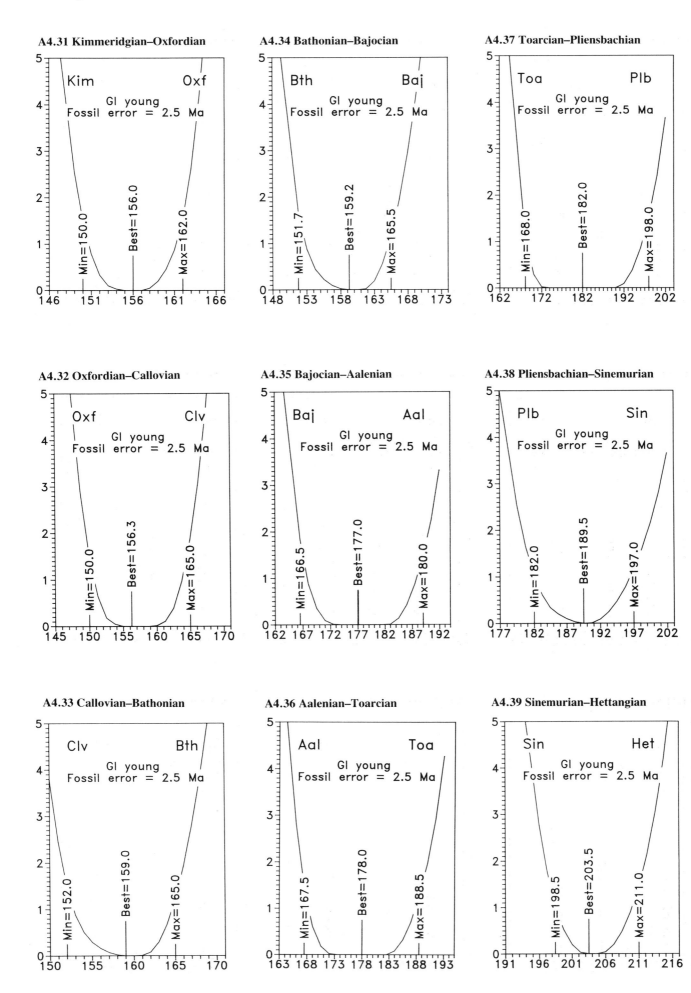

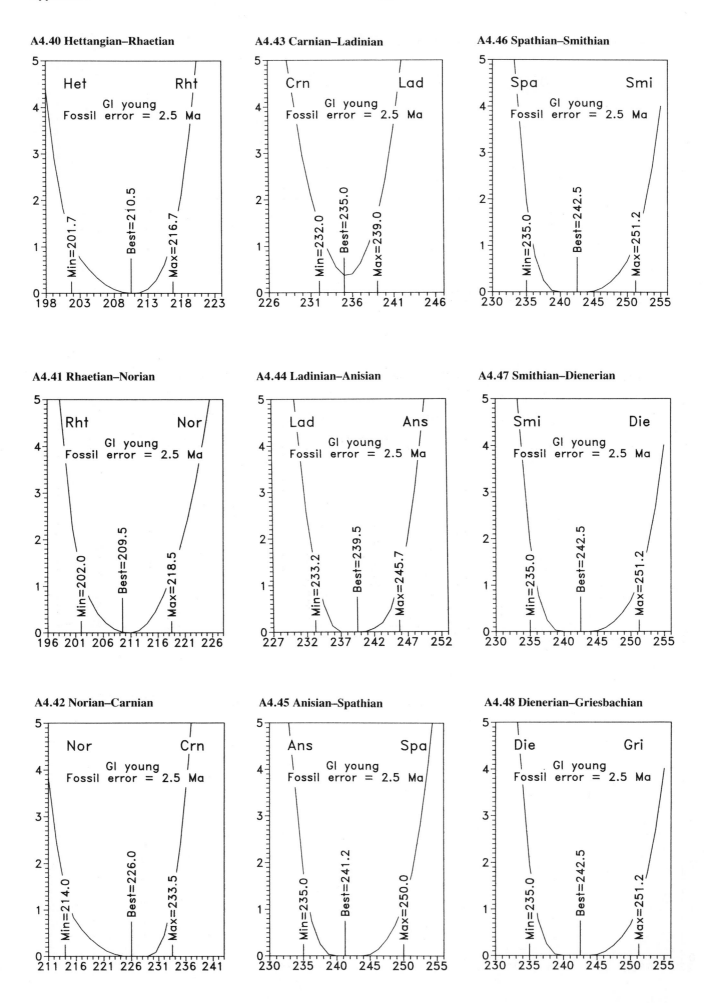

A4.40 Hettangian–Rhaetian

Het   Rht
Gl young
Fossil error = 2.5 Ma
Min=201.7   Best=210.5   Max=216.7

A4.43 Carnian–Ladinian

Crn   Lad
Gl young
Fossil error = 2.5 Ma
Min=232.0   Best=235.0   Max=239.0

A4.46 Spathian–Smithian

Spa   Smi
Gl young
Fossil error = 2.5 Ma
Min=235.0   Best=242.5   Max=251.2

A4.41 Rhaetian–Norian

Rht   Nor
Gl young
Fossil error = 2.5 Ma
Min=202.0   Best=209.5   Max=218.5

A4.44 Ladinian–Anisian

Lad   Ans
Gl young
Fossil error = 2.5 Ma
Min=233.2   Best=239.5   Max=245.7

A4.47 Smithian–Dienerian

Smi   Die
Gl young
Fossil error = 2.5 Ma
Min=235.0   Best=242.5   Max=251.2

A4.42 Norian–Carnian

Nor   Crn
Gl young
Fossil error = 2.5 Ma
Min=214.0   Best=226.0   Max=233.5

A4.45 Anisian–Spathian

Ans   Spa
Gl young
Fossil error = 2.5 Ma
Min=235.0   Best=241.2   Max=250.0

A4.48 Dienerian–Griesbachian

Die   Gri
Gl young
Fossil error = 2.5 Ma
Min=235.0   Best=242.5   Max=251.2

A4.49 Griesbachian–Tatarian (Lopingian)

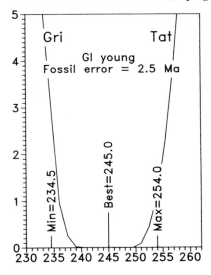

A4.52 Ufimian–Kungurian

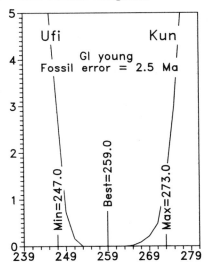

A4.55 Sakmarian–Asselian

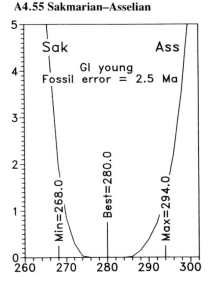

A4.50 Tatarian(Lopingian)–Kazanian (Guadelupian)

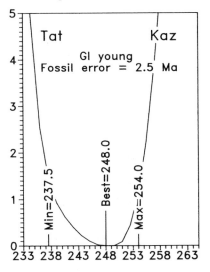

A4.53 Kungurian–Artkinskian

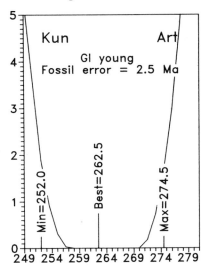

A4.56 Asselian–Noginskian

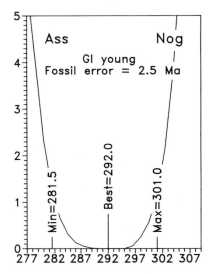

A4.51 Kazanian (Wordian)–Ufimian

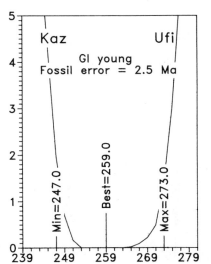

A4.54 Artkinskian–Sakmarian

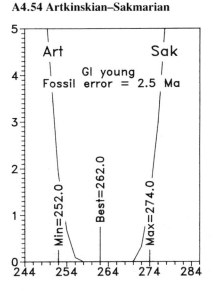

A4.57 Noginskian–Klazminskian

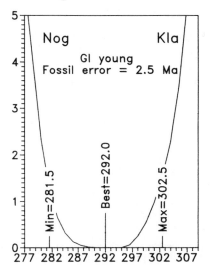

A4.58 Klazminskian–Dorogomilovskian

A4.61 Krevyakinskian–Myachovskian

A4.64 Kashirskian–Vereiskian

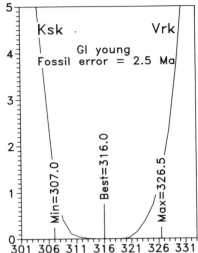

**A4.59 Dorogomilovskian–
Chamovnicheskian**

A4.62 Myachovskian–Podolskian

A4.65 Vereiskian–Melekesskian

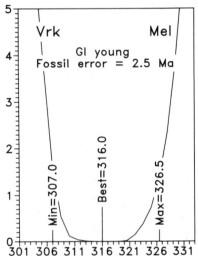

**A4.60 Chamovnicheskian–
Krevyakinskian**

A4.63 Podolskian–Kashirskian

A4.66 Melekesskian–Cheremshanskian

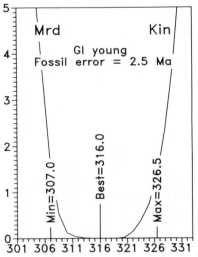

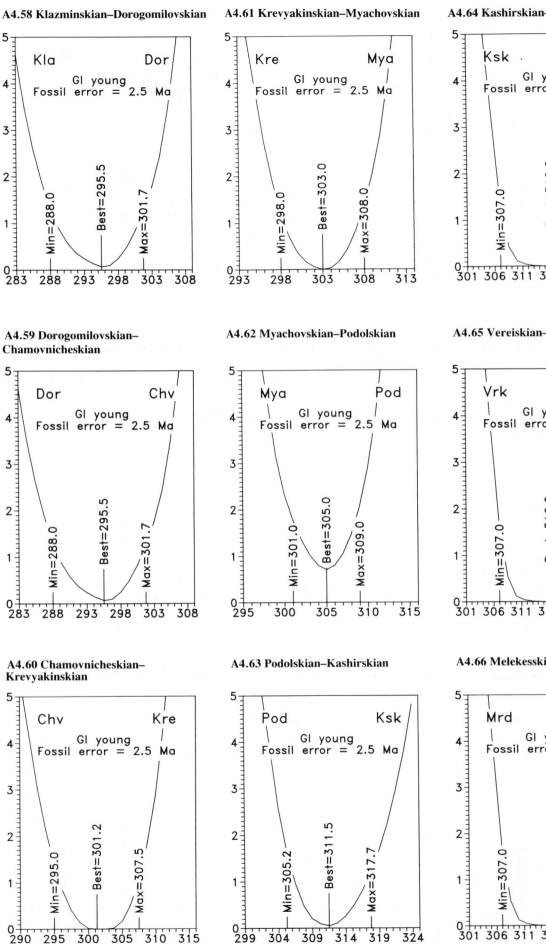

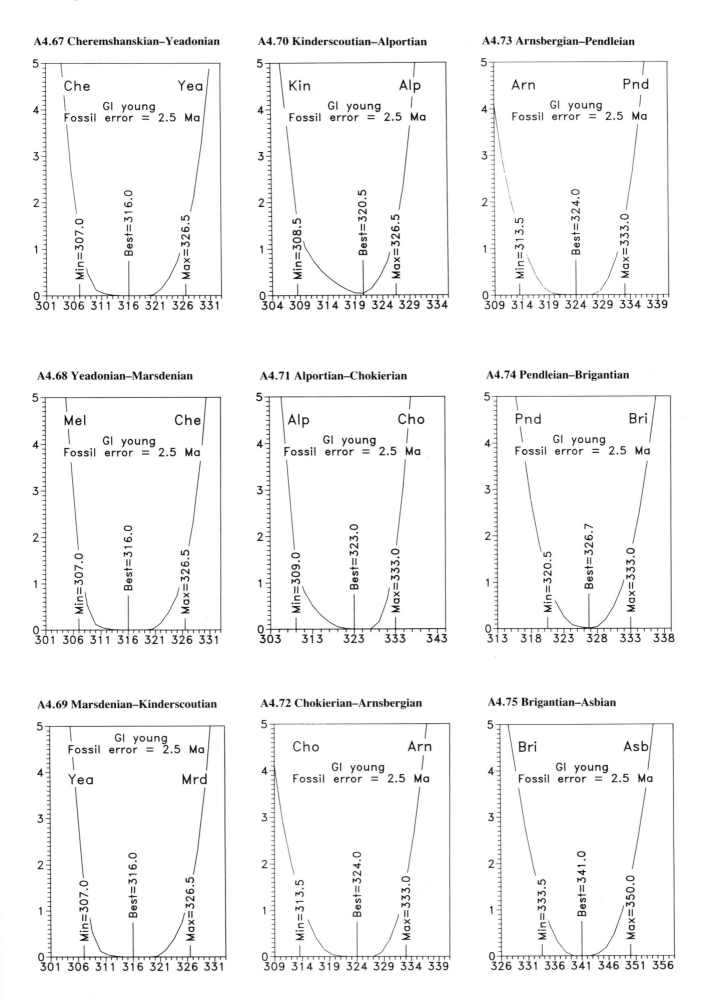

A4.76 Asbian–Holkerian

A4.79 Chadian–Ivorian

A4.82 Famennian–Frasnian

A4.77 Holkerian–Arundian

A4.80 Ivorian–Hastarian

A4.83 Frasnian–Givetian

A4.78 Arundian–Chadian

A4.81 Hastarian–Famennian

A4.84 Givetian–Eifelian

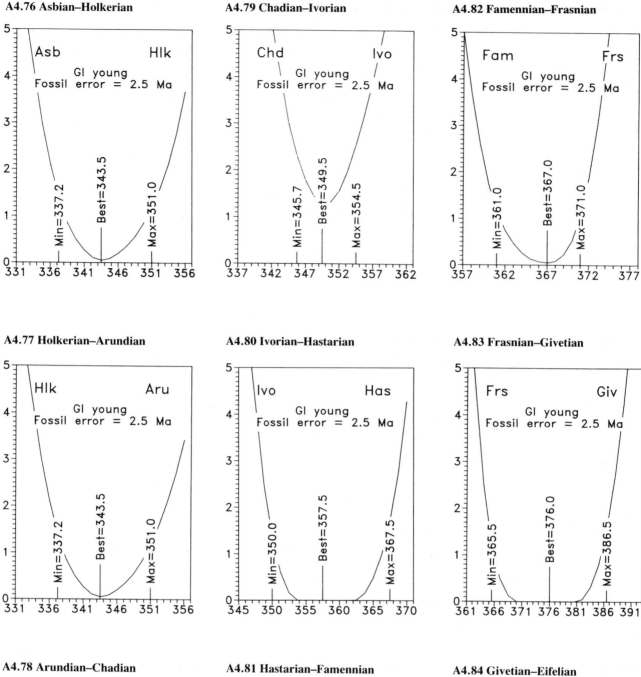

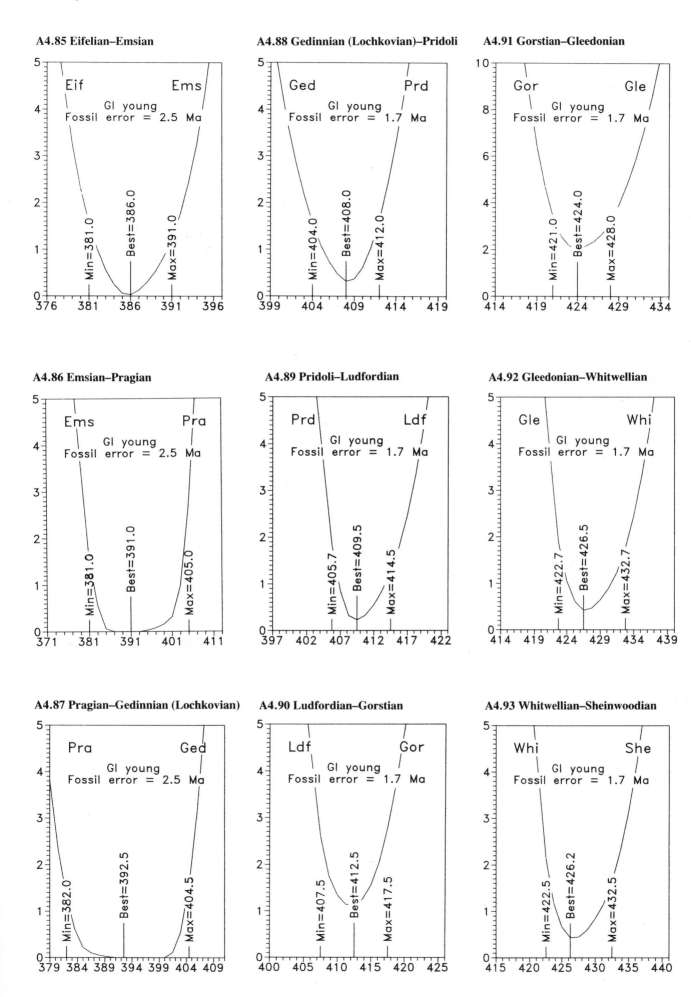

A4.94 Sheinwoodian–Telychian

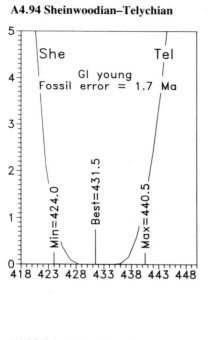

A4.95 Telychian–Aeronian

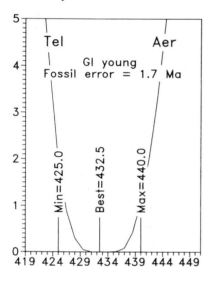

A4.96 Aeronian–Rhuddanian

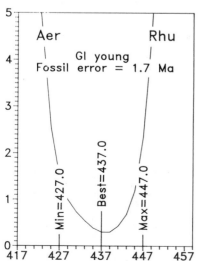

A4.97 Rhuddanian–Hirnantian

A4.98 Hirnantian–Rawtheyan

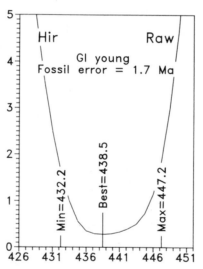

A4.99 Rawtheyan–Cautleyan

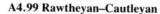
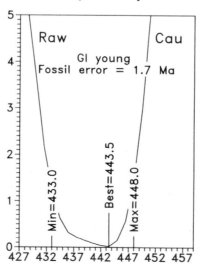

A4.100 Cautleyan–Pusgillian

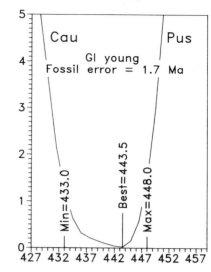

A4.101 Pusgillian–Onnian

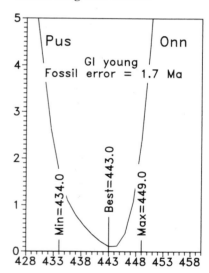

A4.102 Onnian–Actonian

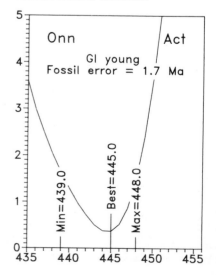

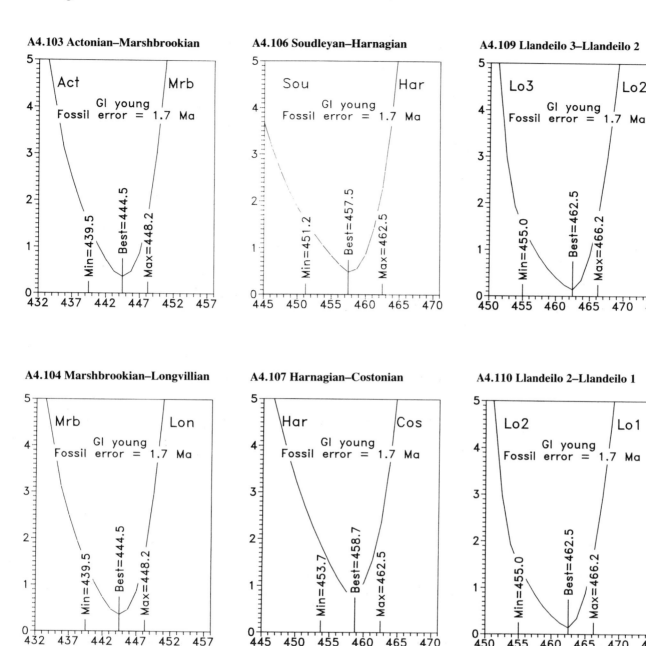

A4.103 Actonian–Marshbrookian

Act · Mrb · GI young · Fossil error = 1.7 Ma
Min=439.5 · Best=444.5 · Max=448.2

A4.106 Soudleyan–Harnagian

Sou · Har · GI young · Fossil error = 1.7 Ma
Min=451.2 · Best=457.5 · Max=462.5

A4.109 Llandeilo 3–Llandeilo 2

Lo3 · Lo2 · GI young · Fossil error = 1.7 Ma
Min=455.0 · Best=462.5 · Max=466.2

A4.104 Marshbrookian–Longvillian

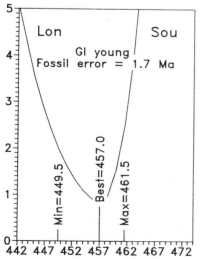

Mrb · Lon · GI young · Fossil error = 1.7 Ma
Min=439.5 · Best=444.5 · Max=448.2

A4.107 Harnagian–Costonian

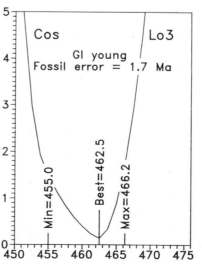

Har · Cos · GI young · Fossil error = 1.7 Ma
Min=453.7 · Best=458.7 · Max=462.5

A4.110 Llandeilo 2–Llandeilo 1

Lo2 · Lo1 · GI young · Fossil error = 1.7 Ma
Min=455.0 · Best=462.5 · Max=466.2

A4.105 Longvillian–Soudleyan

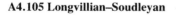

Lon · Sou · GI young · Fossil error = 1.7 Ma
Min=449.5 · Best=457.0 · Max=461.5

A4.108 Costonian–Llandeilo 3

Cos · Lo3 · GI young · Fossil error = 1.7 Ma
Min=455.0 · Best=462.5 · Max=466.2

A4.111 Llandeilo 1–Llanvirn 2

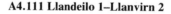
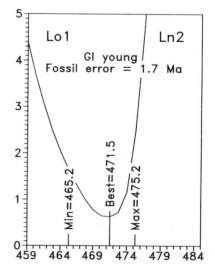

Lo1 · Ln2 · GI young · Fossil error = 1.7 Ma
Min=465.2 · Best=471.5 · Max=475.2

A4.112 Llanvirn 2–Llanvirn 1

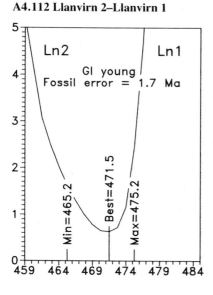

A4.113 Llanvirn 1–Arenig

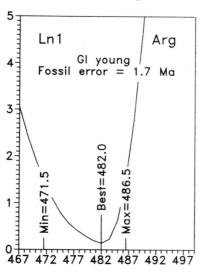

A4.114 Arenig–Tremadoc

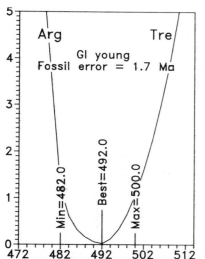

A4.115 Tremadoc–Dolgellian

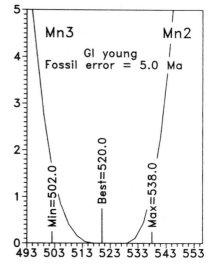

A4.116 Dolgellian–Maentwrogian

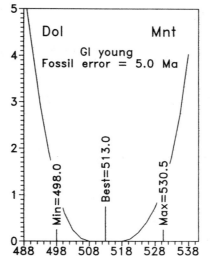

A4.117 Maentwrogian–Menevian 3

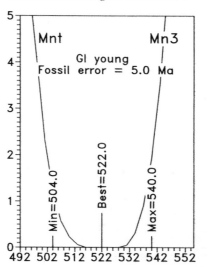

A4.118 Menevian 3–Menevian 2

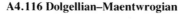

A4.119 Menevian 2–Menevian 1

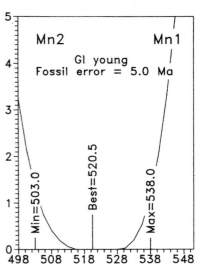

A4.120 Menevian 1–Solvan 2

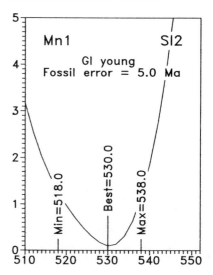

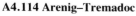

A4.121 Solvan 2–Solvan 1

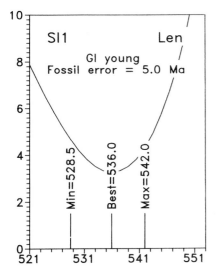

A4.124 Atdabanian–Tommotian

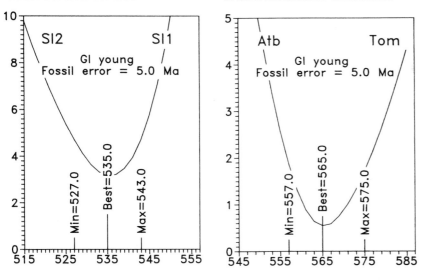

A4.122 Solvan 1–Lenian

A4.125 Tommotian–Poundian

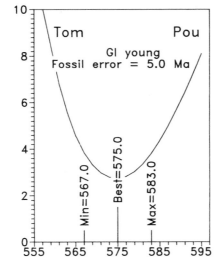

A4.123 Lenian–Atdabanian

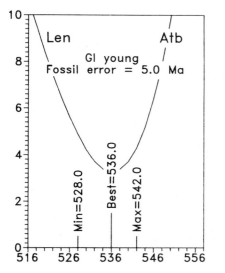

APPENDIX 5

Linear plot of magnetic polarity reversal time scale 0 to 150 Ma

---

The magnetic reversal time scale, as developed in Chapter 6, is summarized in Section 6.6 (Figures 6.9a, b and c). Figures A5.1 to A5.16 in this Appendix plot the same data on a larger scale using a rectangular box convention in which the line is to the right when the field was normal and to the left when the field was reversed. Mixed intervals are shown in the centre. Symbols for the magnetostratigraphic chrons are placed beside each interval with the normally magnetized intervals named on the right-hand side and the reversely magnetized intervals on the left-hand side of each figure. The chronostratic scale is plotted with biostratigraphic correlation to the magnetostratigraphic chrons. Sources of these correlations include Berggren *et al.* (1985a, 1985b) in Snelling (1985) for Cenozoic time and Kent & Gradstein (1985) for Cretaceous and part of Jurassic time. The sixteen figures span 10 Ma each, depicting the time scale back to 160 Ma on a linear scale.

Figures A5.1–A5.16. The Cenozoic and Mesozoic time scale for the past 160 Ma.

These figures display the interrelations between the polarity reversal scale, the GTS 89 numerical time scale and biostratigraphic zonal schemes. The vertical scale is age in Ma. Each figure is divided into columns giving biostratigraphic zones and stratigraphic subdivisions on the left and the polarity reversal time scale on the right. The correlations of biostratigraphic and stratigraphic with the magnetic reversal time scale are taken from BKF85 or from BKV85. The correlation of all these with the numerical time scale is made using the age assignments to the magnetic reversal time scale discussed in Chapter 6.

Where present, the column headings have the following meanings:

a-e. Plankton zones: (a) & (b) Foraminifera, (a) tropical and sub-tropical, (b) sub-tropical and temperate; (c) & (d) Calcareous nannoplankton, (c) tropical and sub-tropical, (d) inter-regional; (e) tropical and sub-tropical Radiolaria [Figures A5.1–A5.3] and northwest European dinoflagellates [Figures A5.4–A5.6].

f-h. Stratigraphic subdivisions.

Magnetic polarity columns: Normal polarities are on the right e.g. C2AN; reversed polarities on the left e.g. C3.1R.

Further notes are provided after the figure number for some of the individual figures.

213

Figure A5.1. The Zanclian/Messinian boundary is given as 5.2 Ma, rather than 5.7 Ma. The biostratigraphic/polarity assignments have not been altered. However, the change in numerical age modifies the biostratigraphic zones assigned to the late Messinian and early Zanclian. For example, N18 is assigned here to the late Messinian, rather than to the early Zanclian.

BKC85 show a subdivision of CN11 into CN11a and CN11b, and CN12 into CN12a and CN12b. For clarity, these subdivisions have been omitted.

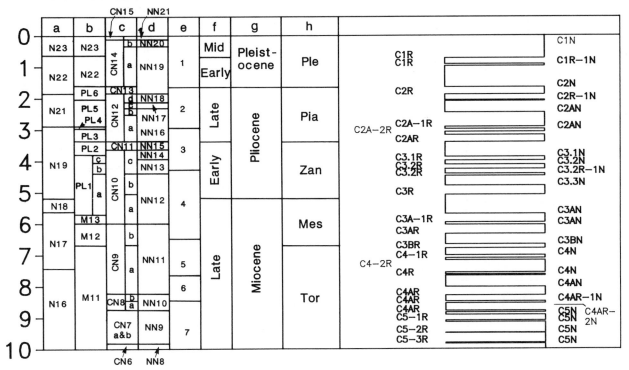

Figure A5.2.

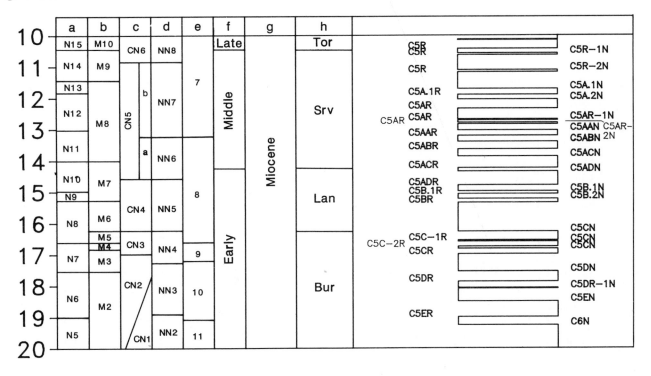

Figure A5.3. BKC85 show a subdivision of CN1 into CN1a, CN1b and CN1c. For clarity, these subdivisions have been omitted.

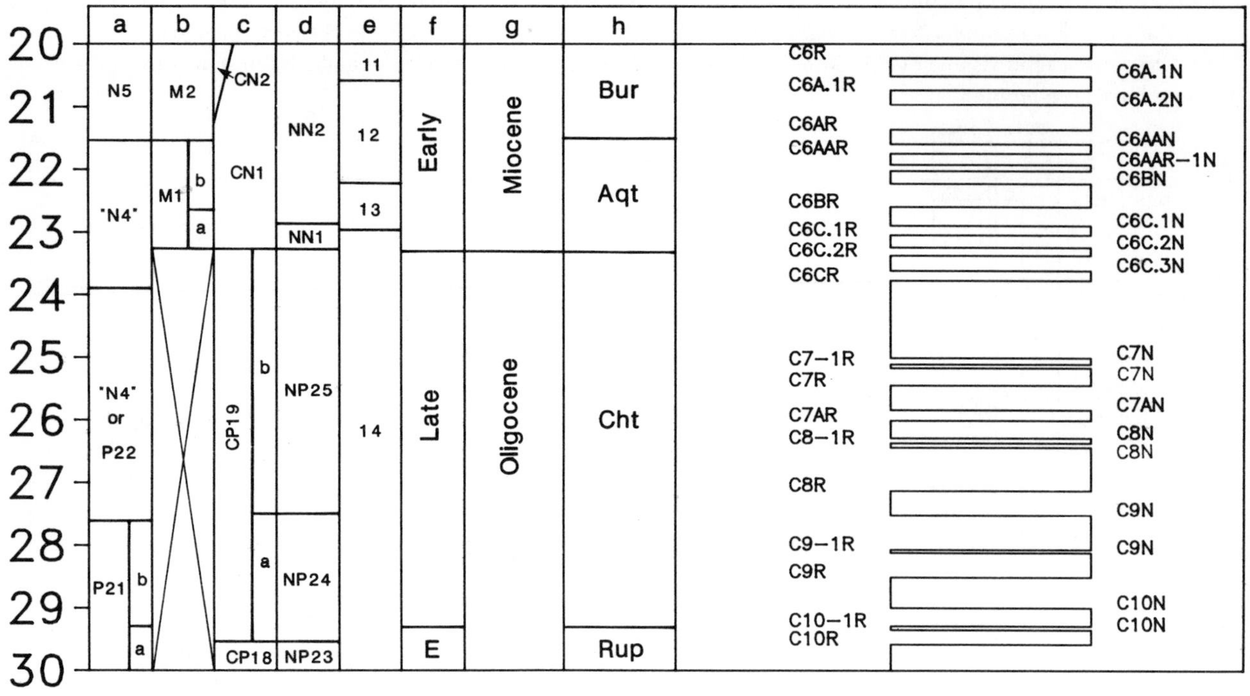

	a	b	c	d	e	f	g	h	
20					11				C6R — C6A.1N
21	N5	M2	CN2			Early	Miocene	Bur	C6A.1R — C6A.2N
22			b	NN2	12				C6AR — C6AAN
	"N4"	M1	CN1					Aqt	C6AAR — C6AAR−1N / C6BN
23			a	NN1	13				C6BR — C6C.1N / C6C.1R — C6C.2N / C6C.2R — C6C.3N
24									C6CR
25	"N4"		b	NP25		Late	Oligocene		C7−1R — C7N / C7R — C7N
26	or P22		CP19		14			Cht	C7AR — C7AN / C8−1R — C8N / C8N
27									C8R
28			a	NP24					C9−1R — C9N / C9R — C9N
29	P21	b							C10−1R — C10N / C10R — C10N
30		a	CP18	NP23		E		Rup	

Figure A5.4. There is no anomaly C14 (see Chapter 6).

Column b is now northwest European dinoflagellates instead of tropical and sub-tropical foraminifera.

The P14/P15 boundary is drawn at 1979 position only, whereas BKF85 shows its 1969 and 1979 positions. The top of subdivision VI is not shown. Premoli-Silva *et al.* (1988) placed the Priabonian/Rupelian boundary at the top of P17 and about 0.2 Ma down into C13R, giving an age in this study of ~35 Ma. See note at end of Postscript (page xv).

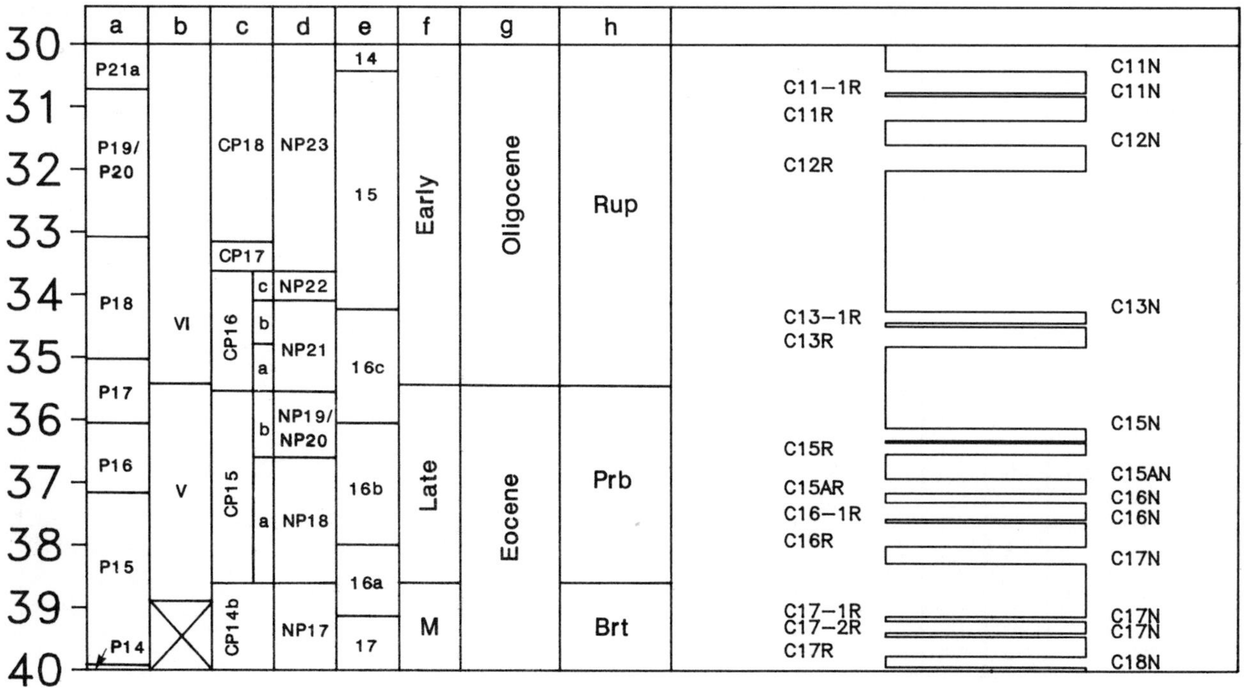

	a	b	c	d	e	f	g	h	
30	P21a				14				C11N
31						Early	Oligocene	Rup	C11−1R — C11N / C11R
32	P19/ P20		CP18	NP23	15				C12R — C12N
33			CP17						
34	P18	VI	c	NP22					C13−1R — C13N / C13R
35			CP16 b	NP21	16c				
	P17		a						
36			b	NP19/ NP20					C15N
37	P16	V	CP15		16b	Late	Eocene	Prb	C15R / C15AR — C15AN / C16−1R — C16N / C16N
38	P15		a	NP18					C16R — C17N
39			CP14b	NP17	16a	M		Brt	C17−1R — C17N / C17−2R — C17N
40	P14				17				C17R — C18N

Figure A5.5.

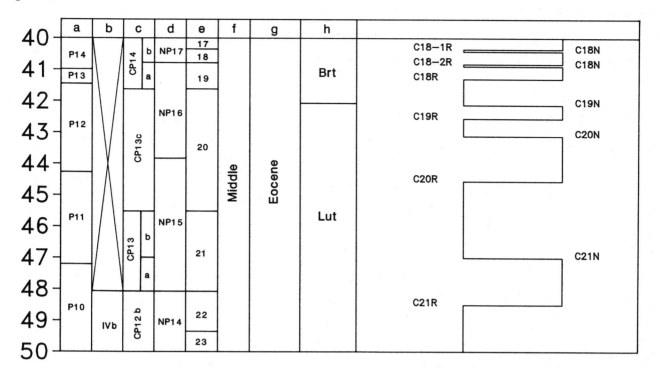

Figure A5.6. Thanetian is used instead of Selandian.

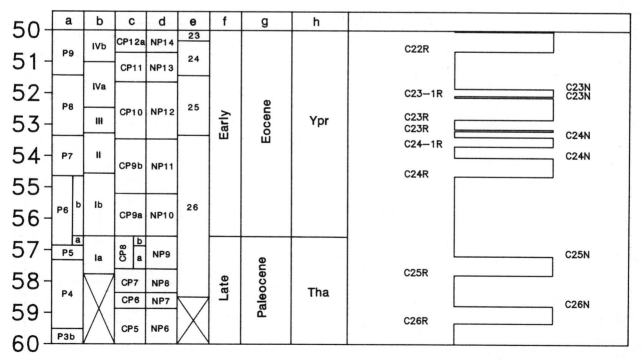

Figure A5.7. As used here, the Thanetian is equivalent to the Selandian of BKF85. The early Thanetian in Figure A5.7 is equated by BKF85 (Figure 3, p. 154) to an unnamed interval below the Thanetian *sensu stricto*. See BKF85 for details.

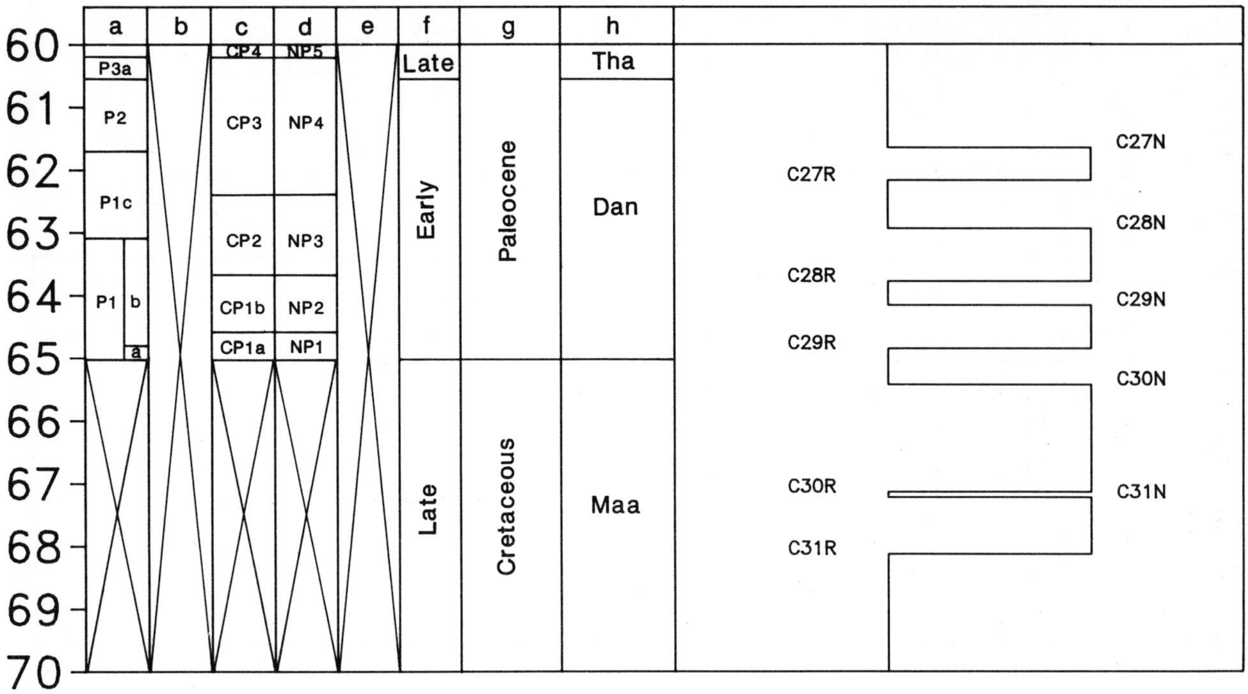

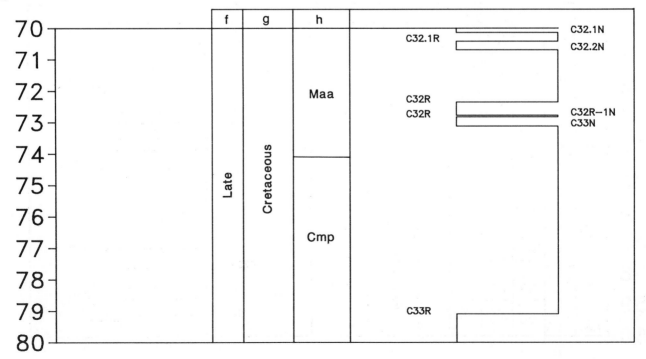

Figure A5.8.

Figure A5.9.

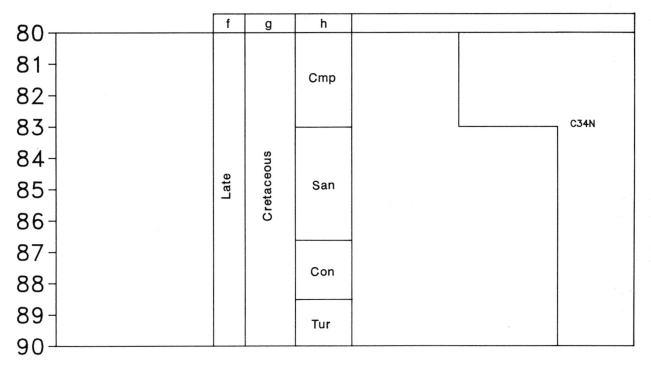

Figure A5.10. C34N should have been placed lower down and the horizontal at 90 Ma continued to the right-hand margin.

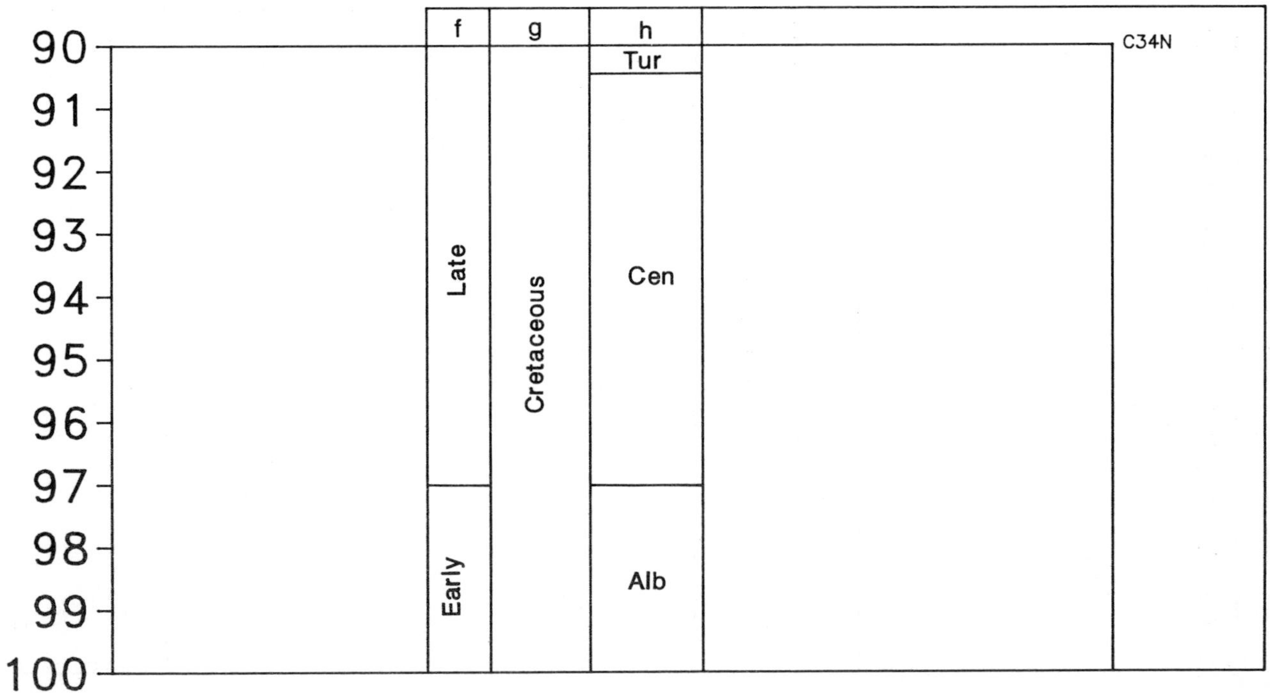

Figure A5.11. Correction as to Figure A5.10.

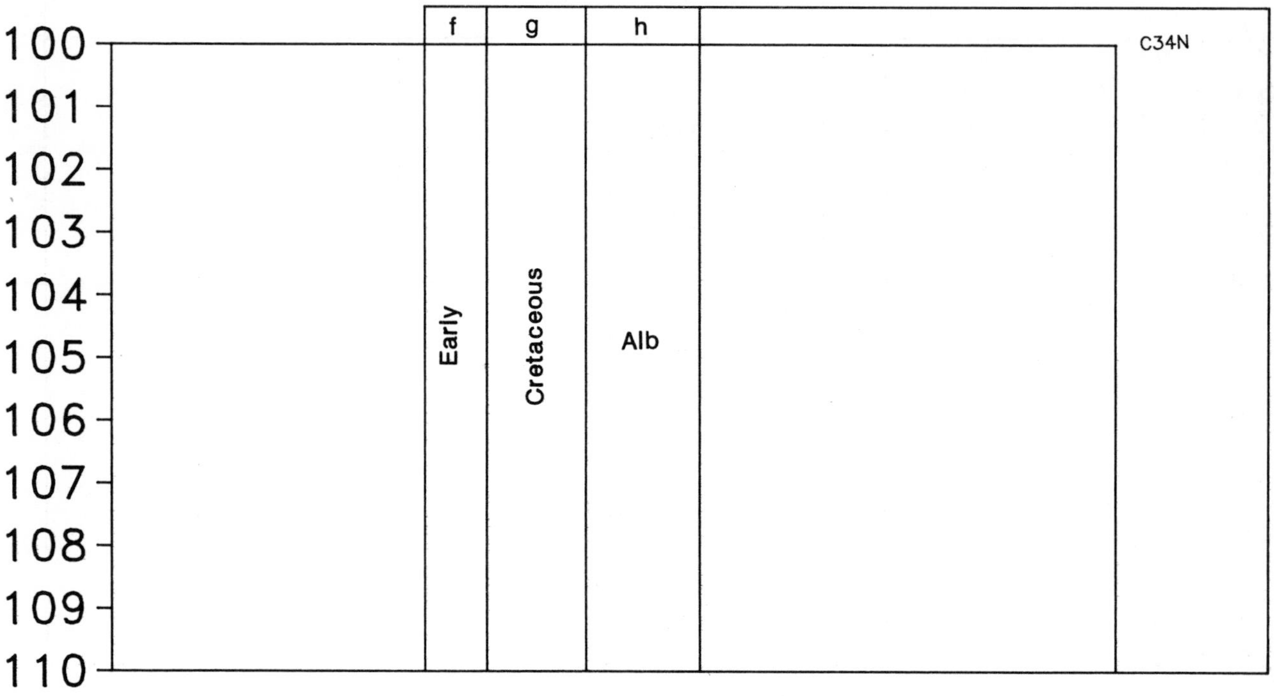

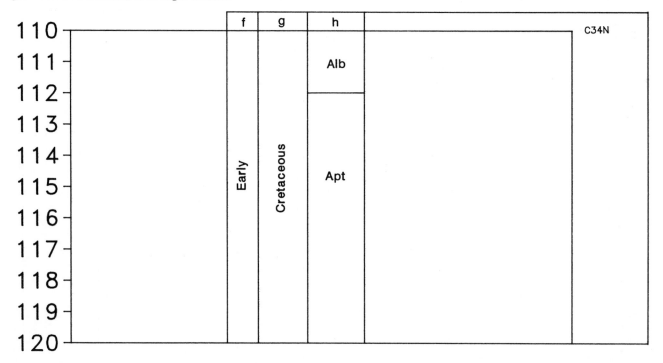

Figure A5.12. Correction as to Figure A5.10.

Figure A5.13. There is currently no M0N: M0 is the reversed interval directly after C34N. In our view it could be named C34Rand M0 suppressed. There is currently no M2R or M3N. M2, which is an interval of normal polarity, is followed by M3, which is of reversed polarity. For simplicity, the sequence could be renamed M2N and M2R, with M3 being suppressed. Correction as to Figure A5.10.

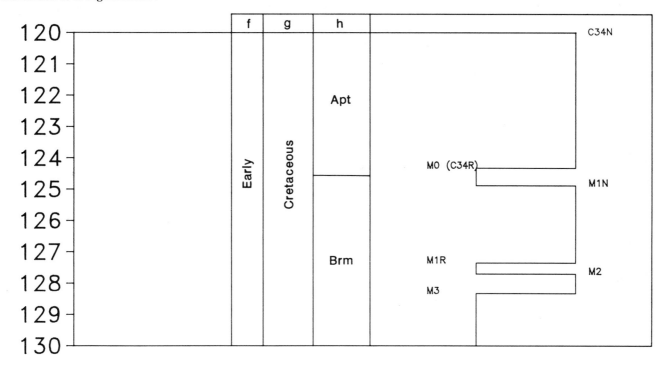

Figure A5.14. There is currently no M4R or M5N. M4 is an interval of normal polarity followed by M5, which is of reversed polarity. For simplicity, the sequence could be renamed M4N and M4R, with M5 being suppressed.

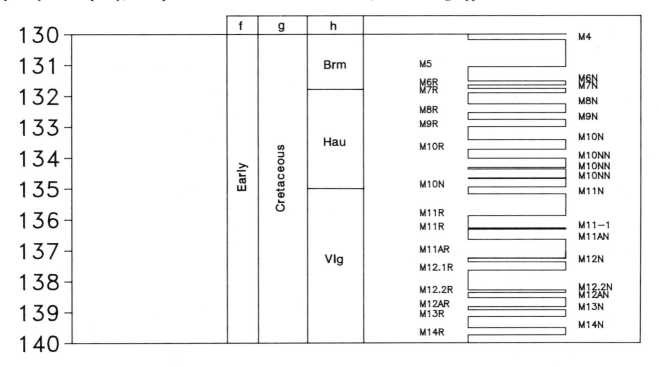

Figure A5.15.

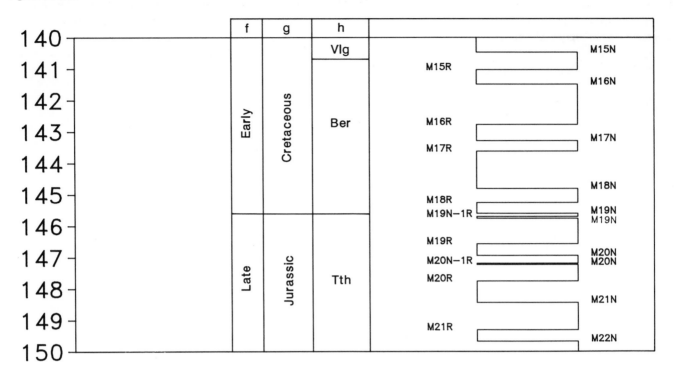

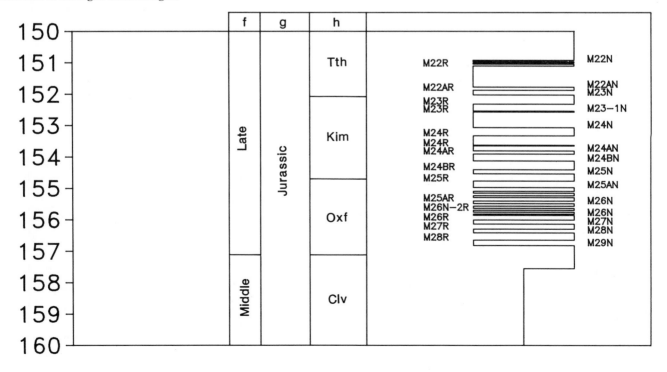

Figure A5.16. M29 is the oldest ocean-floor magnetic anomaly. A very detailed magnetic anomaly sequence has been worked out for Callovian and older Jurassic rocks (Ogg & Steiner in press, Steiner & Ogg in press). The horizontal line at 150 Ma should continue to the right-hand margin.

APPENDIX 6

A geologic time scale wall chart 1989

Introduction

The numerous and changing subdivisions of geologic time and geologic history are difficult to conceive without some visual aid, and the demand for visual aids is demonstrated by the readiness with which suitable charts are acquired and displayed. A wall chart displaying the essential data was an essential companion to GTS 82 (also by Harland *et al.* and published by CUP in 1982). A new wall chart also accompanies GTS 89. As in GTS 82, the GTS 89 wall chart is printed with colours approximating to an international standard (see below). The evidence and reasoning that have contributed to it have been recorded in this book. So that the information on the wall chart is available in this volume we have essentially repeated the different parts of the wall chart as figures in Chapter 7.

Previous chapters have discussed the chronostratic divisions of geologic time, the chromometric dating of their boundaries (and their uncertainties) and paleomagnetic reversal events. The overall dimensions of the wall chart (about 100 cm × 60 cm) showing these features were chosen to allow an adequate level of detail within a sheet of good proportion and manageable size.

International colour scheme

From the outset, we decided to print the chart in colour for both clarity and interest and to restate the desirability of a standard international colour scale for time stratigraphy.

The choice of colours was less easy. We are familiar with several coloured charts, e.g. the Shell Oil Company Legend, a Total Oil Company chart, and Van Esinga's wall chart (which has become perhaps the best known of late), but these all differ in their use of colour.

The Commission for the Geological Map of the World (CGMW) was conceived at the Second International Geological Congress, Bologna, 1881, and the choice of colours for (largely) time–stratigraphic purposes is discussed at length in the Congress Proceedings. Apart from colour plates in the Proceedings volume, an early example of the agreed standard colour scheme is presented by the *Geological map of Europe 1:500 000* prepared in accordance with Congress's resolutions. It seemed appropriate in 1981 to consider adopting the 1881

scheme, but this idea was abandoned at an early stage when we found the original descriptions of the colours to be imprecise while, after the passage of nearly 100 years, the colours in the Proceedings volume and on two separate examples of the European map no longer matched and no 'standard' colours could therefore be determined.

The work of the CGMW has continued, mainly since the Eleventh International Geological Congress in Stockholm, 1910, with interruptions in the war years 1914–18 and 1939–45, and is pursued in Paris where it has been connected with the International Union of Geological Sciences, and with Unesco which publishes its maps. The Unesco *Geological World Atlas* (Choubert & Faure-Muret 1976) represents the culmination of decades of trials and the first sheets were produced in 1974. Vojacek (1979) outlined the history of this project and provided useful background technical information about the printing of the *Atlas*. He also noted (1979) that 'with this project Unesco also hopes to set international standards for the presentation of geological symbols and for the colour designs of geological maps'.

This seemed a laudable aim, and sufficient justification for adopting the Unesco colours, but there proved to be a serious commercial obstacle to this course of action. Unesco used no less than 35 colours (including black) in printing their atlas (Vojacek 1979). This was impractical for the low-cost commercial printing operation envisaged by Cambridge University Press, and represents a standard unlikely to be emulated by many organizations, governmental or private. We are indebted to H. J. Vojacek of Mercury-Walch in Hobart, Tasmania, for providing proof sheets of the Unesco scheme (personal communication with BP).

We have matched the Unesco legend colours as closely as possible and the equivalents between our wall chart and their colours are shown in Table A6.1. Some modifications have been necessary as the Unesco legend has no subdivisions for Cenozoic, Phanerozoic and Priscoan and uses a scheme of Precambrian subdivisions which has been superseded. Also, as explained in Chapter 2, we have used Mississippian and Pennsylvanian in preference to Dinantian and Silesian for the international scale.

Table A6.1. *Wall chart colour equivalents with Unesco international colour scheme for geologic maps*

Our colour block	Unesco
Cenozoic	no equivalent
Quaternary and Pleistogene	Q
Tertiary	TT
Neogene	N
Pliocene	N_2
Miocene	N_1
Paleogene	PG
Oligocene	PG_3
Eocene	PG_2
Paleocene	PG_1
Mesozoic	Mz
Cretaceous	K
Late Cretaceous	K_2
Early Cretaceous	K_1
Jurassic	J
Late Jurassic	J_3
Middle Jurassic	J_2
Early Jurassic	J_1
Triassic	T
Late Triassic	T_3
Middle Triassic	T_2
Early Triassic	T_1
Paleozoic	Pz
Phanerozoic	no equivalent
Permian	P
Late Permian	P_2
Early Permian	P_1
Carboniferous	C
Pennsylvanian	no equivalent
Mississippian	no equivalent
Devonian	D
Late Devonian	D_3
Middle Devonian	D_2
Early Devonian	D_1
Silurian	S
Ordovician	O
Cambrian	€
Proterozoic	PD
Sinian (& Pt_3)	PA
Vendian	PA_1
Sturtian	PA_2
Riphean (& R_3, R_2, R_1)	PC
Huronian (& Pt_1)	PD_1
Pt_2	PB
Archean (Randian, Swazian, Isuan, Hadean, Ar_3, Ar_2, Ar_1)	A
Priscoan	no equivalent

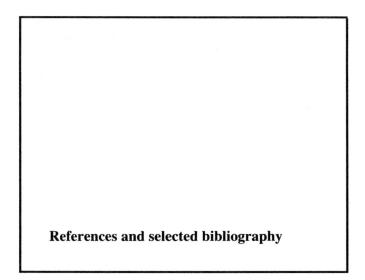

References and selected bibliography

All publications cited in the foregoing pages (except in Appendix 1) should be found in this list. In addition, further key publications that might have been cited here have also been included. This list may thus be useful as a bibliography.

Abele, C. & Page, R. W. 1974. Stratigraphic and isotopic ages of Tertiary basalts at Maude and Airey's Inlet, Victoria, Australia. *Proceedings, Royal Society of Victoria* **86**, 143–150.

Afanasyev, G. D. & Rubinstein, M. M. 1964. Explanatory note to the geochronological scale in the absolute chronological system. In Absolute age of geological formations – papers presented by Soviet geologists at 22nd International Geological Congress (Delhi, India). *Doklady Sovetskitch Geologov* (Papers of the Soviet Geologists) **3**, 187–234 (in Russian).

Afanasyev, G. D. 1987. *Phanerozoic Geochronological Scale and the Problem of Geological Time.* Moscow: Nedra, 144 pp. (in Russian).

Afanasyev, G. D. & Zykov, S. I. 1975. *Phanerozoic Geochronological Time Scale in the Light of Significantly New Decay Constants.* Moscow: Nauka, 100 pp. (in Russian).

Agterberg, F. P., Gradstein, F. M. & Williamson, M. 1984. Biochronology and quantitative stratigraphy. Conference Report. *Episodes* **7**, 63–69.

Aguirre, E. & Pasini, G. 1985. The Pliocene–Pleistocene Boundary. *Episodes* **8**, 116–120.

Alcock, F. J. 1934. Report of the National Committee on Stratigraphical Nomenclature. *Transactions of the Royal Society of Canada*, Series III **28**, Section 4, 113–121.

Alcock, F. J. 1955. American Commission on Stratigraphic Nomenclature, Report 3. *American Association of Petroleum Geologists Bulletin* **39**, 1859–1861.

Alvarez, W. & Lowrie, W. 1978. Upper Cretaceous paleomagnetic stratigraphy at Moria (Umbrian Apennines, Italy): Verification of the Gubbio section. *Geophysical Journal of the Royal Astronomical Society* **55**, 1–17.

Alvinerie, J. *et al.* 1973. A propos de la limite oligo–miocène; résultats préliminaires d'une recherche collective sur les gisements d'Escornebeou (Saint Geours de Maremme, Landes, Aquitaine méridionale). Présence de *Globigerinoides* dans les faunes de l'Oligocène supérieur. *Compte Rendu Société géologique France* **15 (3–4)**, 75–76.

Anderson, J. M. 1981. World Permo–Triassic correlations: their stratigraphic basis. In *Gondwana Five, Proceedings 5th International Gondwana Symposium* Wellington 1980 (eds. M. M. Cresswell and P. Vella), pp. 3–17. Rotterdam: Balkema.

Anderton, O., Bridges, P. H., Leeder, M. R. & Sellwood, B. W. 1979. *A Dynamic Stratigraphy of the British Isles, a Study in Crustal Evolution.* London: George Allen and Unwin, 301 pp.

Anglada, A. O. 1971a. Sur la position du datum à *Globigerinoides* (Foraminiferida) dans la zone N4 (Blow 1967) et la limite oligo–miocène en Méditerranée. *Académie des Sciences, Comptes Rendus* **272**, 1067–1070.

Anglada, A. O. 1971b. Sur la limite Aquitanien–Burdigalien, sa place dans l'échelle des Foraminifères planctoniques et sa signification dans le Sud-Est de la France. *Académie des Sciences, Comptes Rendus* **272**, 1948–1950.

Anhaeusser, C. R. & Wilson, J. F. 1981. The granitic-gneiss greenstone shield. In Hunter, pp. 423–499.

Anonymous, 1882. *Congrès International de Géologie – Comptes Rendus de la 2me Session*, Bologna 1881. Bologna: Fava & Garagni, 661 pp.

Anonymous, 1979. Magnetostratigraphic polarity units. A supplementary chapter of the International Subcommission on Stratigraphic Classification International Stratigraphic Guide. *Geology* **7**, 578–583.

Arkell, W. J. 1933. *The Jurassic System in Great Britain.* Oxford: Clarendon Press, 681 pp.

Arkell, W. J. 1956. *Jurassic Geology of the World.* Edinburgh: Oliver and Boyd, 806 pp.

Armstrong, R. L. 1966. K–Ar dating of plutonic and volcanic rocks in orogenic belts. In Schaeffer & Zähringer, pp. 117–133.

Armstrong, R. L. 1969. K–Ar dating of laccolithic centers of the Colorado Plateau and vicinity. *Geological Society of America Bulletin* **80**, 2081–2086.

Armstrong, R. L. 1970. Geochronology of Tertiary igneous rocks, eastern Basin and Range Province, western Utah, eastern Nevada, and vicinity, U.S.A. *Geochimica et Cosmochimica Acta* **34**, 203–232.

Armstrong, R. L. 1978a. K–Ar dating: late Cenozoic McMurdo Volcanic Group and dry valley glacial history, Victoria Land Antarctica. *New Zealand Journal of Geology and Geophysics* **21**, 685–689.

Armstrong, R. L. 1978b. Removal of atmospheric argon contamination and the use and misuse of the K–Ar isochron method: Discussion. *Canadian Journal of Earth Sciences* **15**, 325–326.

Armstrong, R. L. 1978c. Pre-Cenozoic Phanerozoic time scale – computer file of critical dates and consequences of new and in-progress decay-constant revisions. In Cohee *et al.*, pp. 73–93.

Armstrong, R. L. 1986. Mesozoic and Early Cenozoic magmatic evolution of the Canadian Cordillera. In *Processes in Continental Lithospheric Deformation: a symposium to honour John Rogers* (eds. S. P. Clark, H. C. Burchfield and J. Suppe). *Geological Society of America* Special Paper **218**, 55–91.

Armstrong, R. L., Isachen, C. & Scott, K. (unpublished). Rb–Sr and Sr isotopic study and U–Pb dating of Vancouver Group igneous rocks and related Island Intrusions and of the Coast Plutonic Complex and Early Cenozoic igneous rocks of Vancouver Island.

Armstrong, R. L. & Ramaekers, P. 1985. Sr isotopic study of Helikian sediment and basaltic dikes in the Athabasca Basin, northern Saskatchewan. *Canadian Journal of Earth Sciences* **22**, 399–407.

Armstrong, R. L., Tauberneck, W. H. & Hales, P. O. 1977. Rb–Sr and K–Ar geochronometry of Mesozoic granitic rocks and their Sr isotopic composition, Oregon, Washington and Idaho. *Geological Society of America Bulletin* **88**, 397–411.

Arnold, A. & Jäger, E. 1965. Rb–Sr Altersbestimmungen an Glimmern im Grenzbereich zwischen voralpinen Alterswerten und alpiner Verjungung der Biotite. *Eclogae geologicae Helvetiae* **58**, 369–390.

Ascoli, P. 1976. Foraminiferal and ostracod biostratigraphy of the Mesozoic–Cenozoic, Scotian Shelf, Atlantic Canada. 1st International Symposium on Benthonic Foraminifera of Continental Margins Part B: Paleoecology and biostratigraphy. *Maritime Sediments Special Publication* 1, 653–771.

Ashwal, L. D., Wooden, J. L., Phinney, W. C. & Morrison, D. A. 1985. Sm–Nd and Rb–Sr isotope systematics of an Archean anorthosite and related rocks from the Superior Province of the Canadian Shield. *Earth and Planetary Science Letters* 74, 338–346.

Aubry, M-P. 1983. Calcareous nannofossil biostratigraphy Leg 64. In *Initial Reports of the Deep Sea Drilling Project* 64, 955–972.

Aubry, M-P., Berggren, W. A., Kent, D. V., Flynn, J. J., Klitgord, K. D., Obradovich, J. D. & Prothero, D. R. 1988. Paleogene geochronology: An integrated approach. *Paleoceanography* 3, 707–742.

Baadsgaard, H. & Lerbekmo, J. F. 1982. The dating of bentonite beds. In Odin, pp. 423–440.

Baadsgaard, H. & Lerbekmo, J. F. 1983. Rb–Sr and U–Pb dating of bentonites. *Canadian Journal of Earth Sciences* 20, 1282–1290.

Baadsgaard, H., Lerbekmo, J. F. & Evans, M. E. 1978. Geochronology and magnetostratigraphy of fluvial-deltaic sediments embracing the Cretaceous–Tertiary boundary, Red Deer Valley, Alberta, Canada. In Zartman, pp. 17–18.

Baadsgaard, H., Lerbekmo, J. F. & McDougall, I. 1988. A radiometric age for the Cretaceous Tertiary boundary based on K–Ar, Rb–Sr, and U–Pb ages of bentonites from Alberta, Saskatchewan and Montana. *Canadian Journal of Earth Sciences* 25, 1088–1097.

Baadsgaard, H., Lipson, J. & Folinsbee, R. E. 1961. The leakage of radiogenic argon from sanidine. *Geochimica et Cosmochimica Acta* 25, 147–157.

Backman, J., Morton, A. C., Roberts, D. G., Brown, S., Krumsiek, K. & Macintyre, R. M. 1984. Geochronology of the lower Eocene and upper Paleocene sequences of Leg 81. *Initial Reports of the Deep Sea Drilling Project* 81, 877–881.

Bailey, J. W. 1923. George Frederic Matthew (1837–1923). *Proceedings and Transactions of the Royal Society of Canada* 3rd Series 17 (4), vii–x.

Baksi, A. K. 1974. Isotopic fractionation of a loosely held atmospheric argon component in the Picture Gorge Basalts. *Earth and Planetary Science Letters* 21, 431–438.

Baksi, A. K. 1988. Estimation of lava extrusion and magma production rates for two flood basalt provinces. *Journal of Geophysical Research* 93, 11809–11815.

Bamber, E. W. *et al.* 1970. Biochronology: standard of Phanerozoic time. In *Geology and Economic Minerals of Canada*, 5th edition, (ed. R. J. W. Douglas). *Geological Survey of Canada Economic Geology Report* 1, 591–694.

Bandet, Y. *et al.* 1984. Position du Langhien dans les échelles de chronologie biostratigraphique, radiométrique, et geomagnétique. *Comptes Rendus de Séances – Académie des Sciences* Série II 299, 651–656.

Bandy, O. L. (ed.) 1960. Radiometric dating and paleontologic zonation. *Geological Society of America Special Paper* 124, 247 pp.

Banerjee, S. K., Lund, S. P. & Levi, S. 1979. Geomagnetic record in Minnesota lake sediments – absence of the Gothenburg and Erieau excursions. *Geology* 7, 588–591.

Banks, N. G., Cornwall, H. R., Silberman, M. L., Creasey, S. C. & Marvin, R. F. 1972. Chronology of intrusion and ore deposition at Ray, Arizona. Part 1, K–Ar ages. *Economic Geology* 67, 864–878.

Banks, N. G. & Stuckless, J. S. 1973. Chronology of intrusion and ore deposition at Ray, Arizona. Part II, Fission track ages. *Economic Geology* 68, 657–664.

Banks, P. O. 1973. Permian–Triassic radiometric time scale. In *The Permian and Triassic Systems and their Mutual Boundary* (ed. A. Logan and L. V. Hills) *Canadian Society of Petroleum Geologists Memoir* 2, 669–677.

Banner, F. T. & Blow, W. H. 1965. Progress in the planktonic foraminiferal biostratigraphy of the Neogene. *Nature* 208 (5016), 1164–1166.

Barbier, R. & Thieuloy, J. P. 1965. Etage Valanginien. *Mémoires du Bureau de Recherches Géologiques et Minières* 34, 79–84.

Barnes, C. R., Norford, B. S. & Skevington, D. 1981. *The Ordovician System in Canada.* International Union of Geological Sciences Publication No 8, 27 pp.

Barrell, J. 1917. Rhythms and the measurements of geologic time. *Geological Society of America Bulletin* 28, 745–904.

Basaltic Volcanism Study Project, 1981. *Basaltic Volcanism on the Terrestrial Planets.* New York: Pergamon Press, 1286 pp.

Bassett, M. G. (ed.) 1976. *The Ordovician System – Proceedings of a Palaeontological Association Symposium, Birmingham, September 1974.* Cardiff: University of Wales Press/National Museum of Wales (for Palaeontological Association), 696 pp.

Bassett, M. G. 1984. Lower Palaeozoic Wales – a review of studies in the past 25 years. *Proceedings Geologist's Association* 95, 291–311.

Bassett, M. G. 1985a. Towards a 'common language' in stratigraphy. *Episodes* 8, 87–92.

Bassett, M. G. 1985b. Transition and Grauwacke – The Silurian and Cambrian systems through 150 years. *Episodes* 8, 231–235.

Bassett, M. G., Cocks, L. R. M., Holland, C. H., Rickards, R. B. & Warren, P. T. 1975. *The Type Wenlock Series.* London: Her Majesty's Stationery Office, Institute of Geological Sciences Report 75/13, 19 pp.

Bassett, M. G. & Dean, W. T. (eds.) 1982. The Cambrian–Ordovician boundary: Sections, fossil distribution and correlations. I.U.G.S. Publication 10. *National Museum of Wales Geological Series* 3, 227 pp.

Bates, R. L. & Jackson, J. A. (eds.) 1980. *Glossary of Geology*, 2nd edition. Falls Church Virginia: American Geological Institute, 749 pp.

Bayer, U. 1987. Chronometric calibration of a comparative time scale for the Mesozoic and Paleozoic. *Geologische Rundschau* 76, 485–503.

Beara, J. H., Sangree, J. B. & Smith, L. A. 1982. Quaternary chronology, palaeoclimate, depositional sequences and eustatic cycles. *American Association of Petroleum Geologists Bulletin* 66,158–169.

Bellon, H., Fabre, A., Sichler, B. & Bonhomme, M. G. 1986. Contribution to the numerical calibration of the Bajocian–Bathonian boundary: ^{40}K–^{40}Ar and paleomagnetic data from Les Vignes basaltic complex (Massif Central, France). *Chemical Geology* 59, 155–161.

Benedek, P. N. von & Müller, C. 1974. Nannoplankton–Phytoplankton Korrelation um Mittel und Ober-Oligozän von NW-Deutchland. *Neues Jahrbuch für Geologie und Paläontologie Monatshefte* 7, 385–397.

Benton, M. J. 1987. Do mass extinctions select their targets? *Geology Today* March–April 1987, 48–50.

Benus, A. P. 1988. Sedimentological context of a deep-water Ediacaran fauna (Mistaken Point Formation, Avalon Zone, Eastern Newfoundland). In *Trace Fossils and Small Shelly Fossils and the Precambrian–Cambrian Boundary* (ed. E. Landing, G. M. Narbonne and P. Myrow). Proceedings

University of the State of New York, New York State Museum Bulletin **463**, 8–9.

Berger, A. L. 1977. Long-term variation of the earth's orbital elements. *Celestial Mechanics* **15**, 53–74.

Berger, A. L. 1978. Long-term variations of caloric insolation resulting from the earth's orbital elements. *Quaternary Research* **9**, 139–167.

Berggren, W. A. 1969. Cenozoic stratigraphic, planktonic foraminiferal zonation and the radiometric time scale. *Nature* **224**, 1072–1075.

Berggren, W. A. 1971. Tertiary boundaries and correlations. In *The Micropalaeontology of Oceans* (ed. B. M. Funnell and M. R. Reidel), pp. 693–809. Cambridge: Cambridge University Press.

Berggren, W. A. 1972. A Cenozoic time-scale – some implications for regional geology and paleobiogeography. *Lethaia* **5**, 195–215.

Berggren, W. A. 1973. The Pliocene time-scale: calibration of planktonic foraminiferal and calcareous nannoplankton zones. *Nature* **243**, 391–397.

Berggren, W. A. 1978. Recent advances in Cenozoic planktonic foraminiferal biostratigraphy, biochronology and biogeography: Atlantic Ocean. *Micropaleontology* **24**, 337–370.

Berggren, W. A., Burckle, L. H., Cita, M. B., Cooke, H. B. S., Funnell, B. M., Gartner, S., Hays, J. D., Kennett, J. P., Opdyke, N. D., Pastouret, L., Shackleton, N. J. & Takayanagi, Y. 1980. Towards a Quaternary time scale. *Quaternary Research* **13**, 277–302.

Berggren, W. A., Kent, D. V. & Flynn, J. J. 1985a. Jurassic to Paleogene: Part 2, Paleogene geochronology and chronostratigraphy. In Snelling, pp. 141–195, references pp. 251–260.

Berggren, W. A., Kent, D. V., Flynn, J. J. & Van Couvering, J. A. 1985b. Cenozoic geochronology. *Geological Society of America Bulletin* **96**, 1407–1418.

Berggren, W. A., Kent, D. V. & Van Couvering, J. A. 1985c. The Neogene: Part 2, Neogene geochronology and chronostratigraphy. In Snelling, pp. 211–250, references pp. 251–260.

Berggren, W. A. & Van Couvering, J. A. 1974. The Late Neogene. *Palaeogeography, Palaeoclimatology, Palaeoecology* **16**, 1–216.

Berggren, W. A. & Van Couvering, J. A. 1978. Biochronology. In Cohee *et al.*, pp. 39–55.

Bernard-Griffiths, J. 1976. Essai sur la signification des âges au strontium dans une série métamorphique. Le bas Limousin (Massif central francais). *Annales scientifiques, Clermont-Ferrand* **55**, 1–243.

Berry, W. B. N. 1987. *Growth of a Prehistoric Time Scale based on Organic Evolution*. Palo Alto: Blackwell, 202 pp.

Bianchi, F. R., Driever, B. W. M., Jonkers, H. A. & Kreuzer, H. 1985. K–Ar date and biostratigraphic position of a volcanic ash layer in the marine Pliocene of Crete, Greece. *Geologie en Mijnbouw* **64**, 103–111.

Black, L. P., Bell, T. H., Rubenach, M. J. & Withnall, I. W. 1979. Geochronology of discrete structural–metamorphic events in a multiply deformed Precambrian terrain. *Tectonophysics* **54**, 103–137.

Blackwelder, E. 1912. United States of America. In *Handbuch der Regionalen Geologie* (ed. G. Steinmann and O. Wilkens), **8** (2), 1–258.

Blakely, R. J. 1974. Geomagnetic reversals and crustal spreading rates during the Miocene. *Journal of Geophysical Research* **79**, 2979–2985.

Blakely, R. J. & Cox, A. 1972. Evidence for short geomagnetic polarity intervals in the early Cenozoic. *Journal of Geophysical Research* **77**, 7065–7072.

Blow, W. H. 1969. Late Middle Eocene to Recent planktonic foraminiferal biostratigraphy. In *Proceedings of the First International Conference on Planktonic Microfossils*. Geneva: Leiden Brill **1**, 199–422.

Blow, W. H. 1979. *The Cainozoic Globigerinida*. Geneva 1967: Leiden Brill, 3 volumes, 1413 pp.

Boles, J. R. & Landis, C. A. 1984. Jurassic sedimentary melange and associated facies, Baja California, Mexico. *Geological Society of America Bulletin* **95**, 513–521.

Bond, G. 1978. Speculations on real sea-level changes and vertical motions of continents at selected times in the Cretaceous. *Geology* **6**, 247–250.

Bonhomme, M. G. 1982. The use of Rb–Sr and K–Ar dating methods as a stratigraphic tool applied to sedimentary rocks and minerals. *Precambrian Research* **18**, 5–25.

Bonhommet, N. & Zähringer, J. 1969. Paleomagnetism and potassium–argon age determinations of the Laschamp geomagnetic polarity event. *Earth and Planetary Science Letters* **6**, 43–46.

Bottino, M. L. & Fullagar, P. D. 1968. The effects of weathering on whole-rock Rb–Sr ages of granitic rocks. *American Journal of Science* **266**, 661–670.

Bouche, P. M. 1962. Nannofossiles calcaires du Lutétien du Bassin de Paris. *Revue de Micropaléontologie* **5** (2), 75–103.

Boucot, A. J. 1975. *Evolution and Extinction Rate Controls*. Amsterdam: Elsevier, 427 pp.

Bouroz, A. 1978. Report on isotopic dating of rocks in the Carboniferous System. In Cohee *et al.*, pp. 323–326.

Bowen, D. Q. 1978. *Quaternary Geology*. Oxford: Pergamon Press, 221 pp.

Bralower, T. J. 1987. Valanginian to Aptian calcareous nannofossil stratigraphy and correlation with the upper M-sequence magnetic anomalies. *Marine Micropaleontology* **11**, 293–310.

Bralower, T. J., Monechi, S. & Thiestein, H. R. 1989. Calcareous nannofossil zonation of the Jurassic–Cretaceous boundary interval and correlation with the geomagnetic polarity time scale. *Marine Micropaleontology* **14**, 153–235.

Bramlette, M. N. & Sullivan, F. R. 1961. Coccolithophorids and related nannoplankton of the early Tertiary in California. *Micropaleontology* **7**, 129–188.

Brass, G. W. 1975. The effect of weathering on the distribution of strontium isotopes in weathering profiles. *Geochimica et Cosmochimica Acta* **39**, 1647–1653.

Breistroffer, M. 1947. Sur les zones d'Ammonites de l'Albien de France et d'Angleterre. *Travaux Laboratoire Géologie Faculté Sciences Université Grenoble* **26**, 1–88.

Bridgewater, D., Keto, L., McGregor, V. R. & Myers, J. S. 1976. Archean gneiss complex of Greenland. In *Geology of Greenland* (ed. A. Escher and W. S. Watt). Odense: Geological Survey of Greenland, 19–75.

Bromfield, C. S., Erickson, A. J. Jr, Haddidin, M. A. & Mehnert, H. H. 1977. Potassium–argon ages of intrusion, extrusion and associated ore deposits, Park City mining district, Utah. *Economic Geology* **72**, 837–848.

Brooks, C. & Hart S. R. 1978. Rb–Sr mantle isochrons and variations in the chemistry of Gondwanaland's lithosphere. *Nature* **271**, 220–223.

Brooks, C., Hart, S. R. & Wendt, I. 1972. Realistic use of two-error regression treatments as applied to rubidium–strontium data. *Reviews of Geophysics and Space Physics* **10**, 551–577.

Brooks, C., James, D. E. & Hart, S. R. 1976. Ancient lithosphere: Its role in young continental volcanism. *Science* **193**, 1086–1094.

Brown, L. E. Jr & Fisher, W. L. 1979. *Principles of seismic stratigraphic interpretation*. In *AAPG-SEG Stratigraphic Interpretation of Seismic Data School Notes, Austin Texas*: American Association Petroleum Geologists Education Department.

Bryan, W. B. & Duncan, R. A. 1983. Age and provenance of clastic horizons from hole 516F. *Initial Reports of the Deep Sea Drilling Project* **72**, 475–477.

Buckman, S. S. 1902. The term 'Hemera'. *Geological Magazine* **4**, 554–557.

Bukry, D. 1975. Coccolith and silicoflagellate stratigraphy, northwestern Pacific Ocean. *Initial Reports of the Deep Sea Drilling Project* **32**, 677–686.

Bukry, D. & Kennedy, M. P. 1969. Cretaceous and Eocene coccoliths at San Diego California. *California Division Mines and Geology* Special Report **100**, 33–43.

Busnardo, R. 1965. Le stratotype du Barrémien: 1– Lithologie et macrofaune. *Mémoires du Bureau de Recherches Géologiques et Minières* **24**, 99–116.

Butler, R. F. & Coney, P. J. 1981. A revised magnetic polarity time scale for the Paleocene and Early Eocene and implications for Pacific plate motion. *Geophysical Research Letters* **8**, 301–304.

Butler, R. F., Lindsey, E. H., Jacobs, L. L. & Johnson, M. M. 1977. Magnetostratigraphy of the Cretaceous–Tertiary boundary in the San Juan basin New Mexico. *Nature* **267**, 318–323.

Button, A., Pretorus, D. A., Jansen, H., Stocklmayer, V., Hunter, D. R., Wilson, J. F., Wilson, A. H., Vermaak, C. F., Lee, C. A. & Stagman, J. G. 1981. The cratonic environment. In Hunter, pp. 501–639.

Calder, N. 1983. *Timescale: An Atlas of the Fourth Dimension*. The Viking Press, 288 pp.

Callomon, J. H. 1984a. The measurement of geological time. *Proceedings of the Royal Institution of Great Britain* **56**, 65–69. London: Science Reviews Ltd.

Callomon, J. H. 1984b. Biostratigraphy, chronostratigraphy and all that – again! *International Subcommission on Jurassic stratigraphy, Erlangen*, Symposium Volume **III**, Copenhagen: Geological Survey of Denmark, 612–614.

Cande, S. C., Larson, R. L. & LaBrecque, J. L. 1978. Magnetic lineations in the Pacific Jurassic quiet zone. *Earth and Planetary Science Letters* **41**, 434–440.

Carloni, G. C., Marks, P., Rutsch, R. F. & Selli, R. 1971. Stratotypes of Mediterranean Neogene stages. *Giornale di Geologia* Series 2, **37**, fasc. 2, 1–266.

Carr, P. F., Jones, B. G., Quinn, B. G. & Wright, A. J. 1984. Toward an objective Phanerozoic time scale. *Geology* **12**, 274–277.

Casey, R. 1963. The dawn of the Cretaceous Period in Britain. *South Eastern Union Scientific Societies Bulletin* **117**, 1–15.

Casey, R. 1967. The position of the Middle Volgian in the English Jurassic. *Proceedings of the Geological Society of London* **1640**, 246–247.

Cassignol, C. & Gillot, P.-Y. 1982. Range and effectiveness of unspiked potassium–argon dating: Experimental groundwork and applications. In Odin, pp. 159–179.

Cattell, A., Krogh, T. E. & Arndt, N. T. 1984. Conflicting Sm–Nd whole rock and U–Pb zircon ages for Archean lavas from Newton Township, Abitibi Belt, Ontario. *Earth and Planetary Science Letters* **70**, 280–290.

Chadwick, G. H. 1930. Subdivision of geologic time (abstract). *Geological Society of America Bulletin* **41**, 47.

Challinor, J. 1978. *A Dictionary of Geology*, 5th edition. Cardiff: University of Wales Press, 365 pp.

Chamberlin, T. C. & Salisbury, R. D. 1906. Earth History: Genesis – Paleozoic. *Geology*, volume 2. New York: Henry Holt, 602 pp.

Champion, D. E., Dalrymple, G. B. & Kuntz, M. A. 1981. Radiometric and paleomagnetic evidence for the Emperor Reversed Polarity Event at 0.46 ± 0.05 m.y. in basalt lava flows from the eastern Snake River Plain, Idaho. *Geophysical Research Letters* **8**, 1055–1058.

Champion, D. E., Lanphere, M. A. & Kuntz, M. A. 1988. Evidence for a new geomagnetic reversal from lava flows in Idaho: discussion of short polarity intervals in the Brunhes and late Matuyama polarity chrons. *Journal of Geophysical Research* **93**, 11667–11680.

Channell, J. E. T., Bralower, T. J. & Grandesso, P. 1987. Biostratigraphic correlation of Mesozoic polarity chrons CMI to CM23 at Capriolo and Xausa (Southern Alps, Italy). *Earth and Planetary Science Letters* **85**, 203–221.

Channell, J. E. T. & Grandesso, P. 1987. A revised correlation of Mesozoic polarity chrons and calpionellid zones. *Earth and Planetary Science Letters* **85**, 222–240.

Channell, J. E. T., Rio, D. & Thunell, R. C. 1988. Miocene/Pliocene boundary magnetostratigraphy at Capo Spartivento, Calabria, Italy. *Geology* **16**, 1096–1099.

Chauvel, C., Dupre, B. & Jenner, G. A. 1985. The Sm–Nd age of Kambalda volcanics is 500 Ma too old! *Earth and Planetary Science Letters* **74**, 315–324.

Chen, C-c. 1974. The Triassic System. In *Handbook of the Stratigraphy and Palaeontology in Southwest China*, pp. 58–65. Nanjing: Institute of Geology and Palaeontology Academia Sinica Beijing (in Chinese).

Chen, J. H. & Moore, J. G. 1982. Uranium–lead isotopic ages from the Sierra Nevada batholith, California. *Journal of Geophysical Research* **87**, 4761–4784.

Chen, J. H. & Tilton, G. R. 1976. Isotopic lead investigations on the Allende carbonaceous chondrite. *Geochimica et Cosmochimica Acta* **40**, 635–643.

Chen, J. H. & Wasserburg, G. J. 1981. The isotopic composition of uranium and lead in Allende inclusions and meteoritic phosphates. *Earth and Planetary Science Letters* **52**, 1–15.

Chlupac, I., Jaeger, H. & Zacmundova, J. 1972. The Silurian–Devonian boundary in the Barrandian. *Canadian Petroleum Geologists Bulletin* **20**, 104–174.

Chopin, C. & Maluski, H. 1980. $^{40}Ar–^{39}Ar$ dating of high pressure metamorphic micas from the Gran Paradiso area (Western Alps): Evidence against the blocking temperature concept. *Contributions to Mineralogy and Petrology* **74**, 109–122.

Choubert, G. & Faure-Muret, A. (General co-ordinators), Chanteux, P. (Cartographic art) 1976. (Commission for the Geological Map of the World). *Geological World Atlas* 1/10 000 000. Paris: Unesco.

Chumakov, N. M. 1978. Vendian glaciation of Europe and the North Atlantic in the late Precambrian. *Academy of Science USSR Doklady Earth Sciences Section* **198 (1–6)**, 69–73 (in Russian).

Chumakov, N. M. & Semikhatov, M. A. 1981. Riphean and Vendean of the USSR. *Precambrian Research* **15**, 229–253.

Churkin, M., Jr, Carter, C. & Johnson, B. R. 1977. Subdivision of Ordovician and Silurian time scale using accumulation rates of graptolitic shale. *Geology* **5**, 452–456.

Cita, M. B. 1975. The Miocene/Pliocene boundary: History and definition. In *Late Neogene Epoch Boundaries* (eds. T. Saito and L. H. Burckle), pp. 1–30. Micropaleontology Press: American Museum of Natural History New York.

Cita, M. B. & Elter, G. 1960. La posizione stratigrafica delle marne a Pteropodi della Langhe della Collina di Torino ed il significato cronologico del Langhiano. *Accademia Nazionale Dei Lincei* (Universita Milano Series B **141**) **29 (5)**, 360–369.

Cita, M. B. & Premoli-Silva, I. 1960. Pelagic foraminifera from the type Langhian. International Geological Reports XXI Session Part XXII. *Proceedings International Paleontological Union*, Copenhagen, pp. 39–50.

Clague, D. A. & Dalrymple, G. D. 1973. Age of Koko seamount, Emperor seamount chain. *Earth and Planetary Science Letters* **17**, 411–415.

Clauer, N. 1979a. A new approach to Rb–Sr dating of sedimentary rocks. In Jäger and Hunziker, pp. 30–51.

Clauer, N. 1979b. Relationship between the isotopic composition of strontium in newly formed continental clay minerals and their source material. *Chemical Geology* 27, 115–124.

Clauer, N. 1981. Strontium and argon isotopes in naturally weathered biotites, muscovites and feldspars. *Chemical Geology* 31, 325–334.

CLIMAP Project Members, 1984. The last interglacial ocean. *Quaternary Research* 21, 123–224.

Cloetingh, S., McQueen, H. & Lambeck, K. 1985. On a tectonic mechanism for regional sea level variations. *Earth and Planetary Science Letters* 75, 157–166.

Cloud, P. E. 1972. A working model of the primitive earth. *American Journal of Science* 272, 537–548.

Cloud, P. E. 1973. Possible stratotype sequences for the basal Paleozoic in North America. *American Journal of Science* 273, 193–206.

Cloud, P. E. 1976. Major features of crustal evolution. A. L. du Toit Memorial Lecture Series 14. *Transactions of the Geological Society South Africa* 79, 33 pp.

Cloud, P. E. 1987. Trends, transition and events in Cryptozoic history and their calibration *a propos* recommendations by the subcommission on Precambrian stratigraphy. *Precambrian Research* 37, 257–264.

Cloud, P. E. & Glaessner, M. F. 1982. The Ediacarian Period and System: Metazoa inherit the Earth. *Science* 217, 783–792.

Coats, R. P. 1981. Late Proterozoic (Adelaidean) tillites of the Adelaide Geosyncline. In Hambrey and Harland, pp. 537–548.

Cobb, J. C. & Kulp, J. L. 1961. Isotopic geochemistry of uranium and lead in the Swedish Kolm and its associated shale. *Geochimica et Cosmochimica Acta* 24, 226–249.

Cobban, W. A. & Reedside, J. B. Jr. 1952. Correlation of the Cretaceous formations of the western interior of the United States. *Geological Society of America Bulletin* 63, 1011–1044.

Cocks, L. R. M. & Fortey, R. A. 1982. Faunal evidence for oceanic separation in the Palaeozoic of Britain. *Journal of the Geological Society of London* 139, 467–478.

Cocks, L. R. M., Holland, C. H., Rickards, R. B. & Strachan, I. 1971. A correlation of Silurian Rocks in the British Isles. *Geological Society of London Special Paper* 1, 136 pp.

Cocks, L. R. M. & Rickards, R. B. (eds.) 1988. A global analysis of the Ordovician boundary. *Bulletin of the British Museum (Natural History) – Geology* 43, 394 pp.

Cocks, L. R. M., Toghill, P. & Zeigler, A. M. 1970. Stage names within the Llandovery Series. *Geological Magazine* 107, 79–87.

Cocks, L. R. M., Woodcock, N. H., Rickards, R. B., Temple, J. T. & Lane, P. D. 1984. The Llandovery Series of the type area. *Bulletin of the British Museum (Natural History) – Geology* 38, 131–182.

Cogley, N. G. 1981. Late Phanerozoic extent of dry land. *Nature* 291, 56–58.

Cohee, G. V. 1970. *Generally recognised European stages.* Paper issued April 15 1970 by George V. Cohee, Chairman AAPG Advisory Committee on Stratigraphic Coding.

Cohee, G. V., Glaessner, M. F. & Hedberg, H. D. (eds.) 1978. *Contributions to the Geologic Time Scale*, papers given at the Geological Time Scale Symposium 106.6, 25th IGC Sydney Australia August 1976. Tulsa: American Association of Petroleum Geologists, Studies in Geology 6, 388 pp.

Cohee, G. V. *et al.* 1967. Standard stratigraphic code adopted by AAPG. *American Association Petroleum Geologists Bulletin* 51, 2146–2150.

Colalonga, M. L. 1970. Appunti biostratigrafical sul Messiniano. *Giornale di Geologia* Series 2, 36, 515–542.

Compston, W. & Jeffery, P. M. 1959. Anomalous 'common strontium' in granite. *Nature* 184, 1792–1793.

Compston, W., McDougall, I. & Wyborn, D. 1982. Possible two-stage [87]Sr evolution in the Stockdale Rhyolite. *Earth and Planetary Science Letters* 61, 297–302.

Compston, W. & Pidgeon, R. T. 1986. Jack Hills, evidence of more very old detrital zircons in Western Australia. *Nature* 321, 766–769.

Conil, R., Groessens, E. & Pirlet, H. 1977. Nouvelle charte stratigraphique du Dinantian type de la Belgique. *Annales de la Société géologique du Nord* 96, 363–371.

Conway Morris, S. 1988. Radiometric dating of the Precambrian–Cambrian boundary in the Avalon Zone. In *Trace Fossils, Small Shelly Fossils and the Pre-Cambrian Boundary* (ed. E. Landing, G. M. Narbonne and P. Myrow), pp. 53–58. New York State Museum Bulletin 463.

Conway Morris, S. 1989. South-eastern Newfoundland and adjacent areas (Avalon Zone). In Cowie and Brasier, pp. 7–39.

Conybeare, W. D. & Phillips, W. 1822. *Outlines of the Geology of England and Wales, with an Introductory Compendium of the General Principles of that Science and Comparative View of the Structure of Foreign Countries*, Part 1. London: Phillips, 470 pp.

Cooke, C. W., Gardner, J. & Woodring, W. P. 1943. Correlation of the Cenozoic formations of the Atlantic Coastal Plain and Caribbean region. *Geological Society of America Bulletin* 53, 569–598.

Cooper, G. A., Butts, C., Caster, K. E., Chadwick, G. H., Goldring, W., Kindle, E. M., Kirk, E., Merriam, C. W., Swartz, F. M., Warren, P. S. & Willard, B. 1942. Correlation of the Devonian sedimentary formations of North America. *Geological Society of America Bulletin* 53, 1729–1794.

Cooper, J. A., James, P. R. & Rutland, R. W. R. 1982. Isotopic dating and structural relationship of granitoids and greenstones in the East Pilbara, Western Australia. *Precambrian Research* 18, 199–236.

Cooper, M. R. 1977. Eustacy during the Cretaceous: its implication and importance. *Palaeogeography, Palaeoclimatology, Palaeoecology* 22, 1–60.

Cope, J. C. W., Duff, K. L., Parsons, C. F., Torrens, H. S., Wimbledon, W. A. & Wright, J. K. 1980a. A correlation of Jurassic rocks in the British Isles. Part 2 – Middle and Upper Jurassic. *Geological Society of London Special Report* 15, 109 pp.

Cope, J. C. W., Getty, T. A., Howarth, M. K., Morton, N. & Torrens, H. S. 1980b. A correlation of Jurassic rocks in the British Isles Part 1: Introduction and Lower Jurassic. *Geological Society of London Special Report* 14, 73 pp.

Cope, K. G., Watson, L. K. B. & Smith, A. H. V. 1978. Appendix 1; Coal seams of the Steeple Aston Borehole. In *Stratigraphy of the Steeple Aston Borehole Oxfordshire* (ed. E. G. Poole). *Bulletin of the Geological Survey of Great Britain* 57, 23–28.

Cordani, U. G., Kawashita, K. & Filho, A. T. 1978. Applicability of the rubidium–strontium method to shales and related rocks. In Cohee *et al.*, pp. 93–117.

Corfu, F. & Andrews, A. J. 1986. A U–Pb age for mineralized Nipissing diabase, Gowanda, Ontario. *Canadian Journal of Earth Sciences* 23, 107–109.

Courtillot, V., Besse, J., Vandamme, D., Montigny, R., Jaeger, J.-J. & Capetta, H. 1986. Deccan flood basalts at the Cretaceous/ Tertiary boundary? *Earth and Planetary Science Letters* 80, 361–374.

Cowie, J. W. 1985. Continuing work on the Precambrian–Cambrian boundary. *Episodes* 8, 93–97.

Cowie, J. W., Bassett, M. G., Fritz, W. H., Harland, W. B., Menner, V. V., Palmer, A. R., Rozanov, A. Y., Semikhatov, M. A. & Sokolov, B. S. 1984. Chronostratigraphic definition of the

Precambrian–Cambrian boundary. *Geological Magazine* **121**, 649.

Cowie, J. W. & Brasier, M. D. (eds.) 1989. *The Precambrian–Cambrian Boundary*. Oxford: Oxford University Press, 213 pp.

Cowie, J. W. & Cribb, S. J. 1978. The Cambrian System. In Cohee *et al.*, pp. 355–362.

Cowie, J. W. & Harland, W. B. 1989. Chapter 9 Chronometry. In Cowie and Brasier, pp. 186–198.

Cowie, J. W. & Johnson, M. R. W. 1985. Late Precambrian and Cambrian geological time scale. In Snelling, pp. 47–64.

Cowie, J. W., Rushton, A. W. A. & Stubblefield, C. J. 1972. A correlation of Cambrian rocks in the British Isles. *Geological Society of London Special Report* **2**, 1–42.

Cowie, J. W., Ziegler, W., Boucot, A. J., Bassett, M. G. & Remane, J. 1986. Guidelines and statutes of the International Commission of Stratigraphy (ICS). *Courier Forschungsinstitutt Senkenberg* **83**, 1–14.

Cox, A. V. 1968. Lengths of geomagnetic polarity intervals. *Journal of Geophysical Research* **73**, 3247–3260.

Cox, A. V. 1981. A stochastic approach towards understanding the frequency and polarity bias of geomagnetic reversals. *Physics of the Earth and Planetary Interiors* **24**, 178–190.

Cox, A. V. & Dalrymple, G. B. 1967. Statistical analysis of geomagnetic reversal data and the precision of potassium–argon dating. *Journal of Geophysical Research* **72**, 2603–2614.

Craig, L. E., Smith, A. G. & Armstrong, R. L. 1989. Calibration of the Geologic time scale: Cenozoic and Late Cretaceous glauconite and non-glauconite dates compared. *Geology* **17**, 830–832.

Creer, K. M., Readman, P. W. & Jacobs, A. M. 1980. Paleomagnetic and paleontological dating of a section of Gioia Tauro, Italy: identification of the Blake Event. *Earth and Planetary Science Letters* **50**, 289–300.

Croll, J. 1875. *Climate and Time – in their Geological Relations: A Theory of Secular Changes in Earth's Climate*. London: Daldy, Isbister & Co, 577 pp.

Culver, S. J., Pojeta, J. Jr & Repetski, J. E. 1988. First record of Early Cambrian shelly microfossils from West Africa. *Geology* **16**, 596–599.

Curry, D. 1985. Oceanic magnetic lineations and the calibration of the late Mesozoic time-scale. In Snelling, pp. 269–272.

Curry, D., Adams, C. G., Boulter, M. C., Dilley, F. C., Eames, F. C., Funnell, B. M. & Wells, M. K. 1978. A correlation of Tertiary rocks in the British Isles. *Geological Society of London Special Report* **12**, 72 pp.

Curry, D. K. & Odin, G. S. 1982. Dating of the Palaeogene. In Odin, pp. 607–630.

Curtis, G. H. 1966. The problem of contamination in obtaining accurate dates of young geologic rocks. In Schaeffer and Zähringer, pp. 151–162.

Curtis, G. H. & Reynolds J. H. 1958. Notes on the potassium–argon dating of sedimentary rocks. *Geological Society of America Bulletin* **69**, 151–160.

D'Onofrio, S. 1964. I Foraminiferi del Neostratotipo del Messiniano. *Giornale di Geologia* Series **2 32(2)**, 409–461.

D'Orbigny, A. 1849–1852. Cours élémentaire de Paléontologie et de Géologie Stratigraphique. Paris, 847 pp.

Dallmeyer, R. D. 1979. ^{40}Ar/^{39}Ar dating: Principles, techniques, and applications in orogenic terranes. In Jäger and Hunziker, pp. 77–104.

Dallmeyer, R. D. & Rivers, T. 1983. Recognition of extraneous argon components through incremental-release ^{40}Ar/^{39}Ar analysis of biotite and hornblende across the Grenvillian metamorphic

gradient in southwestern Labrador. *Geochimica et Cosmochimica Acta* **47**, 413–428.

Dallmeyer, R. D. & Van Breeman, O. 1981. Rb–Sr whole rock and ^{40}Ar/^{39}Ar mineral ages of the Togus and Hallowell Quartz Monzonite and Three Mile Pond Granodiorite plutons south central Maine: Their bearing on post-Acadian cooling history. *Contributions to Mineralogy and Petrology* **78**, 61–73.

Dallmeyer, R. D. & Villeneuve, M. 1987. ^{40}Ar/^{39}Ar mineral age record of polyphase tectonothermal evolution in the southern Mauritanide orogen, southeastern Senegal. *Geological Society of America Bulletin* **98**, 602–611.

Dalrymple, G. B. 1964. Argon retention in a granitic xenolith from a Pleistocene basalt, Sierra Nevada, California. *Nature* **201**, 282.

Dalrymple, G. B. 1969. ^{40}Ar/^{36}Ar analyses of historic lava flows. *Earth and Planetary Science Letters* **6**, 47–55.

Dalrymple, G. B. 1979. Critical tables for conversion of K–Ar ages from old to new constants. *Geology* **7**, 558–560.

Dalrymple, G. B. & Clague, D. A. 1976. Age of the Hawaiian–Emperor bend. *Earth and Planetary Science Letters* **31**, 313–329.

Dalrymple, G. B. & Hirooka, M. 1965. Variation of potassium, argon, and calculated age in a late Cenozoic basalt. *Journal of Geophysical Research* **70**, 5291–5296.

Dalrymple, G. B. & Lanphere, M. A. 1969. *Potassium–Argon Dating Principles, Techniques and Applications to Geochronology*. San Fransisco: W.H. Freeman & Co., 258 pp.

Dalrymple, G. B. & Lanphere, M. A. 1971. ^{40}Ar/^{39}Ar technique of K–Ar dating. A comparison with the conventional technique. *Earth and Planetary Science Letters* **12**, 300–308.

Dalrymple, G. B. & Moore, J. G. 1968. Argon-40 excess in submarine pillow basalts from Kilauea Volcano, Hawaii. *Science* **161**, 1132–1135.

Damon, P. E. & Kulp, J. L. 1958. Excess helium and argon in beryl and other minerals. *American Mineralogist* **43**, 433–459.

Dana, J. D. 1872. *Corals and Coral Islands*. London: Sampson Low, Marston Low & Searle, 398 pp.

Dana, J. D. 1875. *Manual of Geology*, 2nd edition.

Dasch, E. J. 1969. Strontium isotopes in weathering profiles, deep sea sediments, and sedimentary rocks. *Geochimica et Cosmochimica Acta* **33**, 1521–1552.

Davis, D. W. 1982. Optimum linear regression and error estimation applied to U–Pb data. *Canadian Journal of Earth Sciences* **19**, 2141–2149.

Davis, D. W., Blackburn, C. E. & Krogh, T. E. 1982. Zircon U–Pb ages from the Wabigoon-Manitou Lakes region, Wabigoon Subprovince, northwest Ontario. *Canadian Journal of Earth Sciences* **19**, 254–266.

de Boer, P. L. & Wonders, A. A. H. 1984. Astronomically induced bedding in Cretaceous Pelagic Sediments near Moria (Italy); In *Milankovitch and Climate*, **Part 1** (eds. A. Berger, J. Imbrie, J. Hays, G. Kukla and B. Saltzman). Dordrecht: D. Reidel, 177–190.

DePaolo, D. J. 1987. Correlating rocks with strontium isotopes. *Geotimes*, **December 1987**, 16–18.

de Rouville, P. G. 1853. *Description géologique des environs de Montpellier*. Montpellier: Boehm, 185 pp.

Debelmas, J. & Thieuloy, J. P. 1965. Etage Hauterivien. *Mémoires du Bureau de Recherches Géologiques et Minières* **34**, 85–96.

Del Moro, A., Puxeddu, M., Radicati di Brozolo, F. & Villa, I. M. 1982. Rb–Sr and K–Ar ages on minerals at temperatures of 300–400°C from deep wells in the Laderello geothermal field (Italy). *Contributions to Mineralogy and Petrology* **81**, 340–349.

Denham, C. R., Anderson, R. F. & Bacon, M. P. 1977.

Paleomagnetism and radiochemical age estimates for Late Bruhnes polarity episodes. *Earth and Planetary Science Letters* **35**, 384–397.

Dewitt, E., Armstrong, R. L., Sutter, J. F. & Zartman, R. E. 1984. U–Th–Pb, Rb–Sr, and Ar–Ar mineral and whole rock isotope systematics in a metamorphosed granitic terrain, southeastern California. *Geological Society of America Bulletin* **95**, 723–739.

Diakow, L. J. 1985. Potassium–argon age determinations from biotite and hornblende in Toodoggone volcanic rocks (94E). *British Columbia Ministry of Energy Mines and Petroleum Resources Paper* **1985–1**, 298–300.

Diener, C. 1908. The fauna of the Traumatocrinus Limestone of Painkhanda. *Memoirs of the Geological Survey of India, Palaeontologia Indica* **15 (6,2)**, 39 pp.

Diener, C. 1912. The Trias of the Himalayas. *Geological Survey of India Memoirs* **36 (3)**, 1–159.

Diener, C. 1926. Die Fossillagerstätten in den Hallstätter Kalken des Salzkammergutes. *Sitsungsberichte Akademie Wissenschaften* **135**, 73–101.

Dietl, G. & Etzold, A. 1977. The Aalenian at the type locality. *Beitrage Naturkunde Stuttgart* Series **30**, 1–13.

Dodson, M. H. 1963. Further argon age determinations on slates from southwest England. *Proceedings of the Ussher Society* **1**, 70–71.

Dodson, M. H. 1973. Closure temperature in cooling geochronological and petrological systems. *Contributions to Mineralogy and Petrology* **40**, 259–274.

Dodson, M. H. 1976. Kinetic processes and thermal history of slowly cooling solids. *Nature* **259**, 551–553.

Dodson, M. H., Compston, W., Williams, I. S. & Wilson, J. F. 1988. A search for ancient detrital zircons in Zimbabwean sediments. *Journal of the Geological Society of London* **145**, 977–983.

Donovan, D. T. & Jones, E. J. W. 1979. Causes of world-wide changes in sea level. *Journal of the Geological Society London* **136**, 187–192.

Douglas, R. J. W. 1980. Proposals for time classification and correlation of Precambrian rocks and events in Canada and adjacent areas of the Canadian Shield. *Geological Survey of Canada Paper* **80–24**, 19 pp.

Drooger, C. W. 1964. Problems of mid-Tertiary stratigraphic interpretation. *Micropaleontology* **10**, 369–374.

Drury, S. A., Francis, P. W., Gass, I. G., Jackson, D. E., Melton, L. R. A., Pearce, J. A., Thorpe, R. S., Williams, D. W. & Wilson, R. C. L. 1976. *Lunar Geology Case Study*. Earth Science Topics and Methods. Milton Keynes: Open University Press, 115 pp.

du Toit, A. L. 1926. *Geology of South Africa*. Edinburgh: Oliver & Boyd, 463 pp.

Dunbar, C. O. 1940. The type Permian: its classification and correlation. *American Association of Petroleum Geologists Bulletin* **24**, 237–281.

Dunbar, C. O. 1942. Artinskian Series (Discussion). *American Association of Petroleum Geologists Bulletin* **26**, 402–409.

Dunbar, C. O. 1960. Correlation of the Permian formations of North America. *Geological Society of America Bulletin* **71**, 1763–1806.

Dunbar, C. O. *et al.* 1942. Correlation charts prepared by the Committee on Stratigraphy of the National Research Council. *Geological Society of America Bulletin* **53**, 429–434.

Dunning, G. R., Kean, B. F., Thurlow, J. G. & Swinden, H. S. 1987. Geochronology of the Buchans, Roberts Arm and Victoria Lake groups and Mansfield Cove Complex, Newfoundland. *Canadian Journal of Earth Sciences* **24**, 1175–1184.

Eardley, A. J., Shuey, R. T., Gvosdetsky, V., Nash, W. P., Picard, M. D., Grey, D. C. & Kukla, G. J. 1973. Lake cycles in the Bonneville Basin Utah. *Geological Society of America Bulletin* **84**, 211–215.

Edwards, M. B. & Føyn, S. 1981. Late Proterozoic tillites in Finnmark North Norway. In Hambrey and Harland, pp. 606–610.

Egyed, L. 1956. Change of earth dimensions as determined from palaeogeographical data. *Geofisica Pura e Applicata* **33**, 42–48.

Eicher, D. L. 1968. *Geologic Time*. New Jersey: Prentice Hall (Foundations of Earth Science Series), 149 pp. 2nd edition 1976, 150 pp.

Elder, W. P. 1988. Geometry of Upper Cretaceous bentonite beds: Implications about volcanic source areas and paleowind patterns, western interior, United States. *Geology* **16**, 835–838.

Elderfield, H. 1986. Strontium isotope stratigraphy. *Palaeogeography, Palaeoclimatology, Palaeoecology* **57**, 71–90.

El-Naggar, Z. R. 1966a. Stratigraphy and planktonic foraminifera of the Upper Cretaceous–Lower Tertiary succession in the Esna-Idfu region, Nile Valley, Egypt, UAR. *Bulletin of the British Museum(Natural History) – Geology* **2**, 1–279.

El-Naggar, Z. R. 1966b. Stratigraphy and classification of type of Esna Group of Egypt. *American Association of Petroleum Geologists Bulletin* **50 (7)**, 455–477.

Emilia, D. A. & Heinrichs, D. F. 1969. Ocean floor spreading: Olduvai and Gilsa events in the Matuyama epoch. *Science* **166**, 1267–1269.

Emiliani, C. 1965. Pleistocene temperatures. *Journal of Geology* **63**, 538–578.

Emiliani, C. 1966. Paleotemperature analysis of Caribbean cores P6304–8 and P6304–9 and a generalized temperature curve for the past 425,000 years. *Journal of Geology* **74**, 109–126.

Emmons, S. F. 1842. *Geology of New York, Part II, comprising the survey of the Second District*. Albany, pp. 4–5 and 136–141.

Emmons, S. F. 1888. Letter to Persifor Frazer, dated May 25, 1887, published in *International Geological Congress, American Committee Reports*. Philadelphia **A58**.

Engebretson, D. C., Cox, A. & Gordon, R. G. 1985. Relative motions between oceanic and continental plates in the Pacific Basin. *Geological Society of America Special Paper* **206**, 59 pp.

Engelhardt, W. von & Zimmermann, J. 1982. *Theorie der Geowissenschaft*. Paderborn: Ferdinand Schoningh, 382 pp.

Evans, A. L. 1970. Geomagnetic polarity reversals in a late Tertiary lava sequence from the Akaroa Volcano, New Zealand. *Geophysical Journal of the Royal Astronomical Society* **21**, 163–183.

Evans, J. W. & Stubblefield, C. J. (eds.) 1929. *Handbook of the Geology of Great Britain*. London: Murby, 556 pp.

Evans, P. 1971. Towards a Pleistocene time-scale. In Harland *et al.*, pp. 123–356.

Evernden, J. F., Curtis, G. H., Kistler, R. W. & Obradovich, J. 1960. Argon diffusion in glauconite, microcline, sanidine, leucite and phlogopite. *American Journal of Science* **258**, 583–604.

Fabre, A. & Bellon, H. 1985. Contribution to the numerical calibration of the Bajocian–Bathonian boundary: A report of ^{40}K–^{40}Ar radiometric data of Les Vignes Basaltic System (French Massif Central). *Terra cognita* **5**, 235.

Fairbridge, R. W. 1961. Eustatic changes in sea-level. In *Physics and Chemistry of the Earth* **4**, 99–185.

Farquharson, R. B. & Richards, J. R. 1975. Isotopic remobilization in the Mount Isa tuff beds. *Chemical geology* **16**, 73–88.

Faure, G. 1977. *Principles of Isotope Geology*. New York: Wiley, 464 pp.

Faure, G. 1986. *Principles of Isotope Geology*, 2nd edition. New York: Wiley, 589 pp.

Faure, G. & Powell, J. L. 1972. *Strontium Isotope Geology*. New York: Springer, 188 pp.

Fechtig, H., Gentner, W. & Kalbitzer, S. 1961. Argonbestimmungen an Kaliummineralien – IX. Messungen zu den verschiedenen Arten der Argondiffusion. *Geochimica et Cosmochimica Acta* **25**, 297–311.

Fewtrell Smith, M. (Compiler) 1981. Open University Handbook and Wall Chart, S364, *Evolution*. Milton Keynes: Open University.

Field, D. & Råheim, A. 1979. Rb–Sr total rock isotope studies on Precambrian charnockitic gneiss from south Norway. Evidence for isochron resetting during a low-grade metamorphic-deformational event. *Earth and Planetary Science Letters* **45**, 32–44.

Fischer, A. G. 1981. Climatic oscillations in the biosphere. In *Biotic Crises in Ecological and Evolutionary Time* (ed. M. H. Nitecki), pp. 102–131. New York: Academic Press.

Fischer, A. G. 1984. The two Phanerozoic supercycles. In *Catastrophes in Earth History* (ed. W. A. Berggren and J. A. Van Couvering), pp.129–150. Princeton: Princeton University Press.

Fischer, A. G. & Schwarzacher, W. 1984. Cretaceous bedding rhythms under orbital control? In *Milankovitch and Climate*, Part 1 (ed. A. Berger, J. Imbrie, J. Hays, G. Kukla and B. Saltzman), pp. 163–175. Dordrecht: D. Reidel.

Fisher, D. E. 1971. Excess rare gases in a subaerial basalt from Nigeria. *Nature, Physical Sciences* **232**, 60–61.

Fitch, F. J., Forster, S. C. & Miller, J. A. 1974. Geological time scale. *Rep. Prog. Phys.* **37**, 1433–1496.

Fitch, F. J., Hooker, P. J., Miller, J. A. & Brereton, N. R. 1978. Glauconite dating of Palaeocene–Eocene rocks from East Kent and the time-scale of Palaeogene volcanism in the North Atlantic region. *Journal of the Geological Society* **135**, 499–512.

Fleischer, R. L., Price, P. B. & Walker, R. M. 1975. *Nuclear Tracks in Solids*. Berkeley: University of California Press, 605 pp.

Flemming, N. C. & Roberts, D. G. 1973. Tectonic-eustatic changes in sea-level and sea-floor spreading. *Nature* **243**, 19–22.

Flint, R. F. 1971. *Glacial and Quaternary Geology*. New York: Wiley, 892 pp.

Flynn, J. J. 1986. Correlation and geochronology of Middle Eocene strata from the western United States. *Palaeogeography, Palaeoclimatology, Palaeoecology* **55**, 335–406.

Forster, S. C. & Warrington, G. 1985. Geochronology of the Carboniferous, Permian and Triassic. In Snelling, pp. 99–113.

Frarcy, M. J. 1981. A provisional standard for correlating the Precambrian rocks of the Canadian Shield: discussion. *Geological Survey of Canada Paper* **81–1C**, 83–88.

Frebold, H. 1953. Correlation of the Jurassic formations of Canada. *Geological Society of America Bulletin* **64**, 1229–1246.

Frith, R. A. 1979. Precambrian division. (Summarized in *Open Earth* **5**, 13).

Froude, D. O., Ireland, T. R., Kinny, P. D., Williams, I. S., Compston, W., Williams, I. R. & Myers, J. S. 1983. Ion microprobe identification of 4100–4200 Myr-old terrestrial zircons. *Nature* **304**, 616–618.

Fullagar, P. D. & Ragland, P. C. 1975. Chemical weathering and Rb–Sr whole rock ages. *Geochimica et Cosmochimica Acta* **39**, 1245–1252.

Funkhouser, J. G., Barnes, I. L. & Naughton, J. J. 1966. Problems in the dating of volcanic rocks by the potassium–argon method. *Bulletin Volcanologique* **30**, 709–717.

Funkhouser, J. G., Fisher, D. E. & Bonatti, E. 1968. Excess argon in deep-sea rocks. *Earth and Planetary Science Letters* **5**, 95–100.

Fyffe, L. R. & Cormier, R. F. 1979. The significance of radiometric ages from the Gulquac Lake area of New Brunswick. *Canadian Journal of Earth Sciences* **16**, 2046–2052.

Gale, N. H. 1982. The physical decay constants. In Odin, pp. 107–122.

Gale, N. H. 1985. Numerical calibration of the Palaeozoic time scale; Ordovician, Silurian and Devonian Periods. In Snelling, pp. 81– 88.

Gale, N. H. & Beckinsale, R. D. 1983. Comments on the paper 'Fission-track dating of British Ordovician and Silurian stratotypes' by R. J. Ross and others. *Geological Magazine* **120**, 295–302.

Gale, N. H., Beckinsale, R. D. & Wadge, A. J. 1979. A Rb–Sr whole rock isochron for the Stockdale Rhyolite of the English Lake District and a revised mid-Palaeozoic time-scale. *Journal of the Geological Society of London* **136**, 235–242.

Gale, N. H., Beckinsale, R. D. & Wadge, A. J. 1980. Discussion of a paper by McKerrow, Lambert and Chamberlain on the Ordovician, Silurian and Devonian time-scale. *Earth and Planetary Science Letters* **51**, 9–17.

Gartner, S. 1977. Calcareous nannofossil biostratigraphy and revised zonation of the Pleistocene. *Marine Micropaleontology* **2**, 1–25.

Gartner, S. & McGuirk, J. P. 1979. Terminal Cretaceous extinction scenario for a catastrophe. *Science* **206**, 1272–1276.

Gebauer, D. & Grünenfelder, M. 1974. Rb–Sr whole-rock dating of late diagenetic to anchimetamorphic Paleozoic sediments in southern France (Montagne Noire). *Contributions to Mineralogy and Petrology* **47**, 113–130.

Gebauer, D. & Grünenfelder, M. 1976. U–Pb zircon and Rb–Sr whole rock dating of low-grade metasediments. Example; Montagne Noire (Southern France). *Contributions to Mineralogy and Petrology* **59**, 13–32.

Gebauer, D. & Grünenfelder, M. 1979. U–Th–Pb dating of minerals. In Jäger and Hunziker, pp. 105–131.

Geological Society of London, 1968. International Geological Correlation Programme – United Kingdom Contribution *Recommendations on Stratigraphical Classification*. London: The Royal Society, 43 pp.

George, T. N. *et al.* 1967. The stratigraphic code – report of the Stratigraphical Code Sub-committee. *Proceedings of the Geological Society of London* **1638**, 75–87.

George, T. N. *et al.* 1968. International Geological Correlation Programme: United Kingdom Contribution. *Royal Society London*, 1–43.

George, T. N. *et al.* 1969. Recommendations on stratigraphical usage. *Proceedings of the Geological Society of London* **1656**, 139–166.

George, T. N., Johnson, G. A. L., Mitchell, M., Prentice, J. E., Ramsbottom, W. H. C., Sevastopulo, G. D. & Wilson, R. B. 1976. A correlation of Dinantian rocks in the British Isles. *Geological Society of London Special Report* **7**, 87 pp.

George, T. N. & Wagner, R. H. 1972. IUGS Subcommission on Carboniferous Stratigraphy, *Septième Congrès International de Stratigraphie et de Géologie du Carbonifère Compte Rendu*, Krefeld 1971, 139–147.

Gerasimov, P., Kuznetsova, K., Mikhailov, N. P. & Uspenskaya, E. A. 1975. Correlation of the Volgian, Portlandian and Tithonian stages. *Mémoires du Bureau de Recherches Géologiques et Minières* (Colloque sur La Limite Jurassique–Crétacé, 1973, Lyon, Neuchâtel).

Geyer, O. F. 1964. Die Typuslokalität des Pliensbachium in Würtemberg (Sudwestdeutschland). *Colloque du Jurassique*. Luxembourg 1962, Luxembourg, 161–167.

Gignoux, M. 1955. *Stratigraphic Geology*. London: Freeman, 682 pp.

Gill, J. E. (ed.) 1957. The Proterozoic in Canada. *Royal Society of Canada, Special Publication* **2**, 191 pp.

Gino, G. F. *et al.* 1953. Studi stratigrafice e micropaleontologiche sull'Apennino Tortonese. In *Observazione geologiche sui Dintorne di Sant'Agata Fossili Tortona Alessandria*. Milan: *Memoria Rivista Italiana di Paleontologia e Stratigrafia* **VI**, 7–24.

Glaessner, M. 1977. The Ediacara fauna and its place in the evolution of the Metazoa. In *Correlation of the Precambrian* (ed. A. V. Sidorenko), Moscow: Nauka **1**, 257–268.

Glaessner, M. F. 1984. *The Dawn of Animal Life*. Cambridge: Cambridge University Press, 244 pp.

Glass, B. P., Hall, C. M. & York, D. 1986. ^{40}Ar/^{39}Ar laser-probe dating of North American tectite fragments from Barbados and the age of the Eocene–Oligocene boundary. *Chemical Geology* (Isotope Geoscience Section) **59**, 181–186.

Gleadow, A. J. W. & Lovering, J. F. 1974. The effect of weathering on fission track dating. *Earth and Planetary Science Letters* **22**, 163–168.

Goldich, S. S. 1968. Geochronology in the Lake Superior region. *Canadian Journal of Earth Sciences* **5**, 715–724.

Goldich, S. S., Baadsgaard, H. & Nier, A. O. 1957. Investigations in Ar^{40}/K^{40} dating. *American Geophysical Union Transactions* **38**, 547–551.

Goldich, S. S. & Gast, P. W. 1966. Effects of weathering on the Rb–Sr and K–Ar ages of biotite from the Morton Gneiss, Minnesota. *Earth and Planetary Science Letters* **1**, 372–375.

Gonovin, D. I., Buyakaite, M. I., Vinogradov, V. I. & Krasheninnikov, V. A. 1985. K–Ar and Rb–Sr dating of glauconite from the base of the Agarinina Pentacamerata Zone, Lower Eocene, Syria. *Terra cognita* **5**, 235.

Goodwin, A. M. 1956. Facies relations in the Gunflint Iron Formation. *Economic Geology* **51**, 565–595.

Gordon, M. Jr & Mamet, B. L. 1978. Committee for IGCP. Moscow: Academy of Sciences USSR. The Mississippian–Pennsylvanian boundary. In Cohee *et al.*, pp. 327–335.

Gosselet, J. 1880. *Esquisse géologique du nord de la France*. 3 volumes, 342 pp.

Gradstein, F. M. 1978. A revision of the Mesozoic-Cenozoic time-scale. (Summarized in *Open Earth* **3**, 13.)

Gradstein, F. M., Agterberg, F. P., Aubrey, M.-P., Berggren, W. A., Flynn, J. J., Hewitt, R., Kent, D. V., Klitgord, K. D., Miller, K. G., Obradovich, J., Ogg, J. G., Prothero, D. R. & Westermann, G. E. G. 1988. Chronology of fluctuating sea levels since the Triassic; A critique. *Science* **241**, 599–601.

Grant, N. K., Freeth, S. J. & Rex, D. C. 1972. K/Ar data on the origin of feldspar megacrysts in basalt: rejoinder to J. B. Wright. *Nature Physical Sciences* **238**, 42–43.

Gray, C. M. & Compston, W. 1978. A rubidium–strontium chronology of the metamorphism and prehistory of central Australian granulites. *Geochimica et Cosmochimica Acta* **42**, 1735–1747.

Green, P. F. 1985. Comparison of zeta calibration baselines for fission-track dating of apatite, zircon and sphene. *Chemical Geology* **58**, 1–22.

Gregory, J. W. & Barrett, B. H. 1931. *General Stratigraphy*. London: Methuen, 285 pp.

Guex, J. 1978. Le trias inférieur des Salt Ranges (Pakistan) problèmes biostratigraphiques. *Eclogae geologicae Helvetiae* **71**, 105–141.

Guidish, T. M., Lerche, I., Kendall, C. G. St. C. and O'Brien, J. J. 1984. Relationship between eustatic sea level changes and basement subsidence. *American Association of Petroleum Geologists Bulletin* **68**, 164–177.

Gulson, B. L. & Krogh, T. E. 1973. Old lead components in the young Bergell Massif, south-east Swiss Alps. *Contributions to Mineralogy and Petrology* **40**, 239–252.

Hailwood, E. A. & Kerth, M. (1989). The role of magnetostratigraphy in the development of Geological Time Scales. *Paleoceanography* **4**, 1–18.

Hall, C. M., Walter, R. C., Westgate, J. A. & York, D. 1984. Geochronology, stratigraphy and geochemistry of Cindery Tuff in Pliocene Hominid-bearing sediments of the Middle Awash, Ethiopia. *Nature* **308**, 26–31.

Hallam, A. 1975. *Jurassic Environments*. Cambridge: Cambridge University Press, 269 pp.

Hallam, A. 1977. Secular changes in marine inundation of USSR and North America through the Phanerozoic. *Nature* **269**, 769–772.

Hallam, A. 1978. Eustatic cycles in the Jurassic. *Palaeogeography, Palaeoclimatology, Palaeoecology* **23**, 1–32.

Hallam, A. 1981. A revised sea-level curve for the early Jurassic. *Journal of the Geological Society of London* **138**, 735–743.

Hallam, A. 1986. The Pliensbachian and Tithonian extinction events. *Nature* **319**, 765–768.

Hallam, A., Hancock, J. M., LaBrecque, J. L., Lowrie, W. & Channell, J. E. T. 1985. Jurassic to Paleogene: Part 1 Jurassic and Cretaceous geochronology, and Jurassic to Paleogene magnetostratigraphy. In Snelling, pp. 118–140.

Hallberg, J. A. & Glikson, A. Y. 1981. Archean granite-greenstone terranes of western Australia. In Hunter, pp. 33–103.

Hamamoto, R., Miyata, Y., Futakami, M. & Tanabe, K. 1980. Isotopic ages of the Cretaceous tuff from the Manji area, Hokkaido. *Proceedings of the Japan Academy* Series **B**, **56**, 545–550.

Hambrey, M. J. 1983. Correlation of Late Proterozoic tillites in the North Atlantic region and Europe. *Geological Magazine* **120**, 209–232.

Hambrey, M. J. 1988. Late Proterozoic stratigraphy of the Barents Shelf. In *Geological Evolution of the Barents Shelf Region* (ed. W. B. Harland and E. K. Dowdeswell). London: Graham & Trotman, 49–72.

Hambrey, M. J. & Harland, W. B. 1981. Stratigraphic time-scale. In Hambrey and Harland (eds), pp. 10–11.

Hambrey, M. J. & Harland, W. B. (eds.) 1981. *Earth's Pre-Pleistocene Glacial Record*. Cambridge: Cambridge University Press, 1004 pp.

Hamilton, E. I. & Farquhar, R. M. (eds.) 1968. *Radiometric Dating for Geologists*. New York: Wiley Interscience, 506 pp.

Hamilton, G. N. G. & Cooke, H. B. S. 1969. *Geology for South African Students*, 4th edition, Johannesburg.

Hancock, J. M. & Kauffman, E. G. 1979. The great transgression of the Late Cretaceous. *Journal of the Geological Society of London* **136**, 175–186.

Hansen, H. J. 1970. Danian foraminifera from Nugssuaq, West Greenland. *Grønlands Geologiske Undersøgelse Bulletin* **93**, 1–132.

Haq, B. U., Berggren, W. A. & Van Couvering, J. A. 1977. Corrected age of the Pliocene/Pleistocene boundary. *Nature* **269**, 483–488.

Haq, B. U., Hardenbol, J. & Vail, P. R. 1987. Chronology of fluctuating sea levels since the Triassic. *Science* **235**, 1156–1167.

Haq, B. U., Hardenbol, J. & Vail, P. R. 1988. Mesozoic and Cenozoic chronostratigraphy and cycles of sea level change. *SEPM Special Publication* **42**, 71–108.

Haq, B. U. & Van Eysinga, F. W. B. 1987. *Geological Time Table*, (4th edition, Wall Chart). Amsterdam: Elsevier.

Hardenbol, J. 1968. The Priabon type section (France). *Mémoires du Bureau de Recherches Géologiques et Minières* **58**, 629–635.

Hardenbol, J. & Berggren, W. A. 1978. A new Paleogene numerical time scale. In Cohee *et al.*, pp. 213–234.

Hardenbol, J., Vail, P. R. & Ferrer, J. 1981. Interpreting

paleoenvironments, subsidence history and sea-level changes of passive margins from seismic and biostratigraphy. *Oceanologica Acta Special Publication*, 33–44.

Harland, W. B. 1964. Critical evidence for a great infra-Cambrian glaciation. *Geologische Rundschau* **54**, 45–61.

Harland, W. B. 1974. The Pre-Cambrian–Cambrian boundary. In *Cambrian of the British Isles, Norden and Spitsbergen, Lower Palaeozoic Rocks of the World* Volume 2 (ed. C. H. Holland), pp. 15–42. London: Wiley Interscience.

Harland, W. B. 1975. The two geological time scales. *Nature* **253**, 505–507.

Harland, W. B. 1978. Geochronologic scales. In Cohee *et al.*, pp. 9–32.

Harland, W. B. 1981. The Late Archean (?), Witwatersrand conglomerates, South Africa. In Hambrey and Harland, pp. 185–187.

Harland, W. B. 1982. The Proterozoic glacial record. *Geological Society of America Memoir* **161**, 279–288.

Harland, W. B. 1983a. Precambrian geochronology in Canada – Essay review. *Geological Magazine* **120**, 195–203.

Harland, W. B. 1983b. More time-scales – Essay review. *Geological Magazine* **120**, 393–400.

Harland, W. B. 1989. Palaeoclimatology. In Cowie and Brasier, pp. 199–204.H

Harland, W. B., Cox, A. V., Llewellyn, P. G., Pickton, C. A. G., Smith, A. G. & Walters, R. 1982. *A Geologic Time Scale*. Cambridge: Cambridge University Press, 131 pp.

Harland, W. B. & Francis, H. (eds.) 1971. *The Phanerozoic Time-Scale – a supplement*. Geological Society of London, Special Publication **5**, 356 pp.

Harland, W. B. & Herod, K. M. 1975. Glaciations through time. In *Ice Ages: Ancient and Modern* (ed. A. E. Wright and F. Moseley), pp. 189–216. *Geological Journal Special Issue* **6**. Liverpool: Steel Horse Press.

Harland, W. B., Holland, C. H., House, M. R., Hughes, N. F., Reynolds, A. B., Rudwick, M. J. S., Satterthwaite, G. F., Tarlo, L. B. H. & Willey, E. C. (eds.) 1967. *The Fossil Record*. London: Geological Society of London, 827 pp.

Harland, W. B., Smith, A. G. & Wilcock, B. (eds.) 1964. *The Phanerozoic Time-Scale*. (A symposium dedicated to Professor Arthur Holmes). *Quarterly Journal of the Geological Society of London* **120s**, 458 pp.

Harper, C. T. 1964. Potassium–argon ages of slates and their geological significance. *Nature* **203**, 468–470.

Harper, C. T. (ed.) 1973. *Geochronology: Radiometric Dating of Rocks and Minerals,* Benchmark Papers in Geology. Stroudsburg, Pennsylvania: Dowden, Hutchison & Ross, 469 pp.

Harper, C. T. & Schamel, S. 1971. Note on the isotopic composition of argon in quartz veins. *Earth and Planetary Science Letters* **12**, 129–133.

Harper, G. D. 1984. The Josephine ophiolite, northwestern California. *Geological Society of America Bulletin* **95**, 1009–1026.

Harris, A. L., Shackleton, R. M., Watson, J., Downie, C., Harland, W. B. & Moorbath, S. 1975. A correlation of Precambrian rocks in the British Isles. *Geological Society of London Special Report* **6**, 136 pp.

Harrison, C. G. A., Brass, G. W., Saltzman, E., Sloan, J., Southam, J. & Whitman, J. M. 1981. Sea level variations, global sedimentation rates and the hypsographic curve. *Earth and Planetary Science Letters* **54**, 1–16.

Harrison, C. G. A., McDougall, I. & Watkins, N. D. 1979. A geomagnetic field reversal time scale back to 13.0 million years before present. *Earth and Planetary Science Letters* **42**, 143–152.

Harrison, T. M. 1983. Some observations on the interpretation of ^{40}Ar/^{39}Ar age spectra. *Chemical Geology* (Isotope Geoscience Section) **41**, 319–338.

Hawking, S. W. 1988. *A Brief History of Time, from the Big Bang to Black Holes*. London: Bantam Press, 198 pp.

Hay, W. W. 1967. Calcareous nannoplankton zonation of the Cenozoic of the Gulf Coast and Caribbean–Antillean area and transoceanic correlation. *Gulf Coast Association Geological Societies* **17**, 428–480.

Hay, W. W. & Mohler, H. P. 1967. Calcareous nannoplankton from early Tertiary rocks at Pont Labau, France and Paleocene–early Eocene correlations. *Journal of Paleontology* **41**, 1505–1541.

Hays, J. D., Imbrie, J. & Shackleton, N. J. 1976. Variations in the Earth's orbit: Pacemaker of the ages. *Science* **194**, 1121–1132.

Hays, J. D. & Pitman, W. C., III 1973. Lithospheric plate motion, sea-level changes and climatic and ecological consequences. *Nature* **246**, 18–22.

Hedberg, H. D. 1976. *International Stratigraphic Guide*. New York: Wiley, 200 pp.

Heirtzler, J. R., Dickson, G. O., Herron, E. M., Pitman, W. C., III & Le Pichon, X. 1968. Marine magnetic anomalies, geomagnetic field reversals, and motions of the ocean floor and continents. *Journal of Geophysical Research* **73**, 2119–2136.

Hellman, K. N. & Lippolt, H. J. 1981. Calibration of the Middle Triassic time scale by conventional K–Ar and ^{40}Ar/^{39}Ar dating of alkali feldspars. *Journal of Geophysics* **50**, 73–88.

Hellman, P. L., Smith, R. E. & Henderson, P. 1979. The mobility of the rare earth elements: evidence and implications from selected terrains affected by burial metamorphism. *Contributions to Mineralogy and Petrology* **71**, 23–44.

Hellsey, C. E. & Steiner, M. B. 1969. Evidence for long intervals of normal polarity during the Cretaceous period. *Earth and Planetary Science Letters* **5**, 325–332.

Herbert, T. D. & Fischer, A. G. 1986. Milankovitch climatic origin of mid-Cretaceous black shale rhythms in central Italy. *Nature* **321**, 739–743.

Hess, J. C. & Lippolt, H. J. 1986. ^{40}Ar/^{39}Ar ages of tonstein and tuff sanidines: new calibration points for the improvement of the Upper Carboniferous time scale. *Chemical Geology* (Isotope Geoscience Section) **59**, 143–154.

Hess, J. C., Lippolt, H. J. & Borsuk, A. M. 1987. Constraints on the Jurassic time-scale by ^{40}Ar/^{39}Ar dating of North Caucasian volcanic rocks. *Geology* **95**, 563–571.

Hickman, A. H. 1983. Geology of the Pilbara Block and its environs. *Geological Survey of Western Australia Bulletin* **127**, 268 pp.

Hickman, M. H. & Glassley, W. E. 1984. The role of metamorphic fluid transport in the Rb–Sr isotopic resetting of shear zones: evidence from Nordre Strømfjord, West Greenland. *Contributions to Mineralogy and Petrology* **87**, 265–281.

Hicks, H. 1881. The classification of the Eozoic and Lower Palaeozoic rocks of the British Isles. *Popular Science Review, New Series* **5**, 5.

Higgins, M. W., Sinha, A. K., Zartman, R. E. & Kirk, W. S. 1977. U–Pb zircon dates from the central Appalachian Piedmont: A possible case of inherited radiogenic lead. *Geological Society of America Bulletin* **88**, 125–132.

Hilgren, F. J. & Langereis, C. G. 1988. The age of the Miocene–Pliocene boundary in the Capo Rossetto area (Sicily). *Earth and Planetary Science Letters* **91**, 214–222.

Hill, D. 1967. Devonian of eastern Australia. In *International Symposium of the Devonian System* (ed. D. H. Oswald), pp. 613–630. Calgary: Alberta Society of Petroleum Geologists.

Hinton, R. W. & Long, J. V. P. 1979. High-resolution ion-microprobe measurement of lead isotopes: variations within single zircons from Lac Seul, northwestern Ontario. *Earth and Planetary Science Letters* **45**, 309–325.

Hofmann, A. 1971. Age measurements on size fractions of two Pennsylvanian underclays. *Carnegie Institute of Washington Yearbook* **70**, 245–248.

Hofmann, A. W. 1979. Rb–Sr dating of thin slabs: an imperfect method to determine the age of metamorphism. In Jäger and Hunziker, pp. 27–29.

Hofmann, H. J. & Patel, I. M. 1989. Trace fossils from the type 'Etcheminian Series' (Lower Cambrian Ratcliffe Brook Formation), Saint John area, New Brunswick, Canada. *Geological Magazine* **126**, 139–157.

Holland, C. H. 1980. Silurian series and stages: decisions concerning chronostratigraphy. *Lethaia* **13**, 238.

Holland, C. H. 1984a. *Lower Palaeozoic Rocks of the World* **4**, *Lower Palaeozoic of north western and west central Africa*. Chichester: Wiley-Interscience, 512 pp.

Holland, C. H. 1984b. Steps to a standard Silurian. *Proceedings of the 27th International Geological Congress* **1**, 127–156.

Holland, C. H. 1985. Series and stages of the Silurian System. *Episodes* **8**, 101–103.

Holland, C. H., Audley Charles, M. G., Bassett, M. G., Cowie, J. W., Curry, D., Fitch, F. J., Hancock, J. M., House, M. R., Ingham, J. K., Kent, P. & 8 others. 1978. *A Guide to Stratigraphical Procedure. Geological Society of London Special Report* **11**, Edinburgh: Scottish Academic Press, 18 pp.

Holland, C. H., Lawson, J. D. & Walmsley, V. G. 1963. The Silurian rocks of the Ludlow District, Shropshire. *Bulletin of the British Museum (Natural History) – Geology* **8**, 93–171.

Holland, C. H., Lawson, J. D., Walmsley, V. G. & White, D. E. 1980. Ludlow stages. *Lethaia* **13**, 268.

Holmes, A. 1913. *The Age of the Earth*. London, New York: Harper, 196 pp.

Holmes, A. 1937. *The Age of the Earth* (new edition, revised). London: Nelson, 263 pp.

Holmes, A. 1947. The construction of a geological time-scale. *Transactions of the Geological Society Glasgow* **32**, 117–152.

Holmes, A. 1960. A revised geological time-scale. *Transactions of the Edinburgh Geological Society* **17**, 183–216.

Hopkins, D. M. 1975. Time stratigraphic nomenclature for the Holocene epoch. *Geology* **3**, 10.

Hopson, C. A., Mattinson, J. M. & Pessagno, E. A. Jr. 1981. Coast Range ophiolite, western California. In *The Geotectonic Development of California* (ed. W. G. Ernst), pp. 418–510. New Jersey: PrenticeHall.

House, M. R. 1985a. Correlation of mid-Palaeozoic ammonoid evolutionary events with global sedimentary perturbations. *Nature* **313**, 17–22.

House, M. R. 1985b. The ammonoid time-scale and ammonoid evolution. In Snelling, pp. 272–283.

House, M. R., Richardson, J. B., Chaloner, W. G., Allen, J. R. L., Holland, C. H. & Westoll, T. S. 1977. A correlation of Devonian rocks of the British Isles. *Geological Society of London Report* **8**, 1–110.

Howarth, M. K. 1955. Domerian of the Yorkshire Coast. *Proceedings of the Yorkshire Geological Society* **30**, 147–175.

Howchin, W. 1901. Preliminary note on the existence of glacial beds of Cambrian age in South Australia. *Transactions of the Royal Society of South Australia* **25**, 10–13.

Hower, J., Hurley, P. M., Pinson, W. H. Jr & Fairbairn, H. W. 1962. The dependence of K–Ar on the mineralogy of various particle size ranges in a shale. *Geochimica et Cosmochimica Acta* **27**, 405–410.

Hsü, K. J., LaBrecque, J. and 11 others. 1984. Numerical ages of Cenozoic biostratigraphic datum levels: Results of South Atlantic Leg 73 drilling. *Geological Society of America Bulletin* **95**, 863–876.

Huang, T. K. 1932a. The Permian formations of southern China. *Memoirs of the Geological Survey of China* Series A **10**, 161 pp.

Huang, T. K. 1932b & 1933. Late Permian brachiopoda of south western China. *Palaeontologica Sinica*, Series B, **9** (1) (1932) 138 pp and (2) (1933) 172 pp.

Hubacher, F. A. & Lux, D. R. 1987. Timing of Acadian deformation in northeastern Maine. *Geology* **15**, 80–83.

Hubbard, R. J. 1988. Age and significance of sequence boundaries on Jurassic and early Cretaceous rifted continental margins. *American Association of Petroleum Geologists Bulletin* **72**, 49–72.

Hughes, N. F. 1989. *Fossils as Information*. Cambridge: Cambridge University Press, 136 pp.

Hughes, N. F., Williams, D. B., Cutbill, J. L. & Harland, W. B. 1967. A use of reference points in stratigraphy. *Geological Magazine* **104**, 634–635.

Humphries, F. J. & Cliff, R. A. 1982. Sm–Nd dating and cooling history of Scourian granulites, Sutherland. *Nature* **295**, 515–517.

Hunter, D. R. (ed.) 1981. *Precambrian of the Southern Hemisphere. Developments in Precambrian Geology* **2**. Amsterdam: Elsevier, 882 pp.

Hunter, D. R. & Pretorius, D. A. 1981. Structural framework. In Hunter, pp. 397–422.

Hurford, A. J. & Green, P. F. 1982. A users' guide to fission track dating calibration. *Earth and Planetary Science Letters* **59**, 343–354.

Hurford, A. J. & Green, P. F. 1983. The zeta age calibration of fission-track dating. *Chemical Geology* (Isotope Geoscience Section) **41**, 285–317.

Hurley, P. M., Cormier, R. F., Hower, J., Fairbairn, H. W. & Pinson, W. H. Jr. 1960. Reliability of glauconite for age measurement by K–Ar and Rb–Sr methods. *American Association of Petroleum Geologists Bulletin* **44**, 1793–1808.

Hurley, P. M., Hunt, J. M., Pinson, W. H., Jr, & Fairbairn, H. W. 1963. K–Ar age values on the clay fractions in dated shales. *Geochimica et Cosmochimica Acta* **27**, 279–284.

Hurst, J. M. 1979. The stratigraphy and brachiopods of the upper part of the type Caradoc of South Salop. *Bulletin of the British Museum (Natural History) - Geology* **32** (4), 183–304.

Imbrie, J. & Imbrie, K. P. 1979. *Ice Ages, Solving the Mystery*. London: Macmillan, 224 pp.

Imlay, R. W. 1944. Correlation of the Cretaceous formations of the Greater Antilles, Central America and Mexico. *Geological Society of America Bulletin* **55**, 1005–1045.

Imlay, R. W. 1952. Correlation of the Jurassic formations of North America, exclusive of Canada. *Geological Society of America Bulletin* **63**, 913–992.

Imlay, R. W. & Reedside, J. B. Jr. 1954. Correlation of the Cretaceous formations of Greenland and Alaska. *Geological Society of America Bulletin* **65**, 223–246.

Ingham, J. K. 1979. Geology of a continental margin 2: Middle and Late Ordovician transgression, Girvan. In *Crustal Evolution in Northwestern Britain and Adjacent Regions* (ed. D. R. Bowes), pp. 163–176. *Geological Journal (Liverpool)* Special Issue **10**.

Ireland, T. R. & Compston, W. 1986. U–Pb age determinations of meteoric perovskites. *Terra cognita* **6**, 174.

Irving, E. & Couillard, R. W. 1973. Cretaceous normal polarity interval. *Nature – Physical Science* **244**, 10–11.

Irving, E. & McGlynn, J. C. 1976. Proterozoic magnetostratigraphy and the tectonic evolution of Laurentia. *Philosophical Transactions of the Royal Society* **A 280**, 433–468.

Irving, E. & Parry, L. G. 1963. The magnetism of some Permian rocks from New South Wales. *Geophysical Journal of the Royal Astronomical Society* **7**, 395–411.

Irving, E. & Pullaiah, G. 1976. Reversals of the geomagnetic field, magnetostratigraphy, and relative magnitude of paleosecular variation in the Phanerozoic. *Earth Science Reviews* **12**, 35–64.

Izett, G. A. 1981. Volcanic ash beds: Recorders of upper Cenozoic silicic pyroclastic volcanism in the western United States. *Journal of Geophysical Research* **86**, 10200–10222.

Izett, G. A. 1982a. Stratigraphic succession, isotopic ages, partial chemical analyses, and sources of certain silicic volcanic ash beds (4.0 to 0.1 m.y.) of the Western United States. *U.S.Geological Survey* Open File Report **81-0763**.

Izett, G. A. 1982b. The Bishop ash bed and some older compositionally similar ash beds in California, Nevada and Utah. *U.S. Geological Survey* Open File Report **82–0582**, 47 pp.

Izett, G. A., Lanphere, M. A., MacLachlan, M. E., Naeser, C. W., Obradovich, J. D., Peterman, Z. E., Rubin, M., Stern, T. W. & Zartman, R. E. 1980. *Major Geochronologic and Chronostratic Units*. U.S. Geological Survey, Geological Names Committee, Reston, Washington D.C.

Izett, G. A., Obradovich, J. D. & Mehnert, H. H. 1988. The Bishop Ash Bed (Middle Pleistocene) and some older (Pliocene and Pleistocene) chemically and mineralogically similar ash beds in California, Nevada and Utah. *U.S. Geological Survey Bulletin* **1675**, 37 pp.

Jacobsen, S. B. & Wasserburg, G. J. 1984. Sm–Nd isotopic evolution of chondrites and achondrites, II. *Earth and Planetary Science Letters* **67**, 137–150.

Jaeger, H. 1980. Silurian series and stages: a comment. *Lethaia* **13**, 365.

Jäger, E. & Hunziker, J. C. (eds.) 1979. *Lectures in Isotope Geology*. Berlin: Springer, 329 pp.

James, H. L. 1972. Subdivision of Precambrian: an interim scheme to be used by US Geological Survey. *American Association of Petroleum Geologists Bulletin* **56**, 11026–11030.

James, H. L. 1979. Precambrian subdivided. *Episodes* **4**, 34.

Jeans, C. V., Merriman, R. J., Mitchell, J. G. & Bland, D. J. 1982. Volcanic clays in the Cretaceous of southern England and northern Ireland. *Clay Minerals* **17**, 105–156.

Jenkins, D. G. 1982. Paleogene planktonic foraminifera of New Zealand and the Austral region. *Journal of Foraminiferal Research* **4**, 155–170.

Jenkins, D. G., Bowen, D. Q., Adams, C. G., Shackleton, N. J. & Brassell, S. C. 1985. The Neogene Part I. In Snelling, pp. 199–210.

Jenkins, R. J. F. 1981. The concept of an 'Ediacaran Period' and its stratigraphic significance in Australia. *Transactions of the Royal Society South Australia* **105**, 179–194.

Johnsen, C. D. & Hills, L. V. 1973. Microplankton zones of the Savik Formation (Jurassic), Axel Heiberg and Ellesmere Islands, District of Franklin. *Canadian Petroleum Geology Bulletin* **21**, 178–218.

Johnsen, S. J., Dansgaard, W., Clausen, H. B. & Langway, C. C. 1972. Oxygen isotope profile through the Antarctic and Greenland ice sheets. *Nature* **235**, 429–434.

Johnson, N. M., McGee, V. E. & Naeser, C. W. 1979. A practical method of estimating standard error of the age in the fission track dating method. *Nuclear Tracks* **3**, 93–99.

Johnson, R. G. 1982. Brunhes–Matuyama magnetic reversal dated at 790,000 yr BP by marine astronomical correlations. *Quaternary Research* **17**, 135–147.

Jones, B. G., Carr, P. F. & Wright, A. J. 1981. Silurian and Early Devonian geochronology – a reappraisal, with new evidence from the Bungonia Limestone. *Alcheringa* **5**, 197–207.

Jones, O. T. 1929. Silurian. In Evans and Stubblefield, pp. 88–127.

Kaemel, T. 1986. Zur Entwicklung und Anwendung der radiogeochronologischen Skala des Phanerozoikums. *Zeitschrift Geologische Wissenschaften* **14** (5), 393–606.

Kaneoka, I. 1972. The effect of hydration on the K/Ar ages of volcanic rocks. *Earth and Planetary Science Letters* **14**, 216–220.

Kayser, E. 1881. Ueber einige neue devonische Brachiopoden. *Zeitschrift der Deutschen geologische Gesellschaft* **33**, 331–337.

Kayser, E. 1883. Oberkarbonische Fauna von Loping. In Richthofen, F.F.V. *China*, Berlin: Dietrich Reimer **4 (8)**, pp. 160–208.

Keller, B. M. 1979. Precambrian stratigraphic scale of the USSR. *Geological Magazine* **116**, 419–429.

Keller, B. M. & Krasnobaev, A. A. 1983. Late Precambrian geochronology of the European USSR. *Geological Magazine* **120**, 381–389.

Kent, D. V. & Gradstein, F. M. 1985. A Cretaceous and Jurassic geochronology. *Geological Society of America Bulletin* **96**, 1419–1427.

Kent, D. V. & Gradstein, F. M. 1986. A Jurassic to recent geochronology. In *The Geology of North America, vol. M. The Western North Atlantic Region* (ed. P. R. Vogt and B. E. Tucholke), pp. 45–50. Boulder, Colorado: Geological Society of America.

Kent, L. E. & Hugo, P. J. 1978. Aspects of the revised South African stratigraphic classification and a proposal for the chronostratigraphic subdivision of the Precambrian. In Cohee *et al.*, pp. 367–379.

Khramov, A. N. 1963. Paleomagnetic study of sections of Upper Permian and Lower Triassic of the northern and eastern Russian platform. In *Stratigraficheskiye Issleduvaniya, Vses Neft Nauchno-Issled Geol Razved Inst. Trudy* **204**, 145–174 (in Russian).

Khramov, A. N. 1967. The earth's magnetic field in the late Paleozoic. *Physics of the Solid Earth* (Akademya Nauka SSSR Izvestiya). **1**, 50–63 (in Russian).

Kilian, W. 1887. Note géologique sur la chaîne de Lure (Basses Alpes). *Feuille Jeune Naturaliste* **17**, 53.

Kimbrough, D. L. & Mattinson, J. M. 1984. Zircon U–Pb age constraints in Middle and Upper Triassic biostratigraphic zones in the Murihiku Supergroup, Southland, New Zealand. *Geological Society of America Abstracts with Programs* **16**, 559.

King, A. F. 1931. Structure and stratigraphy of the Upper Carboniferous Bude Sandstone, north Cornwall. *Proceedings of the Ussher Society* **1**, 1–3.

Kiparisova, L. D. & Popov, Yu. N. 1956. Subdivision of the lower series of the Triassic system into stages. *Doklady Akademy Science USSR* **109 (4)**, 842–845 (in Russian).

Kiparisova, L. D. & Popov, Yu. N. 1964. The project of subdivision of the Lower Triassic into stages: *22nd International Geological Congress, Report of Soviet Geologists*, 91–99 (in Russian).

Kiparisova, L. D., Radchenko, G. P. & Gorskiy, V. P. (eds.) 1973. The Triassic System. In *The Stratigraphy of the USSR* (ed. D. V. Nalivkin). 14 volumes. Moscow: Nedra Press, 557 pp. (in Russian).

Kistler, R. W. 1968. Potassium–argon ages of volcanic rocks in Nye and Esmeralda Counties, Nevada. *Geological Society of America Memoir* **110**, 251–262.

Klapper, G., Ziegler, W. & Mashkova, T. V. 1978. Conodonts and correlation of Lower/Middle-Devonian boundary beds in the Barrandian area of Czechoslovakia. *Geologica et Paleontologica* **12**, 103–115.

Klitgord, K. D., Huestis, S. P., Mudie, J. D. & Parker, R. L. 1975.

An analysis of near-bottom magnetic anomalies. *Geophysical Journal of the Royal Astronomical Society* **28**, 35–48.

Knopf, A., Schuchert, C., Kovarik, A. F., Holmes, A. & Brown, E. W. 1931. *The Age of the Earth*. National Academy of Science – National Research Council Bulletin **80**, 487 pp.

Köhler, H. & Müller-Sohnius, D. 1980. Rb–Sr systematics on paragneiss series from the Bavarian Moldanubicum, Germany. *Contributions to Mineralogy and Petrology* **71**, 387–392.

Kokelaar, B. P., Fitch, F. J. & Hooker, P. J. 1982. A new K–Ar age from uppermost Tremadoc rocks of North Wales. *Geological Magazine* **119**, 207–211.

Köppel, V. 1974. Isotopic U–Pb ages of monazites and zircons from the crust–mantle transition and adjacent units of the Ivrea and Ceneri Zones (Southern Alps, Italy). *Contributions to Mineralogy and Petrology* **43**, 55–70.

Köppel, V. & Grünenfelder, M. 1975. Concordant U–Pb ages of monazite and xenotime from the central Alps and the timing of the high temperature Alpine metamorphism, a preliminary report. *Schweizerische Mineralogische und Petrographische Mitteilungen* **55**, 129–132.

Köppel, V. & Sommerauer, J. 1973. Trace elements and the behaviour of the U–Pb system in inherited and newly formed zircons. *Contributions to Mineralogy and Petrology* **43**, 71–82.

Kotlya, G. V. (ed.) 1977. The Carboniferous and Permian. In *A Stratigraphical Dictionary of the USSR* (ed. V. N. Vereschagin). Leningrad: Nedra Press, 535 pp. (in Russian).

Krasnobayev, A. A. 1986. *Zircons as Indicators of Geological Processes*. Uralian Science Centre of the USSR Academy of Sciences. Moscow: Nauka Publishing House, 147 pp. (in Russian).

Krasnobayev, A. A. & Semikhatov, M. A. 1986. Geochronological scale of the Upper Proterozoic (the Riphean and Vendian) of the USSR: the current state. In *Methods of Isotope Geology and Geochronological Scale* (ed. A. Shukolyukov and E. V. Bilikova), 147 pp. Moscow: Nauka Publishing House (in Russian).

Krogh, T. E. 1982a. Improved accuracy of U–Pb zircon ages by the creation of more concordant systems using an air abrasion technique. *Geochimica et Cosmochimica Acta* **46**, 637–649.

Krogh, T. E. 1982b. Improved accuracy of U–Pb zircon dating by selection of more concordant fractions using a high-gradient magnetic separation technique. *Geochimica et Cosmochimica Acta* **46**, 631–635.

Krogh, T. E. & Davis, G. L. 1971. Zircon U–Pb ages of Archean metavolcanic rocks in the Canadian shield. *Carnegie Institution of Washington Yearbook* **70**, 241–242.

Krogh, T. E. & Davis, G. L. 1974a. Alteration in zircons with discordant U–Pb ages. *Carnegie Institution of Washington Yearbook* **73**, 561–567.

Krogh, T. E. & Davis, G. L. 1974b. The age of the Sudbury Nickel Irruptive. *Carnegie Institution of Washington Yearbook* **73**, 567–569.

Krogh, T. E., McNutt, R. H. & Davis, G. L. 1982. Two high precision U–Pb zircon ages for the Sudbury Nickel Irruptive. *Canadian Journal of Earth Sciences* **19**, 723–728.

Krogh, T. E., Strong, D. F., O'Brien, S. J. & Papezik, V. S. 1988. Precise U–Pb zircon dates from the Avalon Terrane in Newfoundland. *Canadian Journal of Earth Sciences* **25**, 442–453.

Kröner, A. (ed.) 1981. Precambrian plate tectonics. *Developments in Precambrian Geology* **4**. Amsterdam: Elsevier, 781 pp.

Krummenacher, D. 1970. Isotopic composition of argon in modern surface volcanic rocks. *Earth and Planetary Science Letters* **8**, 109–117.

Kryngol'ts, G. Ya. (ed.) 1972. Jurassic System. In *The Stratigraphy of the USSR* (ed. D. V. Nalivkin), 14 volumes. Moscow: Nedra Press (in Russian).

Kukla, G. J. 1977. Pleistocene land–sea correlations, I. Europe. *Earth Science Reviews* **13**, 307–374.

Kulling, O. 1951. Spar av Varangeristiden i Norbotten. Eocambriska Varvskiffrar och tilliter I Nordbottensfjallens ostra rand, I Nordlogaste Sverige (Traces of the Vanger Ice Age in the Caledonides of Norbotten, Northern Sweden). *Sveriges Geologiske Undersokning Afh.* Series C **43**, 1–44.

Kulp, J. L. 1961. Geologic time-scale. *Science* **133**, 1105–1114.

Kummel, B. 1961. *History of the Earth*. San Francisco: Freeman, 610 pp.

Kummel, B. & Teichert, C. 1966. Relations between the Permian and Triassic formations in the Salt Range and Trans-Indus ranges, West Pakistan. *Neues Jahrbuch für Geologie und Paläontologie* **125**, 297–333.

Kunk, M. J. & Sutter, J. F. 1984. ^{40}Ar/^{39}Ar age spectrum dating of biotite from Middle Ordovician bentonites, eastern North America. In *Aspects of the Ordovician System* (ed. D. L. Bruton). Oslo: Universitetsforlaget. *Paleontologic Contributions from the University of Oslo* **295**, pp. 11–22.

Kunk, M. J., Sutter, J., Obradovich, J. D. & Lanphere, M. A. 1985. Age of biostratigraphic horizons within the Ordovician and Silurian systems. In Snelling, pp. 89–92.

LaBrecque, J. L., Hsü, K. J. and 11 others. 1983. DSDP Leg 73: contributions to Palaeogene stratigraphy in nomenclature, chronology and sedimentation rates. *Palaeogeography, Palaeoclimatology, Palaeoecology* **42**, 91–125.

LaBrecque, J. L., Kent, D. V. & Cande, S. C. 1977. Revised magnetic polarity time scale for Late Cretaceous and Cenozoic time. *Geology* **5**, 330–335.

Lamb, J. L. & Stainforth, R. M. 1976. Unreliability of *Globigerinoides* datum. *American Association Petroleum Geologists Bulletin* **60**, 1564–1569.

Lambert, R. St. J. 1971. The pre-Pleistocene Phanerozoic time-scale – a review. In Harland *et al*, pp. 9–31.

Lane, H. R. & Manger, W. L. 1985. The basis for a mid-Carboniferous boundary. *Episodes* **8**, 112–115.

Lang, B. & Mimran, Y. 1985. An Early Cretaceous volcanic sequence in Central Israel and its significance to the absolute date of the base of the Cretaceous. *Journal of Geology* **93**, 179–184.

Lang, W. D. 1928. The Belemnite Marls of Charmouth, a series in the Lias of the Dorset Coast. *Quarterly Journal of the Geological Society of London* **74**, 179–257.

Lanphere, M. A. 1981. K–Ar ages of metamorphic rocks at the base of the Semail ophiolite, Oman. *Journal of Geophysical Research* **86**, 2777–2782.

Lanphere, M. A., Churkin, M. Jr. & Eberlein, G. D. 1976. Radiometric age of the *Monograptus cyphus* graptolite zone in southeastern Alaska – an estimate of the age of the Ordovician–Silurian boundary. *Geological Magazine* **114**, 15–24.

Lanphere, M. A. & Dalrymple, G. B. 1967. K–Ar and Rb–Sr measurements on P–207, the U.S.G.S. interlaboratory standard muscovite. *Geochimica et Cosmochimica Acta* **31**, 1091–1094.

Lanphere, M. A. & Dalrymple, G. B. 1971. A test of the ^{40}Ar/^{39}Ar age spectrum technique on some terrestrial materials. *Earth and Planetary Science Letters* **12**, 359–372.

Lanphere, M. A. & Dalrymple, G. B. 1976a. Final compilation of K–Ar and Rb–Sr measurements on P–207, the U.S.G.S. interlaboratory standard muscovite. *U.S. Geological Survey Professional Paper* **840**, 127–130.

Lanphere, M. A. & Dalrymple, G. B. 1976b. Identification of excess ^{40}Ar by the ^{40}Ar/^{39}Ar age spectrum technique. *Earth and Planetary Science Letters* **32**, 141–148.

Lanphere, M. A., Irwin, W. P. & Hotz, P. E. 1968. Isotopic age of the Nevadan orogeny and older plutonic and metamorphic events in the Klamath Mountains, California. *Geological Society of America Bulletin* **79**, 1027–1052.

Lanphere, M. A. & Jones, D. L. 1978. Cretaceous time scale from North America. In Cohee *et al.*, pp. 259–268.

Lanphere, M. A. & Tailleur, I. L. 1983. K–Ar ages of bentonites in the Seabee Formation, northern Alaska: A Late Cretaceous (Turonian) time-scale point. *Cretaceous Research* **4**, 361–370.

Lapworth, C. 1879. On the tripartite classifications of the Lower Palaeozoic rocks. *Geological Magazine* **6**, 1–15.

Larcher, C., Rat, P. & Malapris, M. 1965. Documents Paleontologiques et stratigraphiques sur l'Albien de l'Aube. *Mémoires du Bureau de Recherches Géologiques et Minières* **34**, 237–253.

Larson, R. L., Golovchenko, X. & Pitman, W. C., III. 1981. Geomagnetic polarity time scale. In *Plate Tectonic Map of the Circum Pacific Region, Northeast Quadrant* (ed. K. J. Drummond). Tulsa: *American Association of Petroleum Geologists*.

Larson, R. L. & Hilde, T. W. C. 1975. A revised time scale of magnetic reversals for the Early Cretaceous and Late Jurassic. *Journal of Geophysical Research* **80**, 2586–2594.

Larson, R. L. & Pitman, W. C., III 1972. World-wide correlation of Mesozoic magnetic anomalies and its implications. *Geological Society of America Bulletin* **83**, 3645–3662.

Laughlin, A. W. 1969. Excess radiogenic argon in pegmatite minerals. *Journal of Geophysical Research* **74**, 6684–6690.

Le Pichon, X. & Heirtzler, J. R. 1968. Magnetic anomalies in the Indian Ocean and sea-floor spreading. *Journal of Geophysical Research* **73**, 2101–2117.

Lerbekmo, J. F. & Coulter, K. C. 1985. Late Cretaceous to early Tertiary magnetostratigraphy of a continental sequence: Red Deer Valley, Alberta, Canada. *Canadian Journal of Earth Sciences* **22**, 567–583.

Likharev, B. K. (ed.) 1966. The Permian System. In *Stratigraphy of the USSR*, 14 volumes (ed. D. V. Nalivkin), Moscow: Nedra Press, 563 pp. (in Russian).

Lippolt, H. J., Hess, J. C. and Burger, K. 1984. Isotopische Alter von pyroklastischen Sanidinen aus Kaolin-Kohlentonsteinen als Korrelationsmarken für das mitteleuropaische Oberkarbon. *Fortschrift Geologie Rheinland und Westfalia* **32**, 119–150.

Logan, W. E. 1863. Geology of Canada. *Geological Survey of Canada* Report of progress from its commencement to 1863, 983 pp.

Lowman, P. D., Jr. 1972. The geologic evolution of the moon. *Journal of Geology* **80**, 125–166.

Lowrie, W. & Alvarez, W. 1981. One hundred million years of geomagnetic polarity history. *Geology* **9**, 392–397.

Lowrie, W. & Alvarez, W. 1984. Lower Cretaceous magnetic stratigraphy in Umbrian pelagic limestone sections. *Earth and Planetary Science Letters* **71**, 315–318.

Lowrie, W., Channell, J. E. T. & Alvarez, W. 1980. A review of magnetic stratigraphy investigations on Cretaceous pelagic carbonate rocks. *Journal of Geophysical Research* **85**, 3597–3605.

Lowrie, W. & Ogg, J. G. 1986. A magnetic polarity time scale for the Early Cretaceous and Late Jurassic. *Earth and Planetary Science Letters* **76**, 341–349.

Ludwig, K. R. 1980. Calculation of uncertainties of U–Pb isotope data. *Earth and Planetary Science Letters* **46**, 212–220.

Lyell, C. 1832. *Principles of Geology*. Volume 1, 586 pp.

Lyell, C. 1833. *Principles of Geology*. Volume 2, 338 pp.

Lyell, C. 1833. *Principles of Geology*. Volume 3, 398 + 189 pp.

Macintyre, R. M. & Hamilton, P. J. 1984. Isotopic geochemistry of lavas from sites 553 and 555. *Initial Reports of the Deep Sea Drilling Project* **81**, 775–781.

Magaritz, M., Bar, R., Baud, A. & Holser, W. T. 1988. The carbon-isotope shift at the Permian/Triassic boundary in the southern Alps is gradual. *Nature* **331**, 337–339.

Manhes, G., Göpel, C. & Allegre, C. J. 1986. Lead isotopes in Allende inclusions: the oldest known solar material. *Terra cognita* **6**, 173.

Mankinen, E. A. & Dalrymple, G. B. 1972. Electron microprobe evaluation of terrestrial basalts for whole-rock K–Ar dating. *Earth and Planetary Science Letters* **17**, 89–94.

Mankinen, E. A. & Dalrymple, G. B. 1979. Revised geomagnetic time-scale for the interval 0–5 m.y. B.P. *Journal of Geophysical Research* **84**, 615–626.

Marks, P. 1967. *Rotalipora* et *Globotruncana* dans la Craie de Theligny (Cenomanien, Dept. de la Sarthe). *Proceedings Koninklijke Nederlandse Akademie van Wetenschappen* Series B **70**, 264–275.

Marr, J. E. 1905. Classification of the sedimentary rocks. *Quarterly Journal of the Geological Society of London* **61**, 81.

Martini, E. 1971. Standard Tertiary and Quaternary calcareous nannoplankton zonation. In *Proceedings of the II Planktonic Conference Roma, 1969*, pp. 739–785. Rome: Edizioni Tecnoscienza.

Martini, E. & Muller, C. 1975. Calcareous nannoplankton from the type Chattian (upper Oligocene). *6th Congress Committee Mediterranean Neogene Stratigraphy Proceedings*, 37–41.

Martinsson, A. 1974. The Cambrian of Norden. In *Lower Palaeozoic Rocks of the World* 2, *Cambrian of the British Isles, Norden and Spitsbergen* (ed. C. H. Holland), pp. 185–283. Chichester: Wiley-Interscience.

Martinsson, A. (ed.) 1977. *The Silurian–Devonian Boundary*: Final report of the Committee of the Siluro-Devonian Boundary within IUGS Commission on Stratigraphy and a state of the art report for Project Ecostratigraphy. Stuttgart: Schweizerbartsche Verlagsbuchhandlung, 347 pp.

Martinsson, A., Bassett, M. G. & Holland, C. H. 1981. Ratification of standard chronostratigraphical divisions and stratotypes for the Silurian System. *Episodes* **1981** (2), 36.

Marvin, R. F., Mehnert, H. H. & Noble, D. C. 1970. Use of Ar^{36} to evaluate the incorporation of air by ash flows. *Geological Society of America* Bulletin **81**, 3385–3392.

Matsuda, J. 1974. A virtual Rb–Sr isochron for an open system. *Geochemical Journal* **8**, 153–155.

Matthew, G. F. 1899. A Palaeozoic terrane beneath the Cambrian. *Annals of the New York Academy of Sciences* **12**, 41–56.

Mattison, J. M. 1975. Early Paleozoic ophiolite complexes of Newfoundland: Isotopic ages of zircons. *Geology* **3**, 181–183.

Mattison, J. M. & Echeverria, L. M. 1980. Ortigalita Peak Gabbro, Franciscan Complex: U–Pb dates of intrusion and high-pressure–temperature metamorphism. *Geology* **8**, 589–593.

McDougall, I. 1966. Precision methods of potassium-argon isotopic age determination on young rocks. In *Methods and Techniques in Geophysics 2* (ed. S. K. Runcorn), pp. 279–304. New York: Interscience.

McDougall, I. 1971. The geochronology and evolution of the young volcanic island of Reunion, Indian Ocean. *Geochimica et Cosmochimica Acta* **35**, 261–288.

McDougall, I. 1978. Revision of the geomagnetic polarity time scale for the last 5 m.y. In Zartman, pp. 287–289.

McDougall, I., Davies, T., Maier, R. & Rudowski, R. 1985. Age of the Okate Tuff Complex at Koobi Fora, Kenya. *Nature* **316**, 792–794.

McDougall, I., Kristjansson, L. & Saemundsson, K. 1984. Magneto-stratigraphy and geochronology of northwest Iceland. *Journal of Geophysical Research* **89**, 7029–7060.

McDougall, I. & Page, R. W. 1975. Towards a physical time scale for the Neogene – Data from the Australian region. In Saito and Burckle, pp. 75–84.

McDougall, I., Polach, H. A. & Stipp, J. J. 1969. Excess radiogenic argon in young subaerial basalts from the Auckland volcanic field, New Zealand. *Geochimica et Cosmochimica Acta* **33**, 1485–1520.

McDougall, I. & Watkins, N. D. 1973. Age and duration of the Reunion geomagnetic polarity event. *Earth and Planetary Science Letters* **19**, 443–452.

McDougall, I., Watkins, N. D., Walker, G. P. L. & Kristjansson, L. 1976. Potassium–argon and paleomagnetic analysis of Icelandic lava flows: limits on the age of anomaly 5. *Journal of Geophysical Research* **81**, 1505–1512.

McDowell, F. W., Wilson, J. A. & Clark, J. 1973. K–Ar dates for biotite from two paleontologically significant localities: Duchesne Formation, Utah and Chadron Formation, South Dakota. *Isochron/West* **7**, 11–12.

McElhinny, M. W. 1971. Geomagnetic reversals during the Phanerozoic. *Science* **172**, 157–159.

McElhinny, M. W. 1978. The magnetic polarity time scale: prospects and possibilities in magnetostratigraphy. In Cohee *et al.*, pp. 57–65.

McElhinny, M. W. & Burek, P. J. 1971. Mesozoic palaeomagnetic stratigraphy. *Nature* **232**, 98–102.

McFadden, P. L. & Merrill, R. T. 1984. Lower mantle convection and geomagnetism. *Journal of Geophysical Research* **89**, 3354–3362.

McGregor, V. R. 1968. Field evidence of very old Precambrian rocks in Godthaab area, West Greenland. *Rapport Grønlands Geologiske Undersogelse* **15**, 31–35.

McGregor, V. R. 1973. The early Precambrian gneisses of the Godthaab district, West Greenland. *Philosophical Transactions of the Royal Society of London* A, **237**, 343–358.

McKerrow, W. S., Lambert, R. St. J. & Chamberlain, V. E. 1980. The Ordovician, Silurian and Devonian time scales. *Earth and Planetary Science Letters* **51**, 1–8.

McKerrow, W. S., Lambert, R. St. J. & Cocks, L. R. M. 1985. The Ordovician, Silurian and Devonian periods. In Snelling, pp. 73–80.

McLaren, D. J. 1977. The Silurian–Devonian Boundary Committee. In Martinsson, 134 pp.

McLaren, D. J. 1983. Bolides and biostratigraphy. *Geological Society of America Bulletin* **94**, 313–324.

McLean, F. H. 1953. Correlation of the Triassic formations of Canada. *Geological Society of America Bulletin* **64**, 1206–1228.

McRae, S. G. 1972. Glauconite. *Earth Science Reviews* **8**, 397–440.

McTaggart, J. McT. E. 1908. The unreality of time. *Mind* **18**, 457–484.

Mellor, D. H. 1981. *Real Time*. Cambridge: Cambridge University Press, 203 pp.

Menning, M. 1986. Zur Dauer des Zechsteins aus magneto-stratigraphischen Sicht. *Zeitschrift Geologische Wissenschaften* **14**, 395–404.

Menning, M. 1989. A synopsis of numerical time scales 1917–1986. *Episodes* **12**, 3–5.

Menning, M., Katzung, G. & Lutzner, H. 1988. Magnetostratigraphic investigations in the Rotliegendes (300–252 Ma) of central Europe. *Zeitschrift Geologische Wissenschaften* **16**, 1045–1063.

Merrihue, C. & Turner, G. 1966. Potassium argon dating by

activation with fast neutrons. *Journal of Geophysical Research* **71**, 2852–2857.

Miall, A. D. 1986. Eustatic sea level changes interpreted from seismic stratigraphy: a critique of the methodology with particular reference to the North Sea Jurassic record. *American Association of Petroleum Geologists Bulletin* **70**, 131–137.

Milankovitch, M. 1920. *Théorie mathématique des phénomènes thermiques produits par la radiation solaire*. Paris: Gauthiers-Villars.

Milankovitch, M. 1930. Mathematische Klimalehre und astronomische Theorie der Klimascherankungen. In *Handbuch der Klimatologie*, I(A) (ed. W. Koppen and R. Geiger), pp. 1–176. Berlin: Borntraeger.

Milankovitch, M. 1941. Canon of insolation and the Ice-Age problem. *Königlich Serbische Akademie, Beograd*. English translation by the Israel Program for Scientific Translations and published for the U.S. Department of Commerce and the National Science Foundation, Washington, D.C.

Miller, K. G. and 5 others. 1985. Oligocene–Miocene biostratigraphy, magnetostratigraphy, and isotopic stratigraphy of the western North Atlantic. *Geology* **13**, 257–261.

Misch, P. 1946. On the discovery of Upper Permian (Lopingian) in Western Yunnan. *Bulletin of the Geological Society of China* **26**, 65–82.

Mitchell, G. F., Penny, L. F., Shotton, F. W. & West, R. G. 1973. A correlation of Quaternary deposits in the British Isles. *Geological Society of London Special Report* **4**, 1–99.

Mitchell, J. G. 1968. The argon–40/argon–39 method for potassium–argon age determination. *Geochimica et Cosmochimica Acta* **32**, 781–790.

Mitchum, R. M. Jr, Vail, P. R. & Thompson, S., III. 1977. Seismic stratigraphy and global changes of sea level, Part 2: The depositional sequence as a basic unit for stratigraphic analysis. In Payton, pp. 53–97.

Moczydlowska, M. & Vidal, G. 1988. How old is the Tommotian? *Geology* **16**, 166–168.

Mohr, P., Mitchell, J. G. & Raynolds, R. G. H. 1980. Quaternary volcanism and faulting at O'A Caldera, central Ethiopian Rift. *Bulletin Volcanologique* **43**, 173–189.

Montanari, A., Drake, R., Bice, D. M., Alvarez, W., Curtis, G. H., Turrin, B. D. & DePaolo, D. J. 1985. Radiometric time scale for the upper Eocene and Oligocene based on K/Ar and Rb/Sr dating of volcanic biotites from the pelagic sequence of Gubbio, Italy. *Geology* **13**, 596–599.

Moore, R. C. and 27 others. 1944. Correlation of Pennsylvanian formations of North America. *Geological Society of America Bulletin* **55**, 657–706.

Moorhouse, W. W. 1957. The Proterozoic of the Port Arthur and Late Nipigon Regions, Ontario. In Gill, pp. 67–76.

Mörner, N.-A. 1976a. The Pleistocene/Holocene boundary: proposed boundary stratotypes in Gothenburg, Sweden. *Boreas* **5**, 193–275.

Mörner, N.-A. 1976b. Revolution in Cretaceous sea-level analysis. *Geology* **9**, 344–366.

Morton, N. 1971. The definition of standard Jurassic Stages. *Mémoires du Bureau de Recherches Géologiques et Minières* **75**, 83–93.

Murchison, R. I. 1839. *The Silurian System Founded on Geological Montgomery, Caermarthen, Brecon, Pembroke, Monmouth, Gloucester, Worcester and Stafford: with Descriptions of the Coal Fields and Overlying Formations*, 2 volumes. London: Murray, 768 pp.

Murchison, R. I., De Verneuil, E. & Von, A. 1845. *Geology of Russia in Europe and the Ural Mountains*, 2 volumes, London.

Murray, G. E. 1961. *Geology of the Atlantic and Gulf Coastal Province of North America*. New York: Harper, 692 pp.

Mussett, A. E. & Barker, P. F. 1983. ^{40}Ar/^{39}Ar age spectra of basalts DSDP Site 516. *Initial Reports of the Deep Sea Drilling Project* **72**, 467–470.

Mutch, T. A. 1972. *Geology of the Moon, a Stratigraphic Point of View*. Princeton: Princeton University Press, 392 pp.

Naeser, C. W. 1979. Fission-track dating and geologic annealing of fission tracks. In Jäger and Hunziker, pp. 154–169.

Naeser, C. W., Ross, R. J. & Izett, G. A. 1978. Fission-track dating of the type Ordovician and Silurian. *U.S. Geological Survey Professional Paper (Research Reports)* **1100**, 1–191.

Nalivkin, D. V. 1973. *Geology of the USSR*. Edinburgh: Oliver and Boyd, 855 pp.

Nance, R. D., Worsley, T. R. & Moody, J. B. 1986. Post-Archean biogeochemical cycles and long-term episodicity in tectonic processes. *Geology* **14**, 514–518.

Napoleone, G., Silva, I. P., Heller, F., Cheli, P., Corezzi, S. & Fischer, A. G. 1983. Eocene magnetic stratigraphy at Gubbio, Italy, and its implications for Paleogene geochronology. *Geological Society of America Bulletin* **94**, 181–191.

Ness, G., Levi, S. & Couch, R. 1980. Marine magnetic anomaly time-scales for the Cenozoic and late Cretaceous: a précis, critique and synthesis. *Reviews of Geophysics and Space Physics* **18**, 753–770.

Nikiforova, K. V. 1978. Status of the boundary between Pliocene and Pleistocene. In Cohee *et al.*, pp. 171–178.

Nisbet, E. G., Wilson, J. F. & Bickle, M. J. 1981. Evolution of the Rhodesian and adjacent Archean terrain: tectonic models. In Kröner, pp. 161–183.

Noble, C. S. & Naughton, J. J. 1968. Deep-ocean basalts – Inert gas content and uncertainties in age dating. *Science* **162**, 265–267.

Noma, E. & Glass, A. L. 1987. Mass extinction pattern: Result of chance. *Geological Magazine* **124**, 319–322.

Norford, B. S., Bolton, T. E., Copeland, L. M., Cumming, L. M. & Sinclair, G. W. 1970. Ordovician and Silurian faunas. In *Geology and Economic Minerals of Canada* (ed. R. J. W. Douglas), 5th edition, pp. 601–613. Geological Survey of Canada Economic Geology Report **1**.

Nunes, P. D. 1981. The age of the Stillwater complex – a comparison of U–Pb zircon and Sm–Nd isochron systematics. *Geochimica et Cosmochimica Acta* **45**, 1961–1963.

Obradovich, J. D. 1984. An overview of the measurement of geologic time and the paradox of geologic time scales. *Proceedings of 27th International Geological Congress*. Stratigraphy **1**, 11–30.

Obradovich, J. D. & Cobban, W. A. 1975. A time-scale for the Late Cretaceous System in the Western Interior of North America. In *The Cretaceous System in the Western Interior of North America*. Proceedings of symposium, Saskatchewan, May 1973 (ed. W. G. E. Caldwell), pp. 31–54. Geological Association Canada, Special Paper **13**.

Obradovich, J. D. & Cobban, W. A. 1978. K–Ar dating of the Albian. *U.S. Geological Survey Professional Paper* **1100** (Research Reports), 191 pp.

Obradovich, J. D., Naeser, C. W., Izett, G. A., Pasini, G. & Bigazzi, G. 1982. Age constraints on the proposed Plio-Pleistocene boundary stratotype at Vrica, Italy. *Nature* **298**, 55–59.

Obradovich, J. D., Sutter, J. F. & Kunk, M. J. 1986. Magnetic polarity chron tie points for the Cretaceous and early Tertiary. *Terra cognita* **6**, 140.

Odin, G. S. 1982a. The Phanerozoic time scale revisited. *Episodes* **5**, 3–9.

Odin, G. S. (ed.) 1982b. *Numerical Dating in Stratigraphy*. Chichester: Wiley-Interscience, 2 volumes, 1040 pp.

Odin, G. S. 1986. Recent advances in Phanerozoic time-scale calibration. *Chemical Geology* (Isotope Geoscience Section) **59**, 103–110.

Odin, G. S. & Curry, D. 1982. L'échelle numérique des temps paléogènes en 1981. *Compte Rendu Hebdomadaire des Séances de l'Académie des Sciences, Paris* **293**, 1003–1006.

Odin, G. S., Curry, D., Gale, N. H. & Kennedy, W. J. 1982a. The Phanerozoic time scale in 1981. In Odin, pp. 957–960.

Odin, G. S., Curry, D. & Hunziker, J. C. 1978. Radiometric dates from N. W. European glauconites and the Paleogene time scale. *Journal of the Geological Society of London* **135**, 481–497.

Odin, G. S., Gale, N. H., Auvray, B., Bielski, M., Dore, F., Lancelot, J.-R. & Pasteels, P. 1983. Numerical dating of Precambrian–Cambrian boundary. *Nature* **301**, 21–23.

Odin, G. S., Gale, N. H. & Dore, F. 1985. Radiometric dating of Late Precambrian times. In Snelling, pp. 65–72.

Odin, G. S., Hunziker, J. C., Jeppsson, L. & Spjeldnaes, N. 1986a. Âges radiométriques K–Ar de biotites pyroclastiques sédimentées dans le Wenlock de Gotland (Suède). *Chemical Geology* (Isotope Geoscience Section) **59**, 117–125.

Odin, G. S., Hurford, A. J., Morgan, D. J. & Toghill, P. 1986b. K–Ar biotite data for Ludlovian bentonites from Great Britain. *Chemical Geology* (Isotope Geoscience Section) **59**, 127–131.

Odin, G. S. & Kennedy, W. J. 1982. Mise à jour de l'échelle des temps mésozoïques. *Compte Rendu Hebdomadaire des Séances de l'Académie des Sciences, Paris* **294**, 383–386.

Odin, G. S. & 35 others. 1982b. Interlaboratory standards for dating purposes. In Odin 1982b, pp. 123–150.

Officer, C. B. & Drake, C. L. 1982. Epeirogenic plate movements. *Journal of Geology* **90**, 139–153.

Ogg, J. G. & Lowrie, W. 1986. Magnetostratigraphy of the Jurassic–Cretaceous boundary. *Geology* **14**, 547–550.

Ogg, J. G. & Steiner, M. B. (in press). Late Jurassic and Early Cretaceous magnetic polarity time scale. Jurassic Symposium II.

Ogg, J. G., Steiner, M. B., Company, M. & Tavera, J. M. 1988. Magnetostratigraphy across the Berriasian–Valanginian stage boundary (early Cretaceous), at Cehegin (Murcia Province, southern Spain). *Earth and Planetary Science Letters* **87**, 205–215.

Ogg, J. G., Steiner, M. B., Oloriz, F. & Tavera, J. M. 1984. Jurassic magnetostratigraphy. 1. Kimmeridgian–Tithonian of Sierra Gorda and Carcabuey, Southern Spain. *Earth and Planetary Science Letters* **71**, 147–162.

O'Keefe, J. A. 1980. The terminal Eocene event: formation of a ring system around the earth? *Nature* **285**, 309–311.

Okulitch, A. V. 1988. Proposals for time classification and correlation of Precambrian rocks and events in Canada and adjacent areas of the Canadian Shield. Part 3: A Precambrian time chart for the Geological Atlas of Canada. *Geological Survey of Canada Paper* **87–23**, 20 pp.

Oliver, W. A., Jr, De Witt, W., Jr, Dennison, J. M., Hoskins, D. M. & Huddle, J. W. 1967. Devonian of the Appalachian Basin, United States. In *International Symposium on the Devonian System* (ed. D. H. Oswald), pp. 1001–1040. Calgary: Alberta Society of Petroleum Geologists.

O'Nions, R. K., Carter, S. R., Evensen, N. M. & Hamilton, P. J. 1979. Geochemical and cosmochemical applications of Nd isotope analysis. *Annual Review of Earth and Planetary Sciences* **7**, 11–38.

Oosthuyzen, E. J. & Burger, A. J. 1973. The suitability of apatite as an age indicator by the uranium-lead isotope method. *Earth and Planetary Science Letters* **18**, 29–36.

Oppel, A. 1856. Ueber einige Cephalopoden der Juraformation Wurttembergs. *Wurttenberg Naturwissenschaftlichen Jahresheften*, 12 pp.

Page, N. J. 1982. The Precambrian diamictite below the base of the Stillwater Complex, Montana. Contribution 34. In Hambrey and Harland, pp. 821–822.

Page, R. W. 1978. Response of U–Pb zircon and Rb–Sr total rock and mineral systems to low-grade regional metamorphism in Proterozoic igneous rocks, Mount Isa, Australia. *United States Geological Survey Open-File Report* **78–701**, 323–324.

Palmer, A. R. (compiler) 1983a. Decade of North American Geology (DNAG). Geologic Time Scale. *Geology* **11**, 503–504.

Palmer, A. R. 1983b. Decade of North American Geology (DNAG). Geologic Time Scale [Wall Chart]. Boulder: *Geological Society of America*.

Paproth, E. 1980. The Devonian–Carboniferous boundary. *Lethaia* **13**, 287.

Paproth, E. & Streel, M. 1985. In search of a Devonian–Carboniferous boundary. *Episodes* **8**, 110–111.

Pareto, M. F. 1865. Note sur la subdivision que l'on pourrait établir dans les terrains tertiaires de l'Appenin septentrional. *Bulletin de la Société Géologique de France*, Series **2**, 210–277.

Paris, F., Peucat, J. J. & Chalet, M. 1985. U–Pb zircon dating of volcanic rocks nearby the Silurian–Devonian boundary in southern Armorican Massif (France). *Terra cognita* **5**, 237.

Parkinson, N. & Summerhayes, C. 1985. Synchronous global sequence boundaries. *American Association of Petroleum Geologists Bulletin* **69**, 685–867.

Parrish, R. & Roddick, J. C. 1984. Geochronology and isotope geology for the geologist and explorationist. *Cordilleran Section. Geological Association of Canada* Short Course **4**, 73 pp.

Patterson, C. & Smith, A. B. 1987. Is the periodicity of extinctions a taxonomic artefact? *Nature* **330**, 248–253.

Payton, C. E. (ed.) 1977. *Seismic Stratigraphy – Applications to Hydrocarbon Exploration*. Tulsa: American Association Petroleum Geologists Memoir **26**.

Pechersky, D. M. & Khramov, A. N. 1973. Mesozoic palaeomagnetic scale of the USSR. *Nature* **244**, 499–501.

Perch-Nielsen, K. 1971. Nannofossilien aus dem Eozän von Danemark. *Eclogae geologicae Helvetiae* **60**, 19–32.

Perch-Nielsen, K. 1972. Les nannofossiles calcaires de la limite Crétacé-Tertiaire (France). *Mémoires du Bureau de Recherches Géologiques et Minières* **77**, 181–188.

Perch-Nielsen, K. 1979. Eocene to Pliocene archaeomonads, ebridians and endoskeletal dinoflagellates from the Norwegian Sea, DSDP Leg 38. In *Initial Reports of the Deep Sea Drilling Project*; Supplement to Volumes **38**, **39**, **11** and **11–1**.

Perch-Nielsen, K. 1985. Cenozoic calcareous nannofossils. In *Plankton Stratigraphy* (ed. H. M. Bolli, J. B. Saunders and K. Perch-Nielsen), pp. 427–554. Cambridge: Cambridge University Press.

Perry, E. A. Jr & Turekian, K. K. 1974. The effects of diagenesis on the redistribution of strontium isotopes in shales. *Geochimica et Cosmochimica Acta* **38**, 929–935.

Peterman, Z. E. 1966. Rb–Sr dating of middle Precambrian metasedimentary rocks of Minnesota. *Geological Society of America Bulletin* **77**, 1031–1043.

Peucot, J. J., Paris, F. & Chalet, M. 1986. U–Pb zircon dating of volcanic rocks close to the Silurian–Devonian boundary, from Vendée (western France). *Chemical Geology* (Isotope Geoscience Section) **59**, 133–142.

Phillips, J. 1841. Palaeozoic fossils of Cornwall, Devon, and West Somerset. *Great Britain Geological Survey Memoir*, 160 pp.

Pigage, L. C. & Anderson, R. G. 1985. The Anvil plutonic suite, Faro, Yukon Territory. *Canadian Journal of Earth Sciences* **22**, 1204–1216.

Pitman, W. C., III 1978. Relationship between eustasy and stratigraphic sequences of passive margins. *Geological Society of America Bulletin* **89**, 1389–1403.

Pitman, W. C., III, Herron, E. M. & Heirtzler, J. R. 1968. Magnetic anomalies in the Pacific and sea floor spreading. *Journal of Geophysical Research* **73**, 2069–2085.

Pomerol, C. 1978. Critical review of isotopic dates in relation to Paleogene stratotypes. In Cohee *et al.*, pp. 235–245.

Pomerol, C. 1981. *The Cenozoic Era*. Chichester: Wiley.

Popenoe, W. P., Imlay, R. W. & Murphy, M. A. 1960. Correlation of the Cretaceous formations of the Pacific coast (United States and northwestern Mexico). *Geological Society of America Bulletin* **71**, 491–1540.

Postuma, J. A. 1971. *Manual of Planktonic Foraminifera*. Amsterdam: Elsevier, 420 pp.

Powell, J. W. 1890. U. S. Geological Survey, Tenth Annual Report, letter of transmittal dated 25th July 1889. Part I, 19–20, 59–61, 66.

Preiss, W. V. (ed.) 1987. Adelaide Geosyncline. Late Proterozoic stratigraphy, sedimentation, palaeontology and tectonics. *Geological Survey of South Australia Bulletin* **53**, 438 pp.

Premoli-Silva, I., Coccioni, R. & Montanari, A. (eds.) 1988. The Eocene–Oligocene boundary in the Manche–Umbria Basin (Italy). *Proceedings of the Eocene–Oligocene Boundary Ad Hoc Meeting, Ancona, Oct. 1987*. Anniballi: Ancona, 250 pp.

Prothero, D. R. 1985. Chadron (Early Oligocene) magneto-stratigraphy of eastern Wyoming: implications for the age of the Eocene–Oligocene boundary. *Journal of Geology* **93**, 555–565.

Prothero, D. R. & Armentrout, J. M. 1985. Magnetostratigraphic correlation of the Lincoln Creek Formation, Washington: Implications for the age of the Eocene/Oligocene Boundary. *Geology* **13**, 208–211.

Prothero, D. R., Denham, C. R. & Farmer, H. G. 1982. Oligocene calibration of the magnetic polarity time scale. *Geology* **10**, 650–653.

Raaben, M. Ye. 1975. The Upper Riphean as a unit of the general stratigraphic scale. *Trudy GIN AN SSSR* **273**, Moscow: Nauka.

Raaben, M. E. (ed.) 1978. *The Riphean Lower Boundary and the Aphebian Stromatolites*. Moscow: Nauka, 198 pp.

Raaben, M. Ye. 1981. *The Tommotian Stage and the Cambrian Lower Boundary Problem*. New Delhi: Amerind Publishing Co. (originally Trudy GIN **206**).

Raasch, G. O. 1961. *Geology of the Arctic*. Proceedings of the First International Symposium on Arctic Geology, Calgary, Jan. 1960. Toronto: Toronto University Press **1**, 1–732; **2**, 733–1196.

Råheim, A. & Compston, W. 1977. Correlations between metamorphic events and Rb–Sr ages in metasediments and eclogite from Western Tasmania. *Lithos* **10**, 271–289.

Rampino, M. R. 1981. Revised age estimates of Brunhes paleo-magnetic events: support for a link between geomagnetism and eccentricity. *Geophysical Research Letters* **8**, 1047–1050.

Rampino, M. R. & Stothers, R. B. 1988. Flood basalt volcanism during the past 250 million years. *Science* **24**, 663–668.

Ramsay, A. C. 1866. The geology of North Wales. *Memoirs of the Geological Survey of Great Britain* **3**, 381 pp.

Ramsbottom, W. H. C. (ed.) 1981. Field guide to the boundary stratotypes of the Carboniferous stages of Britain. *IUGS Subcommission on Carboniferous Stratigraphy*. Leeds.

Ramsbottom, W. H. C., Calver, M. A., Eagar, R. M. C., Hodson, F., Holliday, D. W., Stubblefield, C. J. & Wilson, R. B. 1978.

A correlation of Silesian (Upper Carboniferous) rocks in the British Isles. *Geological Society of London, Special Report* **10**, 81 pp.

Ramsbottom, W. H. C., Saunders, W. B. & Owens, B. (eds.) 1982. *Biostratigraphic Data for a mid-Carboniferous Boundary.* Leeds: Subcommission on Carboniferous Stratigraphy, 156 pp.

Raup, D. M. 1987. Mass extinction: a commentary. *Palaeontology* **30**, 1–13.

Rauser-Chernousova, D. M. & Schegolev, A. K. 1979. The Carboniferous–Permian boundary in the USSR. In Wagner *et al.*, pp. 175–195.

Rawson, P. F. 1983. The Valanginian to Aptian stages – current definitions and outstanding problems. *Zitteliana* **10**, 493–500.

Rawson, P. F., Curry, D., Dilley, F. C., Hancock, J. M., Kennedy, W. J., Neale, J. W., Wood, C. J. & Worssam, B. C. 1978. A correlation of Cretaceous rocks in the British Isles. *Geological Society of London, Special Report* **9**, 70 pp.

Rawson, P. F. & Riley, L. A. 1982. Latest Jurassic–early Cretaceous events and the 'Late Cimmerian Unconformity' in North Sea area. *American Association of Petroleum Geologists Bulletin* **66**, 2628–2648.

Raymo, M. E., Ruddiman, W. F., Backman, J., Clement, B. M. and Martinson, D. G. 1989. Late Pliocene variation in northern hemisphere ice sheets and North Atlantic deep water circulation. *Paleoceanography* **4** (4), 413–416.

Rea, D. K. & Blakely, R. J. 1975. Short-wavelength magnetic anomalies in a region of rapid seafloor spreading. *Nature* **225**, 126–128.

Reeside, J. B. Jr *et al.* 1957. Correlation of the Triassic formations of North America exclusive of Canada. *Geological Society of America Bulletin* **68**, 1451–1514.

Rénevier, E. 1867. Notices géologiques et paléontologiques sur les Alpes vaudoises et les régions environnantes: V. Complément de la faune de Cheville. *Bulletin de la Société Vaudoise des Sciences Naturelles* **9**, 115–208.

Rénevier, E. 1897. Chronologie géologique. *Congrès géologique international*, **IV** Session, Zurich, 1894, 523–695.

Reyment, R. A. & Mörner, N.-A. 1977. Cretaceous transgressions and regressions exemplified by the South Atlantic. *Paleontological Society of Japan Special Paper* **21**, 217–261.

Reynolds, P. H., Zentilli, M. & Muecke, G. K. 1981. K–Ar and ^{40}Ar/^{39}Ar geochronology of granitoid rocks from southern Nova Scotia: Its bearing on the geological evolution of the Meguma Zone of the Appalachians. *Canadian Journal of Earth Sciences* **18**, 386–394.

Rich, J. E., Johnson, G. L., Jones, J. E. & Campsie, J. 1986. Correlation between fluctuations in seafloor spreading rates and evolutionary pulsations. *Paleoceanography* **1**, 85–95.

Richards, J. R. 1978. The length of the Devonian Period. In Zartman, p. 351.

Richter, R. 1942. Geschichte und Aufgabe des Wetteldorfer Richtschnittes. *Senckenbergiana* **25**, 357–361.

Riedel, W. R. & Sanfilippo, A. 1971. Cenozoic radiolaria from the Western Tropical Pacific, Leg 7. *Initial Reports of the Deep Sea Drilling Project* **7**, 1529–1672.

Riley, G. H. & Compston, W. 1962. Theoretical and technical aspects of Rb–Sr geochronology. *Geochimica et Cosmochimica Acta* **26**, 1255–1282.

Rodda, P., McDougall, I., Cassie, R. A., Falvey, D. R., Todd, R. & Wilcoxon, J. A. 1985. Isotopic ages, magnetostratigraphy and biostratigraphy from the early Pliocene Suva Marl, Fiji. *Geological Society of America Bulletin* **96**, 529–538.

Roddick, J. C. 1983. High precision intercalibration of ^{40}Ar/^{39}Ar standards. *Geochimica et Cosmochimica Acta* **47**, 887–898.

Roddick, J. C., Cliff, R. A. & Rex, D. C. 1980. The evolution of excess argon in alpine biotites – a ^{40}Ar–^{39}Ar analysis. *Earth and Planetary Science Letters* **48**, 185–208.

Ross, J. R. P. 1981. Biogeography of Carboniferous ectoproct bryozoa. *Palaeontology* **24**, 313–341.

Ross, R. J. Jr 1984. The Ordovician System, progress and problems. *Annual Review of Earth and Planetary Sciences* **12**, 307–335.

Ross, R. J. Jr & Naeser, C. W. 1984. The Ordovician time scale – new refinements. In *Aspects of the Ordovician System* (ed. D. L. Bruton), pp. 5–10. Oslo: Universitetsforlaget *Palaeontological Contributions* **295**.

Ross, R. J., Jr, Naeser, C. W. & Lambert, R. St. J. 1978a. Ordovician geochronology. In Cohee *et al.*, pp. 347–354.

Ross, R. J. Jr *et al.* 1978b. Fission-track dating of lower Paleozoic volcanic ashes in British stratotypes. In Zartman, pp. 363–365.

Ross, R. J. Jr, Naeser, C. W. & 13 others. 1982. Fission-track dating of British Ordovician and Silurian stratotypes. *Geological Magazine* **119**, 135–153.

Rotai, A. P. 1979. Carboniferous stratigraphy of the USSR: proposal for an international classification. In Wagner, pp. 225–247.

Roth, H. P., Baumann, P. & Bertolino, V. 1971. Late Eocene–Oligocene calcareous nannoplankton from central and northern Italy. *2nd International Conference on Planktonic Microfossils Rome (1970)*, Proceedings, 1069–1097.

Roveda, V. 1961. Contributo allo studio di alcuni macroforaminiferi di Priabono. *Rivista Italiano Paleontologia* **2**, 153–224.

Rozanov, A. Yu., 1984. The Precambrian–Cambrian boundary in Siberia. *Episodes* **7**, 20–24.

Ruddiman, W. F. & McIntyre, A. 1984. Ice-age thermal response and climatic role of the surface Atlantic Ocean, 40°N to 63°N. *Geological Society of America Bulletin* **95**, 381–396.

Ruddiman, W. F., Raymo, M. E., Martinson, D. G., Clement, B. M. & Backman, J. 1989. Pleistocene evolution of northern hemisphere climate. *Paleoceanography* **4** (4), 353–412.

Ruddiman, W. F., Raymo, M. & McIntyre, A. 1986. Matuyama 41,000-year cycles: North Atlantic Ocean and northern hemisphere ice sheets. *Earth and Planetary Science Letters* **80**, 117–129.

Rudwick, M. J. S. 1979. The Devonian: a system born from conflict. In *The Devonian System* (eds. M. R. House, C. T. Scrutton and M. G. Bassett), pp. 9–22. London: Palaeontological Association, Special Papers in Palaeontology **23**.

Rudwick, M. J. S. 1985. *The Great Devonian Controversy*. Chicago: University of Chicago Press, 494 pp.

Rundberg, Y. & Smalley, P. C. 1989. High-resolution dating of Cenozoic sediments from the northern North Sea using ^{87}Sr/^{86}Sr stratigraphy. *American Association of Petroleum Geologists Bulletin* **73**, 298–308.

Rundle, C. C. 1986. Radiometric dating of a Caradocian tuff horizon. *Chemical Geology* **59**, 111–115.

Runnegar, B. 1982. The Cambrian explosion: animals or fossils? *Journal of the Geological Society, Australia* **29**, 395–411.

Rutford, R. H., Craddock, C., Armstrong, R. L. & White, C. M. 1972. Tertiary glaciation in the Jones Mountains. In *Antarctic Geology and Geophysics* (ed. R. J. Adie), pp. 239–243. Oslo: Universitetsforlaget.

Rutland, R. W. R., Parker, A. J., Pitt, G. M., Preiss, V. W. & Murrell, B. 1981. The Precambrian of South Australia. In Hunter, pp. 309–360.

Sachs, V. N. & Strelkov, S. A. 1961. Mesozoic and Cenozoic of the Soviet Arctic. In *Geology of the Arctic. Proceedings of First International Symposium on Arctic Geology* (ed. G. O. Raasch), pp. 48–67. Toronto: University of Toronto Press.

Sadler, D. H. 1968. Astronomical measures of time. *Quarterly Journal of the Royal Australia Society* **9**, 281–293.

Saemundsson, K., Kristjansson, L., McDougall, I. & Watkins, N. D. 1980. K–Ar dating, geological and paleomagnetic study of a 5-km lava succession in northern Iceland. *Journal of Geophysical Research* **85**, 3268–3646.

Saito, T. 1974. A paleomagnetic assignment of Neogene stage boundaries and the development of isochronous data-planes between the Mediterranean, the Pacific and Indian Oceans in order to investigate the response of the world ocean to the Mediterranean 'salinity crisis'. *Rivista Italiana di Paleontologia e Stratigrafia* **80**, 631–687.

Saito, T. & Burckle, L. H. (eds.) 1975. Late Neogene epoch boundaries. *Micropaleontology*, Special Publication **No.1**. Symposium on Late Neogene Epoch Boundaries, 24 I.G.C. Montreal, August, 1972. New York: American Museum of Natural History.

Salop, L. J. 1977. *Precambrian of the Northern Hemisphere.* Amsterdam: Elsevier, 378 pp.

Salop, L. J. 1983. *Geological Evolution of the Earth during the Precambrian.* Berlin, Heidelberg, New York: Springer Verlag, 459 pp.

Salvador, A. 1985. Chronostratigraphic and geochronometric scales in COSUNA stratigraphic correlation charts of the United States. *American Association of Petroleum Geologists Bulletin* **69**, 181–189.

Salvador, A. (chairman) 1987. Unconformity-bounded stratigraphic units. International Subcommission on Stratigraphic units. *Geological Society of America Bulletin* **98**, 232–237.

Sandberg, C. A., Gutschick, R. C., Johnson, J. G., Poole, F. G. & Sando, W. J. 1982. Middle Devonian to Late Mississippian geologic history of the overthrust belt region, western United States. In *Geologic Studies of the Cordillerian Thrust Belt* (ed. R. B. Powers), pp. 691–719. Denver, Colorado: Rocky Mountain Association of Geologists.

Sarjent, W. A. S. 1979. Middle and upper Jurassic dinoflagellate cysts: the world excluding North America. In *Contribution of Stratigraphic Palynology (with Emphasis on North America), Volume 2. Mesozoic Palynology*, American Association of Stratigraphic Palynologists Contribution Series, **5**, 133–157.

Sasajima, S. & Shimada, M. 1966. Paleomagnetic studies of the Cretaceous volcanic rocks in southwest Japan – an assumed drift of the Honshu Island. *Journal of the Geological Society of Japan* **72**, 503–514.

Satir, M. 1974. Rb–Sr-Altersbestimmungen an Glimmern der westlichen Hohen Tauern: Interpretation und geologische Bedeutung. *Schweizerische Mineralogische und Petrographische Mitteilungen* **54**, 213–228.

Satir, M. 1975. Die Entwicklungsgeschichte der westlichen Hohen Tauern und der suedlichen Oetztalmasse auf Grund radiometrischer Altersbestimmung. *Memoire degli Istituti di Geologia e Mineralogia dell'Universita di Padova* **30**, 84 pp.

Schaeffer, O. A. & Zähringer, J. 1966. *Potassium Argon Dating.* Berlin: Springer-Verlag, 234 pp.

Schidlowski, M. 1988. A 3,800-million-year isotopic record from carbon in sedimentary rocks. *Nature* **333**, 313–318.

Schopf, J. W. 1970. Precambrian micro-organisms and evolutionary events prior to the origin of vascular plants. *Biological Review* **45**, 319–352.

Schopf, J. W. (ed.) 1983. *Earth's Earliest Biosphere, Its Origin and Evolution.* Princeton: Princeton University Press, 543 pp.

Schopf, J. W. & Packer, B. M. 1987. Early Archean (3.3-billion to 3.5-billion-year old) microfossils from Warrawoona Group, Australia. *Science* **237**, 70–73.

Schwan, W. 1985. The worldwide active middle/late Eocene geodynamic episode with peaks at 45 and 37 m.y. B.P. and

implications and problems of orogeny and sea-floor spreading. *Tectonophysics* **115**, 197–234.

Schweickert, R. A., Bogen, N. L., Girty, G. H., Hanson, R. E. & Merguerian, C. 1984. Timing and structural expression of the Nevadan orogeny, Sierra Nevada, California. *Geological Society of America Bulletin* **95**, 967–979.

Scott, G. H. 1972. *Globigerinoides* from Escornebeou (France) and the basal Miocene *Globigerinoides* datum. *New Zealand Journal of Geology and Geophysics* **15**, 287–295.

Secord, J. A. 1982. King of Silurian: Roderick Murchison and the Imperial Throne in 19th Century British Geology. *Victorian Studies* **25**, 413–442.

Secord, J. A. 1986. *Controversy in Victorian Geology: the Cambrian–Silurian Dispute.* Princeton: Princeton University Press, 363 pp.

Sedgwick, A. 1838. A synopsis of the English series of stratified rocks inferior to the Old Red Sandstone, with an attempt to determine their successive natural groups and formations. *Proceedings of the Geological Society of London* **2**, 675.

Sedgwick, A. 1852. On the classification and nomenclature of the Lower Paleozoic rocks of England and Wales. *Quarterly Journal of the Geological Society of London* **8**, 136–168.

Sedgwick, A. & McCoy, F. 1851, 1852, 1855. *A Synopsis of the Classification of the British Paleozoic Rocks, with a Systematic Description of the British Paleozoic Fossils in the Geological Museum of the University of Cambridge.* London & Cambridge, 661 pp.

Seguenza, G. 1868. La formation zancléenne, ou recherches sur une nouvelle formation tertiaire. *Bulletin de la Société Géologique de France* Series 3, **25**, 465–486.

Seguenza, G. 1879. La formazioni terziare nella provincia di Reggio Calabria. *Mémoires de l'Académie de la R. Lince. Classe de Sciences de Fis...Mat. Nat.* **6 (6)**, 1–446.

Seiders, V. M. & Zartman, R. E. 1978. U–Pb zircon dates from the central Appalachian Piedmont. A possible case of inherited radiogenic lead. Discussion and reply. *Geological Society of America Bulletin* **89**, 1115–1117.

Seidemann, D. E., Masterson, W. D., Dowling, M. P. & Turekian, K. K. 1984. K–Ar dates and $^{40}A/^{39}Ar$ age spectra for Mesozoic basalt flows of the Hartford Basin, Connecticut and the Newark Basin, New Jersey. *Geological Society of America Bulletin* **95**, 594–598.

Selli, R. 1960. Il Messiniano Mayer-Eymar, 1867. Proposta di neostratotipo. *Giornale di Geologia* Series 2, **28**, 1–33.

Selli, R. 1971. Stratotypes of Mediterranean Neogene stages. *Giornale di Geologia* Series 2, **37**, 1–266.

Selli, R. *et al.* 1977. The Vrica section (Calabria, Italy). A potential Neogene/Quaternary boundary stratotype. *Giornale di Geologia* Series 2, **42**, 181–204.

Semikhatov, M. A. 1966. The suggested stratigraphic scheme for the Precambrian. *Izvestiya Akademia Nauk SSSR. Series Geological* **4**, 70 (in Russian).

Semikhatov, M. A. 1974. Proterozoic stratigraphy and geo-chronology. *Trudy Akademii Nauka SSSR* **256**, 302 (in Russian).

Şengör, A. M. C. 1985. The Cimmeride orogenic system and the tectonics of Eurasia. *Geological Society of America Special Paper* **195**.

Shackleton, N. J. & Hall, M. A. (in press). Stable isotope history of the Pleistocene at ODP site 677. *Proceedings of the Ocean Drilling Program.* Texas A & M University.

Shackleton, N. J. & Opdyke, N. D. 1973. Oxygen isotope and palaeomagnetic stratigraphy of equatorial Pacific core V28–238: oxygen isotope temperatures and ice volumes on a 10^5 and 10^6 year scale. *Quaternary Research* **3**, 39–55.

Shackleton, N. J. & Opdyke, N. D. 1976. Oxygen isotope and paleomagnetic stratigraphy of Pacific core V28–239, Late Pliocene and Latest Pleistocene. *Geological Society of America Memoir* **145**, 449–464.

Shatsky, N. S. 1960. Principles of late Pre-Cambrian stratigraphy and the scope of the Riphean Group. *Report of the International Geological Congress 21 Session*. Norden **8**, 7.

Shaw, A. B. 1964. *Time in Stratigraphy*. New York: McGraw-Hill, 365 pp.

Shell, 1980. *Standard Legend*, Time Stratigraphic Table. Maatschappij, The Hague: Shell International Petroleum.

Sheng Shen-Fu 1980. The Ordovician System in China. Correlation chart and explanatory notes. Ottawa: *International Union of Geological Sciences, Publication* **1**, 7 pp.

Shepard, J. B. 1986. The Triassic–Jurassic boundary and the Manicouagan impact: Implications of $^{40}Ar/^{39}Ar$ dates on periodic extinction models. Senior thesis Princeton University, 38 pp.

Sheridan, R. E. 1987. Pulsation tectonics as the control of long-term stratigraphic cycles. *Paleoceanography* **2**, 97–118.

Sherlock, R. L. 1948. *The Permo–Triassic Formations: a World Review*. London: Hutchinson, 367 pp.

Shibata, K. 1986. Isotopic ages of alkali rocks from Nemuro Group in Hokkaido, Japan: Late Cretaceous time-scale points. *Chemical Geology* (Isotope Geoscience Section) **59**, 163–169.

Shibata, K., Matsumoto, T., Yanagi, T. & Hamamoto, R. 1978. Isotopic ages and stratigraphic control of Mesozoic igneous rocks in Japan. In Cohee *et al.*, pp. 143–164.

Shibata, K. & Miyata, Y. 1978. Isotopic ages of the Cretaceous tuff from the Obira area, Hokkaido. *Bulletin of the Geological Survey of Japan* **29**, 175–180.

Shoemaker, E. M. & Hackman, R. J. 1962. Stratigraphic basis for a lunar time scale. In *The Moon* (eds. Kopal and Michailov). Academic Press – reprinted by Oxford University Press, 1976. *IAU Symposium* **14**, 289–300.

Sibrava, V. 1978. Isotopic methods in Quaternary geology. In Cohee *et al.*, 165–169.

Sibrava, V., Bowen, D. Q. & Richmond, G. M. (eds.) 1986. Quaternary glaciation in the northern hemisphere. *Quaternary Science Review* **5**, 514 pp.

Sigal, J. 1977. Essai de zonation du Crétacé méditerranéen à l'aide des foraminifères planctoniques. *Géologie Mediterranéenne* **4**, 99–108.

Silberling, N. J. & Tozer, E. T. 1968. Biostratigraphic correlation of the marine Triassic in North America. *Geological Society of America Special Paper* **110**, 63.

Sims, P. K. 1980. Subdivision of the Proterozoic and Archean Eons: recommendations and suggestions by the International Subcommission on Precambrian Stratigraphy. *Precambrian Research* **13**, 379–380.

Sissignh, W. 1977. Biostratigraphy of Cretaceous calcareous nanno-plankton. *Geologie en Mijnbouw* **56**, 37–56.

Sloss, L. L. 1963. Sequences in the cratonic interior of North America. *Geological Society of America Bulletin* **74**, 93–114.

Smith, A. G., Hurley, A. M. & Briden, J. C. 1981. *Phanerozoic Paleocontinental World Maps*. Cambridge: Cambridge University Press, 102 pp.

Smith, D. B., Brunstrom, R. G. W., Manning, P. I., Simpson, S. & Shitton, F. W. 1974. A correlation of Permian rocks in the British Isles. *Geological Society of London* Special Report **5**, 45 pp.

Smith, D. G. 1982. Stratigraphic significance of a palynoflora from ammonoid bearing Early Norian strata from Svalbard. *Newsletters in Stratigraphy* **11**, 154–161.

Snelling, N. J. 1982. Chronology of the geological record. *Episodes* **2**, 26–27.

Snelling, N. J. 1985a. An interim time scale. In Snelling, pp. 261–265.

Snelling, N. J. (ed.) 1985b. *The Chronology of the Geological Record*. Geological Society of London Memoir **10**, 343 pp.

Sokolov, B. S. 1952. On the age of the old sedimentary cover of the Russian Platform. *Isvestiya Akademii Nauk SSSR, Seriya geologicheskaya* **5**, 21–31 (in Russian).

Sokolov, B. S. 1984. The Vendian system and its position in the stratigraphic scale. *Proceedings of the 27th International Geological Congress (Stratigraphy)* **1**, 241–269.

Sokolov, B. S. & Fedonkin, M. A. 1984. The Vendian as the terminal system of the Precambrian. *Episodes* **1**, 12–19.

Sokolov, B. S. & Fedonkin, M. A. (eds.) 1985. The Vendian System. Historical–geological and palaeontological basis. *Tom 2, Stratigrafiya i geolgicheskiye protsessy*. AN SSST, Otd Geol., Geofiz i Geokhim. Moscow: Nauka, 238 pp. (in Russian).

Sokolov, B. S. & Ivanovskiy, A. B. (eds.) 1985. The Vendian System. Historical–geological and palaeontological basis. *Tom 1, Paleontologiya* AN SSR, Otd. Geol., Geofiz i Geokhim. Moscow: Nauka, 221 pp. (in Russian).

Souther, J. G., Armstrong, R. L. & Harakal, J. 1984. Chronology of the peralkaline late Cenozoic Mount Edziza Volcanic Complex, northern British Columbia, Canada. *Geological Society of America Bulletin* **95**, 337–349.

Spanglet, M., Brueckner, H. K. & Senechal, R. G. 1978. Old Rb–Sr whole-rock isochron apparent ages from Lower Cambrian psammites and metapsammites, southeastern New York. *Geological Society of America Bulletin* **89**, 783–790.

Spath, L. F. 1923. On the ammonite horizons of the Gault and contiguous deposits. *Summary of Progress of the Geological Survey of Great Britain and the Museum of Practical Geology*, (for 1922). London, 139–149.

Spath, L. F. 1932. The invertebrate faunas of the Bathonian–Callovian deposits of Jameson Land (East Greenland). *Meddelelser om Grønland* **87**, 1–158.

Spath, L. F. 1956. The Liassic ammonite faunas of the Stowell Park Borehole. *Geological Survey of Great Britain Bulletin* **11**, 140–164.

Spjeldnaes, N. 1978. The Silurian System. In Cohee *et al.*, pp. 341–345.

Sprigg, R. C. 1947. Early Cambrian (?) jellyfishes from the Flinders Ranges, South Australia. *Transactions of the Royal Society of South Australia* **71**, 212–224.

Stainforth, R. M. *et al.* 1975. Cenozoic planktonic foraminiferal zonation and characteristics of index forms. *Kansas University Paleontological Contribution* **62**, 425 pp.

Stanley, S. M. 1984. Mass extinctions in the ocean. *Scientific American* **June 1984**, 46–54.

Steiger, R. H. & Jäger, E. 1977. Subcommission on geochronology: Convention on the use of decay constants in geo- and cosmochronology. *Earth and Planetary Science Letters* 36, 359–362.

Steiner, M. B. & Ogg, J. G. (in press). Early and Middle Jurassic magnetic polarity time scale. Jurassic symposium II.

Stephenson, L. W., King, P. B., Monroe, W. H. & Imlay, R. W. 1942. Correlation of the outcropping Cretaceous formations of the Atlantic and Gulf Coastal Plain and Trans-Pecos, Texas. *Geological Society of America Bulletin* **53**, 435–448.

Stern, T. W., Bateman, P. C., Morgan, B. A., Newell, M. F. & Peck, D.L. 1981. Isotopic U–Pb ages of zircon from the granitoids of the Central Sierra Nevada. *U.S. Geological Survey Professional Paper* **1185**, 17 pp.

Stern, T. W., Goldich, S. S. & Newell, M. F. 1966. Effects of weathering on the U–Pb ages of zircon from the Morton Gneiss, Minnesota. *Earth and Planetary Science Letters* **1**, 369–371.

Stevens, C. H., Wagner, D. B. & Sumsion, R. S. 1979. Permian fusilinid biostratigraphy, central Cordilleran miogeosyncline. *Journal of Paleontology* **53**, 29–36.

Stevens, G. R. (compiler) 1980. *Geological Time Scale*. Geological Survey of New Zealand.

Stille, H. 1924. *Grundlagen der vergleichenden Tektonik*. Berlin: Borntrager, 443 pp.

Stockwell, C. H. 1964. Fourth report on structural provinces, orogenies and time classification of rocks of the Canadian Precambrian Shield. *Geological Survey of Canada Paper* **64–17**, 21 pp.

Stockwell, C. H. 1973. Revised Precambrian time-scale for the Canadian Shield. *Geological Survey of Canada Paper* **72–52**, 4 pp.

Stockwell, C. H. 1982. Proposals for time classification and correlation of Precambrian rocks and events in Canada and adjacent areas of the Canadian Shield. Part I. : A time classification of Precambrian rocks and events. *Geological Survey of Canada* Paper **80–24**, 19 pp.

Stubblefield, C. J. 1956. Cambrian palaeogeography in Britain. In *El Sistema Cambrico, su Paleogeografia y el Problema du su Base I*. XX International Geological Congress, Mexico, 1–43.

Sturani, C. 1967. Ammonites and stratigraphy of the Bathonian in the Digne-Barrême area (SE France). *Bolletino della Societa Paleontologica Italiana* **5**, 1–55.

Styles, M.T. & Rundle C.C. 1981. A whole rock Rb/Sr isochron for the Kennack Gneiss (abstract). *Proceedings of the Ussher Society* **5**, 246.

Suggate, R. P., Stevens, G. R. & Te Punga, M. T. (eds.) 1978. *The Geology of New Zealand*. Wellington: Government Printer, 2 volumes, 820 pp.

Summerhayes, C. P. 1986. Sea level curves based on seismic stratigraphy: Their chronostratigraphic significance. *Palaeogeography, Palaeoclimatology, Palaeoecology* **57**, 27–42.

Surlyk, F. 1977. Stratigraphy, tectonics and palaeogeography of the Jurassic sediments of the areas north of Kong Oscars Fjord, East Greenland. *Grønlands Geologiske Undersøgelse Bulletin* **123**, 1–56.

Swartz, C. K. *et al.* 1942. Correlation of the Silurian formations of North America. *Geological Society of America Bulletin* **53**, 533–538.

Sweet, W. C. & Bergstrom, S. M. 1976. Conodont biostratigraphy of the Middle and Upper Ordovician of the United States Midcontinent. In Bassett, pp. 121–151.

Tabor, R. W., Mark, R. K. & Wilson, R. H. 1985. Reproducibility of the K–Ar ages of rocks and minerals. An empirical approach. *United States Geological Survey Bulletin* **1654**, 5 pp.

Takai, F., Matsumoto, T. & Toriyma, R. (eds.) 1963. *The Geology of Japan*. Tokyo: University of Tokyo Press, 279 pp.

Tankard, A. J., Jackson, M. P. A., Eriksson, K. A., Hobday, D. K., Hunter, D. R. & Minter, W. E. L. 1982. *Crustal Evolution of Southern Africa: 3.8 Billion Years of Earth History*. New York: Springer Verlag, 523 pp.

Tappan, H. 1974. Molecular oxygen and evolution. In *Molecular Oxygen in Biology: Topics in Molecular Oxygen Research* (ed. O. Hayaishi), pp. 81–135. Amsterdam: North Holland Publishing Company.

Tappan, H. 1980. *Paleobiology of Plant Protists*. Reading: Freeman, 1028 pp.

Tappan, H. & Loeblich, A. R. 1971. Geobiologic implications of fossil phytoplankton evolution and timespace distribution. *Geological Society of America Special Paper* **127**, 247–340.

Tatsumoto, M., Knight, R. J. & Allegre, C. J. 1973. Time differences in the formation of meteorites as determined from the ratio of lead–207 to lead–206. *Science* **180**, 1279–1283.

Tauxe, L., Monaghan, M., Drake, R., Curtis, G. & Staudigel, H. 1985. Paleomagnetism of Miocene East African Rift sediments and the calibration of the geomagnetic reversal time scale. *Journal of Geophysical Research* **90**, 4639–4646.

Tauxe, L., Opdyke, N. D., Pasini, G. & Elmi, C. 1983. Age of the Pliocene–Pleistocene boundary in the Vrica section, southern Italy. *Nature* **304**, 125–129.

Taylor, S. R. 1975. *Lunar Science: a Post Apollo View*. Oxford: Pergamon Press, 372 pp.

Taylor, S. R. 1982. *Planetary Sciences – a Lunar Perspective*. Houston: Lunar and Planetary Institute, Houston, 481 pp.

Termier, H. & Termier, G. 1949. Les sédiments Antécambriens et leur pauvreté en fossiles. *Revue Scientifique* **87**, 74.

Theyer, F. & Hammond, S. R. 1974. Cenozoic magnetic time scale in deep sea cores: completion of the Neogene. *Geology* **2**, 487–492.

Thompson, G. R. & Hower, J. 1973. An explanation for low radiometric ages from glauconite. *Geochimica et Cosmochimica Acta* **37**, 1473–1491.

Thompson, J. 1871. On the occurrence of pebbles and boulders of granite in schistose rocks in Islay, Scotland. *Report of British Association for the Advancement of Science* 40th meeting, Liverpool. Notes and abstracts 88.

Thompson, J. 1877. On the geology of the island of Islay. *Transactions of the Geological Society of Glasgow* **5**, 200–222.

Tilton, G. R. 1973. Isotopic lead ages of chondritic meteorites. *Earth and Planetary Science Letters* **19**, 321–329.

Tilton, G. R. & Grünenfelder, M. H. 1968. Sphene: Uranium–lead ages. *Science* **159**, 1458–1461.

Tilton, G. R., Hopson, C. A. & Wright, J. E. 1981. Uranium–lead isotopic ages of the Semail ophiolitte, Oman, with applications to Tethyan ocean ridge tectonics. *Journal of Geophysical Research* **86**, 2763–2775.

Ting, V. K. & Grabau, A. W. 1934. The Permian of China and its bearing on Permian classification. *Report of the International Geologic Congress*, Washington 1933.

Toghill, P. 1968. The graptolite assemblages and zones of the Birkhill Shales (lower Silurian) at Dobb's Linn. *Palaeontology* **11**, 654–668.

Tomlinson, C. W., Moore, R. C., Dott, R. H., Cheney, M. G. & Adams, J. E. 1940. Classification of Permian rocks – the Association Subcommittee on the Permian. *American Association of Petroleum Geologists Bulletin* **24**, 337–358.

Toriyama, R. 1963. The Permian. In Takai *et al.*, 43–58.

Toucas, A. 1888. Note sur le Jurassique supérieur et le Crétacé inférieur de la vallée du Rhône. *Bulletin de la Société Géologique de France* **16**, 903.

Tozer, E. T. 1967. A standard for Triassic time. *Geological Survey of Canada Bulletin* **156**, 103 pp.

Tozer, E. T. 1979. Latest Triassic ammonoid faunas and biochronology, western Canada. *Geological Survey of Canada Paper* **79–1B**, 127–135.

Tozer, E. T. 1982. Marine Triassic faunas of North America: their significance for assessing plate and terrane movements. *Geologische Rundschau* **71**, 1077–1104.

Tozer E. T. 1984. The Trias and its ammonoids: the evolution of a time scale. *Geological Survey of Canada Miscellaneous Report* **35**, 171 pp.

Tozer, E. T. 1988. Towards a definition of the Permian–Triassic boundary. *Episodes* **11**, 251–255.

Twenhofel, W. H. *et al.* 1954. Correlation of the Ordovician formations of North America. *Geological Society of America Bulletin* **56**, 247–298.

United States Geological Survey, Geologic Names Committee 1980. Time scale first published on inside covers of the journal *Isochron/West*. New Mexico: Bureau of Mines and Mineral Resources **28**, August 1980.

Vail, P. R. & Hardenbol, J. 1979. Sea-level changes during the Tertiary. *Oceanus* **22**, 71–79.

Vail, P. R., Mitchum, R. M. Jr. 1977. Seismic stratigraphy and global changes of sea level, part 1: Overview. *American Association of Petroleum Geologists Memoir* **26**, 51–52 (in Vail *et al.*).

Vail, P. R., Mitchum, R. M. Jr, Todd, R. G., Widmier, J. M., Thompson, S., III, Sangree, J. B., Bubb, J. N. & Hatlelid, W. G. 1977. Seismic stratigraphy and global changes of sea level (in 11 parts). *American Association of Petroleum Geologists Memoir* **26**, 49–212.

Vail, P. R., Mitchum, R. M. Jr & Thompson, S., III. 1977. Seismic stratigraphy and global changes of sea level, part 4: Global cycles of relative changes of sea level. In Payton, 83–97.

Vail, P. R. & Todd, R. G. 1981. Northern North Sea Jurassic unconformities, chronostratigraphy and sea-level changes from seismic stratigraphy. In *Petroleum Geology of Continental Shelf of North-west Europe* (ed. J. V. Cling and C. D. Hobson), pp. 216–235. London: Institute of Petroleum.

Van Breemen, O. & Dallmeyer, R. D. 1984. The scale of Sr isotopicdiffusion during post-metamorphic cooling of gneisses in the Inner Piedmont of Georgia, southern Appalachians. *Earth and Planetary Science Letters* **68**, 141–150.

Van Couvering, J. A. & Berggren, W. A. 1977. Biostratigraphical basis of the Neogene time-scale. In *Concepts and Methods of Biostratigraphy*. (ed. J. Hasel and E. Kaufmann), pp. 283–305. Stroudsberg, Pa: Dowden, Hutchinson & Ross.

Van Donk, J. 1976. A record of the Atlantic Ocean for the entire Pleistocene Epoch. *Geological Society of America Memoir* **145**, 147–163.

Van Eysinga, F. W. B. (compiler) 1975. *Geological Timetable*, 3rd edition. Amsterdam: Elsevier.

Van Hinte, J. E. 1978a. Cretaceous time scale. In Cohee *et al.*, pp. 269–287, and in 1986. *American Association of Petroleum Geologists Bulletin* **60**, 498–516.

Van Hinte, J. E. 1978b. A Jurassic time scale. In Cohee *et al.*, pp. 289–297, and in 1986. *American Association of Petroleum Geologists Bulletin* **60**, 489–497.

Van Hise, C. R. 1892. Correlation papers, Archean and Algonkian. *U.S. Geological Survey Bulletin* **86**, 549 pp.

Van Hise, C. R. *et al.* 1905. Report of the Special Committee for the Lake Superior Region. *Journal of Geology* **13**, 89–104.

Van Hise, C. R. & Leith, C. K. 1909. Precambrian Geology of North America. *U.S. Geological Survey Bulletin* **360**, 939 pp.

Van Hise, C. R. & Leith, C. K. 1911. The geology of the Lake Superior Region. *U.S. Geological Survey Monograph* **52**, 641 pp.

Van Wagoner, J. C. *et al.* 1987. Seismic stratigraphy interpretation utilising sequence stratigraphy. Part 2. The key definitions of seismic stratigraphy. In *Atlas of Seismic Stratigraphy*, vol. 1 (ed. A. W. Bally). American Association of Petroleum Geologists, Studies in Geology **27**.

Vass, D. 1975. Report of the working group on radiometric age and palaeomagnetism. *6th Congress Regional Commission of Mediterranean Neogene Stratigraphy Proceedings*, 289–297.

Verbeek, J. W. 1977. Calcareous nannoplankton biostratigraphy of Middle and Upper Cretaceous deposits in Tunisia, Southern Spain and France. *Utrecht Micropaleontological Bulletin* **16**, 157 pp.

Verosub, K. L. & Banerjee, S. K. 1977. Geomagnetic excursions and their paleomagnetic record. *Reviews of Geophysics and Space Physics* **15**, 145–155.

Verschure, R. H., Andriessen, P. A. M., Boelrijk, N. A. I. M., Hebeda, E. H., Maijer, C., Priem, H. N. A. & Verdurmen, E. A. Th. 1980. On the thermal stability of Rb–Sr and K–Ar biotite systems: Evidence from coexisting Sveconorwegian (ca 870 Ma) and Caledonian (ca 400) biotites in SW Norway. *Contributions to Mineralogy and Petrology* **74**, 245–252.

Vervloet, C. C. 1966. *Stratigraphical and Micropaleontological Data on the Tertiary of Southern Piemont (Northern Italy)*. Utrecht: Schotanus & Jens.

Vidal, G. 1979. Acritarchs and the correlation of the Upper Proterozoic. *Publications from the Institutes of Mineralogy, Paleontology, and Quaternary Geology, University of Lund, Sweden* **219**, 21 pp.

Vidal, G. 1981. Micropalaeontology and biostratigraphy of the Upper Proterozoic and Lower Cambrian sequence in East Finnmark, northern Norway. *Norges Geologiske Undersøgelse* **362**, 1–53.

Vine, F. J. 1966. Spreading of the ocean floor: New evidence. *Science* **154**, 1405–1415.

Vine, F. J. 1968. Magnetic anomalies associated with mid-ocean ridges. In *The History of the Earth's Crust* (ed. R. A. Phinney), pp. 73–89. Princeton, New Jersey: Princeton University Press.

Visser, J. N. V. 1982. The Mid-Precambrian tillite in the Griqualand West and Transvaal basins, South Africa. In Hambrey and Harland, pp. 180–184.

Vojacek, H. J. 1979. UNESCO geological world atlas. *Cartography* **11**, 32–39.

Wager, L. R. 1964. The history of attempts to establish a quantitative time-scale. In Harland *et al.*, pp. 13–28.

Wagner, R. H., Higgins, A. C. & Meyen, S. V. (eds.) 1979. The Carboniferous of the USSR (Reports to IUGS Subcommission on Carboniferous Stratigraphy and Geology, 1975). *Yorkshire Geological Society Occasional Publication* **4**, 22 pp.

Wang Dongfang, 1983. Variations in isotopic ages of Cretaceous volcanic rocks in eastern China and isotope dating of the lower boundary of the Cretaceous System. *Geochimica* **4**, 426–430 (in Chinese).

Wang Hongzhen & Liu Benpei, 1980. *Textbook of Historical Geology*. Beijing, 352 pp. (in Chinese).

Wang Yigang, 1984. Earliest Triassic ammonoid faunas from Jiangsu and Zhejiang and their bearing on the definition of the Permo–Triassic boundary. *Gushengwu Xuebao* [Acta Palaeontologica Sinica] **23**, 257–269 (in Chinese).

Wang Yuelun, Lu Songnian, Gao Zhenjia, Lin Weixing & Ma Guogan, 1981. Sinian tillites of China. In Hambrey and Harland, pp. 386–401.

Warrington, G., Audley-Charles, M. G., Elliott, R. E., Evans, W. B., Ivmey-Cook, H. C., Kent, P. E., Robinson, P. L., Shotton, F. W. & Taylor, F. M. 1980. A correlation of Triassic rocks in the British Isles. *Geological Society of London* Special Report **13**, 78 pp.

Waterhouse, J. B. 1976a. The significance of ecostratigraphy and need for biostratigraphic hierarchy in stratigraphic nomenclature. *Lethaia* **9**, 317–325.

Waterhouse, J. B. 1976b. A major overthrust within the Maitai Group? Comment. *New Zealand Journal of Geology and Geophysics* **19**, 955.

Waterhouse, J. B. 1978. Chronostratigraphy for the world Permian. In Cohee *et al.*, pp. 299–322.

Waterhouse, J. B. & Gupta, V. J. 1982. An early Djulfian (Permian) brachiopod faunule from upper Shyok Valley, Karakoram Range, and the implications for dating of allied faunas from Iran and Pakistan. *Contributions to Himalayan Geology* **2**, 188–233.

Watson, E. B., Harrison, T. M. & Ryerson, F. J. 1985. Diffusion of Sm, Sr, and Pb in fluorapatite. *Geochimica et Cosmochimica Acta* **49**, 1813–1823.

Watts, A. B. 1982. Tectonic subsidence, flexure and global changes of sea level. *Nature* **297**, 469–474.

Watts, A. B. & Steckler, M. S. 1979. Subsidence and eustasy at the continental margin of eastern North America. In *Deep Drilling Results in the Atlantic Ocean: Continental Margins and Paleoenvironment* (ed. M. Talwani, W. Hay and W. B. F. Ryan), pp. 218–234. American Geophysical Union, Ewing Series **3**.

Watts, A. B. & Thorne, J. 1984. Tectonics, global changes in sea level and their relationship to stratigraphical sequences at the U.S. Atlantic continental margin. *Marine and Petroleum Geology* **1**, 319–339.

Weast, R. C. (ed.) 1969. Definitions and formulas. In *Handbook of Chemistry and Physics*, pp. F65–F103. Cleveland, Ohio: Chemical Rubber Co. Press.

Weaver, C. E. *et al.* 1944. Correlation of the marine Cenozoic formations of western North America. *Geological Society of America Bulletin* **55**, 569–598.

Webb, J. A. 1981. A radiometric time scale of the Triassic. *Journal of the Geological Society of Australia* **28**, 107–121.

Webby, B. D. *et al.* 1981. The Ordovician System in Australia, New Zealand and Antarctica. *International Union of Geological Sciences Publication* **3**, 64 pp.

Weedon, G. P. 1985. Hemipelagic shelf sedimentation and climatic cycles: The basal Jurassic (Blue Lias) of south Britain. *Earth and Planetary Science Letters* **76**, 321–335.

Weller, J. M. *et al.* 1948. Correlation of the Mississippian formations of North America. *Geological Society of America Bulletin* **59**, 91–196.

Westerman, G. 1984. Gauging the duration of stages: a new approach to the Jurassic. *Episodes* **9**, 26–28.

Westphal, M., Montigny, R., Thiuizat, R., Bardon, C., Bossert, A., Hamzeh, R. & Rolley, J. P. 1979. Paléomagnétisme et datation du volcanisme permien, triassique et crétacé du Maroc. *Canadian Journal of Earth Science* **16**, 2150–2164.

Wetherill, G. W. 1956. Discordant uranium–lead ages. *Transactions of the American Geophysical Union* **37**, 320–326.

Wetherill, G. W., Aldrich, L. T. & Davis, G. L. 1955. Ar⁴⁰/K⁴⁰ ratios of feldspars and micas from the same rock. *Geochimica et Cosmochimica Acta* **8**, 171–172.

Whiteman, M. 1967. *Philosophy of Space and Time, and the Inner Constitution of Nature*. London: Allen & Unwin, 436 pp.

Whitney, P. R. & Hurley, P. M. 1964. The problem of inherited radiogenic strontium in sedimentary age determinations. *Geochimica et Cosmochimica Acta* **28**, 325–436.

Whittard, W. F. 1961. Lexique stratigraphique international. Volume I. Europe fasc. 3a, Angleterre, Pays de Galles, Ecosse: V. Silurien. *Centre National de la Recherche Scientifique*, 273 pp.

Whittington, H. B., Dean, W. T., Fortey, R. A., Rickards, R. B., Rushton, A. W. A. & Wright, A. D. 1984. Definition of the Tremadoc Series and the series of the Ordovician System in Britain. *Geological Magazine* **121**, 17–33.

Whittington, H. B. & Williams, A. 1964. The Ordovician Period. In Harland *et al.*, 241–254.

Wilhelms, D. E. 1987. The geologic history of the Moon, with sections by J. F. McCauley and N. J. Trask. *U.S. Geological Survey Professional Paper* **1348**, 302 pp.

Williams, A. 1953. The geology of the Llandeilo district, Carmarthenshire. *Quarterly Journal of the Geological Society* **108**, 177–208.

Williams, A., Strachan, I., Bassett, D. A., Ingham, J. K., Wright, A. D. & Whittington, H. B. 1972. A correlation of Ordovician rocks in the British Isles. *Geological Society of London Special Report* **3**, 74 pp.

Williams, C. A. 1986. An oceanwide view of Palaeogene plate tectonic events. *Palaeogeography, Palaeoclimatology, Palaeoecology* **57**, 3–25.

Williams, G. E. (ed.) 1981. *Megacycles*. Woods Hole, Massachusetts: Hutchinson & Ross Publishing Co., Benchmark Papers in Geology, **57**, 434 pp.

Williams, G. E. 1989. Late Precambrian tidal rhythmites in South Australia and the history of the Earth's rotation. *Journal of the Geological Society of London* **146**, 97–117.

Williams, H. S. 1893. Elements of the geological time scale. *Journal of Geology* **1**, 283–295.

Williams, H. S. 1901. The discrimination of time values in geology. *Journal of Geology* **9**, 570–585.

Williams, H. S. 1983. The Ordovician–Silurian boundary graptolite fauna of Dobb's Linn, southern Scotland. *Palaeontology* **26**, 605–639.

Williams, I. S. 1978. U–Pb evidence for the pre-emplacement history of granitic magmas, Berridale batholith, southeastern Australia. In Zartman, pp. 455–457.

Williams, I. S., Tetley, N. W., Compston, W. & McDougall, I. 1982. A comparison of K–Ar and Rb–Sr ages of rapidly cooled igneous rocks: Two points in the Palaeozoic time scale re-evaluated. *Journal of the Geological Society of London* **139**, 557–568.

Willis, B., Blackwelder, E. & Sargent, R. H. 1907. *Research in China*. Carnegie Institution of Washington. **1** (1), 353 pp.

Wilmarth, M. G. 1925. The geologic time classifications of the United States Geological Survey compared with other classifications accompanied by the original definitions of era, period and epoch terms – a compilation. *United States Geological Survey Bulletin* **769**, 138 pp.

Wilson, D. S. & Hey, R. N. 1981. The Galapagos axial magnetic anomaly: evidence for the Emperor event within the Brunhes and for a two-layer magnetic source. *Geophysical Research Letters* **8**, 1051–1054.

Wilson, J. R. & Pedersen, S. 1981. The age of the synorogenic Fongen-Hyllingen complex, Trondheim region, Norway. *Geologiska Foreningens i Stockholm Forhandlingar* **103**, 429–435.

Wilson, M. R. 1972. Excess radiogenic argon in metamorphic-amphiboles and biotites from the Sulitjelma region, central Norwegian Caledonides. *Earth and Planetary Science Letters* **14**, 403–412.

Windley, B. F. 1977. *The Evolving Continents*. London: Wiley, 385 pp.

Windley, B. F. 1984. The Archean–Proterozoic boundary. *Tectonophysics* **105**, 43–53.

Windrim, D. P., McCulloch, M. T., Chappell, B. W. & Cameron, W. E. 1984. Nd isotopic systematics and chemistry of Central Australian sapphirine granulites: an example of rare earth element mobility. *Earth and Planetary Science Letters* **70**, 27–39.

Wise, D. U. 1974. Continental margins, freeboard and the volumes of continents and oceans through time. In *The Geology of Continental Margins* (ed. C. A. Burk and C. L. Drake), pp. 45–58. New York: Springer-Verlag.

Wood, D. A., Gibson, I. L. & Thompson, R. N. 1976. Elemental mobility during zeolite facies metamorphism of the Tertiary basalts of eastern Iceland. *Contributions to Mineralogy and Petrology* **55**, 241–254.

Woollam, R. & Riding, J. B. 1983. Dinoflagellate cyst zonation of the English Jurassic. *Report of the Institute of Geological Sciences* **83** (2), 42 pp.

Worden, J. M. & Compston, W. 1973. A Rb–Sr isotopic study of weathering in the Mertondale granite, Western Australia. *Geochimica et Cosmochimica Acta* **37**, 2567–2576.

Worsley, T. R., Nance, D. & Moody, J. B. 1984. Global tectonics and eustasy for the past 2 billion years. *Marine Geology* **58**, 373–400.

Worsley, T. R., Nance, R. D. & Moody, J. B. 1986. Tectonic cycles and the history of the earth's biogeochemical and paleoceanographic record. *Paleoceanography* **1**, 233–263.

Wyatt, A. R. 1986. Post-Triassic continental hypsometry and sea level. *Journal of the Geological Society of London* **143**, 907–910.

Xing Yusheng, Ding Qixiu, Luo Huilin, He Tinggui & Wang Yangeng, 1984. The Sinian–Cambrian boundary of China and its related problems. *Geological Magazine* **121**, 155–170.

Yanagi, T., Baadsgaard, H., Stelck, C. R. & McDougall, I. 1988. Radiometric dating of a tuff bed in the middle Albian Hulcross Formation at Hudson's Hope, British Columbia. *Canadian Journal of Earth Sciences* **25**, 1123–1127.

Yang Zunyi, Cheng Yuqi & Wang Hongzhen, 1986. The geology of China. *Oxford Monographs on Geology and Geophysics* **3**, 303 pp.

Ye Boden 1986. The boundary age of Cretaceous–Jurassic period in south China. *Terra cognita* **6**, 221.

York, D. 1967. The best isochron. *Earth and Planetary Science Letters* **2**, 479–482.

York, D. & Farquhar, R. M. 1972. *The Earth's Age and Geochronology*. New York: Pergamon, 178 pp.

Young, G. M. (ed.) 1973. Huronian stratigraphy and sedimentation. *Geological Association of Canada Special Paper* **12**, 271 pp.

Young, G. M. 1982a. Diamictites of the Early Proterozoic Ramsay Lake and Bruce Formations, north shore of Lake Huron, Ontario, Canada. In Hambrey and Harland, pp. 813–816.

Young, G. M. 1982b. The Early Proterozoic Gowganda Formation, Ontario, Canada. In Hambrey and Harland, pp. 807–812.

Zachos, J. C. & Arthur, M. A. 1986. Paleoceanography of the Cretaceous/Tertiary boundary event: Inference from stable isotopic and other data. *Paleoceanography* **1**, 5–26.

Zalasiewicz, J. A. 1984. A re-examination of the type Arenig Series. *Geological Journal* **19**, 105–124.

Zartman, R. E. (ed.) 1978. Short papers of the 4th International Conference, Geochronology, Cosmochronology, Isotope Geology, 1978. *U.S. Geological Survey Open-File Report* **78–701**, 476 pp.

Zeuner, F. E. 1945. The Pleistocene Period, its climate, chronology and faunal successions. *Royal Society of London*, 322 pp.

Zhang Zichao, Ma Guogan & Lee Huaqin, 1984. The chronometric age of the Sinian–Cambrian boundary in the Yangtze Platform, China. *Geological Magazine* **121**, 175–178.

Zhang Zuqi 1989. Proposals concerning an international chronostratigraphic system of the Permian. *Oil and Gas Geology* **10**, 1–8 (in Chinese).

Ziegler, W. 1962. Taxionomie und Phylogenie Oberdevonischer Conodonten und ihre stratigraphische Bedeutung. *Abhandlungen des Hessischen Landesamtes für Bodenforschung* **38**, 1–166.

Ziegler, W. 1971. Conodont stratigraphy of the European Devonian. *Geological Society of America Memoir* **127**, 227–283.

Ziegler, W. 1978. Devonian. In Cohee *et al.*, pp. 337–339.

Ziegler, W. 1979. Historical subdivision of the Devonian. In *The Devonian System* (ed. M. R. House, C. T. Scrutton and M. G. Bassett), pp. 23–47. London: *Palaeontological Association, Special Papers in Palaeontology* **23**.

Ziegler, W. 1984. Conodonts and the Frasnian–Famennian crisis. *Geological Society of America Abstracts with Programs* **16**, 73 pp.

Ziegler, W. & Klapper, G. 1982. Devonian Series of the IUGS Subcommission. *Episodes* **4**, 18–21.

Ziegler, W. & Klapper, G. 1985. Stages of the Devonian System. *Episodes* **8**, 104–109.

Zittel, K. A. von 1901. *History of Geology and Palaeontology to the End of the Nineteenth Century*. London: Walter Scott (translated from 1899 German edition by M. M. Ogilvie-Gordon) 562 pp.

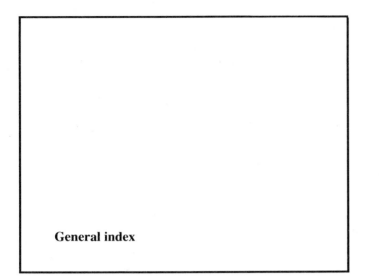

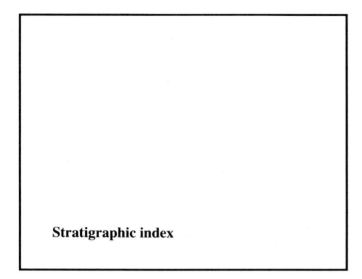

Stratigraphic index

This is intended as a practical rather than a comprehensive index. There should be an entry for each stratigraphic name occurring in this volume. However, there are some exceptions, notably biostratigraphic and magnetostratigraphic names and zonal letters or numbers.

After each entry, the following may be found: (1) the numbers of sections in the text where the name is introduced or discussed; (2) the letter A indicating that the name will be found in Appendix 1; (3) the three-character abbreviations used in the volume and listed in Appendix 2; (4) the letter T with the appropriate number indicating the table in the text where the name occurs; and (5) the letter F with the appropriate number indicating the figure in the text where the name occurs. The names for the selected Phanerozoic divisions used as standard throughout the book (and on its back cover) are printed in bold typeface. Such names are likely to occur in every table and figure where they are relevant, and also in the chronograms of Appendix 4 and the magnetostratigraphic scale in Appendix 5. Therefore, to keep the size of the index manageable, they are only referred where there is some definition or discussion. Where a boundary is indicated the name of the younger overlying division only is entered.

In general, only the distinguishing name of the division is given – the second term (stage, interglacial etc.) and any qualifying term (early, late, etc.) are omitted. Orogenic and some other (rock) names are given where they are illustrated in stratigraphic context. The names of rock units, often somewhat abbreviated, in the main database of Table 4.2 are not listed here.

Supplementary notes added in proof are printed after the Preface on pages xiii to xv and are indicated by a final letter P.

Grey Beds F3.8
Grey Chalk F3.11
Grey Shales F3.9
Griesbachian 3.15.2, 3.15.3, 5.8.4, Gri
Griqualand West 3.5.5
Gshelian (*see* Gzelian) A
Guadalupian 3.14.1, 3.14.3, A, Gua, F3.7
Guan Ling F3.8
Gubbio 6.3.5
Gulf 3.17.1, 3.17.3, Gul, F3.11
Gunflint 3.5.6, Gun, T3.1, F7.8, F7.9
Gunz 3.21.3
Gushan F3.2
Gyliakian F3.11
Gypsum Springs F3.9
Gzelian 3.13.1, 3.13.3, Gze, F3.6b
Gzhelian (*see* Gzelian)

Hackberryfrio F3.13
Hadean 2.4.1, 3.5.1, Hde, T3.1, F1.3, F1.7, F7.8, F7.9
Hadrynian 2.3.1, 2.4.3, Hdy, F2.1
Hafotty F3.2
Hale F3.6b
Halla F3.4
Hammersley 3.5.4, F7.9
Hamra F3.4
Hangeerqiaoke 3.5.8
Hangenberg Kalk 3.13.2
Hangman F3.5
Hannibal F3.6b
Hansborough 3.6
Harbour Main 5.9, F3.1
Harju F3.3
Harlech (Grits) 3.9.1, F3.2
Harnagian 5.8.9, Har
Hasselbachtal 3.13.2
Hastarian 3.13.2, 5.8.6, Has
Hastings Beds F3.11
Haumurian F3.11
Hauptdolomit F3.7, F3.8
Hauterivian 3.17.1, 3.17.2, 5.8.2, A, Hau, T6.11, T6.13
Hayburn F3.9
Heathcotian A
Heersian 3.19.3, A
Heiberg F3.8, F3.9
Heisdorf 3.12.1
Helderbergh 3.9.1
Helderberg(ian) A, F3.5
Helikian 2.3.1, 2.4.3, Hel, F2.1
Helvetian 3.20.2, A
Hemse F3.4
Herangi F3.9
Hercynian F7.7
Heretaungan F3.13
Hervy F3.5
Heterian F3.9
Hetonian F3.11
Hettangian 3.16.2, Het
Hibernian F7.7

Hikawan F3.7
Hirnantian 3.10.4, 5.8.9, Hir
Hod F3.11
Hogklint F3.4
Holkerian 3.13.2, Hlk
Holmia F3.2
Holocene 3.8.4, 3.21, 3.21.6, 5.8.1, Hol
Holsteinian 3.21.3, F3.17
Holyrood 5.9
Hombergian F3.6b
Homerian 3.11.1, Hom, P
Hoogenoeg 3.5.3
Hospital Hill 3.5.4, Hsp, T3.1
Hough Lake 3.5.5, Hou, T3.1, F7.8, F7.9
Houiller 3.13.1
Hoxnian F3.17
Huashiban F3.6b
Hudsonian 3.5.6, F7.9
Hule F3.3
Humorum 3.5.1
Huobachong F3.8
Huronian 2.4.3, 3.5.5, A, Hur, T3.1, F1.3, F1.7, F7.8, F7.9
Hydraulic Limestones F3.9
Hypozoic 2.4.3

Ibexian F3.3
Icenian A
Ichang F3.3
Idamean F3.2
Idavere F3.3
Idwian 3.11.2, Idw
Igualadian A
Ikskiy F3.7
Ilfracombe F3.5
Ilibeyskiy F3.7
Illinoian A, F3.17
Illyrian 3.15.4, Ill, F3.8
Ilychskiy F3.6b
Imbrian 3.5.1, Imn, T3.1, F1.7, F7.8, F7.9
Imbrium 3.5.1, Imm, T3.1
Indigskiy F3.7
Indosinian F7.5
Induan 3.15.3, Ind, F3.8
Inferior Oolite F3.8
Infra-Valanginian (*see* Berriasian) 3.17.2
Inkermanian F3.13
Interglacial III F3.17
Iowan A
Ipswichian F3.17
Irenian Chron F3.7
Irenskiy F3.7
Ironstone Shales F3.9
Ironwood 3.5.6, F7.9
Isuan 3.5.2, Isu, T3.1, F1.3, F1.7, F7.8, F7.9
Iuosuchanskaya F3.8
Ivorian 3.13.2, 5.8.6, Ivo

Jackson F3.13
Jacksonian A